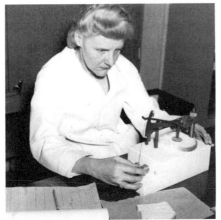

FOOD MATTERS

Critical Histories of Food and the Sciences

EDITED BY

E. C. Spary and Anya Zilberstein

O S I R I S | **35**

A Research Journal Devoted to the
History of Science and Its Cultural Influences

Osiris

On the Cover:

A compressibility apparatus tests the softness of bread made by Miss Josephine Brooks, graduate student in the College of Home Economics at Cornell University, in her research on the thiamine content and eating quality of yeast breads made with enriched flour. Unknown artist, black and white photograph, September 21, 1945. New York State College of Home Economics Records, #23-2-749. Division of Rare and Manuscript Collections, Cornell University Library.

On the Virtues of Historical Entomophagy

by E. C. Spary* and Anya Zilberstein§

"Why *Not* Eat Insects?" inquires a short book, really a pamphlet, first published in London in 1885. Working against the common perception of bugs as pests—at best, an absurdly obvious nonfood, and at worst, a toxin—the author, Vincent M. Holt (who provided no autobiographical details that might establish his credentials) aimed at reversing his readers' general disdain for insects as low and troublesome forms of being, as well as the specific Western objection to entomophagy. Cockchafers, caterpillars, and grubs, he asserted in his opening pages, were "clean, palatable, and wholesome" foods. Indeed, as eaters, these insects were more discerning "than ourselves." It followed, therefore, that eating insects was *not* a form of pica (the mental and physiological disorder of consuming nonfood items); rather, refusing to eat them was merely a provincial prejudice of Europeans, born of ignorance about the consumption of insects, a practice Holt assured his readers was common around the world.[1]

We pose a variant of Holt's deceptively simple question by bringing to the fore the underlying provocation in his manifesto and critique: why *not* study food? Why hasn't food, or the knowledge and practices that surround its production, preparation, distribution, and ingestion, mattered much to historians of science, medicine, and technology? Arguably, the only universal historical constant of human existence (besides death and taxes) is the need to eat and drink. Yet, claims and practices surrounding food and beverages vary widely across time and space. The historicity of food embraces not merely geographic, economic, and political pressures, but also a wide range of claims—theological, legal, medical, traditional—that shape what can, should, or will be consumed by any person or society. Food has long been an object of serious study across the humanities and social

* Faculty of History, University of Cambridge, West Road, Cambridge, CB3 9EF, UK; ecs12@cam.ac.uk.

Many thanks to Anne Secord for giving me a copy of Holt, and Jim Secord for dinosaur advice; to Barbara Orland, Melissa Calaresu, and some of the contributors to this volume for long-standing and fruitful exchange and discussion; and especially to all the brilliant students (you know who you are) who have variously questioned, encouraged, and pushed the boundaries of my own research in this area over the last ten years. Last, but certainly not least, warmest thanks to my coeditor, Anya Zilberstein. I am privileged not only that she invited me to participate in this project (and return to a topic on which I felt the discipline had already heard quite enough from me), but also that it afforded an opportunity to work alongside her over many months of collaborative labor.

§ Department of History, Concordia University, 1455 de Maisonneuve Blvd. West, Montreal, H3G 1M8, Canada; anya.zilberstein@concordia.ca.

Gratitude, first and foremost, to E. C. Spary for agreeing to work with me on the idea for a volume, although we were unacquainted, and teaching me so much in the process. Thanks to Molly Warsh, Danna Agmon, and Daniel Hershenzon for their critical input and encouragement, and to the Rosenfelds and Rivens for coincidentally reminding me of the pleasures of food.

[1] Vincent M. Holt, *Why Not Eat Insects?* (London: Field & Tuer, [1885]).

sciences, especially in anthropology and sociology.[2] Within the historical discipline, by contrast, most professional scholars long regarded food—from its procurement to the crafts of preparing and presenting it—as women's work, existing outside economic and political concerns, and hence low on the scale of serious cultural endeavor and unworthy of scholarly attention. With the important exception of commodity histories of individual ingredients or beverages, historical approaches to food have remained, until quite recently, largely confined to social histories of cooking or diet as aspects of everyday life, or else to cultural analyses or popular treatments of topics like traditional or ethnic foodways, food fads, and colorful chefs.[3] To take food seriously was to be pigeonholed as a scholar of the trivial and mundane.

Yet, from that position of disciplinary marginalization, food history has lately begun to mature into a robust subfield, something attested to by the appearance of edited collections and sourcebooks designed to facilitate teaching the subject; specialized journals such as *Petits Propos Culinaires, Food & History, Food, Culture & Society*, and *Gastronomica*; and international conferences such as the Oxford Symposia on Cookery or the biennial conferences of the Institut Européen d'Histoire et des Cultures de l'Alimentation.[4]

More recent collaboration among scholars with diverse methodological approaches and theoretical commitments—cultural studies, nutritional science, public health, restaurant management, and culinary arts—has generated an interdisciplinarity that makes food studies a thriving new area of inquiry.

From the standpoint of the historian of science, technology, and medicine, however, this very interdisciplinarity has also tended to perpetuate, if unintentionally, the naturalization of different forms of expertise about food. Scientific and medical practitioners' statements

---

[2] Among a large literature, see Bryan S. Turner, "Government of the Body: Medical Regimens and the Rationalisation of Diet," *Brit. J. Sociol.* 33 (1982): 254–69; John Coveney, *Food, Morals and Meaning: The Pleasure and Anxiety of Eating*, 2nd ed. (New York, N.Y.: Routledge, 2006); Jack Goody, *Cooking, Cuisine and Class: A Study in Comparative Sociology* (Cambridge: Cambridge Univ. Press, 1982); Daniel Miller, *Material Culture and Mass Consumption* (Cambridge, Mass.: Blackwell, 1987); Robert J. Foster, *Coca-Globalization: Following Soft Drinks from New York to New Guinea* (Houndmills, UK: Palgrave Macmillan, 2008); Jukka Gronow, *The Sociology of Taste* (New York, N.Y.: Routledge, 1997); and Elisabeth L. Fürst, Ritva Prättälä, Marianne Ekström, Lotte Holm, and Unni Kjærnes, eds., *Palatable Worlds: Sociocultural Food Studies* (Oslo: Solum Verlag, 1991).

[3] Among the best studies in this vein are Jean-Louis Flandrin and Massimo Montanari, eds., *Food: A Culinary History from Antiquity to the Present* (New York, N.Y.: Columbia Univ. Press, 1999); Susan Pinkard, *A Revolution in Taste: The Rise of French Cuisine, 1650–1800* (Cambridge: Cambridge Univ. Press, 2009); Priscilla Parkhurst Ferguson, *Accounting for Taste: The Triumph of French Cuisine* (Chicago: Univ. of Chicago Press, 2004); Paul Freedman, ed., *Food: The History of Taste* (Berkeley and Los Angeles: Univ. of California Press, 2007); Peter J. Atkins, Peter Lummel, and Derek J. Oddy, *Food and the City in Europe since 1800* (Farnham, UK: Ashgate, 2007); Jim Phillips and David F. Smith, eds., *Food, Science, Policy and Regulation in the Twentieth Century: International and Comparative Perspectives* (London: Routledge, 2016); and Adel P. den Hartog, ed., *Food Technology, Science and Marketing: European Diet in the Twentieth Century* (East Linton, UK: Tuckwell, 1997). For scholarship addressing the history of material culture, see, for example, Sara Pennell, *The Birth of the English Kitchen, 1600–1850* (London: Bloomsbury, 2016); on foodways, see Hasia Diner, *Hungering for America: Italian, Irish and Jewish Foodways in the Age of Migration* (Cambridge, Mass.: Harvard Univ. Press, 2003); as well as the journal *Food and Foodways*, published since 1984.

[4] Carole Counihan and Penny Van Esterik, eds., *Food and Culture: A Reader* (New York, N.Y.: Routledge, 1997); Ken Albala, ed., *The Food History Reader: Primary Sources* (London: Bloomsbury Academic, 2014); Raymond Grew, ed., *Food in Global History* (Boulder, Colo.: Westview, 1999); Peter Scholliers, ed., *Food, Drink and Identity: Cooking, Eating and Drinking in Europe Since the Middle Ages* (London: Berg, 2001); Jeffrey M. Pilcher, *Food in World History* (New York, N.Y.: Routledge, 2006); Albala, Joyce E. Chaplin, and Paul Freedman, eds., *Food in Time and Place: The American Historical Association Companion to Food History* (Oakland: Univ. of California Press, 2014).

about food, so far from being neutral, indisputable statements about matters of fact or nature, were always articulated within specific contestations over matters of governance, expertise, resources, and entitlements, and have frequently been overlooked in this burgeoning literature. Even the best cultural histories of food published in recent decades have taken changing claims about what constitutes an "adequate" or "appropriate" diet, or scientific and medical discourse about its content and nature, more or less at face value.[5] The same caveat often applies to work that considers food from the standpoint of the history of stimulants, or in terms of global or imperial commodity flows, the histories of capitalism and development, and their ecological precedents or consequences.[6]

For their part, while historians of science, technology, and medicine have recently confronted the production, application, circulation, or contestation of many forms of knowledge, food, cuisine, and eating have rarely featured among these. Apart from a 2012 forum in *Studies in History and Philosophy of Science* coedited by Spary and Barbara Orland, and a recent special issue devoted to "Food as Medicine, Medicine as Food" in the *Journal of the History of Medicine and Allied Sciences*, leading journals in the field have together published a mere handful of articles concerning the food sciences, and these have mostly been on dietetics, physiology, and metabolism.[7] Steven Shapin and Christopher Lawrence's 1998 edited volume *Science Incarnate* and Harmke Kamminga and Andrew Cunningham's 1995 collection *The Science and Culture of Nutrition* remain among the few works to draw the attention of historians of science, technology, and medicine to the exceptional role of food in the making of natural knowledge.[8] This neglect is the more surprising in that many, if not most, public and policy discussions about food or dietary legislation, regulation, innovation, and marketing rely upon knowledge claims arising out of past scientific and medical research that is often subsequently superseded. Yet, those discussions treat such claims as if they were transhistorical, rather

---

[5] For examples, see Stephen Mennell, *All Manners of Food: Eating and Taste in England and France from the Middle Ages to the Present*, 2nd ed. (Urbana: Univ. of Illinois Press, 1996); Rebecca Lee Spang, *The Invention of the Restaurant: Paris and Modern Gastronomic Culture* (Cambridge, Mass.: Harvard Univ. Press, 2000); and John A. Jakle and Keith A. Sculle, *Fast Food: Roadside Restaurants in the Automobile Age* (Baltimore: Johns Hopkins Univ. Press, 1999). An important exception is Joan Jacobs Brumberg, *Fasting Girls: The Emergence of Anorexia Nervosa as a Modern Disease* (Cambridge, Mass.: Harvard Univ. Press, 1988).

[6] Notably, see Wolfgang Schivelbusch, *Tastes of Paradise: A Social History of Spices, Stimulants, and Intoxicants* (New York, N.Y.: Pantheon, 1992); Nelson Foster and Linda S. Cordell, eds., *Chilies to Chocolate: Food the Americas Gave the World* (Tucson: Univ. of Arizona Press, 1992); Sidney W. Mintz, *Sweetness and Power: The Place of Sugar in Modern History* (New York, N.Y.: Elisabeth Sifton Books and Viking, 1985); David Hancock, *Oceans of Wine: Madeira and the Emergence of American Trade and Taste* (New Haven, Conn.: Yale Univ. Press, 2009); and Akhil Gupta, *Postcolonial Developments: Agriculture in the Making of Modern India* (Durham, N.C.: Duke Univ. Press, 1998).

[7] Julia Adelman and Lisa Haushofer, eds., "Food as Medicine, Medicine as Food," *J. Hist. Med. Allied Sci.* 73 (2018): 127–222; E. C. Spary and Barbara Orland, eds., "Assimilating Knowledge: Food and Nutrition in Early Modern Physiologies," special issue, *Stud. Hist. Phil. Biol. Biomed. Sci.* 43 (2012). Among articles, see Elizabeth A. Williams, "Neuroses of the Stomach: Eating, Gender, and Psychopathology in French Medicine, 1800–1870," *Isis* 98 (2007): 54–79; Vanessa Heggie, "Why Isn't Exploration a Science?," *Isis* 105 (2014): 318–34; Ian Higginson and Crosbie Smith, "'A Magnified Piece of Thermodynamics': The Promethean Iconography of the Refrigerator in Paul Theroux's *The Mosquito Coast*," *Brit. J. Hist. Sci.* 32 (1999): 325–42; Sally Horrocks, "A Promising Pioneer Profession? Women in Industrial Chemistry in Inter-war Britain," *Brit. J. Hist. Sci.* 33 (2000): 351–67; and Anita Guerrini, "The Ghastly Kitchen," *Hist. Sci.* 54 (2016): 71–97.

[8] Christopher Lawrence and Steven Shapin, eds., *Science Incarnate: Historical Embodiments of Natural Knowledge* (Chicago: Univ. of Chicago Press, 1998); Harmke Kamminga and Andrew Cunningham, eds., *The Science and Culture of Nutrition, 1840–1940* (Atlanta, Ga.: Rodopi, 1995).

than the outcome of specific agendas and debates, disciplinary conventions, or professional struggles.[9]

The contributions to *Food Matters*, ranging from early modern dietetics to modern Ayurvedic recipes, from analyses of hungry model organisms to the dining rituals of Silicon Valley entrepreneurs and their patrons, show that such neglect is unwarranted. Our collection seeks to bring the methodological tools developed within the history of science, technology, and medicine over recent decades to bear upon knowledge about food, from the claims of individuals to those asserted by larger collectives, such as government agencies, prisons, armies, cities, or international corporations. Because food ties the body directly to collective life, knowledge about food—as it developed from early modern regimen advice, via the emergence of food chemistry in the eighteenth and nineteenth centuries, into a scientific subdiscipline in its own right—has always raised epistemological, ontological, and definitional questions. These queries have been mediated by macroscopic programs for the management of resources, by taboos and proscriptions, and by individual preferences. We argue that it is only by studying the complex and often contested circumstances under which knowledge claims about food came into being and gained status as authoritative, seemingly transparent reflections of scientific facts about nature that we can begin to understand these past and present debates.

Food Matters opens a broad perspective, one that goes beyond current dietary concerns or political peccadilloes surrounding the food supply. It encourages attention to the importance of a wide range of issues for studying the history of food, including spatiality, disciplinarity, political economy, globalization, translation, gender, practices of cooking and eating, and definitions of "naturalness," "need," or "health." The articles in our collection explicitly reflect upon the methodological potential and problems of food as a central subject for historians of science, technology, and medicine and vice versa, as well as upon the way that our tools, approaches, and preoccupations can be used to investigate the history of food.

TRAJECTORIES OF FOOD GOVERNANCE SINCE 1500

The contributions to this collection form a chronological arc, from early modern medico-theological understandings of ritual foodstuffs or projects of territorial conquest and resource management, via the emergence and elaboration of the food sciences as formal disciplines, to postmodern questioning of the limits of expertise, selfhood, and embodiment. They show that food has long been a focus of inquiry, experimentation, contestation, standardization, quantification, (self-)disciplining, governance, and public concern. As historians well know, it is difficult to overstate the explanatory, institutional, commercial, and broadly political power of modern science, technology, and medicine. Indeed, the prominent role of scientific knowledge claims and expertise in areas as diverse as agriculture, livestock management, the food trade, manufacturing, public health, cooking, eating, and drinking is a prime example of one possible "big picture" we might have of the history of science: a narrative about the increasing purchase of Western scientific knowledge over the daily lives of more and more people across the globe, both in their relationship with governance and in their understanding of themselves.

---

[9] For examples, see Rima D. Apple, *Vitamania: Vitamins in American Culture* (New Brunswick, N.J.: Rutgers Univ. Press, 1996); and Matthew Smith, *Another Person's Poison: A History of Food Allergy* (New York, N.Y.: Columbia Univ. Press, 2015).

Much of the historiography of modern food hinges on a familiar narrative that presents science and technology as progressively discovering solutions to problems of mass consumerism created by industrialization and urbanization—first in Western Europe and the United States, and eventually elsewhere in the world. These solutions included the mechanization of food production, processing, and packaging; the preservation of foodstuffs for long-term storage and long-distance transportation; a greater reliance upon chemical preservatives, colorants, synthetic flavors, and fragrances; and the rise of quantitative nutrition policies, large agribusiness, food conglomerates, globally integrated food chains, fast food, and advertising campaigns.[10] Yet, the contributions to our volume resist any smooth or linear trajectory from "prescientific" or "unscientific" forms of folk knowledge, based on local, familial, or customary practices, to our modern scientific world. Rather, we argue, individual tastes, locally distinctive foodways, and governmental regulation of food have been inseparably entangled with learned knowledge claims from the very earliest scholarly efforts to account for diet.

The extended historical viewpoint adopted by *Food Matters* points to certain crucial relationships between political, economic, and intellectual interventions into the food supply, formulated long before the modern period or well outside conventional arenas of scientific or technical innovation, which made later developments possible. Taken together, the articles in this collection demonstrate the inextricability of the history of food from myriad developments in the sciences more generally. Starting in the late sixteenth century, as Bradford Bouley's article shows, new enterprises of food knowledge emerged from the application of new kinds of natural knowledge to questions of religious dietary proscriptions, agricultural practice, the nature of food and drink, and management of the food supply.[11] The governmentalization of food through the sciences began in this period. From then onward, practitioners of the sciences, engineering, or medicine were prominent in debates between lay, expert, commercial, and governmental food knowledge and practice. Early modern manuals of medicine and rustic economy included medico-culinary recipes serving to feed or cure people and livestock alike, and the science of husbandry embraced food. Rulers and courtiers used food and drink as a way to assert sovereignty or express ownership of lands, both within Europe and in their conquests overseas. The question of how to master and enhance the earth's fertility preoccupied a long series of natural philosophers, beginning in the seventeenth century with Royal Society chymists like Samuel Hartlib, his collaborators, and their princely patrons, as Ted McCormick's article reveals.[12] Such priorities drew scientific practitioners to the problem of how to turn marginal land or wasteland, fallow ground, and swamps into productive spaces. Since the Creator communicated with humanity by blighting a crop or conferring marvelous productivity upon farm animals, techniques to assess soil fertility and fruitfulness were promoted as essential to making

---

[10] Helen Zoe Veit, *Modern Food, Moral Food: Self-Control, Science, and the Rise of Modern American Eating in the Early Twentieth Century* (Chapel Hill: Univ. of North Carolina Press, 2013); Jessica J. Mudry, *Measured Meals: Nutrition in America* (Albany: State Univ. of New York Press, 2010); Michael Pollan, *The Omnivore's Dilemma: A Natural History of Four Meals* (New York, N.Y.: Penguin, 2006); Pollan, *In Defense of Food: An Eater's Manifesto* (New York, N.Y.: Penguin, 2009).

[11] Bradford Bouley, "Digesting Faith: Eating God, Man, and Meat in Seventeenth-Century Rome," in this volume.

[12] Ted McCormick, "Food, Population, and Empire in the Hartlib Circle, 1639–1660," in this volume.

good on the providential contract to husband the earth, and to ensuring more efficient and profitable exploitation of resources.[13]

A century on, early modern cornucopian fantasies of guaranteed abundance had become targets for new social critiques arising in connection with new sciences of population and resources.[14] Enduring practices of governmentality applied to food then assumed new forms as relations of production and consumption altered. In turn, new forms of knowledge emerged to account for the place of food in these relations, as Stefan Pohl-Valero's article on Colombians' shifting perceptions of the social utility of the beverage *chicha* reveals.[15] Attempts to subject food matters to quantification, instrumental logics, discourses of objectivity, and the like were being worked out on shipboard and on land across the colonized spaces of South America, South and East Asia, and Oceania, as the contributions of Joyce E. Chaplin, Rebecca J. H. Woods, Pohl-Valero, Di Lu, and Projit Bihari Mukharji show.[16]

In the nineteenth century, nutrition science emerged alongside, and thanks to, new experimental sciences of physiology, animal chemistry, and metabolism, which coalesced in the later part of the century to form what Corinna Treitel calls "nutritional modernity."[17] Analytical chemistry, in the hands of Giessen chemistry professor Justus von Liebig and his networks of students throughout the world, became an increasingly high-profile form of public expertise from the 1830s. The scientific history of water is exemplary of these shifts in approaches to potable and edible substances, as Chaplin's article shows.[18] For early modern Europeans, water had been a plural thing—"waters"—comprising a range of material and intangible forms or purposes. After the late eighteenth century, chemists, physicians, and dieticians, particularly those working in the service of the British Empire, increasingly began to conceive of water as a singular, pure fluid, or one that could attain an ideal, purified state through chemical interventions, such as distillation. Elemental analysis largely changed the terms of the debate; sanitary officials and medical professionals no longer devoted attention, as their early modern counterparts had, to the question of whether air or water are inherently nourishing.

Cognate changes took place in scientific accounts of the body. The new science of experimental physiology made possible not only experimentation upon hunger, but

---

[13] Louise Hill Curth, *The Care of Brute Beasts: A Social and Cultural Study of Veterinary Medicine in Early Modern England* (Boston: Brill, 2010); Simon Schaffer, "The Earth's Fertility as a Social Fact in Early Modern England," in *Nature and Society in Historical Context*, ed. Mikuláš Teich, Roy Porter, and Bo Gustafsson (Cambridge: Cambridge Univ. Press), 124–47; Richard H. Drayton, *Nature's Government: Science, Imperial Britain, and the "Improvement" of the World* (New Haven, Conn.: Yale Univ. Press, 2000).

[14] Fredrik A. Jonsson, "The Origins of Cornucopianism: A Preliminary Genealogy," *Critical Historical Studies* 1 (2014): 151–68; Jonsson, John Brewer, Neil A. Fromer, and Frank Trentmann, eds., *Scarcity in the Modern World: History, Politics, Society, and Sustainability, 1800–2075* (New York, N.Y.: Bloomsbury Academic, 2019); Alison Bashford and Joyce E. Chaplin, *The New Worlds of Thomas Robert Malthus: Rereading the Principle of Population* (Princeton, N.J.: Princeton Univ. Press, 2016); Emily Pawley, "Feeding Desire: Generative Environments, Meat Markets, and the Management of Sheep Intercourse in Great Britain, 1700–1750," *Osiris* 33 (2018): 47–62.

[15] Stefan Pohl-Valero, "The Scientific Lives of *Chicha*: The Production of a Fermented Beverage and the Making of Expert Knowledge in Bogotá, 1889–1938," in this volume.

[16] Joyce E. Chaplin, "Why Drink Water? Diet, Materialisms, and British Imperialism"; Rebecca J. H. Woods, "The Shape of Meat: Preserving Animal Flesh in Victorian Britain"; Pohl-Valero, "Scientific Lives" (cit. n. 15); Di Lu, "Local Food and Transnational Science: New Boundary Issues of the Caterpillar Fungus in Republican China"; Projit Bihari Mukharji, "Historicizing 'Indian Knowledge Systems': Ayurveda, Exotic Foods, and Contemporary Antihistorical Holisms"; all in this volume.

[17] Corinna Treitel, "Nutritional Modernity: The German Case," in this volume.

[18] Chaplin, "Why Drink Water?" (cit. n. 16).

also laboratory-produced definitions of need, as demonstrated by the articles of Ulrike Thoms, Treitel, Dana Simmons, and Deborah Fitzgerald.[19] Public assertions about the desirability of particular foods can be given a far more rigorous dressing down today—by recourse to experiments, clinical trials, and books of nutritional data and official food standards—than early modern physicians with their humoral exhortations were able to do.[20] The laboratory could become a vantage point from which claims about adequate nourishment were either given or denied scientific standing. It was also a place, from the late eighteenth to the twenty-first century, from which scientific parameters of what counted as "food" could be credibly produced.

The scale of the transformation accomplished by experimentalizing food science is both broad and deep, extending from our self-knowledge as individuals and our personal or familial relations, right up to the way governments engage with questions of the food supply, resource distribution, and foreign aid. Scientific knowledge, expertise, and technological innovation relating to food have often shifted in response to political, economic, or military crises. Such shifts created problems to which new fields of research and new groups of scientific, medical, or technological experts claimed to possess uniquely effective solutions. The long time span covered by *Food Matters* shows how many key features of modern nutrition science were first conceived in earlier moments of profound change or conflict, such as political and religious clashes and colonial encounters, or improvised in response to the exigencies of warfare. Projects for the management of land, resources, and the body enrolled food long before modern nutrition science emerged, and even before what some scholars have dubbed the "nutrition transition" that rendered European eaters dependent upon nonlocal foods.[21] These practices and models flourished over ensuing centuries in a variety of institutional settings, from victualing for the Royal Navy in the eighteenth century, to the New Nutrition research of the Munich School of Metabolism, to the US Army Quartermaster's Subsistence Research Laboratory that was established during the Second World War. They were particularly prevalent in spaces of bodily discipline, from hospitals and poorhouses to prisons and plantations. From the eighteenth century onward, in Europe and North America in particular, scientific food experimentation was also carried out in relation to more ambitious proposals for circumscribing the food entitlement of the hungry poor, always with the end goal of efficiently converting indolent bodies into laboring bodies. One of the best-known manifestations of this phenomenon was the work of American polymath Benjamin Thompson, Count Rumford, whose program of "poor soups" for institutionalized populations was widely taken up around Europe and across the Atlantic world. Based on the principle of minimizing waste, it sought to develop underused materials like animal fodder or bones for human

---

[19] Ulrike Thoms, "The Technopolitics of Food: The Case of German Prison Food from the Late Eighteenth to the Early Twentieth Centuries"; Treitel, "Nutritional Modernity" (both cit. n. 17); Dana Simmons, "Hungry, Thinking with Animals: Psychology and Violence at the Turn of the Twentieth Century"; Deborah Fitzgerald, "World War II and the Quest for Time-Insensitive Foods"; all in this volume.

[20] Elizabeth Neswald, David F. Smith, and Ulrike Thoms, eds., *Setting Nutritional Standards: Theory, Policies, Practices* (Rochester, N.Y.: Univ. of Rochester Press, 2017); John Burnett and Derek J. Oddy, eds., *The Origins and Development of Food Policies in Europe* (London: Leicester Univ. Press, 1994); Alexander Fenton, ed., *Order and Disorder: The Health Implications of Eating and Drinking in the Nineteenth and Twentieth Centuries* (East Linton, UK: Tuckwell, 2000); Thoms, *Anstaltskost im Rationalisierungsprozess: die Ernährung in Krankenhäusern und Gefängnissen im 18. und 19. Jahrhundert* (Stuttgart: Steiner, 2005).

[21] For a definition, see especially Chris Otter, "The British Nutrition Transition and Its Histories," *Hist. Comp.* 10–11 (2012): 812–25.

consumption.[22] Casting the development of useful knowledge about food as a form of social "improvement" was a priority of scientific inquiry among seventeenth-century clerics, eighteenth-century ameliorationist plantation physicians, and the mid-nineteenth-century laboratory of Glasgow physicist William Thomson, Lord Kelvin.[23] Similarly, experimental protocols devised in field, laboratory, or institutional settings during the nineteenth and twentieth centuries became the core of policies designed for the management of publics in the face of new challenges posed by changes in consumption practices, problems in provisioning networks, or programs of population engineering.

Even the sensation of hunger and responses to it have a scientific history, as Simmons argues.[24] Over time, the rise and consolidation of bureaucratic, technocratic forms of statecraft generated new efforts to manage the food supply, determine the "minimum needs" of individual bodies or social groups, and standardize dietary regimens.[25] Expert claims about food have often replicated or supported prevailing social orders, so that inequalities in resource entitlement have been built into the production of scientific knowledge about nutrition. Yet, much mainstream writing, not only in history but also in related disciplines, such as physical anthropology or archaeology, takes "biological need" for granted, accepting categories of analysis developed by and for modern nutrition science (often based on experimentation upon nonhuman model organisms) as tools for constructing accounts of past human embodiment and experience.[26] To take just one influential example from within the fields of demographic and social history, Robert William Fogel's *The Escape from Hunger and Premature Death* uses measures of adult height as a marker for the success of past societies in meeting individual nutritional needs, thereby proceeding as if the cultural definition and experience of hunger are historical constants.[27] Historians of science, technology, and medicine are well positioned to question the epistemological foundations of such approaches by addressing experiences of hunger and appetite as elastic concepts, contingent on complex

[22] Sandra Sherman, *Imagining Poverty: Quantification and the Decline of Paternalism* (Columbus: Ohio State Univ. Press, 2001); E. C. Spary, "Economic Eaters," chap. 1 in *Feeding France: New Sciences of Food, 1760–1814* (Cambridge: Cambridge Univ. Press, 2014); Paul Warde, *The Invention of Sustainability: Nature and Destiny, c. 1500–1870* (Cambridge: Cambridge Univ. Press, 2018); Simon Werrett, *Thrifty Science: Making the Most of Materials in the History of Experiment* (Chicago: Univ. of Chicago Press, 2019).

[23] On Lord Kelvin, see M. Norton Wise, "Work and Waste: Political Economy and Natural Philosophy in Nineteenth-Century Britain (III)," *Hist. Sci.* 28 (1990): 221–61.

[24] Dana Simmons, *Vital Minimum: Needs, Nature and Inequality in Modern France* (Chicago: Univ. of Chicago Press, 2015); see also Carla Cevasco, "Hunger Knowledges and Cultures in New England's Borderlands, 1675–1770," *Early Amer. Stud.* 16 (2018): 255–81.

[25] Wise, "Work and Waste" (cit. n. 23); Anya Zilberstein, "Bastard Breadfruit and other Cheap Provisions: Early Food Science for the Welfare of the Lower Orders," *Early Sci.& Med.* 21 (2016): 492–508; Philip Gibbes, *Instructions for the Treatment of Negroes*, 2nd ed. (1786; London: Shepperson & Reynolds, 1797).

[26] For important exceptions, see E. P. Thompson, "The Moral Economy of the English Crowd in the Eighteenth Century," *Past & Present* 50 (1971): 71–136; and James Vernon, *Hunger: A Modern History* (Cambridge, Mass.: Belknap Press of Harvard Univ. Press, 2007).

[27] Robert William Fogel, "The Persistence of Misery in Europe and America before 1900," chap. 1 in *The Escape from Hunger and Premature Death, 1700–2100* (Cambridge: Cambridge Univ. Press, 2004). See also, for example, Roderick Floud, Kenneth Wachter, and Annabel Gregory, *Height, Health, and History: Nutritional Status in the United Kingdom, 1750–1980* (New York, N.Y.: Cambridge Univ. Press, 1990). The demographic approach to food as a historical object forms its own subgenre, exemplified by works such as Lucile F. Newman, ed., *Hunger in History: Food Shortage, Poverty, and Deprivation* (Cambridge, Mass.: Blackwell, 1995); Robert I. Rotberg and Theodore K. Rabb, eds., *Hunger and History: The Impact of Changing Food Production and Consumption Patterns on Society* (Cambridge: Cambridge Univ. Press, 1985); and Floud, Wachter, and Gregory, *Height, Health and History* (this note).

circumstances—including the experimental milieu within which such concepts were tested—rather than reducing them to anachronistic calculations of body-weight averages, calorific requirements, and vitamin content. The common ground between the history of food and the history of science, from the early modern period to the present day, lies in large part in the ways individual bodies—both of people and of other animals—have repeatedly been mobilized within this emerging complex of knowledge and governance. Their combined methodologies therefore have the potential to transform how food is invoked as a causal factor in much broader historical narratives.

POLICING THE BOUNDARIES OF APPETITE

Bringing the critical apparatus developed within the field of the history of science, technology, and medicine to bear upon historical questions of diet and nutrition allows us to reinsert scientific accounts of food within their specific cultural circumstances of production and reception, and to evaluate what is at stake in universalizing concepts, recipes, and prescriptions stemming from particular historical conjunctures of knowledge and power. Holt's advocacy of insect eating was only one among many past proposals for the radical transformation of what was locally considered adequate or ordinary diet, breaching the divide between "purity" and "pollution" that has been explored by Mary Douglas and others in the anthropological domain.[28] It is no coincidence that Douglas would later turn her attention to problems of consumption, including eating and drinking.[29] Amid the mass transformation of diet in the modern West that has occurred since around 1700—the rise of industrial food manufacturing, the "nutrition transition," and the global application of modern nutrition science—we can discern, in palimpsestic form, the traces of other knowledge systems.[30] Today's culinary practices and preferences still enshrine practices drawn from early modern recommendations for healthy eating: adding spices as warming or preserving elements, frying cold, moist fish, or dressing salads with oil and vinegar are quotidian culinary practices that all have roots in humoral and iatrochemical dietetic recommendations.[31] The transition described by historians of medicine, from early modern humoral theory to modern nutrition science,

[28] Mary Douglas, *Purity and Danger: An Analysis of Concepts of Pollution and Taboo* (London: Routledge and Kegan Paul, 1966).

[29] Mary Douglas and Baron Isherwood, *The World of Goods: Towards an Anthropology of Consumption* (New York, N.Y.: Routledge, 1996); Douglas, "Deciphering a Meal," chap. 18 in Douglas, *Implicit Meanings: Selected Essays in Anthropology* (New York, N.Y.: Routledge, 1999), 231–51; Douglas, ed., *Constructive Drinking: Perspectives on Drink from Anthropology* (New York, N.Y.: Routledge, 2002).

[30] For a model study of such complex hybridizations of food knowledge and practice, see Marcy Norton, *Sacred Gifts, Profane Pleasures: A History of Tobacco and Chocolate in the Atlantic World* (Ithaca, N.Y.: Cornell Univ. Press, 2008).

[31] The literature on these developments is extensive, but key sources include Hans-Jürgen Teuteberg and Günter Wiegelmann, *Der Wandel der Nahrungsgewohnheiten unter dem Einfluß der Industrialisierung* (Göttingen: Vandenhoeck & Ruprecht, 1972); Alexander Fenton, ed., *Order and Disorder: The Health Implications of Eating and Drinking in the Nineteenth and Twentieth Centuries* (East Linton, UK: Tuckwell, 2000); Jack Goody, "Industrial Food: Towards the Development of a World Cuisine," in *Food and Culture: A Reader*, ed. Carole Counihan and Penny Van Esterik (New York, N.Y.: Routledge, 1997); and Stephen Mennell, Anne Murcott, and Anneke H. van Otterloo, *The Sociology of Food: Eating, Diet and Culture* (Newbury Park, Calif.: SAGE, 1992). On the early modern debts of today's cuisine, see Jean-Louis Flandrin, "Assaisonnement, cuisine et diététique," in *Histoire de l'alimentation*, ed. Flandrin and Massimo Montanari (Paris: Fayard, 1996), 491–509; Rachel Laudan, *Cuisine and Empire: Cooking in World History* (Berkeley and Los Angeles: Univ. of California Press, 2013); and Ken Albala, *Eating Right in the Renaissance* (Berkeley and Los Angeles: Univ. of California Press, 2002).

thus turns out to apply only within particular epistemological confines. Forms of old and new knowledge coexist rather than being mutually exclusive, so much is lost in pitting alleged vernacular against expert food knowledge in domains as varied as dietetics, gastronomy, agronomy, biotechnology, chemistry, economics, genetics, physiology, population theory, nutrition, psychology, or thermodynamics. Foods that are deemed to breach boundaries offer particularly interesting cases for historians of science, technology, and medicine to investigate. The caterpillar fungus in Lu's article straddles tradition and innovation; the cultured meat in Benjamin Aldes Wurgaft's account is neither wholly natural nor wholly artificial; and the artificially colored sausages in Carolyn Cobbold's contribution are variously judged to be both authentic and fake.[32]

Historians of food and the sciences should be particularly alert to such processes of exchange and assimilation between localized, scientific knowledge claims and collective cultural practices. Such fusion or "overwriting" of one system of dietary knowledge and practice by others has occurred repeatedly and under varied historical conditions. One of the best-known cases, characterized by the anthropologist Claude Lévi-Strauss as the "raw and the cooked," in fact owes much to Jean-Jacques Rousseau's figurative critique of French courtly society in terms of its diet. A very similar politics of eating is identified by Timothy Morton in the poetics of Percy Bysshe Shelley's opposition to meat eating as a rejection of capitalism.[33] Recombination of earlier dietary systems continues to occur. For example, experimental diets such as the so-called "paleo" or fasting diets, though advertised as "alternative" or "new," are often indebted to Judeo-Christian, Romantic, or Fascist back-to-nature claims about proper regimen. Even Holt's suggestion that Europeans eat insects as a thrifty and rational use of nutritive resources was not new in the late nineteenth century; insect eating had been proposed in the eighteenth century by the astronomer Jérôme de Lalande.[34] Seen as an eccentricity for over two centuries, entomophagy has begun to gain credibility in more recent times, thanks in large part to widespread public concern about the ecological consequences of meat eating. British schoolchildren can now enjoy the odd locust lollipop on a museum field trip, while shuddering at tales of cultures around the world where insects and arachnids are normal dietary components. A new microlivestock industry in the United States and Thailand, developed by a Harvard University-sponsored start-up, cultivates crickets for use in pulverized form as the basis of snack foods.[35] Increasing public commitment to

[32] Lu, "Local Food" (cit. n. 16); Benjamin Aldes Wurgaft, "Meat Mimesis: Laboratory-Grown Meat as a Study in Copying"; Carolyn Cobbold, "The Introduction of Chemical Dyes into Food in the Nineteenth Century"; both in this volume.

[33] Claude Lévi-Strauss, *Introduction to a Science of Mythology*, I: *The Raw and the Cooked*, trans. John Weightman and Doreen Weightman (London: Cape, 1970); Timothy Morton, *Shelley and the Revolution in Taste: The Body and the Natural World* (Cambridge: Cambridge Univ. Press, 1994).

[34] On Lalande, see E. C. Spary, "Eating Beyond Reason" (unpublished manuscript, 2019).

[35] The question of insect eating provoked lively debate in the early 2010s press. See "Why Not Eat Insects?," *Guardian*, 3 December 2010, https://www.theguardian.com/science/punctuated-equilibrium/2010/dec/02/2; Stefan Gates, "Why Not Eat Insects?," Food Blog, BBC, 11 March 2011, http://www.bbc.co.uk/blogs/food/2011/03/why-not-eat-insects.shtml; Joseph Milton, "Why Not Eat Insects? I'll Give You a Couple of Reasons," *Creatology* (blog), *Sci. Amer.*, 29 August 2011, https://blogs.scientificamerican.com/creatology/why-not-eat-insects-ill-give-you-a-couple-of-reasons/; and Krystal D'Costa, "What's Stopping Us from Eating Insects?," *Anthropology in Practice* (blog), *Sci. Amer.*, 24 July 2013, https://blogs.scientificamerican.com/anthropology-in-practice/whats-stopping-us-from-eating-insects/). These followed a public talk by the ecological entomologist Marcel Dicke, "Why Not Eat Insects," filmed on 15 July 2010 at TEDGlobal 2010, Oxford, TED video, 16:34, https://www.ted.com/talks/marcel_dicke_why_not_eat_insects). Former Harvard student Rose Wang's start-up "Chirps" sells insect-based products; see https://eatchirps.com/; and https://innovationlabs.harvard.edu/meet/student-story/rose-wang/.

environmentalism and efforts to address the legacies of colonialism, coupled with the vogue for molecular gastronomy of the early twenty-first century, generated conditions under which a long-lived knowledge claim about the edibility of insects—often discounted in the past as illegitimate because of its associations with food cultures of the southern hemisphere—has now penetrated into cosmopolitan realms of the gustatory-experimental imaginary, and from there into regimes of regulation and commerce.[36]

Because food is both matter taken *into* the body and a medium of relations *between* individuals (within households or wider polities), studying it allows issues of governance to be linked to questions of embodiment and self-fashioning. Taste provides an important locus of resistance to significant transformations in diet or the food supply, something that is particularly apparent when consumers encounter new foods. The many food experiments carried out by scientific and medical practitioners, governments, and businesses, especially in attempts to supplement or substitute customary foods with novel substances, have often met with opposition or indifference, or else have taken decades—or even centuries—to lose the whiff of suspiciousness and inferiority. Well-known examples of this phenomenon include eighteenth-century efforts to reduce dependence on wheat by promoting breadfruit, potatoes, and wild rice across the British Empire; or the use of treacle, vegetable oil, and chicory as *ersatz* foods during wartime shortages of sugar, butter, and coffee. At best, the imposition of such surrogates produced uncertain success. While historians often treat the history of alternative foods as ornamental appendages to more profound transformations in subsistence patterns or provisioning networks, these stories raise the same problems as Holt's pamphlet. Under what circumstances can or should a new food come to be accepted by consumers? What is the legitimate scope of the state's authority to intervene in the food supply, and how far should the law constrain food merchants and manufacturers? These questions, of great interest both historically and in the present day, show why food can be such an inflammatory topic: it is a site of direct encounter between individuals and larger social structures, or transformations, over which they may have little power.[37]

Because no form of matter is more directly relevant to the body and the self than food, dietary choices are indexical and constitutive, and play an integral role in cultures of self-fashioning, as well as in attempts by intellectuals, officials, and medical or religious authorities to classify individuals into groups. Some substances, such as red meat, have been alternately stigmatized and favored for their effects upon physical, mental, national, or spiritual health. Particularly in colonial societies, consuming or abstaining from an unfamiliar food might enhance or undermine one's social status, or worse, threaten the integrity of one's constitution and identity.[38] As new scientific

---

[36] Hervé This, *Molecular Gastronomy: Exploring the Science of Flavor* (New York, N.Y.: Columbia Univ. Press, 2002); Monica Bodirsky and Jon Johnson, "Decolonizing Diet: Healing by Reclaiming Traditional Indigenous Foodways," *Cuizine* 1 (2008), https://doi.org/10.7202/019373ar; for Food and Agricultural Organization (FAO) forecasts about insect farming, see their web page, "Insects for Food and Feed," http://www.fao.org/edible-insects/en/.

[37] Spary, *Feeding France* (cit. n. 22); Anya Zilberstein, "Inured to Empire: Wild Rice and Climate Change," *William Mary Quart.* 72 (2015): 125–56; Rebecca Earle, "Promoting Potatoes in Eighteenth-Century Europe," *Eighteenth-Cent. Stud.* 51 (2017): 147–62; Hans-Jürgen Teuteberg, "The Birth of the Modern Consumer Age," in Freedman, *Food* (cit. n. 3), 233–62.

[38] Rebecca Earle, *The Body of the Conquistador: Food, Race, and the Colonial Experience in Spanish America, 1492–1700* (New York, N.Y.: Cambridge Univ. Press, 2012); Jeffrey M. Pilcher, *The Sausage Rebellion: Public Health, Private Enterprise, and Meat in Mexico City, 1890–1917* (Albuquerque: Univ. of New Mexico Press, 2006); Anita Guerrini, "Health, National Character and the English Diet in 1700," *Stud. Hist. Phil. Biol. Biom. Sci.* 43 (2012): 349–56.

and medical accounts of food emerged in tandem with new agendas for the government of human and other animal bodies, they played a leading role in generating modern economic models of the circulation of resources throughout society, serving to reinforce or recreate social hierarchies. It should not be surprising, then, that scientific models of diet also become enrolled in policing social relations, colonial and national boundaries, and the role of households or consumers within the polity. Dietary choice, where it has been an option, can indicate consumers' priorities for corporeal or mental self-fashioning, as well as the ways commercial agents have trafficked and translated food knowledge between separate domains such as the laboratory, law court, and kitchen, as the articles of Guerrini and Shapin show.[39]

Taste, the Ultima Thule of historical investigation, has been especially distant from the usual themes embraced by historians of science. Yet, it is a topos of scientific concern, both for food manufacturers interested in making their products more appealing to consumers, and for scientists studying the senses and phenomena of embodied cognition. Thoms, Treitel, and Fitzgerald, addressing themes such as the interplay between the emergence of new food categories or standards and the engineering of new processes of food production, show how these were based on contemporary presumptions about the universality of taste or faith in scientists' ability to define a "standard consuming body."[40] Yet, as the Oxford chemist Charles Spence notes, "A growing body of scientific research now suggests that our experience of taste and flavor is determined to a large degree by the expectations that we generate (often automatically) prior to tasting."[41] The vast body of research conducted on different foods and drinks by the chemist Rose Marie Pangborn at the University of California, Davis, from the 1960s onward, for example, has yet to receive any historical attention.[42] Aron's and Cobbold's articles in our collection offer two case studies of how the definition of the taste of *terroir* in French viniculture on the one hand, and the regulation of artificial food additives on the other, became entangled in debates about authenticity, fraudulence, and the reliability of science in appraising or changing food composition.[43] Studying food knowledge, discourse, and praxis allows us to ask questions that are otherwise difficult to approach historically, such as how embodied experience articulates with scientific knowledge claims; how traditions bear upon political or ecological concerns; or how gastronomic experimentation confronts culturally localized senses of danger or disgust.[44]

Sciences of food engage with precisely such questions about boundaries and connections, and so afford a valuable entrée into how big categories such as "nature,"

---

[39] Anita Guerrini, "A Natural History of the Kitchen"; Steven Shapin, "Breakfast at Buck's: Informality, Intimacy, and Innovation in Silicon Valley"; both in this volume.

[40] Thoms, "Technopolitics of Food" (cit. n. 19); Treitel, "Nutritional Modernity" (cit. n. 17); Fitzgerald, "Time-Insensitive Foods" (cit. n. 19); all in this volume.

[41] Charles Spence, "On the Psychological Impact of Food Colour," *Flavour* 4 (2015): 1–16, https://doi.org/10.1186/s13411-015-0031-3; R. Deliza and H. J. H. MacFie, "The Generation of Sensory Expectation by External Cues and Its Effect on Sensory Perception and Hedonic Ratings: A Review," *Journal of Sensory Studies* 11 (1996): 103–28. This kind of investigation began in the late 1950s.

[42] For Panghorn's papers, see https://www.researchgate.net/scientific-contributions/2065837658 _Rose_Marie_Pangborn, accessed 29 September 2018.

[43] Alissa Aron, "Perceptions of Provenance: Conceptions of Wine, Health, and Place in Louis XIV's France"; Cobbold, "Chemical Dyes" (cit. n. 32); both in this volume.

[44] On disgust, see Carolyn Korsmeyer, "Delightful, Delicious, Disgusting," *J. Aesthet. Art Crit.* 60 (2002): 218–25; Korsmeyer, *Savoring Disgust: The Foul and the Fair in Aesthetics* (Oxford: Oxford Univ. Press, 2011); Lauren Janes, "Exotic Eating in Interwar Paris: Dealing With Disgust," *Food & History* 8 (2010): 237–56; and Christopher Forth, "Fat, Desire and Disgust in the Colonial Imagination," *Hist. Workshop J.* 73 (2012): 211–39.

"culture," "knowledge," and "power" have been generated through the manipulation of the material world, epistemic communities, and bodily practices. In the realms of recipe composition, ingredient selection, food processing, preservation and storage, or cookery and baking skill, for example, we might ask about the effects upon food knowledge of the separation of laboratory and industrial spaces from domestic sites of production, about when and how those distinctions emerged, and about how new forms of expertise (and new groups of experts) gained or lost credit. We might also consider the perpetual interplay between what counts as "archaic" and what counts as "modern" food technology. The French company St. Dalfour makes its fruit preserves according to a "Traditional French Recipe," using grape syrup rather than cane sugar as a sweetener. As historians, we can pinpoint this tradition fairly precisely; it originated in efforts by the French emperor Napoleon I to promote indigenous substitutes for cane sugar during the Continental Blockade, which cut French consumers off from colonial trade.[45] Out of a very specific political crisis emerged an experimental innovation—grape syrup—implemented in factory production as an early industrial food during the 1810s, only to be replaced by the beet sugar industry before coming to be recycled as "traditional French cuisine." Studying historical cases points to a common tension in the relationship between food and the sciences: scientific and medical claims are constantly being commodified by food producers, even while the sciences of physiology, food processing, and nutrition are themselves constantly in flux in response to changes in legislation, food technology, or consumer preference. Such feedback loops contrive to blur the separation between "lay" and "expert" food knowledge and practice, forcing us to reconsider category boundaries and recognize the plurality of forms of expertise involved with food, which in turn lead to varied definitions of needs, salubrity, pleasure, or ethical relations.[46]

A similar troubling of boundaries occurs if we consider the circumstances under which particular foods are allocated to (or withheld from) specific groups of eaters, such as "hospital patients," "prisoners," or "children." Likewise, the classification of substances into categories—food/nonfood, healthy/junk, food/medicine, or food/drug—is historically fraught, the outcome of extensive prior work to establish and police such boundaries. The nonobvious and reversible nature of this process is strikingly apparent if we consider substances that have migrated from being seen as waste or nonfood to the realms of the edible, or vice versa. Exploring such changing boundaries between food and nonfood shows why, even as food governance has become ever more tightly coupled to scientific knowledge claims, the reputation of any particular food product or ingredient as "healthy," "safe," or "authentic" has tended to have a limited shelf life. When past experimenters set out to replace one food with another (or otherwise intervene in established dietary practices), whether at home or in colonial situations, they usually needed to take up a position on what the "essence" of a particular ingredient or dish, or indeed, of food in general, was. In making such claims, they provoked debate

---

[45] Spary, "The Empire of Habit," chap. 8 in *Feeding France* (cit. n. 22).

[46] Similar cases of the appropriation of tradition and localism to enhance the value of high-tech, scientifically based food production have been described by Steven Shapin in "Cheese and Late Modernity" (review of *Camembert: A National Myth*, by Pierre Boisard), *London Review of Books*, 20 November 2003; "Hedonistic Fruit Bombs" (reviews of *Bordeaux*, by Robert Parker; *The Wine Buyer's Guide*, by Robert Parker and Pierre-Antoine Rovani; and *Mondovino*, directed by Jonathan Nossiter), *London Review of Books*, 3 February 2005. On tensions between localism and the environmental politics of global food networks, see Anna L. Tsing, *The Mushroom at the End of the World: On the Possibility of Life in Capitalist Ruins* (Princeton, N.J.: Princeton Univ. Press, 2015).

not only about the unique power and qualities of specific foods, but also about their own expertise.

Policies to restrain fraudulence in the food supply, legal prohibitions against particular foods, or advice about proper diet, often presented as self-evident measures by scientific or medical practitioners, almost invariably point, on closer inspection, to the historical circumstances of their coming-into-being. During the shortages of the Napoleonic era, for example, the British government strongly encouraged the consumption of artificial wine manufactured by chemists and druggists using wild fruits. But within a few decades after the conclusion of hostilities, a book by the German chemist Friedrich Accum on the adulteration of commercial food and beverages recast artificial wine as frightful evidence of British grocers' involvement in fraudulent practices.[47] Accum's book was written at a juncture when analytical chemists were struggling to assert the supremacy of their discipline's expertise over the wealth and social authority of grocers. It proved persuasive; its many readers became convinced of the dire state of the food trade and invoked the book in calls for legislation to restrain the food industry. New laws governing food production, so far from being milestones of "progress" in food safety, are complex negotiations, the product of contestations for authority over food production and knowledge. As Alessandro Stanziani brilliantly shows, late nineteenth-century French law banned the sale of admixtures of butter and margarine, even though the sale of each of these foods on its own was permitted.[48] Scientific and expert claims about food, in other words, have often been interventions in, or debates about, a hierarchy of forms of expertise in relation to the public domain. The foods produced and marketed in a given time and place were the artifacts of these contests.

FOOD SCIENCES, IN AND BEYOND THE LABORATORY

These attempts to forge or reform expert knowledge in relation to food and beverages have long escaped scholarly scrutiny, *Food Matters* suggests, in significant part because of the lingering influence of the dualist tradition, which largely excluded matter, practice, and embodiment from the scope of the history of science, technology, and medicine. This has cast an especially long shadow over cooking, eating, and drinking, perhaps because of their associations with pleasure, the baser senses, and the passions, rather than with reason.[49] The conceptual dichotomy between mind and matter—upon which histories of the sciences long rested—was reinforced by the consolidation of laboratory science as a standard means of making natural knowledge. That development further relegated food to the realm of impure, applied, and feminized craft knowledge, as distinct from pure and manly sciences devoted to original discovery. It can seem surprising that

---

[47] Friedrich Accum, *A Treatise on Adulterations of Food, and Culinary Poisons, Exhibiting the Fraudulent Sophistications of Bread, Beer, Wine, Spirituous Liquors, Tea, Coffee, Cream, Confectionery, Vinegar, Mustard, Pepper, Cheese, Olive Oil, Pickles, and Other Articles Employed in Domestic Economy. And Methods of Detecting Them* (London: Longman, Hurst, Rees, Orme, and Brown, 1820).

[48] Alessandro Stanziani, *Histoire de la qualité alimentaire: XIXe—XXe siècle* (Paris: Seuil, 2005); Frederick Filby, *A History of Food Adulteration and Analysis* (London: Allen & Unwin, 1934). On artificial wine projects, see Spary, *Feeding France* (cit. n. 22), 6–7, 161; and Benjamin R. Cohen, *Pure Adulteration: Cheating on Nature in the Age of Manufactured Food* (Chicago: Univ. of Chicago Press, 2020).

[49] Viktoria von Hoffmann, *From Gluttony to Enlightenment: The World of Taste in Early Modern Europe* (Urbana: Univ. of Illinois Press, 2016).

well-known men of science like André-Marie Ampère or Kelvin conducted extensive research into food, even while simultaneously penning the laws of physics. Their involvement is little known to historians, in part because both they and their biographers often
downplayed the significance of such activity.[50] Lalande's habit of eating spiders, originally
undertaken as a corporeal demonstration of the power of reason over prejudice, would later
be appropriated as a stratagem for discrediting and ridiculing him.[51] Ironically, a deliberate
move by professional women scientists in Europe and North America at the turn of the
twentieth century to establish the new discipline of "domestic science" (or home economics, as it was later known) as an avowedly feminine domain of expertise served only to
reinforce cultural and historiographical prejudice about the relatively peripheral place of
food in the history of science.[52]

Yet, as Guerrini's and Shapin's articles make clear, spaces of food procurement,
preparation, and consumption are also spaces of knowledge production and circulation.[53] Close attention to spaces in which recipes or technological ideas circulated can
show how many apparently autonomous scientific principles, practices, and inventions—some seemingly distant from food per se, such as taxonomical descriptions
of birds or pitches for high-tech start-ups—in fact may emerge from cooking experiments in household kitchens or conversations over meals in a diner.[54] One example
that should resonate with historians of science illustrates this ongoing permeability
between the spaces of food and experimentation. The Huguenot physician Denis
Papin, fleeing Louis XIV's sanctions against Protestantism in France, found refuge
in London working as a technician to Robert Boyle, natural philosopher and cofounder
of the Royal Society. Papin's digester has attracted interest within a "Scientific Revolution" historiography as a key device within the experimental tradition surrounding the steam pump.[55] But contemporaries like the French academician Henri Justel
quickly appropriated the digester as a culinary device first and foremost, a way to render matter as hard as an "ivory Ball" soft enough to eat.[56] Experimentation on cuisine

[50] On Ampère, see Spary, *Feeding France* (cit. n. 22), 18; on Kelvin, see Wise, "Work and Waste"
(cit. n. 23).

[51] Spary, "Eating Beyond Reason" (cit. n. 34).

[52] Sarah Stage and Virginia B. Vincenti, eds., *Rethinking Home Economics: Women and the History
of a Profession* (Ithaca, N.Y.: Cornell Univ. Press, 1997); Maresi Nerad, *The Academic Kitchen: A
Social History of Gender Stratification at the University of California, Berkeley* (Albany: State Univ.
of New York Press, 1999); Yuriko Akiyama, *Feeding the Nation: Nutrition and Health in Britain Before World War I* (London: I. B. Taurus, 2008); Carolyn M. Goldstein, *Creating Consumers: Home
Economists in Twentieth-Century America* (Chapel Hill: Univ. of North Carolina Press, 2012).

[53] Guerrini, "Natural History of the Kitchen"; Shapin, "Breakfast at Buck's" (both cit. n. 39).

[54] See also Elaine Leong, *Recipes and Everyday Knowledge: Medicine, Science, and the Household
in Early Modern England* (Chicago: Univ. of Chicago Press, 2018); Leong and Alisha Rankin, eds.,
Secrets and Knowledge in Medicine and Science, 1500–1800 (Burlington, Vt.: Ashgate, 2011). For
domestic experimentation more generally, see also Pamela H. Smith, *The Body of the Artisan: Art
and Experience in the Scientific Revolution* (Chicago: Univ. of Chicago Press, 2004); William Eamon,
Science and the Secrets of Nature: Books of Secrets in Medieval and Early Modern Culture (Princeton,
N.J.: Princeton Univ. Press, 1994); William R. Newman and Anthony Grafton, eds., *Secrets of Nature:
Astrology and Alchemy in Early Modern Europe* (Cambridge, Mass.: MIT Press, 2001); Tara
Nummedal, *Alchemy and Authority in the Holy Roman Empire* (Chicago: Univ. of Chicago Press,
2007); and Werrett, *Thrifty Science* (cit. n. 22).

[55] For example, see David Wootton, *The Invention of Science: A New History of the Scientific Revolution* (New York, N.Y.: HarperCollins, 2015), 492–508.

[56] Henri Justel to John Locke, 10 January 1680, Electronic Enlightenment Scholarly Edition of Correspondence, https://doi.org/10.13051/ee:doc/lockjoOU0020146a1c; see also Spary, "Making More out
of Meat," chap. 6 in *Feeding France* (cit. n. 22).

and on natural processes have rarely, if ever, been distinct enterprises. Conversely, food has often played a central role in the making of scientific, medical, or technological knowledge and authority, often via the locus and manner of its consumption. Eating together in the home, coffeehouse, diner, or lab affords opportunities for sociability, the creation of networks, and the brokerage of knowledge, skill, and credit, whether at the famous Victorian dinner inside the Crystal Palace iguanodon, or among the Silicon Valley venture capitalists studied by Shapin in this volume.[57]

Critically examining the culture and politics of eating further allows historians of science, technology, and medicine to understand how and why certain epistemic developments or technical innovations in food and diet have met with vigorous, organized opposition. When universal standards for nutritional intake, universal accounts of foods' nature and effects, and universal claims about the homogeneity of all matter are exported outside the laboratory, a common reaction has been the reinvention of localism.[58] Whether it be the courtiers of Louis XIV's day (discussed by Aron) who fell back on *terroir* to defend their local power and distinctiveness against the French Crown's attempts at cultural hegemony; the nineteenth-century defenders of regional cuisine against industrialization and urbanization; or today's fears of genetically modified or irradiated produce and embrace of "slow food," these projects of resistance have commonly endorsed forms of food knowledge that were explicitly nonuniversal or antimodern.[59] Such campaigns disclose the food politics that gestures of universalism conceal.[60] The phrase "food miles" began to become current soon after 2000. It emerged directly out of environmentalist calls for a revised economics that included ecological impact within the cost of producing and consuming a given commodity.[61] This move has reshaped the priorities of today's supermarket shoppers, forging a new relationship between food and spatiality. Rising public pressure on food producers to disclose the place of origin of foods has then fused, not always benignly, with exhortations to buy only foods produced within national boundaries, or with attempts to prohibit the consumption of "foreign" foods.[62]

These two trends of localization and globalization have worked dialectically over centuries, producing a repeated remapping of the actual and imagined geography of

[57] J. A. Secord, "Monsters at the Crystal Palace," in *Models: The Third Dimension of Science*, ed. Soraya de Chadarevian and Nick Hopwood (Stanford, Calif.: Stanford Univ. Press, 2004), 138–69, on 150–3. The dinner was described in *Illustrated London News*, 7 January 1854, 22. It copied the genre of Victorian scientific dining clubs such as the X-club; see Roy M. MacLeod, "The X-Club: A Social Network of Science in Late-Victorian England, *Notes Rec. Roy. Soc. Lond.* 24 (1970): 305–22.

[58] Richard Drayton and David Motadel, "Discussion: The Futures of Global History," *J. Glob. Hist.* 13 (2018): 1–21.

[59] Aron, "Perceptions of Provenance" (cit. n. 43).

[60] Julia Abramson, "Legitimacy and Nationalism in the *Almanach des Gourmands* (1803–1812)," *J. Early Mod. Cult. Stud.* 3 (2003): 101–35; Julia Csergo, "La constitution de la spécialité gastronomique comme objet patrimonial en France (fin XVIIIe – XXe siècle)," in *L'Esprit des lieux: Le patrimoine et la cité*, ed. Daniel J. Grange and Dominique Poulot (Grenoble: Presses Univ. de Grenoble, 1997), 183–93; Csergo, "The Emergence of Regional Cuisines," in Flandrin and Montanari, *Food* (cit. n. 3), 500–15.

[61] Stephen Bentley, *Fighting Global Warming at the Farmer's Market: The Role of Local Food Systems in Reducing Greenhouse Gas Emissions*, 2nd ed. (Toronto: FoodShare, 2005), https://foodshare .net/custom/uploads/2015/11/Fighting_Global_Warming_at_the_Farmers_Market.pdf; see also Craig Sams, *The Little Food Book* (Bristol: Alastair Sawday, 2003), 51; William Lockeretz, *Ecolabels and the Greening of the Food Market* (Boston: Friedman School of Nutrition Science and Policy, Tufts University, [2003]), 69; and Giovanni Rebora, *Culture of the Fork: A Brief History of Food in Europe*, trans. Albert Sonnenfeld (New York, N.Y.: Columbia Univ. Press, 2001).

[62] Joanna Bourke, "Pubs Giant JD Wetherspoon to Stop Selling Jägermeister in Drinks Shake-Up Before Brexit," *Evening Standard* (London), 12 September 2018.

food environments. From debates about the components of *terroir* to proprietary practices of selective breeding of livestock or edible plants, the aims and methods employed in experimental laboratory and field programs in the food sciences have been informed by, and have helped to shape, an array of institutions, labor regimes, cultural practices, and ethical commitments. This is particularly visible in Woods's discussion of the sources, quality, and regulation of preserved meat exported from New Zealand in the nineteenth century, and in Wurgaft's exploration of recent investment in the use of stem-cell technology to produce lab-grown meat.[63] A similar politics of resistance has underlain the development of "alternative" diets, which often use radical, scientific knowledge-claims about food to critique the status quo—an example here being the marginal status accorded to vegetarianism over many centuries.[64] Explaining these connections in turn helps to explain how it is that sciences of food have underpinned projects as disparate as purifying individual bodies, reforming the poor, or saving the environment.

CONCLUSION

Eating well, as Bryan S. Turner underlined in a now classic study, has been the subject of scholarly writing for millennia, not just centuries.[65] It is what Claude Fischler terms the "paradox of the omnivore": because what is eaten becomes part of the eater, the alien qualities of food persistently threaten to overwhelm, disfigure, or supplant identity.[66] The historical study of food knowledge provides ample matter for contending that the question of which foods are "good" or "healthy" has never been the subject of universal consensus, from the early modern period right up to the present day. Nor has this judgment ever been free of political significance. Throughout recorded history, diet and eating have been the subject of profound disquiet. During the Renaissance, as Ken Albala shows, food became a particularly charged arena of temptation, offering an unsettling prospect of potentially uncontrolled transformation of the self, which had to be harnessed by right eating, temperance, and self-discipline.[67]

Considering the three and a half centuries from that period to the present day, perhaps the most striking feature of the emergence of the food sciences is the way that the physiological, neurochemical, and biomedical understandings of food that emerged in the nineteenth century, in particular, are now deeply imbricated in the way most people in late modernity understand themselves.[68] The vast shift in self-understandings over the period covered by this collection is brought home to students when they are asked to say whether they have ever weighed themselves or been weighed, and then to reflect

---

[63] Woods, "Preserving Animal Flesh (cit. n. 16); Wurgaft, "Meat Mimesis" (cit. n. 32).

[64] Tristram Stuart, *The Bloodless Revolution: A Cultural History of Vegetarianism from 1600 to Modern Times* (New York, N.Y.: Norton, 2007); Corinna Treitel, *Eating Nature in Modern Germany: Food, Agriculture, and Environment, c. 1870–2000* (New York, N.Y.: Cambridge Univ. Press, 2017); Fenton, *Order and Disorder* (cit. n. 20); Joshua Specht, *Red Meat Republic: A Hoof-to-Table History of How Beef Changed America* (Princeton, N.J.: Princeton Univ. Press, 2019).

[65] Bryan S. Turner, *The Body and Society: Explorations in Social Theory*, 2nd ed. (London: SAGE, 1996).

[66] Claude Fischler, *L'Homnivore. Le goût, la cuisine et le corps* ([Paris]: Editions Odile Jacob, 1993).

[67] Albala, *Eating Right* (cit. n. 31). See also Sandra Cavallo and Tessa Storey, *Healthy Living in Late Renaissance Italy* (Oxford: Oxford Univ. Press, 2013).

[68] Coveney, *Food, Morals and Meaning* (cit. n. 2); Chris Shilling, *The Body and Social Theory* (London: SAGE, 1993); Gronow, *The Sociology of Taste* (cit. n. 2).

on how that self-knowledge now shapes their sense of *who they are*. The rise of the calorie, the language of personal virtue as the default way to talk about dietary choices, and anxieties about the relationship between these choices and one's body mass index, turn out to be the end product of vast arrays of scientific, medical, and technological enterprises that have reshaped selfhood in profound ways, and have become integral to late modern individuals' self-image and self-fashioning.[69]

Bringing together articles written by a group of scholars exploring these and many other related questions, this collection thus sets out to establish the significance of the history of food as a growth area within the history of science, technology, and medicine. It suggests how rich the history of food can be as a subject area for the historian of science, technology, and medicine who is interested in the nexus between material culture, technology, taxonomy, ethics, aesthetics, embodiment, identity, and authority. Yet, until now, few studies have explored the historical processes through which scientific and medical knowledge claims gained such power to shape at once the political, financial, and technical contours of local and international food supply chains, and the cultural and personal dynamics of consumer choice. By highlighting how the historiographies of food and of science, technology, and medicine connect with one another, *Food Matters* should also help readers to identify many other prospective topics that have yet to receive sustained attention within the field. For this collection does not lay claim to exhaustive treatment of a topic of such breadth and depth. Inevitably, many domains remain as fruitful prospects for further inquiry. The culture and politics of labor in the food sciences, including historical changes in farming and cooking practices, as well as in the associated *savoir-faire* or skill, and the gender, ethnic, and socioeconomic dynamics underlying divisions of labor, are all areas that merit further intensive research.[70] Some of the most interesting work in the sciences of food in recent decades has emerged from anthropological methodologies that attend to material culture, ritual, status, and display, offering resources for current explorations of ways in which materiality and spatiality shape the production of natural knowledge, the relationship between gestural or tacit knowledge, the embodiment of gender and social standing, and the extension of "thing theory" to consumed substances.[71]

---

[69] Lucia Dacome, "Living with the Chair: Private Excreta, Collective Health and Medical Authority in the Eighteenth Century," *Hist. Sci.* 39 (2001): 467–500; Anita Guerrini, *Obesity and Depression in the Enlightenment: The Life and Times of George Cheyne* (Norman: Univ. of Oklahoma Press, 2000); Steven Shapin, "Trusting George Cheyne: Scientific Expertise, Common Sense, and Moral Authority in Early Eighteenth-Century Dietetic Medicine," *Bull. Hist. Med.* 77 (2003): 263–97; Michael Stolberg, "'Abhorreas pinguedinem': Fat and Obesity in Early Modern Medicine," *Stud. Hist. Phil. Biol. Biomed. Sci.* 43 (2012): 370–8. For modern weighing programs, see Lawrence T. Weaver, "In the Balance: Weighing Babies and the Birth of the Infant Welfare Clinic," *Bull. Hist. Med.* 84 (2010): 30–57; and Roberta Bivins and Hilary Marland, "Weighting for Health: Management, Measurement and Self-Surveillance in the Modern Household," *Soc. Hist. Med.* 29 (2016): 757–80. On fatness and self-image, see especially Peter N. Stearns, *Fat History: Bodies and Beauty in the Modern West* (New York, N.Y.: New York Univ. Press, 2002); Sander Gilman, *Fat: A Cultural History of Obesity* (Cambridge: Polity, 2008); and Georges Vigarello, *Metamorphoses of Fat: A History of Obesity*, trans. C. Jon Delogu (New York, N.Y.: Columbia Univ. Press, 2013). On the emergence of the calorie, see especially Dietrich Milles, "Working Capacity and Calorie Consumption: The History of Rational Physical Economy," in Kamminga and Cunningham, *Science and Culture of Nutrition* (cit. n. 7), 75–96; Nick Cullather, "The Foreign Policy of the Calorie," *Amer. Hist. Rev.* 112 (2007): 336–64; and Corinna Treitel, "Max Rubner and the Biopolitics of Rational Nutrition," *Cent. Eur. Hist.* 41 (2008): 1–25.

[70] Sophia Roosth, "Of Foams and Formalisms: Scientific Expertise and Craft Practice in Molecular Gastronomy," *Amer. Anthropol.* 115 (2013): 4–16.

[71] Lissa Roberts, ed., *The Mindful Hand: Inquiry and Invention from the Late Renaissance to Early Industrialisation* (Chicago: Univ. of Chicago Press, 2008); Paula Findlen, ed., *Early Modern Things:*

The history of knowledge about food—as well as the knowledge produced in the processes of making, sharing, and arguing about it—has always raised vexing questions about the shifting definition and boundaries of expertise between traditional recipes and experimental protocols; between domestic craft skill and laboratory procedure; and between the distribution of resources throughout the social body on the one hand, and the subjective experiences of individual bodies on the other. At a moment when the authority of science is being questioned by a variety of publics, *Food Matters* is a timely reminder that such tensions were always present in food-related domains of knowledge; indeed, debates over food have expressed the historical circumstances under which modern science became a prevalent force in many areas of public and private life. Appropriately, perhaps, the plan for this volume came into being electronically and transatlantically, but first acquired substance over soup and bread at the ICA Café in London—a space of moody artworks, abstract figurations of dancers' bodies, and healthy food, located a stone's throw from the Royal Society.

Objects and their Histories, 1500–1800 (London: Routledge, 2012); Harry Collins, *Tacit and Explicit Knowledge* (Chicago: Univ. of Chicago Press, 2010); Hjalmar Fors, Lawrence M. Principe, and H. Otto Sibum, "From the Library to the Laboratory and Back Again: Experiment as a Tool for Historians of Science," *Ambix* 63 (2016): 85–97; E. C. Spary and Ursula Klein, eds., *Materials and Expertise in Early Modern Europe: Between Market and Laboratory* (Chicago: Univ. of Chicago Press, 2010).

A Natural History of the Kitchen

by Anita Guerrini*

ABSTRACT

On the evidence of early modern European cookbooks, the wild birds deemed edible around 1600 included cranes, herons, swans, and cormorants. By 1750, none of these were considered palatable. While culinary historians have explored the transition from medieval to modern food marked by François Pierre de La Varenne's *Le cuisinier françois* (1651), less attention has been afforded to another major transition in the European diet—that of eating a large variety of wild and domesticated animals to eating only a few. Historians of science in particular have largely ignored this transition, yet this shift in the definition of what was edible held profound implications for the changing roles of animals in both diets and scientific study. Two kinds of printed sources that are not commonly consulted together—cookery books and works on natural history—afford a new comparative glimpse of early modern animals and their varied meanings. In 1600, cookbooks and natural histories included many of the same animals and talked about some of the same things, since the category of "use" applied to both. By the time Vincent La Chapelle's *The Modern Cook* appeared in 1733, these works had entirely diverged in content. Animals in cookbooks became a means to an end rather than a topic of study, while natural histories ceased to talk about uses and considered instead comparative anatomy and classification. Looking particularly at birds, this article tracks changing meanings surrounding both animals and diets in early modern Europe. The turkey emerges as the critical indicator of these changes in its naturalization from a wild exotic species to a familiar farmyard animal.

INTRODUCTION

One day in the early 1670s, a cormorant wandered into the kitchen at a hostel outside Paris. It attacked the cook, who killed it. It was then dispatched to the Paris Academy of Sciences.[1] The natural history of the cormorant in the Academy's *Mémoires pour servir à l'histoire naturelle des animaux* (1671–76) included this provenance but did

* School of History, Philosophy, and Religion, Oregon State University, Corvallis, OR 97331, USA; anita.guerrini@oregonstate.edu.

I am grateful to the Horning Endowment at Oregon State University and the Descartes Center at the University of Utrecht for financial support. The Linda Hall Library, University of Missouri, Kansas City, provided their usual excellent service for images. For their comments I thank the *Osiris* series editors, the vol. 35 editors, two external referees, and Michael A. Osborne.

[1] "Cet Oiseau fut tué à Sceaux, lors qu'estant entré dans la Cuisine d'vne Hostellerie, il s'acharna sur le Cuisinier, qu'il mordit. Il avoit vne aile rompue & le crane enfoncé, quand on nous l'apporta," from Claude Perrault, ed., *Mémoires pour servir à l'histoire naturelle des animaux* (Paris: Imprimerie Royale, 1676), 107.

not mention culinary uses; a cormorant in the kitchen was clearly out of place. That the bird was dissected rather than butchered and eaten signaled an ongoing and profound change in the Western European diet. On the evidence of printed cookbooks, the wild birds deemed edible around 1600 included cranes, herons, swans, peacocks, and cormorants. Within a century, most of these had disappeared from the table.

Food historian Jean-Louis Flandrin has argued that large wild birds disappeared from European tables by 1650.[2] Ken Albala asserts that wild foods of all sorts, not only birds, faded from aristocratic diets by the late seventeenth century for reasons of aesthetics, fashion, and health.[3] Neither Flandrin nor Albala connect these changes to Rachel Laudan's work on changes in culinary techniques, which she has tied to the prominence of chymical medicine at the French court.[4]

The animals themselves disappear in these historiographical debates. The shift in the definition of what was edible reveals changing meanings surrounding both animals and diets in early modern Europe. As animals came to be studied as themselves and not as symbols or allegories, a number of dichotomies emerged; these included not only edible/inedible, but also domestic and wild, healthful and unhealthful, local and foreign. Herons and other birds that ate food that humans also ate were simultaneously "vermin," yet edible.[5] The proper place of certain animals shifted over time and place; a swan on the dinner table was normal in much of Europe in the sixteenth century but out of place by the eighteenth. This article will track these shifting meanings by means of two print genres that are not commonly consulted together: cookery books and works on natural history.

When François Pierre de La Varenne published his path-breaking *Le cuisinier françois* in 1651, it was not a foregone conclusion that French cuisine would sweep Europe, or that a natural history based on classification and comparative anatomy would supersede a descriptive natural history that included culinary uses.[6] Yet both had occurred by the 1730s. As the aims of natural history changed, the overlapping discourses about animals in cookbooks and natural histories increasingly diverged.

This article will examine the close relationship between the practices of cooking and natural history before 1700, contextualizing culinary history within a larger cultural and intellectual framework while augmenting the historiography of early modern natural history. I will focus on birds, which effectively illustrate these shifts in perception and meaning.

COOKBOOKS AND NATURAL HISTORY TEXTS AS HISTORICAL SOURCES

Bibliographer Henry Notaker defines a cookbook as including both instruction and recipes. Not all early books on food and cooking included recipes, and the recipe

[2] Jean-Louis Flandrin, "Choix alimentaires et art culinaire (xvie–xviiie siècles)," in *Histoire de l'alimentation*, ed. Jean-Louis Flandrin and Massimo Montanari (Paris: Fayard, 1996), 657–81, on 661. The English translation of this work is an abridgment. All translations mine unless otherwise noted.

[3] Ken Albala, *The Banquet: Dining in the Great Courts of Late Renaissance Europe* (Urbana: Univ. of Illinois Press, 2007), viii. See Flandrin, "Choix alimentaires" (cit. n. 2); and Jean-François Revel, *Le festin en paroles* (Paris: Jean-Jacques Pauvert, 1979), trans. by Helen R. Lane as *Culture & Cuisine* (New York, N.Y.: Da Capo, 1982).

[4] Rachel Laudan, "A Kind of Chemistry," *Petits Propos Culinaires* 62 (1999): 8–22.

[5] Mary Fissell, "Imagining Vermin in Early Modern England," *Hist. Workshop J.* 47 (1999): 1–29.

[6] On sixteenth-century natural history and its goals, see Brian Ogilvie, *The Science of Describing* (Chicago: Univ. of Chicago Press, 2008), although his focus is on plants. François Pierre de La Varenne, *Le cuisinier françois* (Paris: Pierre David, 1651).

has its own historiography, which is not entirely about culinary recipes.[7] Notaker identifies one hundred distinct titles between 1470 and 1700 that formed "a distinct genre within so-called 'how-to' books." Simon Varey refers to the cookbook as "an anthology of information about food, medicine, and regimen combined with practical information in the form of recipes."[8] Cookbooks were also technical manuals.[9] A number of subgenres emerge. Books written by (or at least attributed to) cooks in aristocratic households recounted the tastes and diets of the upper classes. Household manuals included general advice on household and estate management as well as culinary recipes, and reflected bourgeois practices that aimed to replicate those of higher ranks. None of these works addressed the laboring classes. Manuscript collections of recipes also addressed, and often were written by, literate housewives of middle-class or higher status.[10] All print cookbooks after 1600 were written in vernacular languages.[11]

The first printed cookbook dates from the 1470s. Medieval manuscript compilations of recipes served several purposes, such as providing medical advice or instruction, preserving ancient knowledge, or showcasing the largesse of a wealthy patron.[12] These aims continued in printed cookbooks, although in the sixteenth century, the dominant genre portrayed the lives and diets of the wealthy. Their content made them irresistible to professional cooks and bourgeois housewives alike.[13]

French, English, German, and Italian constituted the dominant languages of cookbooks in this period, and I confine my account to works in those languages, along

[7] Elaine Leong, "Collecting Knowledge for the Family: Recipes, Gender and Practical Knowledge in the Early Modern English Household," *Centaurus* 55 (2013): 81–103; Leong and Sara Pennell, "Recipe Collections and the Currency of Medical Knowledge in the Early Modern 'Medical Marketplace,'" in *Medicine and the Market in England and its Colonies, c. 1450–c. 1850*, ed. Patrick Wallis and Mark Jenner (London: Palgrave Macmillan, 2007), 133–52; Leong and Alisha Rankin, eds. *Secrets and Knowledge in Medicine and Science, 1500–1800* (Burlington, Vt.: Ashgate, 2011); William Eamon, *Science and the Secrets of Nature: Books of Secrets in Medieval and Early Modern Culture* (Princeton, N.J.: Princeton Univ. Press, 1996); Elizabeth Spiller, "Recipes for Knowledge: Maker's Knowledge Traditions, Paracelsian Recipes, and the Invention of the Cookbook, 1600–1660," in *Renaissance Food from Rabelais to Shakespeare: Culinary Readings and Culinary Histories*, ed. Joan Fitzpatrick (Burlington, Vt.: Ashgate 2010), 55–72; Anne Willan and Mark Cherniavsky, *The Cookbook Library: Four Centuries of the Cooks, Writers and Recipes that Made the Modern Cookbook* (Berkeley and Los Angeles: Univ. of California Press, 2012); Henry Notaker, "Printed Cookbooks: Food History, Book History, and Literature," *Food & History* 10 (2012): 131–59, on 139–41, 153–6.
[8] Henry Notaker, *Printed Cookbooks in Europe, 1470–1700: A Bibliography of Early Modern Culinary Literature* (New Castle, Del.: Oak Knoll Press, 2010), 1–2. See also Notaker, "Printed Cookbooks" (cit. n. 7); Simon Varey, "Medieval and Renaissance Italy, A: The Peninsula," in *Regional Cuisines of Medieval Europe: A Book of Essays*, ed. Melitta Weiss-Adamson (New York, N.Y.: Routledge, 2002), 85–112, on 87. I follow Varey in referring to "cooks" rather than "chefs," a use that he argues is anachronistic.
[9] Pamela O. Long, *Openness, Secrecy, Authorship: Technical Arts and the Culture of Knowledge from Antiquity to the Renaissance* (Baltimore: Johns Hopkins Univ. Press, 2001).
[10] On food and social class, see Allen J. Grieco, "Alimentation et classes sociales de la fin du Moyen Âge à la Renaissance," in Flandrin and Montanari, *Histoire de l'alimentation* (cit. n. 2), 479–90; Anita Guerrini, "The Impossible Ideal of Moderation," in *Lifestyle and Medicine in the Enlightenment: The Six Non-Naturals in the Long Eighteenth Century*, ed. Rina Knoeff and James Kennaway (London: Routledge, 2020), 86–107. On manuscript recipe books, see Leong and Pennell, "Recipe Collections" (cit. n. 7); and *The Recipes Project: Food, Magic, Art, Science and Medicine*, https://recipes.hypotheses.org/.
[11] Notaker, *Printed Cookbooks* (cit. n. 8), 4; the latest entry in the Latin section is dated 1593 (342).
[12] Willan and Cherniavsky, *Cookbook Library* (cit. n. 7), 1; Philip and Mary Hyman, "Imprimer la cuisine: les livres de cuisine en France entre le xve et le xixe siècle," in Flandrin and Montanari, *Histoire de l'alimentation* (cit. n. 2), 643–55.
[13] Albala, *The Banquet* (cit. n. 3), x; Henry Notaker, *A History of Cookbooks* (Berkeley and Los Angeles: Univ. of California Press, 2017), 36–9.

with the most prominent early Latin work.[14] My choice of books is inevitably arbitrary. Notaker observed that more than 45 percent of all cookbooks published between 1480 and 1700 appeared in the second half of the seventeenth century, with few in the first half.[15] Therefore I focus on the beginnings of printing in the 1470s until 1615, and then from 1650 into the early eighteenth century, concluding with Vincent La Chapelle's *The Modern Cook* (1733), which codified haute cuisine and its determination of what was edible for the next two centuries.[16]

Early cookbooks compiled recipes from many sources, leading to overlap and duplication, much like humanist natural histories. Cookbooks sometimes included sample menus and instruction on particular techniques. Culinary knowledge, like other forms of craft knowledge, was passed on by apprenticeship, and culinary recipes and techniques functioned as guild secrets.[17] Cookbooks therefore resembled "books of secrets," and shared recipes with those compilations of natural and magical knowledge.[18] One of the most popular books of secrets, the manual of Alexis of Piedmont or Alessio Piemontese from the mid-sixteenth century, included culinary recipes and enjoyed dozens of editions in numerous languages.[19]

As practitioners of craft knowledge, cooks were of a lower social order than scholars who compiled learned texts of natural history. Many naturalists were physicians who shared a humanist admiration for ancient knowledge. Printed works of natural history, from William Turner, Pierre Belon, and Conrad Gessner in the mid-sixteenth century, to Claude Perrault and Francis Willughby in the last quarter of the seventeenth, addressed a learned audience that expanded as the language of science moved toward vernacular tongues.[20]

Natural histories up to the mid-seventeenth century organized animals into four main groups: quadrupeds (mammals), birds, serpents (reptiles and amphibians, sometimes also including insects), and fish and other aquatic animals. Some authors, such as Belon, wrote on only one or two of these categories. Others, such as Georg Marcgraf and Willem Piso in their *Historia naturalis Brasiliae* (1648), focused on particular places but retained all four categories, as did John Johnstone in the 1650s.[21] Only in the 1670s, in the Paris Academy's *Mémoires pour servir à l'histoire naturelle des animaux*, was this organization abandoned, though it was partially revived in the eighteenth century by Georges-Louis Leclerc, Comte de Buffon's *Histoire naturelle*

---

[14] Notaker, *Printed Cookbooks* (cit. n. 8), 4.

[15] Ibid., 3–4.

[16] Vincent La Chapelle, *The Modern Cook*, 3 vols. (London: for the author by Nicolas Prevost, 1733).

[17] Notaker, *History of Cookbooks* (cit. n. 13), 15, 31.

[18] Ibid.; Eamon, *Science and the Secrets* (cit. n. 7).

[19] Notaker, *Printed Cookbooks* (cit. n. 8); Varey, "Medieval and Renaissance Italy" (cit. n. 8), 94–5; Eamon, *Science and the Secrets* (cit. n. 7), 134–44; Alessio Piemontese, *De' secreti del reuerendo donno Alessio Piemontese, prima parte, diuisa in sei libri* (Venice: Sigismondo Bordogna, 1555).

[20] Michael Gordin, "The Perfect Past that Almost Was," chap. 1 in *Scientific Babel. How Science Was Done Before and After Global English* (Chicago: Univ. Chicago Press, 2015); Sietske Fransen, "Introduction: Translators and Translation in Early Modern Science," in *Translating Early Modern Science*, ed. Fransen, N. Hodson, and Karl A. E. Enenkel (Leiden: Brill, 2017), 1–14; Fransen, "Latin in a Time of Change: The Choice of Language as Signifier of a New Science?," *Isis* 108 (2017): 629–35.

[21] Georg Marcgraf and Willem Piso, *Historiae naturalis Brasiliae* (Leiden: Franciscus Hackius, 1648); John Johnstone [Joannes Jonstonus], *Historiae naturalis de avibus, libri VI* (Heilbrunn, Ger.: Franciscus Josephus Eckebrecht, 1756; first published by Matthäus Merian (Frankfurt am Main, 1650).

(1749–89).[22] While earlier natural histories mentioned culinary uses of the animals they described, later ones did not, apart from domesticated birds.

The best known and most extensive sixteenth-century natural history of animals, the four-volume *Historiae animalium* (1551–58) of Swiss physician and philologist Conrad Gessner (1516–65), compiled knowledge from a number of sources. Unlike cookbook authors, naturalists commonly cited their sources. Gessner quoted Turner, Belon, and the cookbook author Bartolomeo Sacchi (1421–81), known as Platina, as well as ancient authorities including Aristotle, the Romans Pliny the Elder and Aelian, and Apicius, a Roman author on cookery. Production of the multivolume natural history of Ulisse Aldrovandi (1522–1605) extended far past his lifetime, and he referenced all of those named and more. Naturalists generally followed Gessner's model of presentation, which included a picture of the animal in life and information on the name, external and internal parts, habits, and uses in medicine and food, closing with a "philological" section of literary and artistic history.[23] This model echoes the organization of anatomical works in which *historia*, or description of a body part, was followed by *actio*, or how it worked; this was then followed by *usus*, or purpose.[24]

Cookbooks emphasized practical rather than philosophical knowledge. Animals in sixteenth-century cookbooks were those available to the cook-author at the place and time of publication, whereas natural histories included all kinds of animals: local, exotic, even mythical. The first printed cookbook, *De honesta voluptate et valetudine, vel de obsoniis, et Arte coquinaria libri decem* (1474) of Platina, had no illustrations. Its title may be translated as "Of honorable pleasure and health, or of food, and ten books on the art of cooking." In elegant humanist Latin, *De voluptate* combined medicine, natural history, and cookery.[25]

Cooking may have been an art rather than a science, but Sacchi or Platina, the Vatican librarian, began with a Ciceronian treatise on health.[26] He commented on the healthfulness of various foods in terms of humoral medicine. Late medieval cuisine had developed around a Galenic medical theory based on the balance in each individual of the four humors, the Aristotelian qualities, and the so-called "six things nonnatural": food and drink, sleep and waking, exercise, air, evacuations, and the passions.[27] The ideal body was warm (but not hot) and moist, and diet played a critical role in its maintenance, although physicians disagreed on the proper proportions and qualities of particular foods.[28] For example, most fruits and vegetables, particularly raw ones, were deemed cold and wet, and most agreed that they should be consumed

[22] Perrault, *Mémoires* (cit. n. 1); Georges Louis Leclerc, Comte de Buffon, *Histoire naturelle*, 36 vols. (Paris: Imprimerie Royale, 1749–89).
[23] Conrad Gessner, *Historiae animalium*, 4 vols. (Zürich: Christ. Froschauer, 1551–58); Ogilvie, *Science of Describing* (cit. n. 6), 44.
[24] Gianna Pomata, "*Praxis Historialis*: The Uses of *Historia* in Early Modern Medicine," in *Historia: Empiricism and Erudition in Early Modern Europe*, ed. Gianna Pomata and Nancy Siraisi (Cambridge, Mass.: MIT Press, 2005), 105–46, on 117.
[25] Platina, *Platynae de honesta voluptate: & valitudine: vel de obsoniis: & Arte coquinaria libri decem* (Venice: Bernardinus Venetus, 1498); first published by Laurentius de Aquila and Sibylinus Umber (Venice, 1475); Varey, "Medieval and Renaissance Italy" (cit. n. 8), 86–93; Ken Albala, *Eating Right in the Renaissance* (Berkeley and Los Angeles: Univ. of California Press, 2002), 27, 246–8.
[26] Platina, *De voluptate* (cit. n. 25), not paginated; Notaker, *History of Cookbooks* (cit. n. 13), 14–15.
[27] Albala, *Eating Right* (cit. n. 25); Laudan, "A Kind of Chemistry" (cit. n. 4); Peter Niebyl, "The Non-Naturals," *Bull. Hist. Med.* 45 (1971): 386–92; Lelland J. Rather, "The 'Six Things Non-Natural': A Note on the Origins and Fate of a Doctrine and a Phrase," *Clio Medica* 3 (1968): 337–47.
[28] Guerrini, "Impossible Ideal of Moderation" (cit. n. 10).

sparingly, particularly in winter. In this way evolved a cuisine built around the seasons and characterized by spices, sugar, and soft textures; well-cooked food was more digestible, and spices such as ginger, pepper, and cinnamon warmed the body. Physicians and cooks alike valued sugar as ideally warm and moist. In his 1632 treatise on the non-naturals, the French Galenist physician Guy Patin (1601–72) explained that meat was more nutritious than vegetables, except for grains from which bread was made. Raw fruits and vegetables were cold and wet. Most spices, as well as honey and sugar, were hot and acrid, although Patin agreed that sugar was less hot than honey, "more agreeable, less drying, and better for the stomach."[29]

Platina quoted liberally from classical authors; culinary historians compare his commentary to Pliny's *Natural History* in its organization and range of topics and sources.[30] But the second half of the book, while still in Latin, consisted of recipes copied from the manuscript treatise of Martino de'Rossi (ca. 1430–ca. 1490), also known as Martino da Como, cook to the Sforza family in Milan.[31] A few of the recipes came from the manuscript collection attributed to Apicius. Humanists admired Apicius as an avatar of ancient Roman culture.[32] The recipes in *De voluptate* are organized both by cooking methods and by ingredients; they begin with various kinds of meat ("tame and wild," *cicures & silvaticae*), followed by sections on tarts and pies, fritters, and condiments.[33] *De voluptate* was reprinted and translated many times. Although it was atypical of subsequent cookery books, with its humanist Latin and long philosophical preface, it nonetheless constituted a common reference point for naturalists and cooks alike.

NAMING AND TRANSLATING

Because most cooks were not learned men, they wrote recipes in their vernacular languages. Platina translated Maestro Martino's Italian into Latin, often simply inventing words. For example, *torta* means "tart" or "cake" in Italian, but it means "twisted" in Latin; Platina nonetheless used it to mean "tart."[34] Notaker points out that terms such as *Küchenlatein* or *latino maccheronico* came to be employed in the sixteenth century to denote such neologisms in all genres as new knowledge overtook the vocabulary of the Romans.[35] Martino's recipes were copied more often than Platina's

---

[29] Guy Patin, *Traicté de la conservation de santé, par un bon regime & legitime usage des choses requises pour bien & sainement vivre* (Paris: Jean Jost, 1632), 39–40. On early modern theories of digestion, see Bradford Bouley, "Digesting Faith: Eating God, Man, and Meat in Seventeenth-Century Rome," in this volume.

[30] Willan and Cherniavsky, *Cookbook Library* (cit. n. 7), 45; Platina, *De voluptate* (cit. n. 25), books 1–5; Pliny the Elder, *Naturalis historia* (Venice: Johannes de Spira, 1469).

[31] Varey, "Medieval and Renaissance Italy" (cit. n. 8), 88. A modern edition of Martino's fifteenth-century manuscript is Luigi Ballerini, ed., *The Art of Cooking: The First Modern Cookery Book*, trans. Jeremy Parzen (Berkeley and Los Angeles: Univ. of California Press, 2005).

[32] Sally Grainger, "The Myth of Apicius," *Gastronomica* 7 (2007): 71–7; Christopher Grocock and Sally Grainger, *Apicius: A Critical Edition with an Introduction and English Translation* (Totnes, UK: Prospect, 2006); Alberto Capatti and Massimo Montanari, *La cucina Italiana* (Rome: Laterza, 1999), trans. by Aine O'Healy as *Italian Cuisine: A Cultural History* (New York, N.Y.: Columbia Univ. Press, 2003), 98.

[33] Platina, *De voluptate* (cit. n. 25), not paginated.

[34] Notaker, *History of Cookbooks* (cit. n. 13), 73; Varey, "Medieval and Renaissance Italy" (cit. n. 8), 88–9.

[35] Notaker, *History of Cookbooks* (cit. n. 13), 17, 15.

introductory treatise. But translation of Platina's Latin into vernacular languages often resulted in gibberish. Simon Varey observes, "*Macaroni*, for instance, became *exicium siculum* in Sacchi's Latin, which was translated into Italian as 'Sicilian dish.'"[36]

These translations point to the lack of standard names for objects and processes. Printing aided the standardization of vernacular languages, but only in the eighteenth century do we find recognizably modern orthography. By then, the increasingly dominant French vocabulary of cooking encountered its own problems of translation.[37] A good example is the almost ubiquitous term "kickshaw" in seventeenth- and eighteenth-century English cookbooks—a phonetic rendering of the French *quelque chose*, meaning small, made dishes.[38] The difficulties inherent in translation went far beyond spelling; names of animals, as well as of cooking ingredients and techniques, varied widely even within vernacular languages, with local and regional names for common plants and animals.[39] Scientific names of animals also followed no rules in this period.

Gessner listed each name in *Historiae animalium* in German, Spanish, French, English, and Italian, as well as Latin and Greek. He published separate volumes of images and names, underlining the critical importance of naming, and of matching name to image to make accurate identifications.[40] In 1536, the physician Charles Estienne (1504–ca.1564) published a French glossary of Latin and Greek names of trees, fruits, plants, fish, and birds, taken from a long list of classical authors.[41] In his book on birds a few years later, William Turner (1509/10–1568) listed them in alphabetical order according to Aristotle's names (in Greek, Latin, English, and German), with additional information from Pliny. He fit his own observations within this classical framework.[42] In the next decade, Edward Wotton (1492–1555) followed the same format in his *De differentiis animalium* (1552). As Peter Mason has argued, early modern naturalists inhabited "a world with many mansions," with differing registers of response to different phenomena.[43] While the biblical Adam originally named each animal, fixing its

[36] Varey, "Medieval and Renaissance Italy" (cit. n. 8), 88.

[37] Gilly Lehmann, *The British Housewife* (Totnes, UK: Prospect, 2003), 51.

[38] Gervase Markham, in 1615, referred to a "Quelquechose," but this spelling had disappeared by the 1660s; see *The English Hus-Wife* (46) in Gervase Markham, *Countrey Contentments, in Two Bookes* [. . .] *The Second Intituled, The English Huswife* (London: I. B. for R. Jackson, 1615); the two titles are separately paginated.

[39] On local naming, see Keith Thomas, *Man and the Natural World* (New York: Pantheon, 1983), 81–7.

[40] Conrad Gessner, *Icones avium omnium* (Zürich: C. Froschauer, 1560). On Gessner's use of images, see Sachiko Kusukawa, "The Sources of Gessner's Pictures for the *Historia animalium*," *Ann. Sci.* 67 (2010): 303–28; and Kusukawa, *Picturing the Book of Nature* (Chicago: Univ. of Chicago Press, 2012), 139–77 (on plants).

[41] Charles Estienne, *De Latinis et graecis nominibus arborum, fruticum herbarum, piscium, & auium Liber: Ex Aristotele, Theophrasto, Dioscoride, Galeno, Nicandro* [. . .] *cum Gallica eorum nominum appellatione* (Paris: Robert Estienne, 1536).

[42] William Turner, *Avium praecipuarum, quarum apud Plinium et Aristotelem mentio est, brevis & succinta historia* (Cologne: Ioannes Gymnicus, 1544). On this accommodation between classical and modern natural histories, see Laurent Pinon, *Livres de zoologie de la Renaissance: Une anthologie (1450–1700)* ([Paris]: Klincksieck, 1995), 20, 72–3; Anthony Grafton, April Shelford, and Nancy Siraisi, *New Worlds, Ancient Texts: The Power of Tradition and the Shock of Discovery* (Cambridge, Mass.: Harvard Univ. Press, 1992), 161–93; and Anita Guerrini, *Natural History and the New World, 1524–1770* (Philadelphia: American Philosophical Society Library, 1986), 1–9.

[43] Edward Wotton, *De differentiis animalium libri decem* (Paris: Vascosanum, 1552); Peter Mason, *Before Disenchantment: Images of Exotic Animals and Plants in the Early Modern World* (London: Reaktion, 2009), 23.

place in Creation, the multiple names in glossaries and indices established no single correct name.[44]

Moreover, animals used for food had different names in different contexts. English peculiarly has different names for meat and animals; French names for meat (beef/ *boeuf*, pork/*porc*) but English for the animal itself (cattle, pig). An influx of new animals from the New World added further confusion. For example, what did cook to the popes Bartolomeo Scappi (ca. 1500–77) mean in 1570 when he referred to a "*coniglio d'India*" (literally, "Indian rabbit")? Could it have been a guinea pig, a New World animal that Gessner had labelled "*cuniculo vel porcello indico*" (Indian rabbit or small pig) in 1554?[45] Historians have long acknowledged that Europeans believed the New World resembled Europe, and they named New World animals to indicate this relationship. The appellation "Indian" could indicate either east or west, or simply "foreign," highlighting as well continued confusion surrounding the provenance of non-European animals.[46]

Gessner pictured a turkey in the 1550s among wildfowl (*De gallinis sylvestribus*) under the heading of "*De gallopavo*," a neologism combining the Latin words for rooster (*gallus*) and peacock (*pāvō*) (fig. 1).[47] In the 1520s, Gonzalo Fernández de Oviedo (1478–1557) described the Mexican "*pavo real*," noting that it tasted much better than the Spanish variety. Later in the century, Franciscan missionary Bernardino de Sahagún (ca. 1499–1590) listed five Nahuatl names for the bird, none of which crossed the Atlantic, in his *Historia general de las cosas de la Nueva España*.[48] Gessner observed that the bird was also known in Latin as "*gallina Indica*" (Indian chicken) or "*pavo Indica*" (Indian peacock); in Italian as *gallina d'India* (Indian hen); in Spanish as *pavon de las Indias* (Indian peacock); in French as *poulle d'Inde* (Indian hen), and later shortened to *dinde*; in German as "*ein Indianisch oder Kalekuttisch oder Welsch hün*" (an Indian or Calcuttan or Welsch chicken, *Welsch* meaning French or Italian); and in English as "kok of Inde." The Dutch called it *Kalkoen*, also referencing Calcutta.[49] In 1600, Aldrovandi, on the other hand, identified the *gallopavo* with the classical *meleagris*, as had Belon in 1555, who called it a "*coc d'Inde*," although Turner a decade earlier had described the *meleagris* as *Gallina Africana* (adding "a kok of inde" in a marginal note). Gessner believed the turkey and the guinea fowl were different animals. Aldrovandi's illustration of the *gallopavo/meleagris* is clearly a turkey, as is Belon's.[50] This confusion endured for many more years. "Chicken" became

[44] Pinon, *Livres de zoologie* (cit. n. 42), 20.

[45] Bartolomeo Scappi, *Opera* (Venice: Michele and Francesco Tramezzino, 1570), 44; Conrad Gessner, *Appendix historiae quadrupedum viviparorum & oviparorum* (Zurich: C. Froschauer, 1554), 19.

[46] Guerrini, *Natural History* (cit. n. 42), 2–3; Natalie Lawrence, "Assembling the Dodo in Early Modern Natural History," *Brit. J. Hist. Sci.* 48 (2015): 387–408, on 400.

[47] Conrad Gessner. *Historiae animalium, Liber III, qui est de avium natura* (Zurich: Christoph Froschauer, 1555), 464.

[48] Gonzalo Fernández de Oviedo, *De la natural hystoria de las Indias* (Toledo: Ramon Petras, 1526); reprinted as *Sumario de la natural historia de las Indias*, ed. José Miranda (Mexico City: Fondo de Cultura Economica, 1950), 172–4; Bernardino de Sahagún, *Historia general de la cosas de Nueva España*, translated and edited by Charles E. Dibble and Arthur J. O. Anderson as *Florentine Codex: General History of the Things of New Spain*, 13 vols. (Salt Lake City: Univ. of Utah Press, 1963), 11:53–4.

[49] Gessner, *De avium natura* (cit. n. 47), 464. Thanks to Tillmann Taape for help in deciphering the German names; for the Dutch, see Lawrence, "Assembling the Dodo" (cit. n. 46), 400.

[50] Ulisse Aldrovandi, *Ornithologie tomus alter* (Frankfurt am Main: Wolfgang Richter, 1610); first published by Baptista Bellagamba (Bologna: 1600), 19–22 and table 1 (between pages 5 and 6); Alan

Figure 1. Gallopavo *(turkey). From Conrad Gessner,* Historiae animalium, Historiae ani-malium, Liber III, qui est de avium natura, *1555 (cit. n. 23). (Linda Hall Library of Science, Engineering & Technology.)*

a default name for any edible bird of uncertain provenance; in 1605, Charles de L'Écluse (1526–1609) referred to the dodo as *gallinaceus gallus peregrinis*, or "foreign chicken-like cock."[51]

WHAT WAS EDIBLE IN 1600: WILD AND DOMESTIC

If the 1549 banquet celebrating Catherine de'Medici's coronation as Queen of France is any indication, the meat of birds dominated sixteenth-century aristocratic diets. Albala states that "by the sixteenth century, fowl is the food most readily associated with nobility."[52] "Game" is generally defined as wild animals that are hunted for meat, as well as the meat itself. Paula Young Lee explains, "The definition of game has both expanded and contracted" over time, "so that all wild animals are [today] subject to regulation as potential game, but few animals qualify as such."[53] Birds then constituted

---

Davidson, ed., *Oxford Companion to Food* (Oxford: Oxford Univ. Press, 1999), s.v. "Turkey," 809–10 (**hereafter** cited as OCF); Pierre Belon, *L'histoire de la nature des oyseaux* (Paris: Guillaume Cavellat, 1555), 248; Turner, *Avium praecipuarum* (cit. n. 42), 53–7.

[51] Charles de L'Écluse, *Exoticorum libri decem* (Antwerp: Officina Plantiniana Raphelengii, 1605), 99–100.

[52] Albala, *The Banquet* (cit. n. 3), 9; on the Medici banquet, see Stephen Bamforth, "Melons and Wine: Montaigne and *joie de vivre* in Renaissance France," in *Joie de vivre in French Literature and Culture: Essays in Honour of Michael Freeman*, ed. Susan Harrow and Timothy A. Unwin (Amsterdam: Rodopi, 2009), 99–128.

[53] Paula Young Lee, *Game: A Global History* (London: Reaktion, 2013), 8.

a significant proportion of game meat, and continue to today. But not all wild birds constituted game in the sixteenth and seventeenth centuries, since not all were hunted. Many were instead raised in enclosures, indicating an uncertain boundary between domesticated and wild. The article "Game" in the *Oxford Companion to Food* notes the following: "The concept of game has from early times been somewhat blurred by the practice of rearing animals or birds in protected environments in order to provide a stock of game. This conflicts with the general notion that game is wild."[54]

In his *Maison rustique*, Charles Estienne listed three kinds of hunting. Fishing, he said, was suitable for servants; hunting for birds was "*délectable*"; but only hunting for four-footed beasts was an "honest exercise." Among such beasts were red and roe deer, wild boar, and hare.[55] Deer in particular might be kept in enclosed parks with tamed members of their species to lead them to provided food.[56] Wild birds, such as ducks and teals, could be kept in ponds and fed with grain, having first been tamed by pouring lees of wine into the water to make them drunk. Woodcocks, curlews, swans, and cranes, attracted to these ponds, could then be easily caught. "Foreign and wild" birds, such as pheasants, peacocks, and "*poulles d'Inde*" or turkeys, also flourished in enclosures, along with doves, partridges, and quail. Thus, in Estienne's view these animals ceased to be either wild or considered game (meaning subject to hunting), and became tamed if not domesticated in the context of the farm, although they differed from domesticated animals such as chickens.[57] The turkeys that entered Europe from Mexico were already domesticated, but their foreignness led to their identification as wild. By the eighteenth century, European turkeys were considered fully domesticated, and the *History of Birds* (1734) by Eleazar Albin (ca. 1690–ca. 1742) describes a variety of wild turkey in England that was kept in parks to be hunted.[58]

Herons, only eaten when very young, constituted another sort of animal husbandry, as Estienne described. They could never be entirely tamed, and heronries were places where wild herons congregated and were not human-made structures. Herons preferred high trees near water, close to other fish-eating and edible wild birds such as cormorants, bitterns, curlews, and egrets. Landowners used falcons to hunt herons and cranes, thus providing "pleasure and contentment" as well as food. Falcons were trained with young herons that had been partially tamed by feeding them fish.[59] Estienne, a bourgeois city dweller, gained his intimate knowledge of the French countryside by interviewing its inhabitants, and *Maison rustique* is a snapshot of a particular place and time.[60] It went through thirty-four editions by 1600, including translations

[54] OCF, s.v. "Game," 330.

[55] Charles Estienne, *L'agriculture et maison rustique* (Paris: Jacques du Puy, 1570), fols. 231v–232r; first published by Laurens Maury (Rouen, 1558). This book is a translation and abridgment of Estienne's *Praedium rusticum* (Paris: Charles Estienne, 1554).

[56] Estienne, *Maison rustique* (cit. n. 55), fol. 230v.

[57] Ibid., fol. 32r, for "*estranges et sauvages*" (foreign and wild), see fol. 231r.

[58] Eleazar Albin, *A Natural History of Birds*, 3 vols. (London: William Innys, John Clarke, and John Brindley, 1731–38), 2:30.

[59] Estienne, *Maison rustique* (cit. n. 55), fols. 231r–232v; Joop Witteveen, "On Swans, Cranes, and Herons: Part 3, Herons," *Petits Propos Culinaires* 26 (1987), pages not available; reprinted in *The Wilder Shores of Gastronomy*, ed. Alan Davidson and Helen Saberi (Berkeley, Calif.: Ten Speed Press, 2002), 119–128.

[60] Corinne Beutler, "Un chapitre de la sensibilité collective: La littérature agricole en Europe continentale au XVIe siècle," *Annales* 28 (1973): 1280–1301, on 1285–6.

30 ANITA GUERRINI

Table 1. *Wild birds in selected cookbooks, 1470–1733.*

	Bittern or Curlew	Bustard	Cormorant	Crane	Heron	Peacock	Stork	Swan	Turkey
Platina 1476				x		x	x	x	
Grand cuysinier 1543 (1550)		x	x	x	x	x		x	
New Boke 1545	xx	x		x	x	x			
Messisbugo 1549	x	x		x	x	x			
Estienne 1564 (1570)						x			x
Scappi 1570		x		x		x			x
Rumpolt 1581				x	x	x	x	x	x
Murrell 1615	x				x			x	x
Markham 1615	x	x		x		x		x	x
La Varenne 1651					x				x
De Lune 1656		x							x
May 1660	x				x			x	x
Stefani 1662							x		x
Massialot 1691									x
La Chapelle 1733									x

into English, German, Italian, and Dutch.[61] Louis Liger (1658–1717) published *Nouvelle maison rustique* 150 years later, which traded on the continued popularity of Estienne's book.[62]

As table 1 shows, around 1600, cooks considered a number of birds to be edible. Herons, cranes, peacocks, swans, and turkeys most commonly appeared in cookbooks, as well as a long list of smaller birds such as swallows, larks, and blackbirds. Platina, in 1474, which was before turkeys came to Europe, considered storks and cranes edible, as well as partridges, pheasants, and quail. The 1549 *Banchetti* of Christoforo di Messisbugo (d. 1548), a manual for high-end cooks, listed an enormous number of mammals and birds that should be available for noble consumption:

> Beef and cow, veal, wild and domestic pork, stag, deer, roebuck, lamb, kid, wether [gelded sheep], suckling pig, hares, rabbits, dormice, peacocks, wild and domestic pheasants, partridges, *coturnici* [a kind of quail], francolins [a kind of partridge], thrushes, gray partridges, woodcocks, ortolans, figpeckers, quail, crake [a kind of rail], turtle doves, geese, crane, bustard, bittern, herons, snipe, wild and domestic ducks, ducklings, large, medium, and small, *girioli*, plovers, *felizette*, and other birds.
> Capons, fat and meaty, and similar hens, chickens male and female, domestic doves [*casalenghi, o colombara*], or ring-necked doves, eggs, new-laid and not.[63]

[61] Ibid., table 2, page 1299.

[62] Louis Liger, *Oeconomie générale de la campagne, ou nouvelle maison rustique*, 2 vols. (Paris: Charles de Sercy, 1700).

[63] Cristoforo di Messisbugo, *Banchetti, composizioni di vivande e apparecchio generale* (Ferrara: Giovanni Buglhat and Antonio Hucher, 1549), not paginated but page 3 of text. John Florio's 1611 Italian-English dictionary defines *girioli* and *felizette* as "small dainty birds": Florio, *Queen Anna's New World of Words, or Dictionarie of the Italian and English Tongues* (London: Edw. Blount and William Barret, 1611), 183, 212; Capatti and Montanari, *Italian Cuisine* (cit. n. 32), 67, offer a slightly different translation. Dormice had been popular in ancient Rome.

Messisbugo's list evokes a noble clientele who indulged in what Albala has called "Mannerist cuisine," with surprising and sometimes bizarre combinations.[64] But such a variety of wild birds was not confined to the uppermost social tier. Twenty years later, Scappi also listed crane as well as peacock (*"Galline d'India"*) and the more generic *"pollanche d'India,"* or Indian poultry.[65] Scappi's audience included not only cooks who, like him, ministered to the aristocracy, but also the aspiring bourgeoisie who read Estienne's *Maison rustique* and needed instruction in organizing a kitchen and purchasing foodstuffs. The mid-sixteenth-century *Propre Newe Boke of Cookery*, also directed toward the middling sort, included seafowl such as bitterns, curlews, and even gulls, as well as waterfowl—storks, herons, cranes, and bustards. The comprehensive French *Grand cuysinier*, published under various titles beginning around 1543, did not include bitterns but did include cormorants, as well as cranes, herons, peacocks, and swans.[66]

In 1615, Gervase Markham (ca. 1568–1637) appended a household manual *The English Hus-wife* to his *Countrey Contentments*, which was a manual of farriery and hunting. The rural English housewife might cook bittern and shoveler, another large sea bird, as well as gulls and rails. Markham's audience, much like Estienne's, had access to swans, cranes, turkeys, peacocks, and bustards. Bittern supposedly tasted like hare.[67] Swans occupied a peculiar status in England, where they were prized for feasts and heavily managed in the wild. This management may have prevented their extinction, unlike cranes, bustards, and other large birds that were intensively hunted.[68]

But signs of change began to emerge in another book published the same year as Markham's. John Murrell's *New Booke of Cookerie* claimed to "set forth the newest and most commendable Fashion for Dressing or Sowcing, eyther Flesh, Fish or Fowle." His work, he said, was "all set forth according to the now, new, English and French Fashion," with several recipes "in the French fashion." This fashion included the use of fresh herbs alongside, or even substituting for, spices, and the use of wine and especially butter. Murrell had recipes for swan, turkey, curlew, and heron, as well as various wild ducks and smaller birds. But he did not include peacocks, cranes, or bustards. In contrast to Markham, Murrell's book is decidedly urban in orientation, with a section on "London Cookerie" that consisted mainly of domesticated animals. The wild birds available on country estates may have been less accessible in the city.[69]

An earlier work, *Ein new Kochbuch* (1581) by Marx Rumpolt (fl. 1581), cook to the duchy of Mainz, joined natural history to the cookbook as profoundly as had Platina a century earlier, but with Gessner rather than Pliny as its model. Rumpolt

[64] Albala, *The Banquet* (cit. n. 3), 13–20.

[65] Scappi, *Opera* (cit. n. 45), book 2, fols. 60–1.

[66] *A Proper New Booke of Cookery* (London: William How for Abraham Veale, 1575), not paginated; first published in London, 1545; *Le grand cuysinier de toute cuisine* (Paris: Jehan Bonfons, n.d., ca. 1550), fol. xxiii, "Oyseaulx de riviere"; first published as *La fleur de toute cuysine* by Pierre Sergent (Paris, 1542).

[67] Markham, *English Hus-wife* (cit. n. 38), 6. For bittern taste, see OCF, 78. On the relationship between works such as Markham's and English identity, see Wendy Wall, *Staging Domesticity* (Cambridge: Cambridge Univ. Press, 2002).

[68] Arthur MacGregor, "Swan Rolls and Beak Markings: Husbandry, Exploitation and Regulation of *Cygnus olor* in England, c. 1100–1900," *Anthropozoologica* 22 (1996): 39–68, on 64. On the bustard, see Estlin Waters and David Waters, "The Former Status of Great Bustard in Britain," *British Birds* 98 (2005): 275–305.

[69] John Murrell, *A New Booke of Cookerie* (London: John Browne, 1615), title page and 53–88 ("London Cookerie").

focused on meat, poultry, and fish, organized by animal rather than by method of preparation. Rumpolt also included an ostrich as well as a turkey. Following the model of Gessner, he illustrated each animal as it looked in life, even copying some of Gessner's illustrations, such as the turkey and a bird Rumpolt called a *Drappen*, which Gessner called a *Waldrapp*, Belon called a black ibis, and the ancient geographer Strabo called a stork.[70] In contrast to Rumpolt's images of the live animal, the naturalist Adam Lonicer in his 1551 natural history had illustrated his section on animals with images of cuts of meat.[71]

THE NATURALIZATION OF THE TURKEY

Flandrin argued that the turkey largely supplanted other large wild birds in the European diet by 1650. Christopher Columbus noted the animal he called *gallina de la tierra* ("land chicken") in Honduras in 1502.[72] Oviedo, as we have seen, described it in 1526 as a kind of peacock, and it appeared in French naturalist Pierre Gilles's 1533 translation of Aelian as *Gallo peregrino*, "foreign rooster," which noted its New World origins.[73] The first extended description of a turkey by a European appeared in Sahagún's *Historia general* (1569). "It leads the meats," he wrote. "It is tasty, fat, savory."[74]

Neither Belon nor Gessner mentioned culinary uses of the turkey, but there is abundant evidence that the *gallopavo* had fully entered the aristocratic diet by 1600. Catherine de'Medici's 1549 banquet included sixty-six hen turkeys and seven cocks.[75] Aldrovandi's article "*De Gallopavonis*" included *Usus in Cibo* (use as food), remarking that "the flesh of our *gallopavo* as food is most splendid, and worthy of our foremost banquets."[76]

Both cookbooks and natural histories situated the turkey among other large semi-domesticated birds, such as pheasants. Gessner listed birds alphabetically rather than by any classificatory category. Aldrovandi, mainly following Aristotle's classification, placed the *gallopavo* alongside the peacock and other wild (*sylvestris*) but land-dwelling fowl like partridges, quail, bustards, and cranes.[77] Chickens occupied the next category of domesticated birds. Cookbooks followed similar divisions of wild and domestic, although Platina began with the more general category of "edible" (*esculens*).[78]

However, "domestic" is a slippery term, implying a range of animal behaviors from fully subject to human control to a more ambivalent status. Estienne described wild birds, such as ducks, or mammals, such as deer that flourished in enclosures, while

---

[70] Marx Rumpolt, *Ein new Kochbuch* (Frankfurt am Main: Sigmundt Feyerabendt, Peter Fischer, and Heinrich Lacten, 1581), fols. 63v, 65r; Gessner, *Icones avium* (cit. n. 40), 22; and Belon, *Histoire des oyseaux* (cit. n. 50), 199–201 (he cites Strabo's name on page 201).

[71] Rumpolt, *Ein new Kochbuch* (cit. n. 70), fols. 70r, 70v; Adam Lonitzer, *Naturalis historiae opus novum* (Frankfurt am Main: Chr. Egenolph, 1551), 268, 290.

[72] Liliane Plouvier, "Introduction de la Dinde en Europe," *Sci. Hist.* 21 (1995): 13–24, on 15.

[73] Oviedo, *Hystoria* (cit. n. 48); Pierre Gilles, *Ex Aeliani historia per Petrum Gyllium Latini facti* (Lyon: Seb. Gryphium, 1533), 448–9; Andrew F. Smith, *The Turkey: An American Story* (Urbana: Univ. of Illinois Press, 2010), 23–4.

[74] Sahagún, *Florentine Codex* (cit. n. 48), 11:53; Smith, *The Turkey* (cit. n. 73), 9.

[75] Bamforth, "Melons and Wine" (cit. n. 52), 102–3.

[76] Aldrovandi, *Ornithologiae* (cit. n. 50), 21.

[77] Ibid., table of contents. See also J. J. Hall, "The Classification of Birds, in Aristotle and Early Modern Naturalists (II)," *Hist. Sci.* 29 (1991): 223–43. Hall does not address "domestic" as a category.

[78] Platina, *De voluptate* (cit. n. 25), books 5 and 6.

prospering equally outside of human control. Such animals continued to be described as "wild." Estienne distinguished these from chickens, which could not survive without humans.[79] Therefore, early modern writers acknowledged, even if they did not fully articulate, domestication as a lengthy process resulting in an animal with a symbiotic relationship to humans. Only in the eighteenth century were turkeys deemed domesticated as well. Other birds, such as pheasants, might be raised in enclosures, but were never fully domesticated.[80]

The turkey thus occasioned immense confusion among early modern naturalists. Their foreignness in Europe identified them as wild even though they were not. Unlike naturalists, cooks did not confuse the turkey with the guinea fowl. The guinea fowl, designated *gallina indica* by many naturalists (but *gallina Africana* by Turner), originated in Africa and was reintroduced to Europe in the fifteenth century after a long absence following the fall of Rome. Pliny described it and Turner repeated his description.[81] It is smaller than the turkey and lacks the distinctive tail in the males that occasioned comparison to peacocks. In the eighteenth century, Linnaeus perpetuated the confusion between turkey (*gallopavo*) and guinea fowl (*meleagris*) when he gave the name *Meleagris gallopavo* to the turkey, a name it still holds, implying that the two birds were identical.

One indication of European naturalization of the turkey is that it made no appearance in Charles de L'Écluse's 1605 *Exoticorum libri decem*, which catalogued exotic animals and plants.[82] By 1600, the turkey began to lose its status as a wild, exotic animal, joining domesticated ducks, chickens, and geese in the farmyard.[83] Its lighter meat fit well with dietary concerns about red versus white meat, as white meat came to seem healthier and more digestible.[84] To many, turkeys also tasted better than herons or bitterns. Although some wild birds, such as bustards and cranes, were beginning to disappear owing to overhunting, it is not clear that the turkey's increasing prominence was directly due to the growing scarcity of other birds.

Turkeys were an exceptionally versatile ingredient. Aldrovandi, identifying the turkey with the Roman *meleagris*, offered a highly spiced recipe from Apicius. Scappi described the *gallo* and *gallina d'India* as larger than the peacock, with whiter, tenderer meat resembling that of a capon. Scappi roasted capons and turkeys on a spit.[85] Rumpolt listed no fewer than twenty ways to prepare turkey.[86] Murrell, like Scappi, likened a turkey to a capon, and recommended parboiling either bird, larding it, seasoning it with salt and pepper and a few cloves stuck in the breast, enclosing it in a

---

[79] Estienne, *Maison rustique* (cit. n. 55), fols. 25r–30r (chickens), fol. 232r–v (wild animals).

[80] Harriet Ritvo, *The Platypus and the Mermaid* (Cambridge, Mass.: Harvard Univ. Press, 1997), 38–41; Jean-Pierre Digard, *L'homme et les animaux domestiques* (Paris: Fayard, 1990); Nicholas Russell, *Like Engend'ring Like: Heredity and Animal Breeding in Early Modern England* (Cambridge: Cambridge Univ. Press, 1986), 1–10.

[81] Pliny, *Naturalis historia* (cit. n. 30), not paginated; Turner, *Avium praecipuarum* (cit. n. 42), 53–4.

[82] L'Écluse, *Exoticorum libri decem* (cit. n. 51), 94–108.

[83] I have found no discussion among naturalists about the possible exotic or wild origins of the chicken.

[84] Albala, *Eating Right* (cit. n. 25); Guerrini, "Impossible Ideal of Moderation" (cit. n. 10).

[85] Aldrovandi, *Ornithologiae* (cit. n. 50), 21; Scappi, *Opera* (cit. n. 45), book 2, chaps. 140 and 141, fols. 60–1.

[86] Rumpolt, *Ein new Kochbuch* (cit. n. 70), fol. 66, r and v.

"coffin" of dough and then baking it, with plenty of melted butter poured in through a hole in the lid of the coffin.[87]

Turkeys therefore encapsulated the confusions surrounding New World animals. They crossed the boundaries between domesticated and wild; fully domesticated in their homeland, their exotic origins nonetheless situated them among wild birds. Turkeys also revealed differences between the kinds of knowledge held by cooks and naturalists. Cooks readily distinguished turkeys from guinea fowl, and cooked them like other domesticated birds such as capons.

CONSIDERATIONS OF TASTE

At the most basic level, edibility depended upon taste, which encompassed both the physical sense and the cultural imperative. Between 1600 and 1700, taste developed from a classically sanctioned set of experiences to something more subjective. Flandrin attributed the changing meaning of the word "taste" (or *goût*, in French)—from a merely culinary sensation to a broader expression of aesthetic pleasure—to an increased importance of food across all classes.[88] It is obvious that cooks would be concerned with taste in both its meanings.[89] But naturalists also asserted the importance of taste in the culinary sense. In the mid-sixteenth century, Belon extolled the culinary uses of wild birds, presenting his own experience of taste as another piece of sensory evidence and integral to his natural history. As I have argued elsewhere, tasting, like seeing, smelling, and touching, became a standard analytical technique among experimental natural philosophers as well as among cooks in the seventeenth century.[90] As classical ideals receded, taste assumed a new importance in discourses of food, even as discussions of health continued to evolve.

Belon employed his observations to evaluate both ancient precepts and modern popular opinion. He claimed that swans were "the most exquisite birds among French delights." However, the delightfulness of swan meat was due to the animals being raised in semicaptivity.[91] The meat of pelicans was "of the same temperature" (in humoral terms) as that of swans, and comparable in taste to geese, while the flesh of the "*cane de Guinee*" (Guinea duck) "is neither worse nor better than a domestic duck or goose."

Belon continued: "One commonly says, that the Heron is Royal meat. Therefore the noble French make a great case of eating them, but even more the Heronneaux [young herons]: however, foreigners do not recommend them so highly. They are without comparison more delicate than cranes." Nonetheless, "[Pliny] claims that cranes are delicious." On the bustard, "Pliny wrote: *Otidas damnatas in cibis* [bustards are cursed at dinner]." Belon then asked, "But how would it be possible that the Bustard

---

[87] Murrell, *Newe Booke* (cit. n. 69), 28. On pastry "coffins," see Janet Clarkson, *Pie: A Global History* (London: Reaktion, 2009), 18–20.

[88] Jean-Louis Flandrin, "De la diététique à la gastronomie, ou la libération de la gourmandise," in Flandrin and Montanari, *Histoire de l'alimentation* (cit. n. 2), 683–703, on 691.

[89] Laudan, "A Kind of Chemistry" (cit. n. 4); T. Sarah Peterson, *Acquired Taste: The French Origins of Modern Cooking* (Ithaca, N.Y.: Cornell Univ. Press, 1994); Susan Pinkard, *A Revolution in Taste: the Rise of French Cuisine* (Cambridge: Cambridge Univ. Press, 2009); Stephen Mennell, *All Manners of Food*, 2nd ed. (Urbana: Univ. of Illinois Press, 1996), 1–19; Flandrin, "De la diététique à la gastronomie" (cit. n. 88).

[90] Anita Guerrini, "The Ghastly Kitchen," *Hist. Sci.* 54 (2016): 71–97.

[91] Belon, *Oyseaux* (cit. n. 50), 152.

is so bad [tasting], since experience shows that it is a delicious bird, which we now prefer over all others at private banquets?" Cormorants, too, tasted better than their reputation indicated: "they are not as bad as one says."[92]

Belon said little about methods of preparation. Gessner provided recipes for chicken, and Aldrovandi for a number of other birds.[93] But taste, in both sensual and aesthetic terms, became a critical marker of edibility in seventeenth-century Europe, overtaking the relative prestige of wild over domestic. Taste took its place alongside considerations of health.[94] At some point after 1650, certain wild birds went from tasting good to tasting bad, and good taste became identified with a refined, domesticated cuisine. By 1739, the anonymous author of *Les Dons de Comus*, a multivolume compendium of cuisine, could declare that the moderns had carried the art of cuisine and "finesse of taste" far beyond the Romans.[95] Taste had become a subjective criterion rather than a classical imperative.

A GREAT TRANSITION?

While Aldrovandi's students continued to publish volumes based on his voluminous manuscripts into the 1640s, the first new, multivolume natural history of animals began to appear in 1650 from Scottish-Polish physician John Johnstone, known best by his Latinized name Joannes Jonstonus (1603–75). Johnstone's four volumes featured up-to-date research—he cited Marcgraf and Piso, whose volume on Brazil had appeared only two years previously—as well as detailed illustrations.[96]

However, Johnstone largely retained the format of his humanist predecessors. His four volumes covered quadrupeds, birds, fish, and "serpents and dragons"—that is, reptiles, amphibians, and insects. Johnstone compiled material from both ancient and modern authors, but he included new and exotic animals such as the toucan, the rhea, and the dodo. Listing thirty different varieties under the general category of *gallus*, *gallina*, and *gallinula*, he placed the *gallo pavo* as a subcategory of the peacock, equating it like Aldrovandi to the *meleagris*.[97] Johnstone excluded literary and culinary references, and therefore began to demarcate an intellectual place for natural history that diverged from humanist claims to comprehensiveness in order to emphasize close description and rational order.

If Johnstone represented an incremental revision of the previous century's natural histories, the appearance in 1651 of *Rerum medicarum Novae Hispaniae thesaurus* turned the clock back.[98] Based largely on the manuscripts of Spanish physician

[92] Ibid., 190, 202, 237, 164.

[93] Aldrovandi, *Ornithologiae* (cit. n. 50), 2:145–51; Gessner, *De avium natura* (cit. n. 47), 387–90. For an English translation of Gessner's account of chickens in this work, see *Gessnergallus*, transcribed by Fernando Cibardi, trans. Elio Corti (Internet Archive, 2010), https://archive.org/details /TheChickenOfConradGessner.

[94] Flandrin, "De la diététique à la gastronomie" (cit. n. 88); Pinkard, *Revolution in Taste* (cit. n. 89), 3–6; Albala, *The Banquet* (cit. n. 3).

[95] *Les Dons de Comus, ou les delices de la table*, vol. 1 (Paris: Prault, 1739), avertissement, xvi–xvii.

[96] John Johnstone, *Historiae naturalis*, 4 vols. (Frankfurt am Main: Heirs of Matthaus Merian, 1650–53); Marcgraf and Piso, *Historiae naturalis Brasiliae* (cit. n. 21).

[97] Johnstone, *Historiae naturalis de avibus* (cit. n. 21), 57.

[98] Francisco Hernandez et al., *Rerum medicarum Novae Hispaniae thesaurus, seu, plantarum animalium mineraliun Mexicanorum historia* (Rome: Vitalis Mascardi, 1651).

Francisco Hernández (1514–87), who had traveled in Mexico in the 1570s, the *Thesaurus* was a sprawling work, mainly concerning materia medica. But a four hundred-page section in the middle of the book, written by physician Johann Faber (1574–1629) in the 1620s, offered lengthy essays on Mexican animals. Faber did not include the turkey, possibly another indication that by the 1620s, turkeys were no longer considered exotic—or no longer considered Mexican, but instead were a variety of *meleagris*.[99]

Most of the cookbooks published between 1615 and 1650 had reprinted or reshuffled earlier works.[100] Some historians view this less as a sign of waning creativity than of recognition by publishers of new and broader audiences. Piggybacking a book of recipes onto a household manual, as Markham did in 1615, was a way for publishers to save money, and this became an increasingly popular format after 1650. As literacy increased, so did a market for smaller, cheaper cookery books.[101] Celebrity authors also held appeal, such as the Countess of Kent, Elizabeth Grey, supposed compiler of the culinary and medicinal recipes in *A True Gentlewoman's Delight* (1653). The Countess's longtime cook, Robert May, followed a few years later with his own cookbook.[102]

Most culinary historians agree that the publication in 1651 of *Le cuisinier françois* by François Pierre de la Varenne (1615?–78) incited a revolution.[103] The first entirely new French cookbook since 1543, it encapsulated an ongoing shift away from sweet, spiced food. Flandrin calculated that 69 percent of Messisbugo's recipes in the mid-sixteenth century contained sugar, honey, or another sweet ingredient, while 82 percent of his recipes contained some kind of spice.[104] Such ingredients made few appearances in La Varenne's book. His emphasis on simplicity, on food tasting of itself—what his countryman Nicolas de Bonnefons referred to as "*le goût naturel*" ("the natural taste")—put him in the vanguard of a new French cuisine that would dominate European tables.[105] La Varenne described his book as "teaching the manner of preparing & seasoning meats well, following the four seasons of the year, as they are served at present at the tables of the nobility."[106] "Meats" referred to all foods, but meat continued to dominate upper-class tables. However, the animals in La Varenne's "Table des viandes" differed markedly from Messisbugo's list. Although birds continued to predominate, swans, peacocks, cormorants, and bustards made no appearance. Remaining were chickens, turkeys, geese, pigeons raised in dovecotes, and small birds such as thrushes, ortolans, and figpeckers. Game birds consisted of wood pigeons, pheasants,

[99] Johann Faber, "Aliorum Novae Hispaniae Animalium," in Hernández et al, *Rerum medicarum Novae Hispaniae thesaurus* (cit. n. 98), 460–840.

[100] Notaker, *Printed Cookbooks* (cit. n. 8), 3–4.

[101] Willan and Cherniavsky, *Cookbook Library* (cit. n. 7), 129–30.

[102] Ibid., 133, 136; *A True Gentlewoman's Delight. Wherein is Contained All Manner of Cookery: Together with Preserving, Conserving, Drying and Candying. Very necessary for All Ladies and Gentlewomen* (London: William Shears, 1653); Robert May. *The Accomplisht Cook, or The Art and Mystery of Cookery* (London: Nath. Brooke, 1660).

[103] Flandrin, "De la diététique à la gastronomie" (cit. n. 88); Pinkard, *Revolution in Taste* (cit. n. 89); Terence Scully, *La Varenne's Cookery* (Totnes: Prospect, 2006); Peterson, *Acquired Taste* (cit. n. 89); Laudan, "A Kind of Chemistry" (cit. n. 4); Willan and Cherniavsky, *Cookbook Library* (cit. n. 7), 152–63.

[104] Flandrin, "Choix alimentaires" (cit. n. 2), 671.

[105] Pinkard, *Revolution in Taste* (cit. n. 89), 60–4.

[106] La Varenne, *Le cuisinier françois* (cit. n. 6), 1.

partridges, and quail. Only the heron remained of the large number of wild birds deemed edible half a century earlier.[107]

Already, John Murrell's 1615 cookbook distinguished rural from urban cookery, with fewer large wild animals in his "London cookerie." Many of Murrell's recipes "in the French fashion" evidenced a diminution in the use of spices from their peak in Messisbugo. But less spice could have revealed previously masked strong tastes, especially of fish-eating birds. Murrell's recipes for swan and turkey were identical except in their use of spices; both were seasoned with salt and pepper, but while the turkey was merely stuck with a few cloves, seasoning for the swan, as well as for curlews and herons, also included the stronger-tasting ginger.[108]

Changes in medical theories and practices led inevitably to changed dietary recommendations. Galenic humoral theory emphasized warm, moist foods, although physicians were no more consistent in their recommendations than patients were in their observance. Warmth could owe to temperature or to spice. Popular literature on extending the lifespan and preserving health in old age added further nuances to humoral medicine.[109] In *Trattato della vita sobria* (1558); Venetian gentleman Luigi Cornaro (1467?–1566) described a minimalist diet. Light-colored foods such as chicken or milk were to be preferred over dark-colored foods. Cornaro avoided most vegetables and fruit. For meat he recommended poultry, veal, kid, or mutton, or various kinds of white fish, and drinking only "young wine."[110] While Cornaro did not explicitly refer to Galen, humoral rankings of foodstuffs supported his recommendations; that is, red meat was hot and dry, raw vegetables and fruit were cold and wet. Since impinging dryness and coldness characterized aging, such foods were to be avoided.

Guy Patin in 1632 added further commentary on this topic. Young animals, warm and moist, were preferable. Where animals lived and what they ate was also important; domesticated animals had "soft and moist" flesh, while wild animals and those who fed in forests were tougher and drier. On birds he wrote the following:

> The flesh of birds truly nourishes less than that of four-footed animals, although it is digested more easily. Among them the first place of honor belongs to partridges & mountain birds, then to woodcock, to blackbirds, then to squab [baby pigeons], pheasant, & grouse, to which it is necessary to include chickens & capons. The flesh of peacocks is put in the last rank, and is the least prized.[111]

Apart from birds, La Varenne listed only a few mammals, including lambs; rabbits; hares; and in season, deer and suckling pig. He listed the young—*levreaux, pigeonneaux, ramereaux, cailleteaux, oysons*, even *chaponneaux* (leverets, squab, baby quail, goslings, baby capons)—alongside the adult animals, with the list closely corresponding to Patin's recommendations and the emphases of longevity literature.[112]

---

[107] Ibid., "Table des viandes," not paginated.

[108] Murrell, *New Booke of Cookerie* (cit. n. 69), 29–30.

[109] Guerrini, "Impossible Ideal of Moderation" (cit. n. 10); Cynthia Skenazi, *Aging Gracefully in the Renaissance* (Leiden: Brill, 2013); David Gentilcore, *Food and Health in Early Modern Europe* (London: Bloomsbury, 2016).

[110] Luigi Cornaro, *Trattato della vita sobria* (Padua: no publisher given, 1558); translated in Leonard Lessius, *Hygiasticon* (Cambridge: Roger Daniel, 1634), with separate pagination.

[111] Patin, *Traicté* (cit. n. 29), 44.

[112] La Varenne, *Le cuisinier françois* (cit. n. 6), "Table des viandes," not paginated.

The publisher's preface to *Le cuisinier françois* made explicit La Varenne's considerations of health. Pierre David situated the book among the popular genre of maintenance of health, citing "Le Médécin charitable," a reference to Philibert Guybert (1579?–1633), who collected several works on medicine and pharmacy under the title of *Les oeuvres charitables* (1633). Guybert's volume included recipes for medicines, healthy drinks, and preserves.[113] David commended Guybert (without naming him) for wishing to "conserve and maintain health in a good state," though his works taught only how to "corrupt the vicious qualities of meats by contrary and diverse seasonings." Rather than spices, David stated, "it is much sweeter to make an honest and reasonable consumption of ragouts and other delicacies in proportion to one's faculties, to make life and health subsist."[114]

In his list of meats, La Varenne followed the precepts of Galenic dietary rules and the longevity literature; but in rejecting most spices, he also acknowledged a different set of rules. Rachel Laudan has convincingly argued that La Varenne's emphasis on local and seasonal foods, light meats, and vegetables, and his rejection of spices and sugar, owed much to the chymical physicians surrounding Louis XIV at his court. Chymical medicine viewed digestion as a chemical process rather than being based on heat, as Galenists argued. The tendency of vegetables and fruits to putrefy therefore indicated their higher digestibility. Oils and fats provided the foundation for new kinds of sauces based on emulsions of the Paracelsian *tria prima* of sulfur, salt, and mercury, and corresponding to fats, salts, and meat and vegetable essences.[115] But La Varenne did not state this directly, leaving discussion of medical uses to the publisher's preface.

La Varenne's *Cuisinier françois* saw fifty-two printings before 1700 and a number of translations. Although cookbook authors gradually adopted his techniques, the kinds of animals deemed edible remained in flux. Nicolas de Bonnefons's cookbook *Délices de la campagne* (1654), a follow-up to his 1651 *Le jardinier françois*, billed itself as a comprehensive guide for "*Dames ménagères*." He considered peacocks together with pheasants, grouse, and "other similar animals," and devoted a separate chapter to the turkey, but omitted herons and other large wild birds.[116] Pierre de Lune's 1656 *Le cuisinier* followed La Varenne's instructions for bouillons, essences, and ragouts, but continued to recommend bustards and large seabirds.[117]

While peacocks no longer appeared on English tables, swans, curlews, and herons still held their place, according to the cookbooks published by Robert May and William Rabisha in the early 1660s. In *L'arte di ben cucinare* by Bartolomeo Stefani (1661), the list of edible birds included turkey, duck, pheasant, and many smaller wild birds, but none of the larger ones. However, his lengthy section on banquets told a different story, with, among other delicacies, "*due pavoni revestiti*," that is, two peacocks reclothed with their feathers, topped with sugar crowns.[118]

---

[113] Philibert Guybert, *Toutes les oeuvres charitables* (Paris: Jean Jost, 1633); cited in Notaker, *Printed Cookbooks* (cit. n. 8), 161.

[114] La Varenne, *Cuisinier francois* (cit. n. 6), "Le Libraire au Lecteur," not paginated; this preface by David does not appear in the English translation: *The French Cook* (London: Charls [sic] Adams, 1653).

[115] Laudan, "A Kind of Chemistry" (cit. n. 4).

[116] [Nicolas de Bonnefons], *Les délices de la campagne*, 2nd ed. (Amsterdam: Raphael Smith, 1655), 230–4; first published by Pierre Des-Hayes (Paris, 1654).

[117] Pierre de Lune, *Le cuisinier ou il est traitté de la veritable method* (Paris: Pierre David, 1656), 77–8, 80–1.

[118] May, *The accomplisht cook* (cit. n. 102); William Rabisha, *The Whole Body of Cookery Dissected* (London: Giles Calvert, 1661); Bartolomeo Stefani, *L'arte di ben cucinare, et instruire* (Mantua: Stampatore Ducali, 1661), 36–42, 97.

Some natural histories also continued to express this ambivalence between the old and the new. The English edition of Francis Willughby's 1676 *Ornithologiae*, published in 1678, included treatises on falconry and fowling, both adapted from Markham. While Willughby took much of his information from Aldrovandi and other humanist authors, he added his own observations and employed a rational system of classification. His finely detailed illustrations, gathered at the back of the book, borrowed Johnstone's format. Willughby did not specifically denote "use," but he discussed the meat of animals he called "of the poultry kind," which encompassed domesticated and wild birds. Both the turkey and the peacock were domesticated, and he described the bustard's flesh as "delicate and wholesome."[119]

In contrast, the *Mémoires pour servir à l'histoire naturelle des animaux* (1671–76), published by the Paris Academy of Sciences under the editorship of Claude Perrault (1613–88), deliberately rejected any system of classification and presented animals in apparently random order. The descriptions themselves, however, set forth principles by which classification could proceed, with particular attention to naming.[120] Birds included both exotic species and such familiar birds as the cormorant, eagle, and bustard. Perrault definitively identified the *meleagris* with the guinea hen or *peintade*, and took pains to distinguish the *cocq-Indien* (curassow) from the *coq d'Inde* or turkey (fig. 2).[121] But the academy took no interest in culinary or medical uses. In ignoring such uses, Perrault and the Academy of Sciences demarcated the boundaries of natural history around comparative anatomy, precise morphological description, and, although they claimed to eschew it, classification. At the same time, authors of cookbooks demarcated their boundaries to exclude specific medical recommendations, focusing instead on taste.

CONCLUSION

Vincent La Chapelle (fl. 1733–36), who published his three-volume *The Modern Cook* in 1733, combined French and English cooking. La Chapelle, a Frenchman, had long worked in England. *The Modern Cook*, which he issued in French a few years later, borrowed many of its recipes from the *Cuisine roïal et bourgeois* (1691) of François Massialot (1660–1733).[122] Both had cooked at royal courts; Massialot began his book with menus from aristocratic dinners, while at the same time insisting that his book "will not be without use in bourgeois households."[123] La Chapelle similarly directed his work toward both professional cooks and householders. Both continued to retreat from heavy spicing, and to reflect a largely urban sensibility. La Chapelle added recipes for woodcock, snipe, teal, and other game birds that continued to be part of the British diet into the twentieth century. Both continued to use a large variety of small songbirds. But their main emphasis was on domesticated birds, including chickens, capons, ducks, geese, and turkeys. Massialot and La Chapelle thus demonstrate the

[119] Francis Willughby, *Ornithology*, ed. John Ray (London: A. C. for John Martyn, 1678), 159, 179.
[120] Anita Guerrini, *The Courtiers' Anatomists: Animals and Humans in Louis XIV's Paris* (Chicago: Univ. of Chicago Press), 153–6; Perrault, *Mémoires* (cit. n. 1).
[121] Perrault, *Mémoires* (cit. n. 1), 134–9, 146–9.
[122] La Chapelle, *The Modern Cook* (cit. n. 16), 1:ii; François Massialot, *Le cuisine royal et bourgeois*, 2nd ed. (Paris: Charles de Sercy, 1693); first published by Charles de Sercy (Paris, 1691).
[123] Massialot, *Le cuisine royal et bourgeois* (cit. n. 122), preface, not paginated.

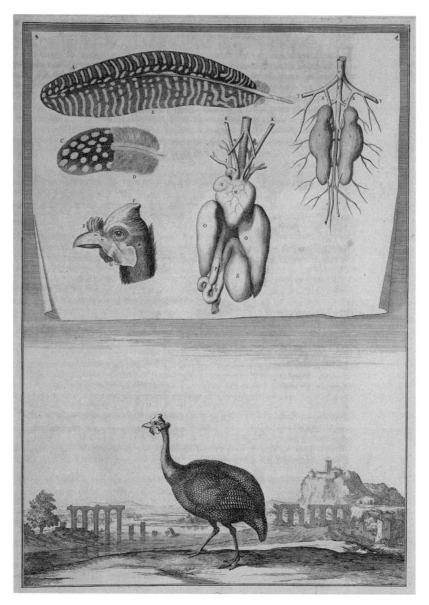

Figure 2. *Peintade (guinea fowl). From Claude Perrault, ed.*, Mémoires pour servir *(cit. n. 1).*
(Linda Hall Library of Science, Engineering & Technology.)

continuing shift of culinary literature toward town and away from court and country
that we first saw in Murrell's 1615 *Newe Booke of Cookerie*.[124]

La Chapelle, like sixteenth-century naturalists and like Buffon, whose *Histoire
naturelle* began publication in 1749, aimed to be comprehensive as well as up to date.
His cookbook codified what came to be known as French cuisine, which became the
cuisine of elites across Europe and which heavily influenced more modest cuisines as

---

[124] Mennell, *All Manners of Food* (cit. n. 89), 102–33.

well. Many of the recipes in Hannah Glasse's 1747 *The Art of Cookery Made Plain and Easy* came straight from La Chapelle.[125]

Already in 1700, Louis Liger's update of Estienne's *Maison rustique* advised city dwellers how to build their weekend house in the country rather than displaying existing country life. Liger only listed farmyard birds, omitting the wild and half-wild birds that Estienne had detailed.[126] La Chapelle prefaced his work with "Directions for a House-Steward," which were instructions for managing a kitchen. Animals came from the butcher or poulterer, not from the field or the pond, and the concern was with price, not with provenance.[127] The links between the kitchen and the natural history of animals, which had become increasingly tenuous by the end of the seventeenth century, had been entirely broken. In cookbooks at least, animals became a means to an end rather than a topic of study.

[125] Buffon, *Histoire naturelle* (cit. n. 22); La Chapelle, *The Modern Cook* (cit. n. 16); Hannah Glasse, *The Art of Cookery Made Plain and Easy* (London: published by author, 1747).
[126] Liger, *Oeconomie générale* (cit. n. 62), vol. 1.
[127] La Chapelle, *The Modern Cook* (cit. n. 16), 1:ii–vii.

Digesting Faith:
Eating God, Man, and Meat
in Seventeenth-Century Rome

by Bradford Bouley*

ABSTRACT

In seventeenth-century Rome, the consumption of meat was on the rise. By the 1630s, Romans were eating double the amount of meat they had consumed fifty years previously, even accounting for growth in population. At the same time that all this meat was being consumed, the papacy came to fiercely defend another comestible: the wafer eaten in the Eucharist. These two products came to be at the center of papal reform in Rome. Eating meat, especially at Easter, and regularly partaking in the body of Christ signaled one's adherence to Catholicism and obedience to the Pope. But the matter was not that simple; accusations of cannibalism in Rome—both real and imagined—led to lengthy medical and theological discussion over how the body digests food. Furthermore, most contemporary medical advice did not recommend heavy consumption of meat. This article thus explores how an alliance between the medical community and the papacy sought to remake alimentary and anatomical ideas related to digestion and healthy eating in early modern Rome. Various sections will detail evolving theories of digestion in the papal capital; how such theories were applied to theological and practical issues such as giving the Eucharist to the sick; whether cannibals could gain sustenance from human flesh; and physician commentary on rising meat consumption in the city. In the end, medical expertise allied with Church authority to defend the aims of the Counter-Reformation papacy.

INTRODUCTION

According to a report recorded by the official in charge of taxing meat coming into Rome, in 1598 the amount of meat imported into the city was sufficient to provide

* Department of History, HSSB 4000, University of California, Santa Barbara, Santa Barbara, CA 93106, USA; bouley@ucsb.edu.

I would like to thank a number of people who have offered helpful comments on this research in general, and this article in particular, including Emma Spary, Paula Findlen, Anya Zilberstein, Tom Cohen, Libby Cohen, Hannah Marcus, Maria Pia Donato, Paolo Savoia, Brian Brege, Hilary Bernstein, Elena Aronova, Utathya Chattopadhyaya, Christina Roberts, and Sheila Dean. The research has also been generously supported by a research year at Villa I Tatti, the Harvard Center for Renaissance Studies, the UCSB Regents Junior Faculty Fellowship, and a Vatican Film Library Mellon Grant. All translations are my own, unless otherwise specified. All errors are, of course, also my own.

three-quarters of a pound per person per day on non-fast days.[1] This is an impressive statistic for the sixteenth century, since the period has largely been seen as one in which food supplies in Europe generally decreased. Furthermore, Rome was no small town, but a large urban area with a population of about 110,000 people.[2] As noteworthy as those statistics for 1598 are, they pale in comparison with the amount of meat brought into the city by the 1630s. The per capita per day consumption of meat by 1629 topped more than a pound on non-fast days.[3] This daily consumption of meat had been growing steadily for fifty years, despite occasional bumps due to famines, and reports generated during this time suggest this increase was the practical realization of a major project initiated by the papal government.[4]

Among the various meats that Romans ate—the tax records reveal sizeable numbers of cattle, veal, pig, and Italian water buffalo—the animal they consumed most was sheep. By 1629 Romans were devouring 165,970 head of sheep per annum. That was three times the amount that Romans had consumed just thirty years earlier and

[1] Archivio di Stato di Firenze (**hereafter** ASF), Carte Strozziane, ser. prima, 233, fol. 61r. These details come from an anonymous document entitled "Relazione distintis[si]ma di Roma, Anime, entrate, Chiese, Palazzi, Casali con molte piante di Ville, e altre minuzie particolari." However, the author has numerous specific details about Rome and lists the name of the tax farmer for the Grascia—Franchino Pistolese—as his source for the information that he recounts here. Furthermore, the document survives in several copies, including one from the files of Paul V's family: Archivio Segreto Vaticano (**hereafter** ASV), Fondo Borghese, ser. IV, no. 285. This source gives specific numbers of animals brought into the city at this moment in time (1598). As there is not space in this article to discuss the various numbers, I have chosen to aggregate into a total the meat consumed. I have estimated the weight of the butchered meat for each animal based on a combination of Ivana Ait's discussion of it for an earlier period (Ivana Ait, "Le carni della città," in *Banchetti e vivande nel rinascimento a Roma*, ed. Ivana Ait (Rome: Roma nel Rinascimento, 2017), 174n22; and a document issued by the Congregation of the Grascia (Rome's government agency in charge of meat import) that allowed a discount on tax if owners brought animals to market over a certain weight: see Archivio di Stato di Roma (**hereafter** ASR), Grascia, b. (*busta*) 1, fasc. 3, doc. 57. I have combined my estimates of butchered meat with more recent studies on animal husbandry related to animal weight and weight brought to market, including Tamara Scully, "Raising Veal: Alternatives to Conventional Models," *Farming Magazine*, 1 August 2014, https://www .farmingmagazine.com/dairy/raising-veal/; and Heather Hamilton, "The Relationship Between Cow Size and Production," *Beef Magazine*, 27 December 2011, http://www.beefmagazine.com/cow-calf /relationship-between-cow-size-production.

[2] Jacques Revel, "A Capital City's Privileges: Food Supplies in Early Modern Rome," in *Food and Drink in History: Selections from the Annales*, trans. Elborg Forster and Patricia M. Ranum, ed. Robert Forster and Orest Ranum (Baltimore: Johns Hopkins Univ. Press, 1979), 37–49; Jean Delumeau, *Vie économique et sociale de Rome dans la seconde moitié du XVIe siècle*, vol. 2 (Paris: Éditions E. de Boccard, 1959), 521–640; Arturo de Sanctis Mangelli, *La pastorizia e l'alimentazione di Roma nel medio evo e nell'età moderna* (Rome: P. Maglione & C. Strini, 1918), 57–89; Ramón A. Banegas López, "Consumption of Meat in Western European Cities during the Later Middle Ages: A Comparative Study," *Food & History* 8 (2010): 63–86, on 77–9.

[3] Archivio Storico Capitolino di Roma Capitale (**hereafter** ASC), cred. 10, tomo 47, fol. 174v also gives population numbers for each year recorded. This document records only sheep production, but the estimate of one pound per person is based on the following pieces of evidence: (1) Fausto Piola Caselli, "Merci per dogana e consumi alimentari a Roma nel seicento," in *La popolazione italiana nel seicento*, ed. Maria Norberta de Simas Bettencourt Amori (Bologna: CLUEB, 1999), 387–410. On page 392 Caselli states that Urban VIII's government said that a pound of meat per day would be the optimal consumption for the Roman citizenry. Unfortunately, Caselli writes that he found this information in ASR, Grascia 1. I have also examined this document but have been unable to corroborate the detail he cites. (2) Nevertheless, if we assume that the production of other animals given in 1598 and the one other year recorded in ASF (Carte Strozziane, ser. prima, 233, fol. 61r) remain the same, and only the amount of sheep consumed by Romans increased to the numbers reported in ASC, cred. 10, tomo 47, fol. 174v, then the daily consumption would indeed top a modern pound per capita on non-fast days.

[4] Jean Delumeau, *Vie économique et sociale de Rome dans la seconde moitié du XVIe siècle*, vol. 1 (Paris: Éditions E. de Boccard, 1957), 124.

six times the consumption documented fifty years before.[5] In contrast, Florence consumed 13,770 head of sheep in the same year (1629).[6] Although Florence was a smaller city (about 75,000 residents to Rome's 110,000 in the period), the per capita difference in consumption is striking. Similar comparisons could be made with other cities, with the much larger city of Paris (population about 400,000 in 1630), consuming 80,000 head of sheep each year in the 1630s—in other words, half of Rome's consumption for a population four times as great.[7] Comparisons with other cities yield similar results; as one anonymous chronicler noted in 1641, "all Rome consumes twice as much meat and wine as Naples, though the latter city has twice as many people."[8]

Romans consumed so much meat, and especially sheep, in these years for reasons that were not just economic or political—that is, the consumption was due to more than the fact that sheep herds were common in the Papal States or the general desire for centralizing rulers to make sure their cities were well supplied. Rather, the heavy Roman consumption of sheep was linked to the religious and political aims of popes responsible for reforming the Church. As Antonio Serra, president of the Grascia, the Roman commission in charge of assuring the supply of meat to the city, noted in a 1630 document, "the City of Rome and the other locations in the Ecclesiastical States must be well supplied with meat necessary for human life, especially during the next Easter festival of the Resurrection . . . and we should be supplied with lambs particularly [for this holiday]."[9] In the wake of the Reformation, the festivals of Easter and Corpus Christi came to especially be celebrated in Rome as signs of the strength of faith and a statement of Catholic unity.[10] During these festivals the rite of the Eucharist was venerated as partaking of the actual flesh and blood of Christ through transubstantiation, and nearly all believers would receive the consecrated bread. Following these events, in the week after Easter, another festival would celebrate the *Agnus Dei*—that is, the lamb of God, which symbolized Christ's sacrifice on the Cross.[11] It was during this time that large amounts of sheep were consumed in Rome, symbolically linking the rite of the Eucharist with the consumption of lamb or sheep on the tables of nearly every Catholic Roman citizen.

[5] For food consumption in 1598, see ASF, Carte Strozziane, ser. prima, 233, fol. 61r; for 1629, see ASC, cred. 10, tomo 47, fol. 174v. For 1579, see Delumeau, *Vie économique et sociale*, vol. 1 (cit. n. 4), 124.

[6] Andrea Zagli, "Da Beccai a Macellai nella Firenze dei Medici," in *Maladetti beccari: Storia dei macellai fiorentini dal Cinquecento al Duemila*, ed. Zagli, Francesco Mineccia, and Andrea Giuntini (Florence: Edizioni Polistampa, 2000), 21.

[7] Sydney Watts, *Meat Matters: Butchers, Politics, and Market Culture in Eighteenth-Century Paris* (Rochester, N.Y.: Univ. of Rochester Press, 2006), 1.

[8] Biblioteca Apostolica Vaticana (**hereafter** BAV), Codices Vaticani Latini (Vat. Lat.) 8354, fol. 187v.

[9] ASV, Miscellanea Armadi (Misc. Arm.), vol. 50, fol. 18r: "Volendo noi, e Monsignor Reverendiss. Serra Presidente della Grascia, provedere, che la Città di Roma, & altri luoghi dello Stato Ecclesiastico siano abbondanti de Carni per il vitto humano, massime nella Pasqua di Resurrettione prossima, & a questo effetto desiderando havere certa notitia di tutti gl'animali, e particolarmente de gl'Agnelli, che si trovano nel distretto di Roma."

[10] Margaret A. Kuntz, "Liturgical, Ritual, and Diplomatic Spaces at St. Peter's and the Vatican Palace: The Innovations of Paul IV, Urban VIII, and Alexander VII," in *A Companion to Early Modern Rome*, ed. Pamela M. Jones, Barbara Wisch, and Simon Ditchfield (Boston: Brill, 2019), 75–98.

[11] Ibid.

This article examines these two foodstuffs that were central to the Reformation in Rome—the bread of the Eucharist, and meat—and argues that both medicine and theology in the Holy City were affected by and contributed to changing consumption patterns. Such a study makes a critical intervention both in the history of medicine and in the history of religion for this period by demonstrating how the act of eating was linked to theological, anatomical, and political debates in the Holy City. Scholars such as Tessa Storey, Sandra Cavallo, David Gentilcore, and Ken Albala have suggested that the late sixteenth and early seventeenth centuries represent a critical juncture in the development of nutritional ideas, since it was during this time that diets became medicalized, with physicians increasingly being looked to as the authorities on what one should eat. However, such eating was closely wedded to the political and cultural context in which it took place; expertise in nutrition and eating habits of the populace were reflections of the society in which they existed.[12]

On a related, religious note, scholars have begun to argue that examining eating practices is central to understanding how the Reformation unfolded in the early modern period. Albala, Gentilcore, Christopher Kissane, Anselm Schubert, and others have argued that food was at the center of both the break of Protestants with Rome and the Roman response to the Reformation.[13] Feasting and fasting were at the root of church holidays, and, by some estimates, as many as 150 days in the pre-Reformation calendar may have consisted of one form of fast or another. Reformation cities like Zwingli's Zurich marked their break with Rome by consuming meat on fast days. As Albala notes, "Food, in this respect, was at the core of the average person's conception of religiosity, and devotion was defined in terms of the things one does, not necessarily the things one believes."[14] This study therefore demonstrates that changed eating practices in Rome can be seen as a direct consequence of the Counter-Reformation. This religious reform in turn sparked a change in medical ideas about healthy eating, and even digestion, in Rome.

The sources used to investigate these changed ideas about eating and nutrition are primarily medical in nature, since a number of physicians in the Holy City produced treatises on the functioning of digestion, diet, cannibalism, and even the absorption of the Eucharistic elements by the body. Rome was a vibrant medical center with a wide range of practitioners, and the evidence points to close collaboration between papal officials and medical professionals in creating new dietary advice.[15] The works of

---

[12] Sandra Cavallo and Tessa Storey, *Healthy Living in Late Renaissance Italy* (New York, N.Y.: Oxford Univ. Press, 2013), 11; David Gentilcore, *Food and Health in Early Modern Europe* (London: Bloomsbury, 2016), 1–2, 13–14; Ken Albala, *Eating Right in the Renaissance* (Berkeley and Los Angeles: Univ. of California Press, 2002), 1–2.

[13] Ken Albala, "The Ideology of Fasting in the Reformation Era," in *Food and Faith in Christian Culture*, ed. Albala and Trudy Eden (New York, N.Y.: Columbia Univ. Press, 2011), 41–57; Gentilcore, *Food and Health* (cit. n. 12), 95–103; Christopher Kissane, *Food, Religion and Communities in Early Modern Europe* (London: Bloomsbury, 2018); Anselm Schubert, *Pasto divino: Storia culinaria dell'eucaristia*, trans. Alice Barale (Rome: Carocci, 2019).

[14] Albala, "Ideology of Fasting" (cit. n. 13), 41–4, quote on 44.

[15] Elisa Andretta, *Roma medica: Anatomie d'un système médical au XVIe siècle* (Rome: École Française de Rome, 2011); Bradford Bouley, *Pious Postmortems: Anatomy, Sanctity and the Catholic Church in Early Modern Europe* (Philadelphia: Univ. of Pennsylvania Press, 2017); Maria Pia Donato, "La medicina a Roma tra sei e settecento: Una proposta di interpretazione," *Roma moderna e contemporanea rivista interdisciplinare di storia* 13 (2005): 99–114; Daniela Mugnai Carrara and Maria Conforti, "L'insegnamento della medicina dall'istituzione delle università al 1550," in *Il Rinascimento Italiano e l'Europa*, vol. 5: *Le Scienze*, ed. Antonio Clericuzio, Germana Ernst, and Maria Conforti

Antonio Brassavola (1500–55), Alessandro Petronio (1510–81), Realdo Colombo (1515–59), Leonardo Fioravanti (1518–88), Andrea Cesalpino (1519–1603), Giovanni Faber (1574–1629), and Paolo Zacchia (1584–1649) are of special relevance because all these authorities worked in Rome and published treatises dealing with diet, digestion, or Roman food more generally. Although no record explicitly shows that the Church looked for the aid of the lay medical group in putting forward new consumption practices, the links between physicians who will be considered here and various papacies are evident. First, more than a few physicians served as chief papal physicians—Cesalpino, Petronio, Zacchia, Brassavola; others were invited to Rome at the behest of a pope—Colombo and Fioravanti; while still others were resident at the Santo Spirito Hospital, which was closely allied to papal aims—Petronio and Faber. Furthermore, most of these physicians knew each other or were familiar with each other's work. Colombo listed Petronio and Brasavola as friends with whom he had worked;[16] Cesalpino was Colombo's student in Pisa; Fioravanti referred to Colombo as a medical competitor in Rome; and Zacchia learned from Faber and performed dissections with him.[17] In short, the writers surveyed here represent a group of physicians working together and for the papacy in early modern Rome. It was they who outlined theories of digestion that found a way to marry old theological ideas on the body with new anatomical studies. These writers also questioned nutritional theories that threatened papal authority, and they argued that the land around Rome was fertile and that animals that grazed in the area provided a healthy diet for all. Together they endorsed a papal food program designed to enhance Catholic unity and highlight papal strength following the Reformation.

In pursuing this argument, this paper begins with a discussion of digestion as it was understood and modified in papal Rome, before turning to the stakes of such anatomical debates. That Catholicism came to be symbolized by an act (the consuming of the Eucharist) that had come to be decried as cannibalism by Protestant reformers meant that how the body interacted with what it consumed mattered immensely to medical writers and Catholic officials. After looking at how ideas on digestion were inextricably linked to theological concerns about how the consecrated bread of the Eucharist moved through a body, we will turn to a discussion of how ideas about animal flesh and its consumption led to a dramatic change and justification of meat-heavy alimentary policies in papal Rome. In the early seventeenth century, a series of popes decided that Catholicism would be defended through the bellies of the faithful.

(Treviso: Fondazione Cassamarca, Angelo Colla Editore, 2008), 475; Silvia de Renzi, "'A Fountain for the Thirsty' and a Bank for the Pope: Charity, Conflicts, and Medical Careers at the Hospital of the Santo Spirito in Seventeenth-Century Rome," in *Health Care and Poor Relief in Counter-Reformation Europe*, ed. Ole Peter Grell, Andrew Cunningham, and Jon Arrizabalaga (New York, N.Y.: Routledge, 1999), 99–130; De Renzi, "Medical Competence, Anatomy and the Polity in Seventeenth-Century Rome," *Renaiss. Stud.* 21 (2007): 551–67.

[16] Robert J. Moes and C. D. O'Malley, "Realdo Colombo: 'On Those Things Rarely Found in Anatomy': An Annotated Translation from the 'De Re Anatomica' (1559)," *Bull. Hist. Med.* 34 (1960): 508–28, on 523 and 526.

[17] Augusto de Ferrari, "Cesalpino (Caesalpinus), Andrea," in *Dizionario biografico degli Italiani*, vol. 24 (Rome: Istituto della Enciclopedia Italiana, 1980), 122–5, http://www.treccani.it/enciclopedia /andrea-cesalpino; Leonardo Fioravanti, *De capricci medicinali* (Venice: Appresso gli Heredi di Melchior Sella, 1582), 81r–v; Silvia de Renzi, "Per una biografia di Paolo Zacchia: Nuovi documenti e ipotesi di ricerca," in *Paolo Zacchia alle origini della medicina legale, 1584–1659*, ed. Alessandro Pastore and Giovanni Rossi (Milan: Franco Angeli, 2008), 50–73, on 53–4.

THE ROMAN BELLY

The combination of a large community of medical practitioners and the religious and political importance of keeping Rome well fed helped create a city for which the topic of food and its digestion was of overweening importance. The starting point was the classic understanding of digestion, which was put forward by the ancient Roman physician Galen and only underwent slight modification in the medieval curriculum. In this conception, digestion of food took place in several stages. First, according to Galen, consumed food traveled to the stomach, where it was cooked by the bodily heat until it became a milky liquid called chyle. This substance was then filtered in the small intestine before it moved to the liver, where it was broken down into the various humors. In Galen's conception of anatomy, the liver was of critical importance not just for digestion but for the entire functioning of the human body. The liver served as a conduit in which nutrient-rich chyle passed through fine channels, each so small as to be invisible. These channels were important for Galen, who saw in the liver the origin of blood, which needed a route from the *vena porta* (portal vein) to the veins that traveled from the liver to the rest of the body.[18] Once reaching the various parts of the body, the blood was absorbed, ultimately becoming human flesh.[19] Thus, the liver became the connection point between the digestive and the circulatory systems.

This conception of digestion was, in the late Middle Ages, wedded to a theory of "moistures" to help explain the theological issues raised by cannibalism and communion. According to this theory, the body maintained life and health through two moistures; these included the nutritive moisture that could be resupplied by food, and the radical moisture that could not be resupplied by food. Each moisture was consumed by the body's natural heat. The radical moisture was what amounted to a person's life essence and was frequently likened to the oil in a lamp. When the radical moisture was exhausted a person died, so the amount of radical moisture a person began with at birth indicated the span of a life, barring, of course, an unforeseeable event. The nutritive moisture, in contrast, was the building block for the flesh that would require continual resupply via food throughout life.[20]

This theory of using moistures to explain how the body assimilated food was important to theologians chiefly because of its bearing on several issues, including resurrection, original sin, and digestion in the Eucharist. Adam's original sin would be carried from generation to generation via the radical moisture, which was passed on unchanged in the semen from father to children. This sin was thereby physically present in the human body. When one consumed the Eucharistic elements, though, the body was thought to absorb Christ into the nutritive moisture, thus allowing this salvation

---

[18] Owen Powell, "Introduction," in *Galen: On the Properties of Foodstuffs*, ed. and trans. Powell (New York, N.Y.: Cambridge Univ. Press, 2003), 1–19, on 17.

[19] Albala, *Eating Right* (cit. n. 12), 56–64; Robert P. Multhauf, "J. B. van Helmont's Reformation of the Galenic Doctrine of Digestion," *Bull. Hist. Med.* 29 (1955): 154–63, on 154–5. This theory of digestion appears in a number of other places, but these two authors provide a good overview of the classical theory and subsequent challenges to it. For Galen's view, which is basically the theory described here, see Powell, "Introduction" (cit. n. 18), 13–18.

[20] Albala, *Eating Right* (cit. n. 12), 53; Michael McVaugh, "The 'Humidum Radicale' in Thirteenth-Century Medicine," *Traditio* 30 (1974): 259–83; Joseph Ziegler, "*Ut Dicunt Medici*: Medical Knowledge and Theological Debates in the Second Half of the Thirteenth Century," *Bull. Hist. Med.* 73 (1999): 208–37, on 217.

to become part of one's very flesh for a time.[21] This was a theory of digestion, then, that allowed the battle for salvation to be constantly waged in the bodies of parishioners.

This theory also solves another problem, that of cannibalism. From the early days of the Church, theologians worried that if you eat another person you might destroy their soul by forcing their flesh to become your flesh.[22] The theory of moistures solved this problem by asserting that a body's radical moisture could never be absorbed into another body. Therefore, some essence of the consumed body always remained and thus could be resurrected.[23]

These medico-theological notions of digestion survived into the early modern period, at which point they were challenged by both new theories of digestion and renewed theological concern about communion and cannibalism.[24] In Rome, in particular, a succession of prominent anatomists in papal employ sought to preserve what they could of the classical system. The first professor of anatomy at the University of Rome, La Sapienza, Realdo Colombo, was at the vanguard of these efforts. He was a friend, and then later a rival, of Andreas Vesalius (1514–64), and the chair of anatomy at the University of Rome from 1548 until 1559, as well as a pioneer of research into cardiac anatomy.[25] In 1559, Colombo produced an anatomical manual, *De re anatomica*, that summarized the latest discoveries dealing with the human body. This included a somewhat new view, informed by anatomical work, of human digestion.

In his explanation of human digestion, Colombo agreed with earlier writers that chyle is produced in the stomach through a sort of cooking—as in classical theory. However, he argued for a role of the intestine other than just as a filter: "Chyle is the alteration of food [in the stomach] into a material similar to drinkable milk which afterward leaves the stomach and descends through the curved intestine until the intestines expel the full juice, supplying not four or even ten veins, but springing forth into innumerable veins."[26] That is, in Colombo's scheme, the chyle leaves the intestines to travel to the rest of the body via innumerable little veins. Food is absorbed into the bloodstream at various moments, when it is ready to be made into flesh in the body. That which is left descends into the colon to be excreted.[27] Colombo modified Galen by removing some of the liver's necessity, arguing that blood could pass directly from the intestine to the rest of the body. He also more carefully charted the movement of blood from the gastric system to the liver, noting that it travels from the spleen into the portal vein. He did not find Galen's invisible channels.[28]

---

[21] Philip Lyndon Reynolds, *Food and the Body: Some Peculiar Questions in High Medieval Theology* (Boston: Brill, 1999), 105–6, 6–8, 392–3, 431.

[22] Caroline Bynum, *The Resurrection of the Body in Western Christianity, 200-1336* (New York, N.Y.: Columbia Univ. Press, 1995), 55; Bart Wagemakers, "Incest, Infanticide, and Cannibalism: Anti-Christian Imputations in the Roman Empire," *Greece and Rome*, 2nd ser., 57 (2010): 337–54.

[23] Reynolds, *Food and the Body* (cit. n. 21), 234, 392–3; Ziegler, "*Ut Dicunt Medici*" (cit. n. 20), 218–22.

[24] On some of the new theories of digestion see Multhauf, "J. B. van Helmont's Reformation" (cit. n. 19), 154–63. For the popularity of these theories in early modern Rome, see De Renzi, "Paolo Zacchia" (cit. n. 17), 50–73, on 54.

[25] Carlo Colombero, "Colombo, Realdo," in *Dizionario biografico* (cit. n. 17), vol. 24, http://www .treccani.it/enciclopedia/realdo-colombo_%28Dizionario-Biografico%29/.

[26] Realdo Colombo, *De re anatomica Libri XV* (Venice: Ex Typgraphia Nicolai Bevilacquae, 1559), 165.

[27] Ibid.

[28] Ibid.

Colombo's work in general comprised some of the partial challenge to Galenic ideas that was common among sixteenth-century anatomical innovators and undertaken by other contemporary anatomists such as Vesalius and Charles Estienne.[29] As Barbara Orland has recently shown, real change in understanding how chyle was absorbed in the body was slow in coming, but began with Gaspare Aselli (1581–1626), the professor of medicine in Padua who discovered the "milk veins," or small veins that seemed to carry chyle through the body. His discoveries were not published until after his death, and further elaboration of new anatomical ideas on digestion really took off after 1640.[30] Thus, in the early seventeenth century, at the same moment when the papacy began fiercely promoting the Eucharist while also encouraging more meat consumption, Galenic theories of nutrition were breathing one last gasp. Indeed, although Colombo made use of the latest anatomical theory to modify the path of digestion, he seems to have stuck to a largely traditional view of how food is incorporated into nutrients, thus allowing his theory to agree with Christian theology on this matter. This seems to have been appreciated by theologians. For example, Giorgio Polacco (b. 1574), a Venetian inquisitor and theologian, continued to cite medical theories of moistures in a treatise in which he defended the eating of the real presence of Christ in the Eucharist.[31]

Despite the careful discussion of anatomy, Colombo did not specify in his 1559 account the precise mechanism whereby food is absorbed or rejected by the body. Such information can be inferred, though, from case studies that he recounted in other parts of *De re anatomica*. In particular, he told the story of "Lazaro, commonly called the glass-eater." Lazaro "ate glass, stones, rocks, wood, live animals, coals, fish drawn still wriggling from the fish pond. He ate mud, linen and woolen cloths, hay, straw and in short whatever men or living things ate."[32] Lazaro, it seems, was easily able to pass these items that he ate. Those items that did not nourish the body passed out without any digestion. When Colombo eventually performed an autopsy on Lazaro, he found that the root of Lazaro's ability to eat all these strange things was a failure of his nerves to transmit the sensation of taste, rather than any oddity with his gastrointestinal system.[33]

What is implicit in Colombo's theory—that the body only takes nourishment from items that can be made into flesh and does not digest others—is made explicit in the later writings of Alessandro Petronio, Colombo's contemporary and colleague. Petronio was personal physician to several popes and directed the physicians at the

---

[29] Numerous scholars have written on the rising prestige of anatomy and its importance for medicine in the early sixteenth century. See, for example, Roy Porter, "Medical Science," in *The Cambridge History of Medicine*, ed. Porter (New York, N.Y.: Cambridge Univ. Press, 2006), 136–75; Cynthia Klestinec, "Practical Experience in Anatomy," in *The Body as Object and Instrument of Knowledge: Embodied Empiricism in Early Modern Science*, ed. C. T. Wolfe and O. Gal (New York, N.Y.: Springer, 2010), 33–57; Charles O'Malley, *Andreas Vesalius of Brussels, 1514–1564* (Berkeley and Los Angeles: Univ. of California Press, 1964); and Nancy Siraisi, *Medieval and Early Renaissance Medicine: A Guide to Knowledge and Practice* (Chicago: Univ. of Chicago Press, 1990), 190–3.

[30] Barbara Orland, "The Fluid Mechanics of Nutrition: Herman Boerhaave's Synthesis of Seventeenth-Century Circulation Physiology," *Stud. Hist. Phil. Biol. Biomed. Sci.* 43 (2012): 357–69, on 362–3.

[31] Giorgio Polacco, *Trattato divoto, et fruttuoso dove si difende frequente communione da gl'impugnatori di essa con vive dotte ragioni cavate dalla Sacra Scrittua, da. SS.Padri, e Dottori Catolici* (Venice: Appresso Trivisan Bertolotti, 1606), 51–2.

[32] Moes and O'Malley, "Realdo Colombo" (cit. n. 16), 522.

[33] Ibid.

large Santo Spirito hospital in Rome for several decades.[34] He produced a treatise on healthy living in Rome, entitled *Del viver delli romani et di conservar la sanità*, in which he described more clearly which foods could be well digested by the human body and why: "We begin by stating first that everything that nourishes the body necessarily has in it a fat and a slow juice [*un succo grosso & lento*]." That is, Petronio thought that when the body broke down foods into chyle, there were fatty elements and torpid elements in food that could be digested. The fatty juice was what allowed the chyle to nourish the human body as it was this part that would "restore and join on to the substance and soft-ness of our body"—that is, the fatty part of the chyle was what allowed it to become the flesh.[35] The significance of the "slow juice" was that as the gastric system made food into chyle it could not be too thin or it would be passed quickly into the stool. This qual-ity of the slow juice was of importance to the retention of what was ingested in the Eu-charist, and Petronio explicitly commented on the easy digestibility of the included wheat.[36] Items consumed that did not possess enough fattiness and torpidity were simply passed out of the body. He explained this was because it was necessary that food was "converted into our substance . . . and such a thing cannot be done if it does not take part in fatness and slowness [*tal cosa non si può fare se non participa di grossezza & di lentore*]."[37] This reflected generally accepted medical wisdom about how the body digested food, but with a little more specific detail about what was in food that allowed it to be digested.[38] Such specificity was important for understanding how the body di-gested the Eucharistic elements as well as both human and animal flesh—foodstuffs that had become important and contested in contemporary Rome.

EATING HUMANS AND EATING GOD

Although the sacrament of Communion had become very important in the late me-dieval period, in the period after the Reformation it came to be understood as the cen-tral act of the Catholic faith; adherence to the real presence of Christ in the bread and wine via transubstantiation had come to mark the key dividing line between Catholics and all other Christian denominations. However, both prior to the Reformation and after, the laity were only given the bread and not the wine. Nevertheless, Frederick McGinniss has argued that participating in the regular taking of the Eucharist was seen as engaging fully in the reform efforts of the Church. The Eucharist was so im-portant that the entire reform effort in Rome was organized around this sacrament. Far from being merely an annual practice—as it had been for most Christians prior to the Reformation—Catholic reformers were now urging regular Communion. De-votions in which the Host, or Eucharistic wafer, was venerated for forty hours without pause became a regular event in Rome, with Pope Clement VIII even decreeing there needed to be ceaseless devotion to the Eucharist.[39] Robert Bellarmine touted the

---

[34] Colombero, "Colombo, Realdo" (cit. n. 25); Moes and O'Malley, "Realdo Colombo" (cit. n. 16), 523.

[35] Alessandro Petronio, *Del viver delli Romani et di conservar la sanità, libri cinque* (Rome: Domenico Basa, 1592), 92.

[36] Ibid., 144.

[37] Ibid., 92.

[38] Albala, *Eating Right* (cit. n. 12), 66–7.

[39] Frederick J. McGinness, "'Roma Sancta' and the Saint: Eucharist, Charity, and the Logic of Re-form," *Hist. Reflect.* 15 (1988): 99–116.

Eucharist as the most important Catholic sacrament, one with far-reaching benefits for the soul.[40] To consume Christ in the form of wine and the baked wafer was to be a Catholic.[41]

Part of the vehement defense of, and strict adherence to, this sacrament was a response to vigorous Protestant attacks. In antiquity, Romans had accused Christians of devouring their God when attending mass. These accusations resurfaced during the Reformation, as reformers again cast the Catholic belief in transubstantiation as a doctrine of cannibalism.[42] Catholic reformers were especially sensitive to such accusations because since at least the eleventh century, theologians worried that if any of the Eucharistic elements were not totally absorbed by a person's body, Jesus might, quite literally, end up in a person's feces.[43] Such a desecration would have been abhorrent and therefore required resolution through theories of digestion that assured total absorption of the Eucharistic elements. This renewed concern seems to have been taken seriously on the part of medical writers in Rome in the early modern period.

Although medical writers tended to not be explicit about the theological meaning of their theories, that Roman medical writers were concerned about absorption of the Eucharistic wafer can be inferred from a number of treatises. Petronio, for example, stated that wheat grain was one of the most easily digested of all foods—thus implying that the wafer, which canon law demanded be made from wheat, would quickly and easily be absorbed by the body.[44]

A later Roman physician, Paolo Zacchia, was more explicit and tried to examine carefully how quickly the Eucharistic elements would be absorbed by the body. Zacchia was trained in Rome, becoming one of the preferred expert witnesses of a number of Roman courts as well as personal physician to Pope Innocent X (1644–55).[45]

In his widely published *Medical Legal Questions*, in which he dealt with medical issues related to canon and civil law, Zacchia took up the issue of whether there are moments in which dying patients should not be administered the Eucharist.[46] His concerns seem to have been that undigested hosts—that is, consecrated wafers, which in Catholic

[40] Robert Bellarmine, *Spiritual Writings*, ed. and trans. John Patrick Donelly, S. J., and Roland J. Teske (New York, N.Y.: Paulist, 1989), 286; Polacco, *Trattato divoto* (cit. n. 31), 63.

[41] Schubert, *Pasto divino* (cit. n. 13), 10. Schubert has observed that even the composition of the bread—Eucharistic wafer, thick loaf, and so forth—varied across assorted confessions during the Reformation. So, not just taking the Eucharist and believing in transubstantiation, but also the material form of it, denoted one's faith.

[42] Regina M. Schwartz, "Real Hunger: Milton's Version of the Eucharist," *Religion & Literature* 31 (1999): 1–17. For a general overview of controversies surrounding the Eucharist in the Reformation period, see Lee Palmer Wandel, *The Eucharist in the Reformation: Incarnation and Liturgy* (New York, N.Y.: Cambridge Univ. Press, 2006).

[43] Reynolds, *Food and the Body* (cit. n. 21), 5; Gary Macy, "The Medieval Inheritance," in *A Companion to the Eucharist in the Reformation*, ed. Lee Palmer Wandel (Boston: Brill, 2014), 15–37, on 23.

[44] Petronio, *Del viver delli Romani* (cit. n. 35), 144. On the necessity of the Eucharist including wheat bread and the lengths that people went to in the colonial context to import wheat for this purpose, see Rebecca Earle, "'If you Eat Their Food . . .': Diet and Bodies in Early Colonial Spanish America," *Amer. Hist. Rev.* 115 (2010): 688–713, on 699–700.

[45] De Renzi, "Paolo Zacchia" (cit. n. 17), 50–73.

[46] Paolo Zacchia, *Quaestionum Medico-Legalium Tomus Poterior* (Lyon, Fr.: Ioan. Ant Huguetan & Marci Amnt, 1671), liber 9, titulus 9, 88: "De morbis impedientibus Eucharistia Assumptionem." On the wide circulation and frequent reprinting of Zacchia's *Quaestiones*, see Maria Gigliola and Renzo Villata, "Paolo Zacchia, la medicina come sapere globale e la 'sfida' al diritto," in Pastore and Rossi, *Paolo Zacchia alle origini* (cit. n. 17), 9–49.

theology are now actually the body of Christ—should not be expelled, either as vomit or in diarrhea. In one passage, Zacchia discusses *lienteria*, which the contemporary English translator John Florio explained was a "flux of the stomacke which can keepe nothing but presently voydeth the same undigested [sic]."[47] Zacchia's explanation is that a patient with *lienteria* should not be given the Eucharist since "with lienteria the danger is certain, because the holy bits have been neither absorbed nor altered but at once are cast out with the great horror and the most unimaginable indecency."[48] That is, the general concern that the host not end up as feces is here repeated in a seventeenth-century medical manual. Zacchia argues, though, that in all but the most extreme cases of diarrhea this problem is avoided because the host is so easily absorbed by the body. When he is discussing another sort of diarrhea, *caeliaca passio*, Zacchia notes that food only gets "partially cooked" by the stomach before being evacuated.[49] This, according to a slightly later account, meant that the disease leads to the expulsion of chyle, rather than the completely undigested food.[50] Such an evacuation, however, should mean that the food has not been absorbed by the body, since both modern and classical theory held that the chyle only provides nourishment as it moves through the small intestine or the liver.[51] Zacchia, however, insists that it is safe to give the host to someone suffering from this disease since "the sacred particles have stayed in the belly for enough time" for the essence of the host to be absorbed.[52] That is, in Zacchia's estimation the host is so quickly absorbed by the body that it only needed to be partially digested to effectively pass its spiritual benefits on to the patient.

Roman medical writers in the late sixteenth and early seventeenth centuries thus incorporated new medical theories of digestion into an old solution to the question of how the body absorbs other ingested material. The Eucharistic bread is absorbed quickly and easily; however, such a claim led back to the old theological and medical issue. That is, if the body quickly absorbs Jesus's flesh in the Eucharistic wafer, what happens if one consumes another person? Does the cannibal destroy the victim's identity, since in the nutritional theory outlined above, things consumed became the flesh of the eater?[53] Although these were long-standing questions, and had been answered by the theory of moistures, they became urgent once again in Counter-Reformation Rome, where emphasis on consumption of the Eucharistic elements sparked new accusations of, and concerns about, cannibalism in Rome and Catholicism more generally.

That contemporaries were intensely concerned with the stakes of such debates can be seen by the fact that tales of both real and imagined cannibalism appeared with surprising frequency in sixteenth- and seventeenth-century Rome. Pamphlets circulated in the 1630s and 1640s in Rome averring that during times of famine several

---

[47] John Florio, *Queen Anna's New World of Words, Or Dictionarie of the Italian and English Tongues* (London: Melch. Bradwood, 1611), 283, http://www.pbm.com/~lindahl/florio/299small .html.

[48] Zacchia, *Quaestionum Medico-Legalium* (cit. n. 46), 91.

[49] Ibid.

[50] William Cockburn, *The Nature and Causes of Looseness Plainly Discovered* (London: B. Bartker, 1701), 7.

[51] Galen stated, however, that the stomach itself gained nourishment from the chyle that was cooked therein. So perhaps the idea was that even this level of absorption was sufficient for the Eucharist. See Powell, *Galen* (cit. n. 18), 16.

[52] Zacchia, *Quaestionum Medico-Legalium* (cit. n. 46), 91.

[53] Bynum, *Resurrection of the Body* (cit. n. 22), 31–3.

butchers had slain fellow Romans and ground their flesh with pork to make sausages.[54] In a similarly unpleasant recounting earlier in the sixteenth century, Pope Innocent VIII was accused of drinking the blood of children in an attempt to prolong his life.[55] These narratives appear to have been popular in Rome and circulated alongside travel and confessional literature whose authors regularly sought to use the cannibalism trope to cast enemies in a depraved light.[56]

In Rome, however, cannibalism was no mere trope; rather, there was some truth amidst the horror stories. As several modern scholars have noted, there was a growing enthusiasm across Europe in the sixteenth and early seventeenth centuries for mumia—dried human flesh that was an ingredient in medicines for various ailments. Although originally derived from Egyptian mummies, by the late sixteenth century, recipes that described how to make mumia from fresh human corpses were circulating throughout Europe.[57] Antonio Brasavola, who served as a physician to Pope Paul III and resided in Rome for much of the 1540s, commented on the effectiveness of mumia for the treatment of a range of diseases, thereby essentially advocating the consumption of human flesh in papal Rome.[58] Domenico Auda, chief apothecary to the Santo Spirito hospital in mid-seventeenth-century Rome, advised the drinking of a solution distilled from human cranial bones to fight off the falling sickness (epilepsy); he suggested another solution made from human blood for the restoration of health and vigor.[59] These pieces of medical advice strayed from the theory of moistures and suggested that the cannibal gains some of the life essence from the bodies they consume.

This idea that consumption of the human body, or parts of it, might sometimes be beneficial was not far removed from either religious or medical practice. For example, shortly after the death of Fra Felice of Cantalice in Rome in 1587, two noblewomen broke into his tomb and collected a fluid issuing from his corpse, which they then sold

---

[54] BAV, Codici Urbinati Latini (Urb. Lat.) 1647, fols. 529r–543v; a slightly different version of this also appears in BAV, Vaticani Latini (Vat. Lat.) 13658, fols. 346–355; and a third appears in the British Library, Egerton Manuscripts (MS. Eg.) 1101, fols. 93r–98r.

[55] Stefano Infessura, *Diario della città di roma di Stefano Infessura scriba senato. Nuova edizione a cura di Oreste Tommasini* (Rome: Forzani E. C. Tipografi del Senato, 1890), 275–6.

[56] Catalin Avramescu, *An Intellectual History of Cannibalism*, trans. Alistair Ian Blyth (Princeton, N.J.: Princeton Univ. Press, 2009), 87. Avramescu notes that in this period those who consumed other human flesh were considered not only to be outside the bounds of civilization, but also to be allied with Satan, the ultimate devourer of humans. In addition to Avramescu, there is a wealth of literature on the circulation and meaning of cannibalism tropes in early modern and modern European history. For a sampling of some of this literature see William Arens, *The Man-Eating Myth: Anthropology and Anthropophagy* (New York, N.Y.: Oxford Univ. Press, 1979); Francis Barker, Peter Hulme, and Margaret Iverson, eds., *Cannibalism and the Colonial World* (New York, N.Y.: Cambridge Univ. Press, 1998); Anthony Pagden, *The Fall of Natural Man* (New York, N.Y.: Cambridge Univ. Press, 1982), 80–7; Frank Lestringant, "Catholiques et cannibales: Le thème du cannibalisme dans le discours Protestant au temps des guerres de religion," in *Pratiques et discours alimentaires a la Renaissance*, ed. Jean-Claude Margolin and Robert Sauzet (Paris: Maisonneuve et Larose, 1979), 233–45; Janet Whatley, "Food and the Limits of Civility: The Testimony of Jean de Lery," *Sixteenth Cent. J.* 15 (1984): 387–400.

[57] Karl H. Dannenfeldt, "Egyptian Mumia: The Sixteenth Century Experience and Debate," *Sixteenth Cent. J.* 16 (1985): 163–80, on 163–70.

[58] Ibid., 172. For Brasavola's biography, see Vivian Nutton, "The Rise of Medical Humanism: Ferrara, 1461–1555," *Renaiss. Stud.* 11 (1997): 11–16; Giuliano Gliozzi, "Brasavola, Antonio, detto Antonio Musa," in *Dizionario biografico* (cit. n. 17), vol. 14, http://www.treccani.it/enciclopedia /brasavola-antonio-detto-antonio-musa_(Dizionario-Biografico)/.

[59] Domenico Auda, *Breve compendio di maravigliosi secreti approvati con felice sucesso nelle indispositioni corporali* (Venice: Giacomo Zattoni, 1668), 157–61, 178.

to other Romans as a healing tonic.[60] In another instance occurring in Rome in 1600, pieces of linen soaked in the blood from Angelo del Pas's heart were put in and on sick bodies to cure ailments ranging from battle wounds to difficult childbirth.[61] Such stories of ingesting parts of saints or prospective saints are found regularly in hagiographic accounts of the period.[62] It may have been that this consumption seemed an extension of the Eucharist, wherein one partakes of holiness by eating it. But, if so, the practice seemed to stray from the carefully laid out theory of moistures and invited renewed concerns about cannibalism.

This deviation from the carefully constructed medical theology of moistures came to full fruition in several medical treatises in the sixteenth and early seventeenth centuries, when two Roman practitioners directly addressed what happens to the human body when other humans are consumed. Fioravanti and Cesalpino, each of whom resided in Rome in the service of the papacy in the late sixteenth century, devoted sections in their medical treatises to exploring the effect that cannibalism had on the perpetrator's body. In a treatise published in 1561, Fioravanti recounted a series of unusual experiments he undertook for this purpose:

> The first experiment was the following: I took a little pig and began to feed it at my house. In all the foods I gave it I included a little pig fat. In a few days the pig became totally bald and full of sores due to eating its own kind. Not content with only one example, I decided to try a second: I took a puppy and I kept it in a room. I fed it only dog meat for two months, after which time the poor dog became covered in sores, hairless, and developed a great deal of pain, so that he cried like a human being. I wanted to see a number of examples of the same experiment [*la medesima esperientia volsi vedere in un nebbio*], so I fed a bird of prey its own kind, which had the same effect as it did on the other animals. From these experiments I came to understand that feeding on your own kind creates the kind of corruption or illness as I have described.[63]

Fioravanti likely performed these experiments in Rome, where he had been residing and treating the *mal francese* (syphilis) as recently as 1559.[64] Fioravanti's experiments confirmed what he had already concluded—consuming the flesh of one's own kind destroyed the essence of a creature. In Fioravanti's estimation, this sort

---

[60] Mariano D'Alatri, ed., *Processus Sixtinus Fratris Felicis a Cantalice* (Rome: Institutum Historicum O. F. M. Cap., 1964), 335.

[61] ASV, Congregazione dei Riti Processus (RP) 2812, fols. 1v–2v, 8r, 34r.

[62] Cases were also taking place outside of Rome. See, for example, use of the healing fluid issuing from the corpse of Teresa of Avila noted in ASV, RP 3156, 269 (unpaginated, but numbers are my count of pages); or the woman who bit off one of Francis Xavier's toes to take as a relic, noted in *Relatio facta in Consistorio Secreto coram S. D. N. Gregorio Papa XV* [. . .] *die XIX Ianuarii MDCXXII super vita, sanctitate, actis canonizationis, & miraculis Beati Francisci Xavier e Societate Iesu* (Rome: apud Haerede Bartholomaei Zannetti, 1622), 48–9.

[63] Leonardo Fioravanti, *De Capricci Medicinali* (Venice: Appresso gli Heredi di Melchior Sella, 1582), 50v–51r. William Eamon translates the same passage slightly differently in "Cannibalism and Contagion: Framing Syphilis in Counter-Reformation Italy," *Early Sci. & Med.* 3 (1998): 1–31, on 11. In particular, he suggests that the word "*nebbio*" in Fioravanti's text "probably means *gheppio*, a kind of sparrow hawk." Although such a translation is possible, given that Florio's contemporary dictionary gives a translation for *nebbio* as "a multitude together," and that Fioravanti wants to emphasize that he made repeated tests of the same experiment, I think a more accurate reading of the passage might be as I have rendered it above. See Florio, *Queen Anna's New World* (cit. n. 47), 330, http://www.pbm.com/~lindahl/florio/345small.html.

[64] Fioravanti, *De capricci medicinali* (cit. n. 63), fols. 80v–81r. Fioravanti mentions being in Rome for the death of Realdo Colombo, who died in 1559.

of spiritual destruction led to the advent of syphilis. That is, he saw the most egregious crime of his age as causing the most feared disease of his age. Cesalpino, chief physician to several popes, doctor to at least one saint, and a pioneer of cardiac anatomy, also saw cannibalism as the cause of syphilis.[65] However, he does not seem to have drawn on Fioravanti for this conclusion; in fact, Cesalpino cites an entirely different example of cannibalism in Italy as the cause of syphilis.[66] That two Roman medical men in the employ of the papacy separately would seek to understand the effects of cannibalism on the body suggests that such concerns were common in early modern Rome; it also certainly hints at how important it was to the medical community to understand the effect of consuming certain foods. This issue must have seemed additionally pressing because it was also at this moment that large amounts of animal flesh began to be consumed in Rome.

EATING MEAT

How flesh—whether divine or human—was digested became a central issue of debate among physicians in Rome in the late sixteenth and early seventeenth centuries. Although medieval digestion theories had solved the problem, early modern religious controversy reignited convictions that what one consumed dramatically affected the eater. That early modern popes therefore began dramatically increasing the supply of meat, especially Easter lamb, to Rome suggests that they saw this meat as part of the evangelization of the faithful. Roman bodies became Catholic ones through eating Christ in the Eucharist and Easter lamb, an act that symbolized a connection with the papacy providing this food.

That this eating campaign was connected with spirituality and politics is suggested by the fact that regular and heavy meat eating was not necessarily the recommended dietary advice at the time. As Albala has noted, early Christians had generally abstained from meat eating because they thought that it aroused the passions, especially the libido. According to Galenic theories of digestion, meat resulted in increased production of blood in the body, which in turn led to feelings of lust.[67] There also was not generalized dietary advice for all people, so although eating large amounts of beef could be good for a person with a hot complexion, it would have been terrible for a person with a more sedentary lifestyle and therefore a more frigid complexion.

Despite the lack of nutritional wisdom supporting a meat-heavy diet, meat was regarded as a highly desirable, quasi-luxury product. As Gentilcore has argued, "access to fresh meat defined early modern elites, in their own eyes as much as in the eyes of others, [and served] as a point of social distinction."[68] Allen Grieco has argued that although meat was not prohibitively expensive, it was expensive enough that it was routinely out of the reach for most people in early modern society, making it something of a luxury for most. Finally, Massimo Montanari has argued that beginning

[65] On the importance of Cesalpino to the development of cardiac anatomy, see Frederick A. Willius and Thomas Dry, *A History of the Heart and Circulation* (Philadelphia: Saunders, 1948), 292; and De Ferrari, "Cesalpino" (cit. n. 17), 122–25.

[66] Andrea Cesalpino, *Praxis universae artis medicae* (Treviso: Sumptibus Roberti Meietti, 1606), 349.

[67] Albala, *Eating Right* (cit. n. 12), 148.

[68] Gentilcore, *Food and Health* (cit. n. 12), 57.

in medieval Europe, eating meat was clearly associated with strength and power.[69] It was surely these connotations that Roman pontiffs meant to draw upon when they started increasing the meat supply to the city.

Aware that such a luxury diet generally was at odds with contemporary dietary advice, Roman physicians associated with the papacy defended in print the meat-heavy diet Rome enjoyed. Giovanni Faber came to Rome in 1600 to begin a training program at the Santo Spirito Hospital, a pious institution closely connected to the papacy. After his training there, he remained in papal employ until his death in 1629.[70] Faber noted in a later publication that the residents of Santo Spirito had begun in the seventeenth century to enjoy a large supply of oxen and buffalo, which was used to feed its sick and poor. In this discussion, he explained that such food was beneficial to the residents, because the Roman soil and fodder on which these animals fed was "more fertile and more noble" than other soils and therefore led to healthier animals that were better for eating. He even added that farmers from lands outside the Papal States attempted to pasture their animals inside the pope's lands because it would enhance their products.[71] That is, the animals brought into Rome for use at the Santo Spirito were of better quality than those eaten elsewhere, and therefore were suitable for all people to eat.

Alessandro Petronio, lead physician at the Santo Spirito at the end of the sixteenth century, contributed to this discussion. He specifically argued against those who claimed that the difficulty of digesting beef led to health problems, especially an overproduction of black bile, or melancholy. He remarked, "many have said that [beef] generates the melancholy humor: [but] those who have this opinion have not judged from experience, because in truth one does not find this." Petronio, furthermore, claimed that poultry was the easiest food to digest, which therefore made it beneficial for consumption even by the sick and poor in Santo Spirito. He argued that although sheep had largely been a food for the lower classes, the sheep produced in the hills around Rome were of such high quality that "frequently they also were eaten by the elites."[72]

The writings of Petronio and Faber repeatedly emphasize that the location where the food was produced makes a great difference as to whether certain kinds of foods are healthy. Petronio states it clearly: "All meats, even if they are from the same types of animals, are good or bad depending on the variety of places where they feed."[73] Both authors further make it clear that the lands around Rome had recently improved so wonderfully that foods produced there could be consumed by everyone. The anonymous chronicler of 1641 mentioned above clearly drew on such ideas when he contended that the meat in Rome was of better quality than before and "did not have the same deformities as it once had."[74] The newly bountiful nature of the Roman countryside,

---

[69] Allen J. Grieco, "Food and Social Classes in Late Medieval and Renaissance Italy," in *Food: A Culinary History from Antiquity to the Present*, ed. Jean-Louis Flandrin and Massimo Montanari (New York, N.Y.: Columbia Univ. Press, 1999): 303; Montanari, *The Culture of Food*, trans. Carl Ipsen (Cambridge, Mass.: Blackwell, 1994), 13–14.

[70] Gabriella Belloni Speciale, "Faber, Giovanni," in *Dizionario biografico* (cit. n. 17), vol. 43, http://www.treccani.it/enciclopedia/giovanni-faber_%28Dizionario-Biografico%29/.

[71] Francisco Hernandez and Johannes Faber, *Rerum medicarum Novae Hispaniae thesauraus seu plantarum animalium mineralium Mexicanorum historia* (Rome: Typographia Vitalis Mascardi, 1651), 594.

[72] Petronio, *Del viver delli Romani* (cit. n. 35), 171.

[73] Ibid., 168.

[74] BAV, Vat. Lat. 8354 (cit. n. 8), fols. 112v–113r.

he further claimed, was the cause of such plenty that Rome's countryside "could feed an entire kingdom."[75]

Such a food program, in which large quantities of meat were considered to be good for people of all walks of life, was clearly in evidence at the Santo Spirito Hospital in the early seventeenth century. The Santo Spirito was one of the most prominent charitable institutions in a city known for dozens of charities and was closely allied with the papacy for centuries. It was therefore a natural place for successive popes to unfurl the alimentary changes that were part of their reforming plan. After all, here the improvements would be on display for visitors to the capital.

Apostolic visits for these years provide evidence of a startling level of supply to the hospital; foundlings housed there were given more than a third of a pound of meat daily, three small loaves of bread, about fourteen ounces of wine, and salad—a large amount of meat, and a generous diet by most standards. The twenty-four wet nurses on permanent staff ate even better; they received three-quarters of a pound of meat per day, two quarts of wine, broth in the morning, three small loaves of bread, and salad twice a week.[76] It is likely, given Faber's anecdote and the Roman butchering cycle that was mandated by the papacy, that these people in the care of the Santo Spirito ate a great deal of beef and buffalo. Such a conclusion is supported by the diet that the hospital supplied for over four hundred unmarried women without additional means of support; these women were given a third of a pound of oxen or beef every day.[77] The ill, in contrast, were given a paste made from at least six chickens, as well as broth.[78] Thus, all the people cared for at the Santo Spirito Hospital were fed a large amount of meat by contemporary standards.

Most medical personnel outside Rome would not have approved of this program for the poor or the sick, though both were treated at the Santo Spirito.[79] Mutton and beef—two meats supplied in large quantities in early seventeenth-century Rome—were frequently described as tough and only fit for those with the most robust constitutions.[80] The sick, children, and single women—all of whom were given these foods at the Santo Spirito Hospital—certainly were not among the populations for which contemporary medical counsel prescribed beef. The connection between heavy meat eating and promiscuity would have marked the unmarried women as unsuitable candidates for regular consumption of beef. Finally, consumption of fowl was supposed to be reserved for the upper classes and was not considered healthy for those lower on the social scale.[81] Thus, the food served at the Santo Spirito seemed to fly in the face of contemporary, non-Roman dietary advice. Yet, the physicians on staff not only supplied this diet, but they wrote treatises in which such decisions were defended.

---

[75] Ibid., fol. 112r–v
[76] BAV, Vat. Lat. 7941, fols. 187r–188r.
[77] BAV, Vat. Lat. 7941, fol. 188v.
[78] ASF, Carte Strozziane, ser. prima, 233, fols. 63v–64r.
[79] Few beds were reserved for nobles at the Santo Spirito, while the regular beds were the most numerous and reserved for average Romans; see ASF, Carte Strozziane, ser. prima, 233, fols. 62v–63r.
[80] Albala, *Eating Right* (cit. n. 12), 68; Albala, *Food in Early Modern Europe* (London: Greenwood, 2003), 63; Grieco, "Food and Social Classes" (cit. n. 69), 311.
[81] Grieco, "Food and Social Classes" (cit. n. 69), 305–10.

CONCLUSION

This article has focused on two items consumed by early modern Romans—the Eucharist breadstuff and meat—to demonstrate how medical dietary advice responded to the demands of the Counter-Reformation in Rome. After all, food and its effect on the human body were at the center of the Reformation in Rome. The Eucharist, which had been an important part of the Catholic faith even in the late Middle Ages, came to be vigorously defended and supported as the key sacrament. But, as it was attacked by Protestant reformers and, in particular, decried as being an act of cannibalism, medical professionals and theologians in Rome worked to craft a view of digestion in which consuming the body of Christ could not be considered cannibalism.

Such discussions, however, reignited old concerns about the eating of flesh, both human and animal. The specter of cannibalism lurked strongly enough that Roman physicians even undertook experimental programs in an attempt to understand its effect on the human body.

Meanwhile, as part of the papal efforts to defend the faith against attack, meat consumption in Rome skyrocketed. Romans across the social spectrum apparently had regular access to generous meat supplies, and this program of heavy meat consumption, with the most indulgent moments happening around Easter, was intended to display the rightness of the Catholic faith and the just rule of the popes. As other scholars have noted when discussing the architectural and artistic innovations in Rome during the same period, the city itself was a weapon in the arsenal of the Counter-Reformation.[82] Nowhere was that weapon more prominently deployed than in remaking the daily lives of average citizens through the food that they ate. Still, such an ambitious program brought detractors. The specter of cannibalism, which seemed always to be poorly hidden in Rome, most clearly hinted at dissent toward papal food programs. To help refute this and other dietary concerns, ecclesiastical officials allied with Roman physicians to create new nutritional standards.

Rome's meat moment was, however, short lived. Although it began in the 1590s and reached its peak in the 1630s, the heavy supply of meat evaporated by the end of the 1640s. The disastrous war of the Castro (1641–44) that Pope Urban VIII waged and then lost to an unruly vassal, Odoardo Farnese, demonstrated the fragility of Rome's supplies and the mythical nature of the bountiful Papal States. As a result of the conflict, 50,000 fewer sheep entered Rome in 1649 than two decades before, during the trade's heyday; the farmers supplying the city had declined by a third, and Roman efforts to requisition food from neighboring states now led to riots, such as one in Fermo that had to be quelled by papal troops in 1648.[83] In a dramatic display of the changing times, when Gian Lorenzo Bernini's Four Rivers Fountain was unveiled in Piazza Navona in 1651, the population was said to have shouted "pane non fontane"—that is, "bread, not fountains."[84]

---

[82] Stefano Andretta, "Religious Life in Baroque Rome," in *Rome, Amsterdam: Two Growing Cities in Sixteenth-Century Europe*, ed. Peter van Kessel and Elisja Schjulte (Amsterdam: Amsterdam Univ. Press, 1997), 168–74.

[83] ASC, cred. X, tomo 47, fols. 174v–176r; ASC, cred. XIV, tomo 9, fol. 46r–v.

[84] Theodore K. Rabb, "Play not Politics: Who Really Understood the Symbolism of Renaissance Art?" *Times Literary Supplement*, 10 November 1995, 19.

The meat moment in Rome faded, as did the dietary solution worked out by Roman medical men and theologians. Even as Roman supply lines were being disrupted, new anatomical understanding of digestion, along with changing notions of healthy eating, meant that the ideal of heavy meat consumption did not continue in Rome.[85] Nevertheless, meat's persistence as a luxury product meant that other European cities, including some in Italy, would flirt with their own meat moments during the eighteenth century.[86]

[85] Orland, "Fluid Mechanics of Nutrition" (cit. n. 30), 357–69.
[86] Karl Appuhn, "Ecologies of Beef: Eighteenth-Century Epizootics and the Environmental History of Early Modern Europe," *Environ. Hist.* 15 (2010): 268–87; Watts, *Meat Matters* (cit. n. 7).

Food, Population, and Empire in the Hartlib Circle, 1639–1660

by Ted McCormick*

ABSTRACT

The idea of population control is often associated with Malthusian views of scarcity and their twentieth-century political and technological legacies. Though sixteenth- and seventeenth-century political thinkers and scientific projectors often described human multiplication in religious—especially biblical and providentialist—terms, they similarly understood population to be constrained by the capacity of limited resources to feed growing numbers, and they sought ways to manage this relationship by "improvements" that combined technological and political innovations in both metropolitan and colonial settings. This article examines how these efforts engaged with population, focusing on several projects relating to food connected with Samuel Hartlib (1660–62) and the Hartlib Circle: Gabriel Plattes's manifold agricultural improvements for domestic use, Hugh L'Amy and Pierre Le Pruvost's promotion of colonial trade and fisheries, Cressy Dymock's corn-setting and "perpetual motion" machines for use in England and Barbados, and John Beale's promotion of fruit trees and cider. While the Hartlibians developed no theory or doctrine of population and made scant use of demographic quantification, their projects framed the problem of feeding populations central to the management of human multiplication, both as a global, historical concern and as a key problem of colonial empire. They thus shed light not only on the emergence after 1660 of new discourses of demographic quantification, and the background to sustained demographic growth after 1750, but on the origins of population as an object of scientific-*cum*-political intervention through the medium of food.

"The idea of controlling the population of the world," writes a recent historian, "is a modern phenomenon."[1] It began, in this account, with Thomas Malthus's *Essay on the Principle of Population* (1798), which contrasted finite increases in food supply with the boundless propensity of people to procreate. Malthus influenced Darwin's portrayal of a "struggle for existence," and Darwin inspired Francis Galton's interest in selective breeding in human populations, giving rise to the eugenics movement of the twentieth century.[2] Yet, if eugenics depended on modern theories and technologies,

* Department of History, Concordia University, LB-1001.01, 1455 De Maisonneuve Blvd. W., Montreal, QC H3G 1M8, Canada; ted.mccormick@concordia.ca.

I would like to thank the volume editors for the invitation to contribute, and Keith Pluymers for his comments on a draft of this article.

[1] Matthew Connelly, *Fatal Misconception: The Struggle to Control World Population* (Cambridge, Mass.: Belknap Press of Harvard Univ. Press, 2008), 7.

[2] Ibid., 2.

the emergence of population as an object of scientific and technological intervention did not. Early modern thinkers and projectors knew that human numbers depended on limited resources. Rather than restraining population, however, they sought to augment natural wealth, facilitate multiplication, and redistribute populations in Europe and across colonial empires.[3] Exploring projects associated with the "Hartlib Circle" in England under the Parliamentary and Cromwellian regimes of the 1640s and 1650s, in particular, reveals complex engagements with the links between population, subsistence, and territory. Central to these were questions about food, including how to produce it, how to improve it, and for whom. The answers projectors proffered held implications not only for lands and people in England, but also for the scope and nature of colonial empire and the organization of different populations within it. They also suggest that the history of projects may have more to tell us about the origins of population as an object of control than the history of canonical demographic thought.

Despite its name, the Hartlib Circle was not a tight-knit group but an affiliation of like-minded participants in overlapping correspondence and patronage networks.[4] Its heart was the German-born "intelligencer" Samuel Hartlib (1600–62).[5] Hartlib's expansive correspondence with Protestant thinkers and reformers across Europe, his extensive political connections within the Commonwealth and Cromwellian regimes, and his tireless circulation and publication of his contacts' ideas all made him a fulcrum in the pursuit of social, economic, educational, medical, and spiritual "improvement" in the revolutionary atmosphere of the 1640s and 1650s.[6] From Hartlib's perspective, agricultural innovation was but a part—though a fundamental part—of a multifaceted reformation touching virtually all areas of knowledge and life. In this context, he and his associates looked to agriculture not only as a source of vital sustenance but also as a locus of employment, a driver of trade, a justification of private

[3] See, for examples, Abigail L. Swingen, *Competing Visions of Empire: Labor, Slavery, and the Origins of the British Atlantic Empire* (New Haven, Conn.: Yale Univ. Press, 2015); Leslie Tuttle, *Conceiving the Old Regime: Pronatalism and the Politics of Reproduction in Early Modern France* (Oxford: Oxford Univ. Press, 2010); Sarah Lloyd, *Charity and Poverty in England, c.1680–1820: Wild and Visionary Schemes* (Manchester: Manchester Univ. Press, 2009); Carol Blum, *Strength in Numbers: Population, Reproduction, and Power in Eighteenth-Century France* (Baltimore: Johns Hopkins Univ. Press, 2002); and Joyce E. Chaplin, *Subject Matter: Technology, the Body, and Science on the Anglo-American Frontier, 1500–1676* (Cambridge, Mass.: Harvard Univ. Press, 2001). On the importance of colonial spaces to Malthus's demographic thought, see Alison Bashford and Joyce E. Chaplin, *The New Worlds of Thomas Robert Malthus: Rereading the* Principle of Population (Princeton, N.J.: Princeton Univ. Press, 2016).

[4] See Mark Greengrass, Michael Leslie, and Timothy Raylor, "Introduction," in *Samuel Hartlib and the Universal Reformation: Studies in Intellectual Communication*, ed. Greengrass, Leslie, and Raylor (Cambridge: Cambridge Univ. Press, 1994), 1–14; and Thomas Leng, *Benjamin Worsley (1618–1677): Trade, Interest and the Spirit in Revolutionary England* (Woodbridge, UK: Royal Historical Society/Boydell Press, 2008), 4–7.

[5] An important early study is Dorothy Stimson, "Hartlib, Haak and Oldenburg: Intelligencers," *Isis* 31 (1940): 309–26; see also Hugh Trevor-Roper, "Three Foreigners: The Philosophers of the Puritan Revolution," in *The Crisis of the Seventeenth Century: Religion, the Reformation, and Social Change*, ed. Trevor-Roper (New York, N.Y.: Harper & Row, 1967), 219–71. More recent scholarship is cited in n. 6, below.

[6] See Paul Slack, *The Invention of Improvement: Information and Material Progress in Seventeenth-Century England* (Oxford: Oxford Univ. Press, 2015). The most comprehensive study of the Hartlib Circle remains Charles Webster, *The Great Instauration: Science, Medicine, and Reform, 1626–1660* (London: Duckworth, 1975); see also Greengrass, Leslie, and Raylor, *Samuel Hartlib* (cit. n. 4); and Slack, *From Reformation to Improvement: Public Welfare in Early Modern England* (Oxford: Oxford Univ. Press, 1999), 77–101.

property, a topos of moral reflection, an object of administrative reform, a proving ground for new mechanical contrivances, or "engines," and a point of contact between divine Providence, bounteous nature, and human art. Individual projects could involve any of these things, or all of them. Both food and the numbers it supported, consequently, resonated with many meanings. There was no single, coherent Hartlibian doctrine of population.

Paradoxically, this dearth of discernible demographic thought reflects not the poverty but the richness of Hartlibian engagements with food and population between 1639 (when Gabriel Plattes's *Discovery of Infinite Treasure* appeared in print) and 1660 (when the Restoration of the Stuart monarchy ended the prospect of state-supported reformation). It may also be a consequence of the structure of the Hartlib Circle and of the nature of projecting, which tended to transgress boundaries between different areas of knowledge and practice and to emphasize the broad, practical, and transformative ramifications of specific technological changes.[7] Projectors typically stood outside or at the margins of learned disciplines and institutions and straddled the boundaries between state and society, purporting (not always convincingly) to align public good and private interest. The structure of the Hartlib Circle likewise militated against doctrinal consensus; though they shared commitments to experiment, invention, and improvement, and faced common challenges, Hartlib's contacts nursed rivalries and resentments as they competed for esteem and support, and their schemes often conflicted. Moreover, projects did not all operate on the same level, and their consequences were not mapped out in consistent ways. Those geared to improve soil or introduce new crops were often local in focus; as Sir Richard Weston wrote in *A Discours of Husbandrie Used in Brabant and Flanders* (1650), "the chiefest and fundamentallest point in *Husbandrie*, is, to understand the nature and condition of the Lands that one would Till."[8] New engines, by contrast, might have almost universal application. At every level, change, not theory, was the point.

Indeed, scholars have recognized practical links between Hartlibian projecting and population, exploring projects ranging from the introduction of new food crops, soil preparations, and rotations; to experiments with seed drills, milling mechanisms, and food preservation techniques; to the promotion of orchards, gardens, irrigation, and fen drainage.[9] Some argue that fear of overpopulation (reflecting real demographic expansion between 1500 and about 1650) motivated projectors; others suggest that

---

[7] Vera Keller and Ted McCormick, "Toward a History of Projects," *Early Sci.& Med.* 21 (2015): 423–44. Recent studies of projecting include Vera Keller, *Knowledge and the Public Interest, 1575–1725* (Cambridge: Cambridge Univ. Press, 2015); Eric H. Ash, *The Draining of the Fens: Projectors, Popular Politics, and State Building in Early Modern England* (Baltimore: Johns Hopkins Univ. Press, 2017); and Koji Yamamoto, *Taming Capitalism before Its Triumph: Public Service, Distrust, and "Projecting" in Early Modern England* (Oxford: Oxford Univ. Press, 2018).

[8] [Richard Weston], *A Discourse of Husbandrie Use in Brabant and Flanders, Shewing the Wonderful Improvement of Land There; and Serving as a Pattern for our Practice in this Common-Wealth* (London: printed by William Du-Gard, 1650), 1. Commentary on the best land for particular improvements abounds; for examples pertaining to food crops see Walter Blith's compendium, *The English Improver Improved or the Survey of Husbandry Surveyed Discovering the Improveableness of All Lands* [. . .] (London: printed for John Wright, 1652), 235 (regarding hops), 244 (saffron), 246 (licorice).

[9] Joan Thirsk, ed., *Chapters from the Agrarian History of England and Wales, 1500–1750*, vol. 3, *Agricultural Change: Policy and Practice, 1500–1750* (Cambridge: Cambridge Univ. Press, 1990); Thirsk, *Alternative Agriculture: A History* (Oxford: Oxford Univ. Press, 1997), 23–71; Webster, *Great Instauration* (cit. n. 6), 465–83.

seventeenth-century projects paved the way for gains in agricultural productivity and for unprecedented population growth after 1750.[10] Still others have noted the Hartlibian roots of new approaches to demographic quantification after 1660, and shown that the Hartlibian alchemists' attempts to concentrate the vegetative energies and nutritive powers at work in the earth—in part through improved fertilizers—incubated ideas of limitless growth.[11] Such improvements were not just productive appropriations of material resources but also intellectual and spiritual engagements with a natural world made hostile by human sin, yet governed by a beneficent God.[12]

Despite the demographic significance of their projects, however, it remains unclear how mid-seventeenth-century projectors understood the demographic effects of their plans. This question matters not only for our picture of the Hartlibian world, but also for understanding subsequent efforts at demographic quantification and schemes for the maintenance and manipulation of various subpopulations over the long eighteenth century—including the Enlightened optimism and revolutionary radicalism Malthus attacked. Looking at projects touching the production or supply of food reveals consequential engagements with population that histories of formal doctrines or statistical techniques miss.[13] Analogous to recent work on the political and social contexts of later population thought, a focus on projects shows that the mid-seventeenth-century English shift from anxiety about overpopulation to fear of underpopulation was tied to new ideas about the promises and perils of colonial empire (associated with "mercantilist" ideas) and to the legacy of Baconian ideas about the purposes of natural

[10] On the demographic background to agricultural improvement, see Ann Kussmaul, *A General View of the Rural Economy of England, 1538–1840* (Cambridge: Cambridge Univ. Press, 1990), 76–102; Joan Thirsk, "Enclosing and Engrossing, 1500–1640," chap. 2 in *Agrarian History of England and Wales* (cit. n. 9), 54–109; and Thirsk, *Alternative Agriculture* (cit. n. 9), 23–71. On the legacy of Hartlibian projecting, see Webster, *Great Instauration* (cit. n. 6), 482–3; see also Mark Overton, *Agricultural Revolution in England: The Transformation of the Agrarian Economy 1500–1850* (Cambridge: Cambridge Univ. Press, 1996), 8–10; but cf. Joshua Lerner, "Science and Agricultural Progress: Quantitative Evidence from England, 1660–1780," *Agr. Hist.* 66 (1992): 11–27, which considers Hartlib only as an influence on the Royal Society. On the long-term significance of gains in agricultural productivity prior to 1750, see E. A. Wrigley, *Poverty, Progress, and Population* (Cambridge: Cambridge Univ. Press, 2004), 44–67, 225–6, 258.

[11] Ted McCormick, "Who Were the Pre-Malthusians?," in *New Perspectives on Malthus*, ed. Robert J. Mayhew (Cambridge: Cambridge Univ. Press, 2016), 25–51, at 31–5; Carl Wennerlind, "Credit-Money as the Philosopher's Stone: Alchemy and the Coinage Problem in Seventeenth-Century England," *Hist. Polit. Econ.* 35 (2004): 234–61; Wennerlind, *Casualties of Credit: The English Financial Revolution, 1620–1720* (Cambridge, Mass.: Harvard Univ. Press, 2011), 44–80. See also Ted McCormick, "Alchemy into Economy: Material Transmutation and the Conceptualization of Utility in Gabriel Plattes (c. 1600–1644) and William Petty (1623–1687)," in *"Eigennutz" und "gute Ordnung": Ökonomisierungen im 17. Jahrhundert*, ed. Guillaume Garner and Sandra Richter (Wiesbaden, Ger.: Harrassowitz Verlag, 2016), 339–52.

[12] On "improvement" as a theme in seventeenth- and eighteenth-century economy, politics, and society, see Slack, *Invention of Improvement* (cit. n. 6); Richard W. Hoyle, ed., *Custom, Improvement, and the Landscape in Early Modern Britain* (New York, N.Y.: Routledge, 2011); Lloyd, *Charity and Poverty* (cit. n. 3); James Livesey, *Civil Society and Empire: Ireland and Scotland in the Eighteenth-Century Atlantic World* (New Haven, Conn.: Yale Univ. Press, 2009); and Toby Barnard, *Improving Ireland? Projectors, Prophets and Profiteers, 1641–1786* (New Haven, Conn.: Yale Univ. Press, 2008). On the prior history of the concept, see Paul Warde, "The Idea of Improvement, c. 1520–1700," in Hoyle, *Custom, Improvement, and the Landscape* (cit. n. 12), 127–48; and Joan Thirsk, "Agricultural Innovations and Their Diffusion, 1640–1750," chap. 7 in *Agrarian History of England and Wales* (cit. n. 9), 263–319.

[13] For a critique of earlier literature, see Philip Kreager, "Histories of Demography: A Review Article," *Population Studies* 47 (1993): 519–39; more recently, see Yves Charbit, *The Classical Foundations of Population Thought: From Plato to Quesnay* (Dordrecht: Springer, 2011), 1–12.

knowledge.[14] Even in the absence of a clear doctrine or concept of population, as-sumptions about the relationship between nature and human multitude—and about the power of innovation, helped by Providence, to reshape this relationship through the medium of food—encouraged seventeenth-century projectors to frame interven-tions in the size, health, and distribution of populations in Britain and across the Atlantic. Unlike Malthus, but like his eugenicist heirs, they saw the control of popula-tion as a political task made feasible by technology working upon a pliant nature—with God's assistance. Unlike both Malthus and the neo-Malthusians, they saw Prov-idence guiding human art to sustain, not suppress, multiplication.

MANAGING POPULATION FROM BOTERO TO MALTHUS

The observation that food supply constrains population, and the idea that the fruits of the earth are limited while the human propensity to reproduce is not, were two views already in circulation by 1600. Both had roots in sacred history and biblical exegesis. Adam's fall from grace had introduced not only human mortality but also the need to toil for sustenance; the Great Flood had further abbreviated human longevity, while the ensuing covenant included the injunction to multiply.[15] Chronologists and exe-getes duly set about calculating the rate at which Noah's progeny must or might have grown, with one eye on the emerging historical record and the other on the demands of Scripture. But these ideas also framed discussions of politics and population in the present. The Italian diplomat and reason-of-state theorist Giovanni Botero articulated them in his influential work *On the Causes of the Greatness and Magnificence of Cit-ies* (1588). "It is beyond doubt," he wrote, "that the procreative force has remained constant, at least for the last three thousand years . . . so that if there were no other obstacle, human beings would have propagated endlessly." This implied that "if the population fails to increase . . . [it] is the result of a lack of food and subsistence." Since "a city's provisions are obtained either from its territory, or from abroad," an expanding city would inevitably come to depend upon distant sources of food. While trade could feed growth for a time, "obstacles finally become so many and so grave that they defeat all human effort and diligence."[16] This had been the fate of ancient Rome. Multiplication (Botero put the Roman population at two million at its zenith) had outpaced the productive capacity of the surrounding land and exhausted the city's ability to make shortfalls good through commerce. Stagnation, decline, and disper-sion had followed.[17]

---

[14] Recent examples dealing with the seventeenth and eighteenth centuries include Susan E. Klepp, *Revolutionary Conceptions: Women, Fertility, and Family Limitation in America, 1760–1820* (Chapel Hill: Univ. of North Carolina Press, 2009); and Lloyd, *Charity and Poverty* (cit. n. 3). Studies con-cerned with later periods include Juanita de Barros, *Reproducing the British Caribbean: Sex, Gender, and Population Politics after Slavery* (Chapel Hill: Univ. of North Carolina Press, 2014); and Karl Ittman, *A Problem of Great Importance: Population, Race, and Power in the British Empire, 1918–1973* (Oakland: University of California Press, 2013).

[15] Gen. 3:19, 9:7. On declining longevity, see Peter Harrison, *The Fall of Man and the Foundations of Science* (Cambridge: Cambridge Univ. Press, 2007), 163–5. On biblical chronology and ideas of human multiplication, see Jed Z. Buchwald and Mordechai Feingold, *Newton and the Origin of Civ-ilization* (Princeton, N.J.: Princeton Univ. Press, 2013), 164–94.

[16] Giovanni Botero, *On the Causes of the Greatness and Magnificence of Cities*, trans. Geoffrey Symcox (Toronto: Univ. of Toronto Press, 2012), 73–4.

[17] Ibid., 74, 78–80.

In Botero's rendering, this was a problem for statesmen to ponder, but it was not one that they could permanently solve. Rather than a technological challenge, the tie between food and population was subsumed under the ancient political question of whether cities should limit their size for the sake of order (as Aristotle counseled in the *Politics*), or expand for the sake of strength (as the Romans had done).[18] Botero plumped for the latter: "Experience, which teaches us that because human nature is corrupt, power prevails over reason and force of arms over law, also teaches us that the Romans' policy is to be preferred to that of the Greeks."[19] Corruption was inevitable, but greatness was not. Through growth, Rome had enjoyed a season of glory (before its decline) that had eluded the Greek city-states, for all their internal order. Stagnation, however, was the inescapable result of tension between the "the procreative power of human beings," which did not abate, and "the cities' power to sustain them," which could not be augmented indefinitely. This conflict, by which "eternal Providence cause[d] the few to multiply, and place[d] a limit on the many," was beyond the power of human ingenuity to resolve.[20] In emulating their ancient models, modern city-states would inevitably face the same problems of order and of logistics. Eventually, these challenges would prove insurmountable.

Yet, if expansion could not circumvent nature entirely, it did furnish a means of managing growth for a time. Botero emphasized, in particular, the role of colonies in alleviating the pressures of expansion in the context of city-states. Exporting colonists allowed metropoles to adjust to the needs of growing numbers without setting bounds to their increase: "Just as plants cannot grow and multiply as well in the nursery . . . as in the open ground where they are transplanted," he wrote, "human beings do not propagate as successfully when enclosed within the walls of the city where they are born, as they would in different places to which they are sent."[21] Colonization mitigated the effects of hunger and of plague—and even, paradoxically, of war.[22] Botero's English readers took the argument to heart, adjusting for circumstances. Alarmed by the perceived effects of overpopulation—urban crowding, unemployment, vagrancy, crime, and disease—Elizabethan and Jacobean writers promoted trans-Atlantic colonization as a panacea for the kingdom's social ills, while emphasizing its potential to make the idle industrious and augment national strength.[23]

---

[18] In Book II of *The Politics*, Aristotle remarked, "One would have thought that it was even more necessary to limit population than property" (II.6) for the sake of preventing poverty; at VII.4, he described optimal population as a mean between dependency and disorder: "A state when composed of too few is not, as a state ought to be, self-sufficient; when of too many, though self-sufficient in all mere necessaries, as a nation may be, it is not a state, being almost incapable of constitutional government." See Aristotle, *The Politics and the Constitution of Athens*, trans. Jonathan Barnes, ed. Stephen Everson (Cambridge: Cambridge Univ. Press, 1996), 41, 173.

[19] Botero, *Causes* (cit. n. 16), 71.

[20] Ibid., 71–3. Despite this apparent natural limit to population, Vera Keller has compellingly emphasized Botero's stress on the power of art to exploit nature, creating an "above-ground mine" of wealth without expansion. See Vera Keller, "Mining Tacitus: Secrets of Empire, Nature and Art in the Reason of State," *Brit. J. Hist. Sci.* 45 (2012): 189–212.

[21] Botero, *Causes* (cit. n. 16), 33. See Mildred Campbell, "'Of People Either Too Few or Too Many': The Conflict of Opinion on Population and Its Relation to Emigration," in *Conflict in Stuart England: Essays in Honour of Wallace Notestein*, ed. William Appleton Aiken and Basil Duke Henning (London: Jonathan Cape, 1960), 169–201, on 183.

[22] Botero, *Causes* (cit. n. 16), 33–4.

[23] See, for example, Richard Hakluyt, *A Particuler Discourse Concerninge the Greate Necessitie and Manifolde Commodyties that Are Like to Growe to This Realme of Englande by the Westerne*

With metropolitan population growth stagnating by midcentury, attitudes changed. A decade filled by civil strife in England, Scotland, and Ireland, and another marked by political dislocation, foreign war (notably with the Dutch), and colonial misadventure (with Cromwell's failed "Western Design" to capture the Spanish colony of Hispaniola), culminated in the Restoration of Charles II in 1660. The ensuing decades saw the creation of "political arithmetic," and with it a series of populationist projects designed to augment English numbers in the face of Dutch and French commercial and military competition. In this context, colonial emigration from England came to many to seem undesirable if not hazardous; at the same time, colonial plantations promised wealth in the form of marketable foodstuffs, notably sugar.[24] As Abigail Swingen has shown, this tangle of concerns eventuated in a debate over imperial governance between planters and state authorities in which both sides converged on the enslavement of Africans as the solution to colonial labor supply. Population management—harmonizing the dual imperatives of colonial production and metropolitan reproduction—was central to the business of empire.[25] Over the course of the Restoration and through the eighteenth century, schemes for augmenting and improving populations proliferated at every level; they were generated by the state and emerged from civil society, in metropole and colony alike. Though varying in their social aims and political implications, they shared a confidence in the capacity of human agency, art, or policy to shape the conditions in which numbers of people multiplied.

Malthus, famously, attacked this confidence. Yet, his assault on ideas about human perfectibility and equality had—like those ideas—roots in seventeenth-century scientific and technological engagement with nature. Even if Malthus, like Botero, contrasted an unlimited "passion between the sexes" with the limited "power in the earth to produce subsistence"—and even if both saw the hand of Providence in the perpetual oscillation of human numbers that resulted—Malthus still expressed the problem more firmly as a universal and perpetual clash of human nature with the nature of the earth.[26] He encapsulated this conflict, further, by contrasting the "arithmetical" growth of agricultural productivity and the "geometrical" increase of humankind—implying a need for calculation in probing the limits of demographic growth.[27] (Only in the second, 1803, edition of the *Essay* would statistics play a significant role.) In asserting insurmountable natural barriers, Malthus argued not just about the scope of policy but also about the limits of technological and social improvement.[28] Malthus's reply to Enlightenment optimism was also a response to improvement's promise—a

---

Discoueries Lately Attempted, Written in the Yere 1584 [. . .] *Known as Discourse of Western Planting*, ed. David B. Quinn and Alison M. Quinn (London: Hakluyt Society, 1993), 4, 28–32, 115–6. On the links between overpopulation, vagrancy, and colonization schemes, see Abbot Emerson Smith, *Colonists in Bondage: White Servitude and Convict Labor in America, 1607–1776* (1947; repr., New York, N.Y.: W. W. Norton, 1971), 3–25; Campbell, "'Of People Either Too Few'" (cit. n. 21); Paul Slack, *Poverty and Policy in Tudor and Stuart England* (Harlow, UK: Longman, 1988); and Timothy Sweet, *American Georgics: Economy and Environment in Early American Literature* (Philadelphia: Univ. of Pennsylvania Press, 2002), esp. 12–28; see also Slack, *Invention of Improvement* (cit. n. 6), 15–76.

[24] Swingen, *Competing Visions of Empire* (cit. n. 3), 105.

[25] Ibid.; Abigail L. Swingen, "Labor: Employment, Colonial Servitude, and Slavery in the Seventeenth-Century Atlantic," in *Mercantilism Reimagined: Political Economy in Early Modern Britain and Its Empire*, ed. Philip J. Stern and Carl Wennerlind (Oxford: Oxford Univ. Press, 2013), 46–73.

[26] Thomas Robert Malthus, *An Essay on the Principle of Population, as It Affects the Future Improvement of Society* (London: J. Johnson, 1798), 11–16.

[27] Ibid., 13.

[28] Ibid., iii, 144–52, 217–18, 271–2, 346.

promise based in a providential view of nature that Malthus shared—to transform the material conditions of multiplication.[29]

It was during the 1640s and 1650s that this promise first became a rallying cry for Hartlibians. Their projects opened new spaces for state and private action to create different, even utopian futures—for England, for Protestant Europe, and ultimately for all humankind.[30] The production and supply of food were central to this. What follows here, accordingly, is an examination of the Hartlib Papers and related sources, with a focus on engagements with populations' dependence on food.[31] It omits schemes designed to employ rather than to feed the poor, and those touted more for their profitability than for their impact on population—though projects often claimed to reconcile private and public gain. Projects establishing new trades or reorganizing old ones have been left out where questions of food were not raised. What remains reveals close and conscious links between the history of technology and the capacity of the earth; between plantations, fisheries, and the qualities of the human populations; between engines, land use, and the distribution of people; between fruit cultivation, diet, and health; between thinking about food production and envisioning populations in quantitative terms; and, above all, between technological improvements in the production or supply of food and prospects for increasing or redistributing population in England and across a far-flung colonial empire.

TECHNOLOGY, MULTIPLICATION, AND PROVIDENCE: GABRIEL PLATTES'S FRAMEWORK FOR IMPROVEMENT

The broadest statement of the relationship between technology, food, and population came at the beginning of the Hartlibian engagement with agriculture. Its author was Gabriel Plattes (ca. 1600–44), an alchemical, mineralogical, and agricultural writer whose best-known work, the utopian *Description of the Famous Kingdome of Macaria* (1641), was attributed to Hartlib until the twentieth century.[32] Plattes was an obscure figure even in life, though Richard Weston knew him.[33] Charles Webster credits him with stoking the Hartlib Circle's interest in agriculture through his *Discovery of Infinite Treasvre* (1639), which Hartlib judged "worthy to bee translated and set on foot by praxis."[34] A would-be patron, Sir Cheney Culpeper, pestered Hartlib for copies or news of Plattes's works, but learned of his death only months after the event.[35] Yet

---

[29] On Malthus's providential beliefs, see A. M. C. Waterman, *Revolution, Economics, and Religion: Christian Political Economy 1798–1833* (Cambridge: Cambridge Univ. Press, 1991).

[30] See Webster, *Great Instauration*; and Slack, *From Reformation to Improvement*, 77–101 (both cit. n. 6). Hartlibian utopias influenced by Francis Bacon's *New Atlantis* include [Gabriel Plattes], *A Description of the Famous Kingdome of Macaria* (London: printed for Francis Constable, 1641); and William Petty, *The Advice of W. P. to Mr. Samuel Hartlib for the Advancement of Some Particular Parts of Learning* (London: [publisher unknown], 1648).

[31] M. Greengrass, M. Leslie, and M. Hannon, eds., *The Hartlib Papers* (Sheffield, UK: Digital Humanities Institute, University of Sheffield, 2013), http://www.hrionline.ac.uk/hartlib (**hereafter** HP).

[32] Charles Webster, "The Authorship and Significance of Macaria," *Past & Present* 56 (1972): 34–48.

[33] Samuel Hartlib, "Ephemerides 1639, Part 3," HP 30/4/18b–27b, on 26a.

[34] Webster, *Great Instauration* (cit. n. 6), 471; Gabriel Plattes, *A Discovery of Infinite Treasvre, Hidden since the World's Beginning* (London: printed by I[ohn] L[egat], 1639); Samuel Hartlib, "Ephemerides 1639, Part 1," HP 30/4/1a–9a, on 4a; Hartlib, "Ephemerides 1640, Part 1," HP 30/4/37a–44b, on 39b.

[35] See Sir Cheney Culpeper to Hartlib, 13 and 20 November 1644, 4 and 28 January 1645, 18 March 1645, 20 May 1645, 2 July 1645, and 12 November 1645 (HP 13/52a–53b, 55a–55b, 59a–60b, 69a–70a, 78a–79b, 88a–89b, 92a–93b, 121a–122b). On the problem of self-interest in projecting, and on

Plattes's ideas and example outlived him. The farmer Peter Smith, returning copies of Plattes's works to the clergyman and projector John Beale in 1656, lauded Plattes's "vsefull knowledge & ingenuity" and deplored that "the ingratitude of his age" had stopped him from sharing more "experiments of great vse & benefit."[36]

Macaria is a useful point of entry into Plattes's thinking, both for its directness and because it was addressed to Parliament as a realizable plan.[37] On one level, this dialogue between a Traveler and a Scholar (recalling Thomas More's *Utopia*) proposed a reorganization of central government based on parliamentary reform and the creation of five "under Councills"; the first among them was a "Council of Husbandry" empowered to enforce the improvement of land, with others for fisheries, land and sea trade, and plantations.[38] On another level, however, *Macaria* (closer in this regard to Francis Bacon's *New Atlantis*) presented improvement as a way of avoiding the choice between stasis and expansion. At the end of the work, the Traveler promises to "propose a book of Husbandry"—evidently the *Discovery of Infinite Treasvre*—"to the high Court of Parliament, whereby the Kingdome may maintaine double the number of people, which it doth now." Before perusing the book in question, the Scholar remarks that "if a Kingdome may be improved to maintaine twice as many people as it did before, it is as good as the conquest of another Kingdome, as great, if not better."[39] Like an aboveground mine—an image Botero and Bacon both employed, as Vera Keller has explored—improvement could increase the capacity of the kingdom's existing territory, removing a key obstacle to growth and obviating the demographic price of conquest.[40]

The Discovery of Infinite Treasvre had spelled out the anticolonial (or, more properly, antiexpansionist) implications of this in greater detail, proposing improved husbandry as "a better cure for an over-peopled Common-wealth, then to make violent incursions upon others territories"—especially as "the finding of new worlds, is not like to be a perpetuall trade."[41] The specific changes Plattes called for indicated many directions that subsequent Hartlibian efforts would take: fruit trees (later advocated by Beale, Ralph Austen, and John Evelyn), new crops and rotations (Weston's focus), refined implements (taken up by Walter Blith), fertilizers (part of Benjamin Worsley's saltpeter scheme), machines such as seed drills (pursued by Cressy Dymock and William Petty), enclosure (advocated by Dymock), sluices and "Persian wheels" for irrigation, drainage pools, and more—as well as measures against mildew and sheep rot.[42] But Plattes was unusual in presenting them as a package—though Hartlib and Blith produced compendious collections.[43] More explicitly than most, Plattes based

---

Plattes's strategy, see Koji Yamamoto, "Reformation and the Distrust of the Projector in the Hartlib Circle," *Hist. J.* 55 (2012): 375–97.

[36] Peter Smith to [John Beale?], 7 April 1656, in "Copy Letters in Scribal Hand A, Peter Smith to Beale?," HP 67/23/1a–17b, on 17a–b. See Gabriel Plattes, *A Discovery of Subterraneall Treasure* (London: imprinted by I. Okes for Iasper Emery, 1639).

[37] [Plattes], *Macaria* (cit. n. 30), sig. A2r–A2v.

[38] Ibid., 3–4. Plattes cites More and Bacon as exemplars at sig. A2r–A2v.

[39] Ibid., 11.

[40] See Francis Bacon's essay "Of Seditions and Troubles," in Bacon, *The Essayes or Covnsels, Civill and Morall* (London: printed by Iohn Haviland for Hannah Barret, 1625), 76–90, on 84. On art as an alternative to conquest in Botero, Jakob Bornitz, and Bacon, see Keller, "Mining Tacitus" (cit. n. 20).

[41] Plattes, *Infinite Treasvre* (cit. n. 34), sigs. A4r, C3v.

[42] Ibid., 9–15, 23–46, 54–74.

[43] See Samuel Hartlib, *Samuel Hartlib His Legacie: Or an Enlargement of the Discourse of Husbandry Used in Brabant and Flaunders* [. . .] (London: printed by H. Hills for Richard Wodenothe,

this package on an alchemical theory of fertility according to which the products of art and nature alike—all manifestations of wealth—were combinations of terrestrial and celestial vapors, congealed and fused in the bowels and on the surface of the earth.[44] He was unlike his successors, too, in seeing improvement as an alternative to conquest—a matter of internal colonization—rather than a tool of empire.[45]

If Plattes's vision of improvement rested on an alchemical theory of wealth, it also reflected a providential view of the history of multiplication. Human "improvements" and "innovations" had already played vital roles in averting disaster and fostering growth:

> Such innovations as these have beene accustomed in all ancient times; as the people grew more and more numerous, to be put in practise: for three severall times the people growing too numerous for their maintenance, God hath given understanding to men to improve the earth in such a wonderfull manner, that it was able to maintaine double the number . . . for when there were but few, they were maintained by Fish, Fowle, Venison, and Fruits; freely provided by Nature: but when they grew too numerous for that food, they found out the Spade and used industry to augment their food by their indeavours: then they growing too numerous againe, were compelled to use the plough, the chiefest of all engines, and happily found out: whereby all Commonwealths have ever since been maintained.[46]

The providential-historical role of technology in expanding capacity by transforming food production was clear. Inasmuch as these improvements redressed the degraded state and circumstances of humankind since the biblical fall of man, Plattes's presentation of them paralleled familiar Baconian claims about the capacity of new instruments to augment the degenerated powers of the senses.[47] At key points, new tools—first the spade, then the plough—had relieved the pressure of growing numbers by introducing new foods, food sources, and levels of productivity. Human ingenuity grappled with the powers as well as the limits of nature and, working with and upon them, transformed the conditions of life. This process continued in the seventeenth century. New skills, crops, treatments, and rotations were needed as old technology lost ground against population:

> At length [the plough] would not serve the turne neither without new skill in the using of it: for at the first they used to till the Land till the fatnesse thereof was spent, and so to let

---

1651); and Blith, *English Improver Improved* (cit. n. 8). Benjamin Worsley, whose enthusiasm for alchemy exceeded Plattes's, also linked a range of improvements, without connecting them to population. See [Benjamin Worsley?], "Copy 'Profits Humbly Presented to This Kingdom' in Hand B," HP 15/2/61a–64b.

[44] Plattes, *Infinite Treasvre* (cit. n. 34), sig. A4v–A5r. On the Hartlibians' alchemical approach to fertility, see Simon Schaffer, "The Earth's Fertility as a Social Fact in Early Modern Britain," in *Nature and Society in Historical Context*, ed. Mikuláš Teich, Roy Porter, and Bo Gustafsson (Cambridge: Cambridge Univ. Press, 1997), 124–47, on 127–8.

[45] But see Sarah Hogan, *Other Englands: Utopia, Capital, and Empire in an Age of Transition* (Stanford, Calif.: Stanford Univ. Press, 2018), which argues (164) that "Plattes conceives of agrarian improvement in colonial terms," at least in *Macaria*. This reading can be reconciled with Plattes's presentation of improvement as an alternative to expansion if we distinguish colonial conquest as a political strategy (which Plattes disavowed) from the expropriation of putative wasteland or unimproved land (which Plattes enjoined).

[46] Plattes, *Infinite Treasvre* (cit. n. 34), sig. C3r.

[47] See Peter Harrison, *The Bible, Protestantism, and the Rise of Natural Science* (Cambridge: Cambridge Univ. Press, 1998), 226–35; and Harrison, *Fall of Man* (cit. n. 15).

it lye a long time to gather fatnesse againe of it selfe; and in the mean time to till fresh Land: but when they grew too numerous for the food gotten that way, they were compelled to find out the fallowing and manuring of Land: by which invention the Land recovered more fatnesse in one yeare, then before in many yeares; and so a Countrey would maintaine double the number of people more then before.[48]

Art met and underpinned the "doubling" of population by improving nature, introducing revolutionary changes with national and global demographic effects.

Improvement was no mere *deus ex machina*, appearing as required at critical moments of demographic plenitude; it was the fruit of human effort cooperating with divine Providence to counteract the corrupting consequences of original sin. The continual process of doubling implied a continual need for refinements, if not for epoch-making inventions—all the more so as innovation was resisted, and its promise threatened by the husbandman's attachment to custom and suspicion of novelties:

> Now the people are growne numerous againe, requiring new improvements discovered in this little Booke, and shall be showed . . . that by the Common course of Husbandry used at this day, the barrennesse doth by little and little increase, and the fertilitie decrease every yeare more and more, which in regard that the people doe increase wonderfully, must needs at length produce an horrible mischiefe, and cause the Common-wealth to be oppressed with povertie and beggery.[49]

The authority Plattes gave *Macaria*'s "Council of Husbandry" to force landowners to adopt improvements reflected this sense of sluggishness in the face of creeping disaster. Certainly, Hartlibians lamented popular resistance to new ways of doing things, and conservatism may have been particularly apparent in agriculture, rooted as that was in local knowledge and conditions.[50] One solution to this, even before the creation of a Commonwealth committed to "reformation," was to appeal to the power of a sympathetic lawgiver. Some projects, as John Dury wrote, "can not bee done but by a State."[51] For Plattes himself, "Husbandry is the very nerve and sinew, which holdeth together all the joynts of a Monarchy."[52] Improvement was a political concern.[53]

It was also a popular one. Plattes's last printed work, a pamphlet entitled *The Profitable Intelligencer* (1644), started from the same demographic and alchemical premises as *The Discovery of Infinite Treasvre*: "as God is infinite, and men are infinite by propagation, so the fruits of the Earth for their food, and cloathing are infinite, if men will consent to put their helping hands to this commendable Designe."[54] The agency Plattes now invoked, however, was not the coercive power of the state but the voluntary effort of the people; poor women, maidservants, and ordinary householders were

[48] Plattes, *Infinite Treasvre* (cit. n. 34), sig. C3r.

[49] Ibid., sig. C3r–v.

[50] Sir Cheney Culpeper referred to the "Irishe humor, of keepinge theire olde barbarous custome of plowinge by horses tayles," as a metaphor for English skepticism about Hartlibian schemes; see Culpeper to Hartlib, [1646?], HP 13/284a–285b, on 284b. Cressy Dymock to Hartlib, undated, HP 62/9/2a–b, on 2b, contrasted "reason" and "reall experience" with attachment to the "old customes" of "the bores [i.e., boors] of England."

[51] John Dury to Sir Cheney Culpeper (copy), undated, HP 12/72a–75b, on 72b.

[52] Plattes, *Infinite Treasvre* (cit. n. 34), sig. C3v.

[53] See Hogan, *Other Englands* (cit. n. 45), 158.

[54] Gabriel Plattes, *The Profitable Intelligencer, Communicating His Knowledge for the Generall Good of the Common-wealth and All Posterity* (London: printed for T. U., 1644), sig. A2r.

now, like so many alchemists, to capture and control the fertile powers of the earth. Plattes advised them to collect all kinds of waste material—including dung, urine, and rags; as well as hair, horn, and leather from barbers, butchers, and tanners. Locked up in such leavings was "the vegetable spirit of the world, by which all things do encrease and multiply."[55] While the sale of this fertile waste would generate dowries for young women, its use on the land would compensate for extra mouths by augmenting and pre-serving the soil's fertility: "the excrements, and materials, which any family produceth, being well contrived, will produce yeerly as much bread, and drink, as that family spendeth for ever."[56] Improvement recruited the virtues of thrift and industry to liber-ate the fertile powers of waste and put these to work through commercial exchange, agricultural technology, and marital union—returning vegetable spirit to the land as fertilizer while promoting the fertility of the women involved.[57] An alchemically in-flected Protestant ethic would augment food production and transform the conditions of multiplication: "why may not the same thing be done by Art, which was formerly done by Nature, and accident?"[58]

PLANTATION, PRESERVATION, AND POPULATION: HUGH L'AMY AND PETER LE PRUVOST'S IMPERIAL PROJECTS

No Hartlibian historicized the relationship between food, technology, and population as boldly as Plattes did, but other projects sketched improvement's demographic im-plications in imperial terms. Beginning in the mid-1640s, two Huguenot exiles, Hugh L'Amy and Peter Le Pruvost, pursued innovations in husbandry, fisheries, and colo-nial plantation designed to promote English power and advance the Protestant cause in Europe and across the Atlantic. Helped (and translated) by the Scottish reformer John Dury, L'Amy and Le Pruvost hoped to procure an act of Parliament linking new and more productive husbandry to the supply of men for a colonial settlement; this, in turn, would supply labor, ships, and commodities for an expanded fishery. Be-tween 1645 and 1649, L'Amy and Le Pruvost refined their proposals to counter re-sistance from those suspicious of monopolies—and from fish merchants whose priv-ileges their project threatened.[59] These efforts failed.[60] Yet while their ideas are rightly seen as part of a fleeting pan-Protestant political moment, they were also designed to affect the quantity, qualities, and distribution of population across a British Atlantic empire.[61]

---

[55] Ibid., sig. A3r.

[56] Ibid., sig. A2v.

[57] See Wennerlind, "Credit-Money" (cit. n. 11).

[58] Plattes, *Profitable Intelligencer* (cit. n. 54), sig. A3v.

[59] Letters from Culpeper and John Dury highlight the challenges these projectors faced in navigat-ing interests and in managing their own reputations. For Culpeper's thoughts, see his letters to Hartlib of November 1645(?), 9 April 1647, 30 August 1648, and undated (arguing for a patent), on HP 13/277a–278b, 171a–172b, 241a–242b, and 279a–283b, respectively. For Dury's, see Dury to [Peter Le Pruvost?], 20 July 1646, HP 12/20a–b; Dury to Culpeper (undated copy), HP 12/72a–75b (arguing against a patent). On fish merchants' resistance, see Le Pruvost to [Dury?], 6 December 1646, HP 12/18a–19b, and an exchange of letters between Dury and Culpeper (Dury to Culpeper, 25 September 1648, and Culpeper to Dury, 26 September 1648), HP 12/23a–26b.

[60] As the projectors complained in mid-1649, "Lamy a esté a londres y a enuiron quatre ans et tarde la trois mois sans aucun effect sinon promesses pour ladvenir"; see "Note on the Proposals of Hugh L'Amy & Peter Le Pruvost," 20 July 1649, HP 12/110a–b, on 110a.

[61] See Jeremy Fradkin, "Protestant Unity and Anti-Catholicism: The Irenicism and Philo-Semitism of John Dury in Context," *J. Brit. Stud.* 56 (2017): 273–94, on 273. See also Leng, *Benjamin Worsley* (cit. n. 4), 35.

L'Amy and Le Pruvost's husbandry proposals were limited in detail—possibly because participation entailed contribution to the colonization scheme—but familiar in content. They included "dunging and fattning" of the soil, seed preparations to protect corn from "being black" (i.e., black rust) or "blasting," and a modified schedule of sowing and reaping corn and fodder (perhaps a rotation similar to Weston's), besides advice on feeding sheep to improve their wool and make their flesh "better tasted and more healthfull."[62] The positive connection between increased corn production and colonial expansion, however, stood in stark contrast to Plattes. Estimating a productivity increase of one-fifth, besides gains from reduced wastage, L'Amy and Le Pruvost proposed the following:

> All such as will practise or shall follow the waye of improoving husbandrie which shall bee shewed shall bee obliged allwaies in time to come (namely soe long as they shall follow that way) to furnish one man of their kindred or freinds for the service of the Plantation according to the order etc. namely for every 70 acres of Land thus Husbanded one man and that a penaltie of a 1000 lb sterling and confiscation &c shall bee inflicted upon all such as transgresse the order.[63]

In contrast, too, to what Plattes had envisioned in *Macaria*, L'Amy and Le Pruvost—perhaps with Culpeper's warning against expropriative legislation in mind—emphasized that participation was voluntary, a matter of "free Contract."[64] Their improvement was not a utopian substitute for territorial expansion but an interest-driven engine of it. Improvement could respect property at home all the better as it entailed expansion overseas.

Agriculture reformed by the application of new fertilizers, preparations, and rotations would supply men for colonial settlement, "enlarg[ing] the borders of Great Brittaine" and projecting improvement across the sea.[65] Echoing such hostile sources as Bartolomé de Las Casas and Richard Hakluyt, L'Amy and Le Pruvost distinguished a potentially endless British empire of industrious Protestant husbandmen from its brutally avaricious and extractive Iberian competitors.[66] A rising empire of food and free contract could leave an empire of gold and slavery to itself:

> Great Brittaine can with better order and so better purpose erect a Potent plantation by Land and Sea and alwaies encrease the same without end namly by good husbandrie of that new Land without grudging the Spaniard his mines of Gold or Silver or the Trade

---

[62] Hugh L'Amy and Peter Le Pruvost, "Copy Proposals in Hand B," undated, HP 12/61a–63b, on 61a; Le Pruvost, "Draft Petition in Dury's Hand, Peter Le Pruvost to House of Lords," undated, HP 12/150a–b, on 150a. See also "Proposals Of Hugh L'Amy & Peter Le Pruvost," undated, HP 12/93a–98b, partially translated as Hugh L'Amy, "Copy Proposals for Husbandry in Hand B," HP 12/64a–65b.

[63] L'Amy and Le Pruvost, "Copy Proposals," HP 12/62b; see also L'Amy, "Copy Proposals for Husbandry" (both cit. n. 62), HP 12/64a. Compare L'Amy and Le Pruvost, "Copy, Propositions of Hugh L'Amy & Peter Le Pruvost to House of Commons," HP 12/142a–143b, where improvers are to supply men for the navy on similar terms.

[64] L'Amy and Le Pruvost, "Copy, Propositions" (cit. n. 63), HP 12/142a.

[65] L'Amy and Le Pruvost, "Copy Proposals" (cit. n. 62), HP 12/61b.

[66] See Edmund Valentine Campos, "West of Eden: American Gold, Spanish Greed, and Discourses of English Imperialism," in *Rereading the Black Legend: The Discourses of Religious and Racial Difference in the Renaissance Empires*, ed. Margaret R. Greer, Maureen Quilligan, and Walter D. Mignolo (Chicago: Univ. of Chicago Press, 2008), 247–69.

of spices and sugar &c. and without burdening the Plantation with Negros for many reasons.[67]

Like subsequent colonial efforts—the subjugation and resettlement of Ireland from 1649 to 1656, the Western Design of 1655, and the ensuing conquest and settlement of Jamaica—this plantation would also occupy the restless New Model Army (created by Parliament to fight the king in 1645, but a political liability once the English Civil War was over) and absorb other disaffected, displaced, or impoverished groups.[68] Much as earlier seventeenth-century colonial schemes had promised, L'Amy and Pruvost's plantation would alleviate poverty, vagrancy, and overpopulation. England's surplus people might produce valuable commodities while spreading true religion.[69]

It would also contribute to L'Amy and Le Pruvost's other proposals, especially improvements to the fishery. Here, too, the projectors set new elements in a familiar framework. The idea of the fishery as a "nursery of seamen" was well established, and their promotion of fishing was typical in promising to produce seamen for service in peace and war.[70] More distinctive was their concern with the "manner" of fishing and, in particular, the preservation or drying and "Dressing of Fishes." They distinguished "4 cheif kindes of Fishes which may bee dressed to a great deall more benefit and advantage of the Public and off particullars then ordinarily." The first was "the Fish of Newfoundland." For these, they wrote the following: "The Plantation shall furnish gray salt for the salting of fishes; wherby they will bee whiter and consequently more salliable. For the white Salt doth burne the fish and makes it yellowish and unsavourie." Better-dressed fish meant employment for colonists and savings for merchants: "The Plantation shall furnish the third of the men to helpe the Drying of the fishes &c. by which meanes the merchants that send their shippes to the fishing will bee eased of great charges and expences of wages and victualls." Second was herring, "wherin the benefit will come also from a better forme of Nets and better manner of fishing." Third was "the Fish of the North Sea which by this dressing shall bee made both more wholesome to the stomacke, and better tasted to the mouth. Fourth, finally, was "salmon which in like manner shall bee made better."[71] Improved husbandry would facilitate colonial plantation; and plantation, absorbing excess numbers of metropolitan poor, would supply the salt and labor essential to improving fisheries.

---

[67] Hugh L'Amy and Pierre Le Pruvost, "Copy Memo In Hand? Hugh L'Amy [and] Peter Le Pruvost To House Of Commons," HP 12/7a–8b, on 7b.

[68] Ibid., HP 12/7b–8a; L'Amy and Le Pruvost, "Copy Proposals" (cit. n. 62), HP 12/62a. On the Irish settlement, see, most recently, John Cunningham, *Conquest and Land in Ireland: The Transplantation to Connacht, 1649–1680* (Woodbridge, UK: Royal Historical Society/Boydell, 2011); on the Western Design, see Swingen, *Competing Visions of Empire* (cit. n. 3), 32–55.

[69] L'Amy and Le Pruvost, "Copy Proposals," HP 12/62a; L'Amy, "Copy Proposals for Husbandry" (both cit. n. 62), HP 12/65a.

[70] The idea appears in "Polices to Reduce this Realme of Englande vnto a Properus Wealthe and Estate" (1549), in *Tudor Economic Documents: Being Select Documents Illustrating the Economic and Social History of Tudor England*, ed. R. H. Tawney and Eileen Power, 3 vols. (London: Longmans, Green, 1951), 3:311–45, on 336. The phrase occurs in a paper headed "The Detriment which England suffers by prohibiting Cattle" (undated; probably ca. 1667), in Trinity College Dublin Library, Robert Southwell Papers, MS 1180, fols. 27–30, on fol. 29. For later examples, see Anna Gambles, "Free Trade and State Formation: The Political Economy of Fisheries Policy in Britain and the United Kingdom Circa 1780–1850," *J. Brit. Stud.* 39 (2000): 288–316, on 297; and Julian Hoppit, *Britain's Political Economies: Parliament and Economic Life, 1660–1800* (Cambridge: Cambridge Univ. Press, 2017), 121.

[71] L'Amy and Le Pruvost, "Copy Proposals" (cit. n. 62), HP 12/61a–b.

Improvements in the volume and quality of fish—as commodities and as nourishment—
would augment numbers of seamen.

Changes to the production of food on land and sea were understood both to require
and to facilitate a redistribution of population between the British metropole and its
Atlantic colonies. This was more than just an efficient reallocation of laboring units; it
was a qualitative transformation of people, of the land they settled and the sea they
traversed, and of the fruits of both. Huguenot exiles, Cromwellian soldiers, the poor,
and the self-seeking would move, and as they did so, they would change. Refugees
would settle, soldiers would become colonists, and the idle would turn productive.
Meanwhile, "Great Brittaine will bee much mor peopled then ever it hath beene . . .
without any vagabons and debauched people and without being incommodated with
poore as now it is."[72] The implication was not just that colonial expansion was consistent with the maintenance of sufficient metropolitan numbers, but also that (as Botero
had argued) it would create more favorable conditions for the multiplication of those
people—economically, physically, and morally. L'Amy and Le Pruvost went so far as
to suggest that "marriage will be better celebrated and the use of prostitutes abolished"
if their proposals were adopted, while fish and mutton ("which breedeth consumptions") would become healthier.[73] Better food production and better food were crucial
to reorganizing, augmenting, and improving the population of Britain and its empire.

FOOD AND LABOR IN ENGLAND AND BARBADOS: CRESSY DYMOCK'S ENGINES

Just as L'Amy and Le Pruvost were losing hope that Parliament would enact their proposals, Cressy Dymock began promoting a pair of inventions that connected provisioning, population, and colonial empire in related ways. Little is known about Dymock.
He may have been the son of Sir Thomas Dymock, and hence from the Lincolnshire-
Nottinghamshire border, and he appears to have died shortly after the Restoration.
What is clear is that he entered Gray's Inn in 1629, became an enthusiastic supporter
of Hartlib about twenty years later, and advocated enclosure as the basis of various
other agricultural improvements.[74] Among these were new fodder crops (his notes on
Weston's *Discourse* survive), fruit trees, rabbit breeding, and a plan for setting out
and dividing up farmland on a more rational basis.[75] The innovations for which he

---

[72] L'Amy and Le Pruvost, "Copy Memo" (cit. n. 67), HP 12/8a. Disputing the view that colonies
would drain the nation of people, Le Pruvost replied, "Ie proposeroy ordre et moyens pour ne point
manquer d'hommes . . . pour la guerre, et pour l'agriculture, et pour plantations, en rendant
l'Angleterre plus abondante en viures, et en manufactures, et commerce, par l'augmentation de
peuple, et sans aucuns pauures"; see Le Pruvost to Dury and Hartlib, 10 September 1649, HP 12/
28a–29b, on 28b.
[73] "[L]e mariage sera mieux celebre et la putacerie abolye"; L'Amy and Le Pruvost, "Note on the
Proposals" (cit. n. 60), HP 12/110a. On mutton and "consumptions," see L'Amy, "Copy Proposals for
Husbandry" (cit. n. 62), HP 12/64a.
[74] Cressy Dymock, "Memorandum on the Advantages of Enclosure," 20 July 1649, HP 64/18/1a–
2b; Mark Greengrass, "Dymock, Cressy (fl. 1629–1660), agriculturist," *Oxford Dictionary of National Biography* (2004), accessed 13 June 2018, https://www.oxforddnb.com/.
[75] Cressy Dymock, "Notes on a Discourse on Husbandry" (undated but likely 1650 or later), HP 62/
4/1a–2b; Dymock, "Proposition for Planting Quinces" (undated), HP 62/41a–b; Dymock, "Memorandum on the Husbandry of Rabbits" (undated), HP 62/32/1a–2b; Dymock to [Hartlib?], undated but
likely 1653 or earlier, HP 62/29/1a–4b, printed in Samuel Hartlib, *A Discoverie for Division or Setting
out of Land, as to the Best Form* (London: printed for Richard Wodenothe, 1653), 1–11; Weston, *Discourse of Husbandrie* (cit. n. 8).

was best known, however, were his two "engines"—one for "perpetuall motion," the other for "the setting of corn."[76]

Like the philosopher's stone, perpetual motion now has the ring of fantasy; like alchemical transmutation, however, it was a longstanding philosophical *desideratum* that motivated significant conceptual, experimental, and technological effort in the seventeenth century. Among those in or close to Hartlib's network, such diverse figures as the Dutch inventor Cornelius Drebbel, the Moravian philosopher Jan Amos Comenius, the agricultural improver Gabriel Plattes, and the polymath William Petty all pursued it. So, later, did other fellows of the Royal Society.[77] For Dymock, as for others, perpetual motion was an abstract challenge; it was also a source of energy that might drive the machinery of improvement in a variety of settings. Many of its applications had little to do with food or population. Dymock's promotional pamphlet, *An Invention of Engines of Motion Lately Brought to Perfection* (1651), listed thirty-two uses, ranging from grinding brick for plaster and bark for tanning, to weighing ship anchors, to boring holes in wood, metal, or stone.[78] But several uses of perpetual motion impinged on the production of food in ways that affected the land and labor required to support population—and, in colonial contexts, the character of the population itself.

Agriculture was on Dymock's mind when he announced his "instrument or engine" for "perpetuall motion" to Hartlib late in 1648:

> The vses to which this is appliable are turning of grind stones winnowing of Corne, churning of butter, or the like for smale ones, And for great ones (the greater the better) drayning of waters, Iron mills, Corne mils, sugar mils, oyle mils & alsoe for the mooving of carriages of all sorts either about home, or the high wayes, to sutch proportions as may abate of the present charge, & advance trafique verry mutch.[79]

Further work—and the quest for financial support—expanded Dymock's sense of the engine's uses, which now included "to draw the Plough (or rather Spade-plough which is .3. times better, and the grand or second roulers, and the New harrowes. etc.)"[80] It also tamed Dymock's claim to have created a "perfect" perpetual motion, and led him to present his invention as a labor-saving rather than labor-eliminating device: "Such a Motion," he put it, "as shall want little of mooving constantly, and if it faile in that, shal make amends in the strenght it goes by."[81] This was a significant shift, but not because it was a climb down from utopian to modest aspirations. If anything, Dymock ramped up his rhetoric, describing himself as "by [God's] appointment . . . impregnate with an issue which by . . . devine blessing is now brought forth," and his engine as "a living

[76] See Cressy Dymock, *An Invention of Engines of Motion Lately Brought to Perfection* (London: printed by I. C. for Richard Woodnoth, 1651).

[77] See Keller, *Knowledge and the Public Interest* (cit. n. 7), 81, 86, 177–8, 254. On Petty's perpetual motion engine, see Samuel Hartlib, "Ephemerides 1649 Part 1," HP 28/1/1a–13b, on 9b.

[78] Dymock, *Invention of Engines* (cit. n. 76), 8–9. An anonymous, undated paper, likely associated with Dymock's engine and discussed further below, lists even more uses: "Notes on Uses of Perpetual Motion in Scribal Hand B," HP 58/21a–22b.

[79] Dymock to Hartlib, 25 December 1648, HP 62/50/2a–b, on 2a.

[80] Dymock to [?], 25 May 1649, HP 62/50/3a–4b, on 3a–b. On Hartlib's and others' financial contributions to Dymock's perpetual motion project—which ended in bitter disappointment by 1654—see Yamamoto, "Reformation and the Distrust of the Projector" (cit. n. 35), 392–3.

[81] Cressy Dymock, "Copy Extract in Hartlib's Hand, Cressy Dymock on Perpetual Motion," 16 March 1649, HP 62/50/11a–b, on 11a.

male childe cal'd the Marriage of Strength and tyme."[82] The Holy Spirit had fertilized Dymock's invention; art's work upon nature was divinely ordained.

The shift of emphasis from perpetual motion to the marriage (or "wedding") of strength and time tied Dymock's machine more explicitly to the particular labor and land requirements of food production in specific locales.[83] Central to this was the engine's relationship to animal labor. In March 1649, Dymock hoped "to doe the Worke of 3. or 4. Horses with one."[84] The significance of this emerges from an undated "Memo Concerning a Utility Engine," in which Dymock linked his claim "to Marrye strength and tyme togather, or bring them closer each to other then vsuall engines doe" to the existing needs of human- and horse-powered sawmills and waterworks, as well as wind- and water-mills.[85] He noted the following:

> As to ploughing/Vsually men haue in their teemes to plough 4 -5- or -6 horses & 6 or 8 oxen, & to guide these and the plough two men or a man & a boy att least, with which they plough about one acre a day I hope without any horse with 6 men onely or lesse to plough with two ploughs att once of the same strength or bignes. . . two acres a day of the same or the lyke ground
>   As to the grinding of Corne Malt &c. . . . Horse mills to moue them effectually require 3- or -4 horses to worke att a tyme & some must bee kept to releiue them/But I hope to grind -30-or -40- quarters of malt . . . or towards 20 quarters of hard corne in a day without any of the foresaid vsuall helpes perticulerly without horses & with onely two men in their turnes.[86]

Dymock's engine would reduce or remove the need for animal labor in a range of settings, but his specific target reflected his observation of the horsepower that plowing and milling grain typically required. When he advertised shares in his machine, he estimated the costs of completing the engine for use in grinding wheat or malt, but not for other purposes.[87]

In a general sense, what this implied for population was obvious: more food could be grown with less work on the same land. But the reduction of animal labor to a lower minimum, or its replacement by human effort, had more precise implications—and different ones, depending on context. As an anonymous paper likely by Dymock put it, sparing "the labour of horses" meant sparing "the fruites of a great part of the grounds which are now spent to maintaine those horses." These "will serve to feed & maintaine men" instead, so that "by this Invention the Land will maintaine on 3d part of men at the least more then now it doth."[88] Perpetual motion would put more land under corn and employ more people in producing it. But the same technology would mean something quite different in the plantation colonies of the Caribbean:

> If my engine bee made vse of in the Barbados for the grinding of sugar there will nessesarily follow (besids all private benifitts) this publique advantage that whereas they

---

[82] Dymock to [Hartlib?], 25 October 1649, HP 62/50/9a–10b, on 9a. The same providentialism and metaphors shape the account in Dymock, *Invention of Engines* (cit. n. 76), 1–3.

[83] See Cressy Dymock, "Memo on Perpetual Motion," HP 67/17/1a–2b, on 1a.

[84] Dymock, "Copy Extract . . . On Perpetual Motion," HP 62/50/11a.

[85] Cressy Dymock, "Memo Concerning a Utility Engine" (undated), HP 67/13/1a–2b, on 1a–b.

[86] Ibid.

[87] Dymock, "Memo on Perpetual Motion," HP 67/17/1a.

[88] "Notes on Uses of Perpetual Motion," HP 58/21a–22b. The document's contents (including thirty-three uses) resemble that of works by Dymock on the same subject.

are now forced to lett many acres ly for fother [fodder] for those draught cattle winter & somer the proffitt thence arrising beeing farre short of what the same land would yeild if planted with sugar canes, cotton, Indico, or the lyke, by this meanes all that land may bee converted to those more beneficiall vses, to the great increase & trade of those more staple comodityes.[89]

If reducing animal labor would allow a better-fed and better-employed population in England, in Barbados it would free land for sugar and other cash crops, creating commercial opportunities for planters—and work for the indentured or the enslaved. Another paper, almost certainly Dymock's, spelled this out:

For that in the plantations horses etc are dearer, & shorter lived then Negroes, & more troublesome and charegeable to keep then Negroes, it is generally Concluded, that, could an Invention bee found, wher by the hands of 4 or 6 Negroes at a spell the same worke could bee dispatched, both as to strength and time, which is now done by 4 horses or 8 cattle; it would bee a noble usefull designe, particularly for sugar workes, & merite a large reward.[90]

By minimizing the use of animal energy in the production of food, the perpetual motion engine would alter the balance not only between animal and human populations in the metropole, but also between free and enslaved populations in plantation colonies. Dymock's perpetual motion engine animated a mercantilist vision of empire in which the cost of feeding free, unfree, and animal labor was a central problem, and in which imperial relationships and economic efficiency, mediated by divinely inspired technology, dictated a clear hierarchy of labor. While English population would expand at home, the colonial demand for labor would be met neither by English emigration nor by the costly and land-intensive use of animals, but by slavery.[91]

The demographic significance of Dymock's "engine" for setting corn was consistent with this imperial program. As Petty (who produced a similar machine) observed, the instrument's purpose was to save seed.[92] Dymock worked on it alongside the perpetual motion machine in 1649, and attracted "Engagers (tho not many nor deepe)."[93] By 1653, Swedish agents were apparently interested, though Dymock felt "bound to prefere the good of mÿ owne Natives Countrÿe aboue all others."[94] A series of letters trumpeted the prototypes' successes at Wadworth in Yorkshire:

[89] Cressy Dymock, "Memorandum about Engines" (undated), HP 62/8a–b, on 8a.

[90] "Memo on Types of Mills" (undated), HP 67/8/1a–2b, on 1b. Besides the use of "strength and time" and the consistency of the numerical estimates, the document echoes Dymock's comparisons of the efficiency of various mills in his "Memo Concerning a Utility Engine" (cit. n. 85).

[91] See Swingen, *Competing Visions of Empire* (cit. n. 3).

[92] Quoted in Samuel Hartlib, "Ephemerides 1648 Part 3," HP 31/22/27a–28b, 21a–24b, 33a–40b, on 22a. On Petty's "Instrument for setting of Corne," "Agriculture instrument," or "corn engine," see Hartlib, "Ephemerides 1648 Part 2," HP 31/22/14a–20b, 29a–32b, 25a–26b, on 31a; "Ephemerides 1649 Part 1," HP 28/1/1a–13b, on 8a; "Ephemerides 1649 Part 2," HP 28/1/14b–26a, on 24b–25a; "Ephemerides 1649 Part 3," HP 28/1/26b–38b, on 28b–29a; [William Petty?] to [?], 7 August 1651, HP 67/11a–b.

[93] Dymock to [?] (copy), 25 May 1649, HP 62/50/3b. See Yamamoto, "Reformation and the Distrust of the Projector" (cit. n. 35), 392.

[94] Dymock to Hartlib (copy), 26 February 1653, HP 62/28/1a–4b, on 1a–b. Dymock also voiced concern for his family's welfare, but was willing to sell upon "tearmes Convenient" and hoped that "in due tyme" his invention might serve "all mankind." John Dury, who spent time in Sweden and enthused about its success under Axel Oxenstierna, may have played a role; see Dury to Hartlib, 30 November 1645, HP 12/66a–68b, on 66b; [Dury?] to Hartlib, [1645?], 12/69a–71b; Dury to Culpeper (copy), undated, HP 12/72a–75b, on 73b.

The Cattle like them well, by reason they goe easier, and now the seruants like them also, and Confesse that they are fare easier to guide and hold then their old ones, and they now say (:seeing that wee even in their owne Iudgements sow or set thicke enough, & yet spent but 4 or 5 peckes at most for that same quantitye of ground whereon they sow (:and yet sow too thin:) about 5 bushells) they can now say that a great deale of Corne is throwne away in a yeare in England.[95]

The key point was that the engine used a peck of seed where customary methods required a bushel. Whereas "the Manner of ploughing & sowing now usuall" left seed "either buried so deepe as not to come up, couered so thinly as not bee haue sufficient Earth," exposed to birds, worms, or weeds, Petty explained, Dymock's "New way of tillage" meant the following:

1. Ground may be tilled in the same tyme this New as the common, & with fewer horse <cattle> and Mens help.
2. Lesse then a third part of the seed will suffice to sow the same parcell of ground this New then the old way.
3. Whereas the Encrease in the old way of One bushell of seed is seldome 5 <for 1>, this way it will commonly bee 24 for One.[96]

In reducing inputs and increasing yields, the engine had the same basic implications for food supply as the perpetual motion machine.

Concern with wastage led Dymock and his colleagues to connect improvement to population in two other ways. One related to health. As a paper on "Smutty corn" noted, the fungal infection made wheat "most unwholesome for the Bodyes of men," but given the difficulty of cleaning it, "many poore Tenents will bee compelled to sell" and "many thousands of poore people will be necesitated to eate it soe, to the great impairing of their healthes & be getting fowle diseases." One answer was a new way of threshing, perhaps involving the perpetual motion engine—a technological intervention to benefit the health of large numbers of people.[97] A second link lay in the use of numbers themselves. We have seen that assessments of the corn engine involved calculations of yields. Similarly, Dymock's arguments for the enclosure of farmland—which included curbing the destruction of crops by animals—led him to estimate the damage in quantitative terms. In particular, he set losses to pigeons in the context of England's human population: "One paire of olde pidgeons," he wrote, "eates of one sort or another of corne or graine in the yeare att least –6 bushells & . . . there is . . . allmost if not all-togather as <many> sutch paires of pidgeons as there is men woemen & children in England."[98] An anonymous paper, possibly by Dymock, even suggested

[95] Dymock, "Copy Letters In Scribal Hand G, Dymock To Hartlib? On Husbandry," 22 March–23 April 1653, HP 64/17/1a–6b; quotation from Dymock to Hartlib (copy), 26 March 1653, HP 64/17/2a–3a.

[96] William Petty, "Copy in Petty's Hand, Cressy Dymock's Husbandry Design" (undated), HP 62/50/17a–18b, at 17a–b. Angle brackets in *The Hartlib Papers* (HP) represent insertions in the manuscripts by the author or a scribe; the hand a document appears in, if other than the author's, is indicated in the citation.

[97] [Cressy Dymock?], "Copy Proposition In Scribal Hand B About Smutty Corne" (undated), HP 64/15a–b. Dymock listed threshing as one of his engine's uses; see Dymock, *Invention of Engines* (cit. n. 76), 9.

[98] Dymock to Hartlib, undated, HP 62/29/1b; printed in Plattes, *Discovery* (cit. n. 34), 4–5.

that the destruction of corn wrought by pigeons nationwide be expressed in terms of the number of poor families the corn would feed for a fixed number of weeks.[99] Here was a clear vision of population as a number of mouths to be fed, and of improvement as tasked with feeding them.

FRUIT, DRINK, AND THE HEALTH OF ENGLISH BODIES: JOHN BEALE'S FRUIT TREES

The final effort to be considered in this article is John Beale's promotion of fruit trees and cider (as well as perry and other fermented fruit drinks). Beale, a Church of England clergyman and political moderate from a well-off Herefordshire family, is the best documented of the projectors discussed here.[100] His correspondence with Hartlib was frequent and copious through the later 1650s, ranging from the grafting of trees to the search for a political settlement, to prophecy and the interpretation of dreams; also, he printed a work on orchards.[101] He enjoyed education, travel, connections, and a long life—he was buried in April 1683, almost 75 years to the day from his baptism. His projects benefited from this longevity; unlike Plattes, L'Amy, Le Pruvost, Dymock, and Hartlib, whose hopes of reform and support died at the Restoration or before, Beale adjusted to the return of monarchy with relative ease. He became a fellow of the Royal Society in 1663, maintained a correspondence with its secretary, Henry Oldenburg, and contributed to the *Philosophical Transactions*. Beale's writing thus allows us to follow some of the threads connecting improvement to population into a new period.

Most of Beale's writing on fruit trees in the 1650s concerned technical aspects of grafting, orchard keeping, and cider making, as well as the virtues of different apples, rather than their significance for feeding large numbers.[102] The main line he drew between improvement and population concerned the benefits of cider consumption for health:

> You shall find it as briskely danceing in the cup, as much corroborating the stomac, comforting the heart, preserving the balsome of nature, purging corrupt humours, clensing the veines & kidneyes, dissolving the formed stone in the bladder, & all tartarous matter . . . That more excellent qualityes cannot bee attributed to any kind of Wine, that I can heare of.[103]

Much as later proponents of improvement would do, but without the numerical data they prized, Beale cited local health and longevity as empirical evidence of his claims. "The salubrity of the best of thiese liquors is sufficiently attested," he wrote, "by the health & long life of many Thousands, That for some ages in this neighbourhood have

---

[99] "Copy Memo on Destruction of Corn by Pigeons in Hand?," undated, HP 25/3/4a–b. The paper extrapolates national estimates from numbers for Cambridgeshire, making Dymock's authorship unlikely.

[100] Patrick Woodland, "Beale, John (*bap.* 1608, *d.* 1683), Church of England clergyman and writer on agriculture and natural philosophy," *Oxford Dictionary of National Biography* (2004), accessed 18 June 2018, https://www.oxforddnb.com/.

[101] I[ohn] B[eale], *Herefordshire Orchards, a Pattern for All England* (London: printed by Roger Daniel, 1657).

[102] Beale held to the supremacy of Redstreak apples for cider (which he preferred to "the better sort of White Wines, & claret that are found in most Provinces of France") against another proponent of orchards, the religious radical Ralph Austen; see John Beale to Hartlib, 8 May 1658, HP 52/26a–43b, quotation on 30a. The dispute remained amicable.

[103] Ibid., HP 52/38b–39a.

beene accustomed to this drinke, more than to any other."[104] He further speculated that "blossoming trees" might "sweeten & purify the vernall ayre epidemically."[105] Diet and environment connected the cultivation of fruit trees to the health and longevity of local populations.

Beale did not speak of the national population. Nor did he quantify the goals of his project in relation to demographic figures. But it is not clear that the benefits he outlined required this, or that he could have done it had he wished. Rural Herefordshire, unlike London, did not produce bills of mortality, and although Beale might have used parish registers of births, marriages, and burials, he made no claim to have done so. An appeal to local experience, expressed in round numbers, sufficed.[106] He did, nevertheless, push for the adoption of his improvements on a national scale, and he described their potential effects in proportional terms. Like Dymock, he promised massive growth in yields:

> I shall shewe by what art a tree shall grow & beare as much in one yeare, as ordinarily in two; as much in five yeares (by the course which I shall prescribe) as usually in ten yeares; not by violence, or præcipitation, but by naturall ayde, & true improvement.[107]

Yet Beale, like Plattes, interpreted this less in terms of present-tense geopolitical struggle than in relation to a providential-historical dynamic of art working upon nature to undo the effects of the Fall.[108] While versed in classical literature on agriculture, Beale thought that "wee have far exceeded the ancient, & have advanced very far to obliege all posterity."[109] The fruit of this "Taming art" would be "the Resemblance of the Originall Obediance," a reproduction of Eden.[110] Much as for Plattes, improvement would alter the conditions of human multiplication more deeply than any fleeting conquest. "Wee neede not rayse Wars to destroy one another, or eate up one another," he insisted. For "in a short time, Wee may bee provided of food enough for another World as big as this, & soe make this a true paradyse."[111]

Beale's correspondence with Oldenburg after 1660 built on the agricultural concerns of his letters to Hartlib.[112] It was, however, marked by more explicit attention to wider

---

[104] Ibid., HP 52/39a. On later interest in measuring "salubrity" through demographic data, see Andrea Rusnock, *Vital Accounts: Quantifying Health and Population in Eighteenth-Century England and France* (Cambridge: Cambridge Univ. Press, 2002), 29–30, 163–7; and Ted McCormick, "Governing Model Populations: Queries, Quantification, and William Petty's 'Scale of Salubrity,'" *Hist. Sci.* 51 (2013): 179–98.

[105] Beale to Hartlib, 22 May 1658, HP 52/65a–66b, on 66b.

[106] Projecting future cider production, Beale remarked: "It must not bee expected, That my accompt should bee otherwise then of round numbers at large, as an old soldier may <at a viewe> say, howe many Regements of Horse or Foote are drawne foorth into the field." Beale to Hartlib, 22 May 1658, HP 52/65a.

[107] Beale to Hartlib, 8 May 1658, HP 52/42b.

[108] See Harrison, *Fall of Man* (cit. n. 15).

[109] Beale to Hartlib, 8 May 1658, HP 52/42a. On classical authors, see Beale to Hartlib, 15 November 1659, HP 62/25/1a–4b.

[110] Beale to Hartlib, 15 November 1659, HP 62/25/4a. On the goal of rediscovering or re-creating the Garden of Eden in the early modern period, see John Prest, *The Garden of Eden and the Re-Creation of Paradise* (New Haven, Conn.: Yale Univ. Press, 1981).

[111] Beale to Hartlib, 8 May 1658, HP 52/43b; see also Beale to Hartlib, 22 May 1658, HP 52/66a–66b.

[112] See Beale to Henry Oldenburg, 4 and 31 January 1662/3, in *The Correspondence of Henry Oldenburg*, ed. A. Rupert Hall and Marie Boas Hall, 9 vols. (Madison: Univ. of Wisconsin Press, 1965–75), 2:3–5, 16–20.

geographical contexts, especially the increasingly populous Atlantic empire—which now figured much more positively in Beale's thinking about improvement. Beale encouraged the setting up of nurseries in Ireland (a conquered country in the midst of a new land settlement), discussed the effects of imported crops on diet and health, and pondered the technical demands and providential significance of the transplantation of plant species.[113] But in a 1668 letter he also linked these concerns to the size and the quality of colonial populations. This was a longstanding concern of colonial promoters and planters. Early writing on colonial plantation in Virginia and New England dwelt on the suitability of the American soil, water, air, native crops, and climate for English settlement.[114] Domestic critics and traveling observers, meanwhile, complained of the insalubrious colonial environment's degenerative physical and moral effects—a theme that had also loomed large in late Elizabethan and Jacobean writing on English settlement and plantation in Ireland.[115] Whether change operated directly on English bodies through climate or indirectly via adaptations of diet and agricultural practices or interaction with indigenous people, colonial settlement was understood to be a perilous and physically transformative process. Even before political arithmeticians sought to measure it, navigating such change was integral to managing population on an imperial scale.

Beale began by remarking that New England, "though ye poorest, & in ye Coldest Climate, & on ye most barren Lands, yet is ye Granary, & affords ye English Diet to all our American Plantations," whose total population "cannot be so few as a Million of People." Such was the power of improvement to fit the environment to the dietary needs of English bodies. New England's agriculture, "regular Government," and "Fear of God" all stood in contrast, however, to the southern and Caribbean colonies. There, "vast Numbers of ye English are become as wild as ye savages; They destroy all accommodations wherever they come, & so remove from place to place as disorderly as ye wild Tartars." Part of the solution to this cycle of failure to improve, physical transience, and moral decay lay in "ye Transplantation of Spices, & ye Plantations of Gardens & Orchards, Vineyards & Groves of Olives, & other usefull Plants." Cultivation

[113] Beale to Oldenburg, 1 April 1664, in ibid., 2:151–61. Beale reproduces verbatim (without citation) substantial parts of an earlier "Phytologicall Letter" currently attributed to Petty, copies of which are in British Library, Additional MS 4292, fols. 141r–142v, and HP 8/22/1a–4b. See Tony Aspromourgos, "The Mind of the Oeconomist: An Overview of the 'Petty Papers' Archive," *History of Economic Ideas* 9 (2001): 39–102, on 84–5; and Aspromourgos, "The Invention of the Concept of Social Surplus: Petty in the Hartlib Circle," *European Journal of the History of Economic Thought* 12 (2005):1–24. Given Beale's use of the "Phytologicall Letter" and the presence of related material in his correspondence, Petty's authorship appears doubtful. See Beale to [Hartlib], undated, HP 25/6/1a–4b; and Oldenburg, "Extract of Mr. Beale of ye transplanting of spices," in Hall and Hall, *Correspondence of Henry Oldenburg* (cit. n. 112), 2:161–2.

[114] See, for example, Thomas Harriot, *A Briefe and True Report of the New Found Land of Virginia of the Commodities and of the Nature and Manners of the Naturall Inhabitants* (Frankfurt am Main: Typis Ioannis Wecheli, 1590), 13–21, 31–3; William Wood, *Nevv Englands Prospect* (London: printed by Tho. Cotes for Iohn Bellamie, 1634), 3–14. On climate and colonization schemes, see Karen Ordahl Kupperman, "The Puzzle of the American Climate in the Early Colonial Period," *Amer. Hist. Rev.* 87 (1982): 1262–89; and Anya Zilberstein, *A Temperate Empire: Making Climate Change in Early America* (Oxford: Oxford Univ. Press, 2016), 19–52.

[115] On degeneration in Ireland, see [William Gerrard], "Lord Chancellor Gerrard's Notes of His Report on Ireland," *Analecta Hibernica* 2 (1931): 93–291; Edmund Spenser, *A View of the State of Ireland*, ed. Andrew Hadfield and Willy Maley (1596; repr., Oxford: Blackwell, 1997), 54–84; and Nicholas Canny, *Making Ireland British, 1580–1650* (Oxford: Oxford Univ. Press, 2000), 59–120. On degeneration and the early American colonies, see Chaplin, *Subject Matter* (cit. n. 3), 116–56.

of both salable commodities and foodstuffs "would fixe & setle [the colonists] in convenient habitations," and ultimately foster "a numerous people . . . born & bred to agree with ye Air, & soyle, & too strong to be supplanted by their Enemyes."[116] While Beale's appeal to the civilizing effect of plantation resembled Tudor arguments about Ireland (and Jacobean hopes for the draining of the Fens), his vision of colonists taming a savage continent with farms and towns anticipated the projections of Benjamin Franklin and Ezra Stiles.[117] The application of art to the production of food and, by extension, properly English settler populations, was central to both.

CONCLUSION: AGRICULTURAL IMPROVEMENT AND THE IDEA OF POPULATION

Hartlibian projects connected food and population in several ways. New tools, fertilizers, crops, or rotations might mean cheaper, more abundant, longer-lasting, or healthier food, and hence the capacity to sustain greater numbers on the same metropolitan territory. Or, innovation might drive overseas colonization, redistributing population between the metropole and its American colonies, with consequent effects for trade and fisheries. Innovations in milling might alter labor requirements in England and in the Caribbean, replacing fodder crops with sugar cane, and livestock with the enslaved. Improvement might augment Protestant numbers in the context of confessional and imperial conflict, or underpin the multiplication of humankind on its progress through history from a lost paradise of God's making to a new one of its own. The common thread was the conviction that, if industry and ingenuity were applied where Providence directed, art could work upon nature to lasting demographic effect; technology could decisively reshape the material context of multiplication and exert a degree of control over the size, qualities, and prospects of human populations on every scale.

In the near term, the most significant legacy of this conviction was in the "political arithmetic" of the Restoration—the work of Petty and another Hartlibian contact, John Graunt.[118] Political arithmetic has rightly been seen as a pioneering effort of demographic quantification.[119] But a historiographical obsession with the origins and spirit of its quantitative methods has distracted attention from its purpose, which was *qualitative* transformations of engagements with the environment, and of populations, which were understood as living, breathing, eating, and multiplying objects of government. Restoration political arithmetic promised not merely to monitor or even maximize

---

[116] Beale to Oldenburg, 29 August 1668, in Hall and Hall, *Correspondence of Henry Oldenburg* (cit. n. 112), 3:28–33, on 29–30.

[117] See I. B. Smith and Thomas Smith, *A Letter Sent by I. B. Gentleman Vnto His Very Frende Mayster R. C. Esquire, Vvherin Is Conteined a Large Discourse of the Peopling & Inhabiting the Cuntrie Called the Ardes* [. . .] (London: printed by Henry Binneman for [Anthony Kitson], 1572); Benjamin Franklin, *The Interest of Great Britain Considered, with Regard to Her Colonies, and the Acquisitions of Canada and Guadeloupe* (London: printed for T. Becket, 1760), 17–23; and Ezra Stiles, *A Discourse on the Christian Union* [. . .] (Boston, Mass.: printed by Edes and Gill, 1761), 95–109. On fen drainage, see Ash, *Draining of the Fens* (cit. n. 7).

[118] On Graunt, see Margaret Pelling, "Far Too Many Women? John Graunt, the Sex Ratio, and the Cultural Determination of Number in Seventeenth-Century England," *Hist. J.* 59 (2016): 695–719; and Pelling, "John Graunt, the Hartlib Circle and Child Mortality in Mid-Seventeenth-Century London," *Contin. & Change* 31 (2016): 335–59. On Petty, see Ted McCormick, *William Petty and the Ambitions of Political Arithmetic* (Oxford: Oxford Univ. Press, 2009).

[119] See, most recently, William Deringer, *Calculated Values: Finance, Politics, and the Quantitative Age* (Cambridge, Mass.: Harvard Univ. Press, 2018), 1–78.

numbers but to give the state power, through projects, over those numbers' qualities (restraining metropolitan depopulation and colonial degeneration, for example), as well as their distribution within the Three Kingdoms and across the British Atlantic. The use of quantitative arguments was new, but the ambitions look little greater and often little different from those of the Hartlibian schemes of the 1640s and 1650s.

This suggests that the Hartlib Circle played an important role in the formation of a new idea of population as an object of technological intervention, an idea now more closely associated with the emergence of biopolitics in the later eighteenth century.[120] Modifications of this idea would manifest themselves in innumerable improvement schemes over the long eighteenth century, including attempts to secure cheaper, better, or more durable food sources for laboring populations, free and unfree, in Britain and elsewhere.[121] Some of the same ideas would provoke Malthus and David Ricardo to theorize afresh the perils of human multiplication—and the limits of human art—now in the face of a nature conceived as iron fisted, inflexible, and intolerant of political meddling.[122] As the projects explored above also show, however, the roots of this idea of population lie as much in the practice and promotion of agricultural improvements and the sketchy promises and speculative calculations of projectors as in the canon of early modern political thought or the history of quantification. The idea that there might be technological levers of population control was abroad in the early modern period; those levers were first tested not in the court or the study, but in the field and the mill.

---

[120] See Michel Foucault, *The Will to Knowledge: The History of Sexuality Volume 1*, trans. Robert Hurley (New York: Pantheon, 1978), 25, 137–40.

[121] See, for example, Anya Zilberstein, "Bastard Breadfruit and Other Cheap Provisions: Early Food Science for the Welfare of the Lower Orders," *Early Sci. & Med.* 21 (2016): 492–508.

[122] Wrigley, *Poverty, Progress, and Population* (cit. n. 10), 70–3.

Perceptions of Provenance:

Conceptions of Wine, Health, and Place in Louis XIV's France

by Alissa Aron*

ABSTRACT

During the reign of Louis XIV, influences arising from the Galenic and iatrochemical medical traditions collided with changing notions of Frenchness to shape understandings of the healthfulness and quality of a wine in relation to its provenance. Specific locations were believed to impart particular qualities to the people, plants, animals, and waters that originated there. Thus, wines from a particular locale would be marked by the properties of that place, and were thought capable of transmitting their characteristics to a drinker. Some land was seen as inherently suited for producing superior wines, and proponents of wines from particular regions often grounded their arguments on the basis of the proclaimed health benefits. However, this was a period of tension between French and regional identities, and the promotion of prestigious wines from specific regions, like Champagne or Burgundy, can be interpreted as an impediment to the goal of Louis XIV and his minister Jean-Baptiste Colbert to equate Frenchness with quality and distinction.

INTRODUCTION: *TERROIR* AS A GUARANTEE OF WINE QUALITY

> A wine from Champagne is preferred to a wine from Beaune,
> and the wines from Beaune are preferred to those of Tonnerre;
> so it goes for the rest: & all of these differences [are] because
> of the salts filling the soils that these vines are planted in; [with
> some soils] having a few degrees more or less of perfection,
> giving these different qualities to the grapes.[1]

For Louis Liger, the French author of agricultural handbooks who wrote the above passage in 1700, a wine's quality was closely tied to the provenance of the grapes used to make it, a supposition that allowed him to rank viticultural regions and attribute their distinguishing features to the chemical characteristics of the soils in each location. Thus, the place of origin, and more specifically the nature of the soil itself, already guaranteed

* 3150 Richards Road, Suite 200, Bellevue, WA 98005, USA; alissa.aron@gmail.com.
  All translations are the author's.

[1] Louis Liger, *Oeconomie générale de la campagne, ou nouvelle maison rustique*, vol. 2 (Paris: C. de Cercy, 1700), 332, http://gallica.bnf.fr/ark:/12148/bpt6k6546310s/f283.vertical.r = OEconomie %20générale%20de%20la%20campagne.

the characteristics and the quality one could expect in a wine well before this concept was formalized with the introduction of the *appellation contrôlée* system in the early twentieth century; that moment is often pinpointed by modern social scientists as the dawn of the modern conception of *terroir*.[2] This French loanword, attributing a wine's aroma, flavor, and mouthfeel to particular characteristics of the environment the grapes were grown in, has become ubiquitous in English wine parlance, but has also been hotly contested as wine peddlers and writers ceaselessly accuse each other of misusing, over-using, and misunderstanding the term.[3] To some, *terroir* simply encompasses the features, such as soil and climate, that influence the growth and nature of grapes at that locale. To others, *terroir* is the actual character that is conferred from these features to the grapes and the wine made from them. Compounding the confusion, the exact mechanism by which *terroir* is conferred to a wine remains mysterious in scientific circles specializing in viticulture, geology, and chemistry. Studies in the fields of viticulture and oenology have investigated the role of such diverse factors as grape variety, soil, microbes, topography, climate, and landscape characteristics in determining the final distinctive aroma and flavor profile of grapes and wines, with each found to have significant effects. Taken together, such research suggests that *terroir* results from a complex blend of all such factors, yet many winemakers and wine enthusiasts seek to simplify the concept into a straightforward cause and effect relationship. Others have a tendency to employ the term *terroir* to refer exclusively to the effects of a vineyard's soil type. However, the official definition of *terroir*, as published by the International Organization of Vine and Wine, encompasses all of these elements as well as local cultivation and wine-making practices, and emphasizes the interplay between physical, biological, and cultural aspects involved in growing grapes and converting them into wine.[4]

Despite the muddy picture regarding what *terroir* is and how it is conferred to wines, the concept that wines from different places have distinct and reproducible characteristics is codified in the world's appellation systems; these classify wines according to their specific region of origin, requiring the wines of each appellation to conform to particular standards of production, aroma, and taste. However, as Liger's example suggests, French writers consistently classed wines by their provenance well before the birth of the appellation system. Indeed, the first edition of the *Dictionnaire de l'Académie Française*—published in 1694—defined *terroir* both as agricultural land and as the taste or aroma of a wine resulting from that land's particular characteristics.[5]

[2] See Marion Demossier, "Beyond *Terroir*: Territorial Construction, Hegemonic Discourses, and French Wine Culture," *Journal of the Royal Anthropological Institute* 17 (2011): 685–705, on 689; Amy Trubeck, *The Taste of Place: A Cultural Journey into Terroir* (Berkeley and Los Angeles: Univ. of California Press, 2008), 21–2; and Robert Ulin, "Terroir and Locality: An Anthropological Perspective," in *Wine and Culture: Vineyard to Glass*, ed. Rachel Black and Ulin (London: Bloomsbury, 2013), 67–84.

[3] See Eric Asimov, "Terroir," *Diner's Journal: The New York Times Blog on Dining Out*, 10 May 2007, https://dinersjournal.blogs.nytimes.com/2007/05/10/terroir/?searchResultPosition = 1; Dave McIntyre, "Studying Wine from the Ground up," *Washington Post*, 5 November 2012; and James E. Wilson, *Terroir: The Role of Geology, Climate, and Culture in the Making of French Wines* (London: Wine Appreciation Guild, 1998), 55.

[4] Sarah Daynes, "The Social Life of Terroir among Bordeaux Winemakers," in Black and Ulin, *Wine and Culture* (cit. n. 2), 15–32, on 25; International Organization of Vine and Wine (OIV), "Definition of Vitivinicultural 'Terroir,'" Resolution OIV/VITI 333/2010 (2010), http://oiv.int/public/medias/379 /viti-2010-1-en.pdf; Wilson, *Terroir* (cit. n. 3), 55.

[5] "Terroir," *Dictionnaire de l'Académie françoise, dedié au Roy* (Paris: Vve J. B. Coignard et J. B. Coignard, 1694), 522–3, http://gallica.bnf.fr/ark:/12148/bpt6k50398c/f553.vertical.r = Dictionnaire %20de%20l'Académie%20françoise.

Thus, it is clear that by the seventeenth century, geographic origin was already an important factor in the equation that determined the quality and traits of a wine.

However, throughout much of the sixteenth and seventeenth centuries, the term *terroir* carried strongly negative connotations. Thomas Parker has thoroughly explored the intellectual history of the term itself, arguing that complex social, philosophical, and political reasons underlie the attribution of negative connotations to this term around the time of Louis XIV's reign. Parker draws upon Chandra Mukerji's exploration of the gardens of Versailles, which aims to understand why the concept of territoriality became a material practice in the seventeenth century. Mukerji argues that at the time, territoriality went beyond a philosophical conception of how to feel about a particular zone, and came to define how one acted upon the land itself, as evidenced by the earth-moving display at Versailles.[6] For Mukerji, the principal driver of this shift in the understanding and actualization of territoriality was the political and cultural instability of the noble classes, whose status was tied to the land and land ownership. The expansion of trade had opened up unprecedented possibilities for social mobility, threatening to decouple the aristocracy from its holdings. The monarchy's stake in this is clear in that Louis XIV used his display at Versailles to aid consolidation of the power of the crown against any development of an alternative political base. The construction of the gardens at Versailles served to demonstrate his near-divine authority; the architects implemented changes to the earth that had previously been solely under the purview of almighty powers.[7] For Parker, then, the rise of Louis XIV and the Parisian court engendered a shunning of all things provincial among the French elites, including any products with a residual taste or smell of the earth.[8] By extension, artifice was elevated over nature in the eyes of the aristocracy, causing products characterized by their terroir to take on a pejorative status.

Because the negative connotation of *terroir* is at odds with its desirability in modern parlance, other authors have chosen to begin their analysis at a later historical moment. Amy Trubek, for example, attributes the origins of the modern concept of *terroir* to early twentieth-century "tastemakers," such as journalists, cookbook authors, chefs, and taste producers (cheesemakers, winemakers, bakers, and cooks), who shaped how people experienced food and wine. In this later period, Trubek asserts, the term *terroir* took on the multifaceted signification that it holds today as "a sensibility, a mode of discernment, a philosophy of practice, and an analytic category."[9] Trubek's analysis traces the history of the modern concept of *terroir*, then, while Parker's work follows the history of the term itself. Trubek's work, however, shows the dangers of anachronism associated with the use of the term *terroir*, and therefore use of the term in this article will be limited.

In this article, however, I am concerned primarily with one particular facet of *terroir*, whose roots reach back to the period when the term held principally negative connotations—that of geographical provenance and how it related to the understanding and differentiation of wine quality during the reign of Louis XIV. At that time, concerns about how locality influenced wine quality provide a window into the entanglements

---

[6] Chandra Mukerji, *Territorial Ambitions and the Gardens of Versailles* (Cambridge: Cambridge Univ. Press, 1997), 8–9.

[7] Ibid., 18–20.

[8] Thomas Parker, *Tasting French Terroir: The History of an Idea* (Oakland: Univ. of California Press, 2015), 54, 78–83.

[9] Amy Trubek, *The Taste of Place* (cit. n. 2), 21–2.

between geographic, climatic, medical, and gustatory knowledge, and the resistance of monarchical power. I aim to understand why provenance would have been associated with a wine's characteristics in the seventeenth century at all. The answer is multifaceted and related, in the first instance, to contemporary understandings of the relationship between the earth, the plants that subsisted on it, and the human body. We must therefore look to the predominating medical theories of the time, and their debts to the work of Galen and Paracelsus. Second, while still two centuries away from the rise of the modern concept of nationalism in France, the reign of Louis XIV (1643–1715) foreshadowed this shift with pronounced changes in the conception of what it meant to be "French," creating an identity crisis with social, political, and economic implications.[10] This shift was related to the economic program of financial minister Jean-Baptiste Colbert, and was designed to increase the monarchy's wealth in order to finance Louis XIV's enthusiasm for war. A key aspect of what came to be known as his mercantilist policy was the unification of the French economy by eliminating interior customs controls and private tolls on transport routes within France. Additionally, Colbert aimed to improve the quality of French products, increasing their appeal for both export and domestic markets.[11] When this emphasis on building a united French economy is considered alongside the drive to improve the quality of French products, it appears to be no accident that these policies also coincided with a reevaluation of the relationship between provenance and wine quality. Mukerji has suggested that in striving to unify the extremely diverse French state, Colbert's policies instituted "a way of acting on the land that helped to make it seem like France."[12] For this reason, wines—seen as reflections of the land from which they came—would become divided between those seen as "French," such as wines of quotidian consumption in Paris, and the more prestigious wines that retained their regional or local designations. In this article, I will first use the lens of contemporary medical traditions to explore how the French understood the transmutation of land into wine. I will then analyze the evolving relationship between wine and "Frenchness" in the seventeenth century. Throughout the article, I will emphasize that medical, chemical, social, and economic ideas about land and taste were inextricably linked, and shaped contemporary knowledge about the relationship between a wine's provenance and its virtues.

KNOWLEDGE ABOUT SOIL, PLANTS, AND HEALTH IN LOUIS XIV'S FRANCE

Medical treatises and cookbooks published during the reign of Louis XIV described wine quality in terms that blended salubrity and gustatory appeal. Nicholas Lémery, the French chemist and member of the *Académie Royale des Sciences*, explained that "the goodness of a wine . . . consists in a certain proportion and natural liaison of its components, which make a pleasant impression on the tongue's nerve, & which accelerate

---

[10] Chandra Mukerji has argued that the construction of the Canal du Midi—in the 1660s through 1680s—inadvertently became a precursor to French national identity by uniting political territoriality, large-scale engineering, and identity; see Mukerji, "The New Rome: Infrastructure and National Identity on the Canal du Midi," *Osiris* 24 (2009): 15–31. In her work on Versailles, Mukerji has demonstrated how Louis XIV used the ability to control nature as a tactic to consolidate power by making the countryside feel like "France"; *Territorial Ambitions* (cit. n. 6). Hans Kohn posits that the modern centralized state as implemented by French monarchs was a prerequisite for nationalism; see Kohn, "The Nature of Nationalism," *American Political Science Review* 33 (1939): 1001–21, on 1002.

[11] Philippe Minard, *La fortune du Colbertisme: État et industrie dans la France des Lumières* (Paris: Fayard, 1998).

[12] Mukerji, *Territorial Ambitions* (cit. n. 6), 8–9.

the movement of the animal spirits, delight the stomach, the heart & the brain."[13] Lémery numbered among the chemical philosophers who, inspired by the work of Paracelsus, subscribed to a chemical understanding of the universe, where substances were composed of a trinity of elements, including salt, sulfur (or oil), and mercury (or liquor), in addition to, or in place of, the four humors of Galenic medicine (blood, phlegm, black bile, and yellow bile).[14] Chemical philosophers who followed this philosophy, known as iatrochemistry, varied in the degree to which they integrated their ideas with those of Galenic medicine, where all substances, including soils, plants, nonhuman animals, and humans, were understood to be characterized by the four humors.[15] The above passage demonstrates that Lémery drew upon both of these medical philosophies, because the "components" he referred to were an amalgam of the humors of classical medicine and the iatrochemical elements.

Iatrochemical ideas were overlain onto Galenic principles in interpreting the relationship between locality and health. As Andrew Wear has explained, Galenic medical philosophy relied on the view, derived from the Hippocratic treatise *Airs, Waters, Places*, that particular locales could "shape both the physical and mental nature of people" directly and also through the consumption of the plants, animals, and waters that came from them.[16] Within this philosophical framework, then, the geographical origins of medicines and other consumables influenced their healing properties. In general, people believed that products that came from a similar environment, or "climate," that an individual came from would be appropriate and healthful for that individual's constitution. For example, products from hot regions such as Arabia were thought to cause overheating in Europeans by disrupting the equilibrium of their bodies, which were characterized by a relatively cool composition. Others argued that by counteracting the properties of a person's native climate, foreign materials might also be able to restore humoral equilibrium in diseased individuals. By the eighteenth century, debate on the merits of exotic substances had become commonplace.[17] However, before this time, many French medical men had already been arguing in support of the salubrity of local medicines. Writing in 1712, the botanist Pierre-Jean-Baptiste Chomel defended the medical merits of indigenous plants, asking whether it is "so unreasonable to believe that the plants from our climate are more suited to our temperaments than those that are born, so to speak, under another sun?"[18] Soil was considered to be the liaison

---

[13] Nicolas Lémery, *Traité universel des drogues simples* (Paris: Chez Laurent d'Houry, 1699), 810, accessed 17 December 2015, http://www.biusante.parisdescartes.fr/histoire/medica/resultats/index.php?p=830&cote=20798&do=page.

[14] Allen Debus, *The Chemical Philosophy: Paracelsian Science and Medicine in the Sixteenth and Seventeenth Centuries*, vol. 1 (New York, N.Y.: Science History Publications, 1977), 78–84.

[15] Allen Debus, "The Chemical Philosophers: Chemical Medicine from Paracelsus to van Helmont," chap. 3 in *Chemistry, Alchemy and the New Philosophy, 1550–1700* (London: Variorum Reprints, 1987), 235–59, on 238–41.

[16] Andrew Wear, "Place, Health, and Disease: The Airs, Waters, Places Tradition in Early Modern England and North America," *J. Medieval Early Mod. Stud.* 38 (2008): 443–65, on 444.

[17] Alix Cooper, *Inventing the Indigenous: Local Knowledge and Natural History in Early Modern Europe* (Cambridge: Cambridge Univ. Press, 2007), 29, 39; E. C. Spary, "'Peaches Which the Patriarchs Lacked': Natural History, Natural Resources, and the Natural Economy in France," *Hist. Polit. Econ.* 35 (2003): 14–41, on 22.

[18] Pierre-Jean-Baptiste Chomel, *Abregé de l'histoire des plantes usuelles. Dans lequel on donne leurs noms differens, françois & latins. La maniere de s'en servir, la dose, & les principales compositions de pharmacie, dans lesquelles elles sont employées. Avec quelques observations de leurs usages* (Paris: Chez Charles Osmont, 1712), "Discours Préliminaire," https://gallica.bnf.fr/ark:/12148/bpt6k65569549/f36.item.r=bresil.zoom.

between earth and plants, such that plants would be imbued with the qualities of the soil they grew in, and these properties would, in turn, be transferred to anyone who consumed those plants.[19]

Iatrochemistry or chemical philosophy provided a complementary explanation with a theory for how particular properties were transferred from soil to plant. In this view, soil, like all substances, was understood to be composed of varying proportions of the "chemical principles," including salt, sulfur, and mercury, which plants absorbed through their roots. Through the processes of "a diversity of fermentations and other natural elaborations," Lemery believed the soil's elemental composition would be transferred into the plant, thereby influencing its vegetal composition.[20] From a perspective that combined tenets of both Galenic and iatrochemical approaches, it thus followed that wine types "vary by the different natures of the grapes from which they are made; by the different climates in which they grew; according to whether they received more or less heat from the Sun; by their odors, their consistency, their taste, their virtues."[21] Other authors also demonstrated the combined influence of both philosophies. Louis Liger, in his *Œconomie génerale de la Campagne*, identified the soil as the most important trait of a vineyard, and essential to determining the quality of a wine. Secondary factors included slope orientation, grape variety, pruning, and cultivation methods. He described many soil types, including a particularly virtuous yellow-brown soil dotted with small white stones:

> Anyone with vines on such soils is always assured of finding a merchant for his wine . . . The grape grown on such soils always ripens to perfection; because the salt that it accumulates & the substance that it sucks from those salts are of such great character to make the wine delicious, that however cold the year that the grape must endure, the wine that it produces is always of an admirable taste; I mean as long as these soils are situated on slopes with good exposure to the sun. Such are the slopes from which come the great wines of Beaune, Chablis, Tonnerre, Auxerre, Collanges & Champagne.[22]

Liger emphasized that the chemical composition of the soil can result in the differentiation between standard and "admirable" wines, and is thus responsible, in conjunction with the orientation of vineyard slopes with respect to the sun, for the reputation of the greatest wine regions. Despite recognizing the connection between soil and wine quality, the mechanism of the influence remained mysterious to Liger, who wrote that "this yellowish soil is naturally filled with a substance that has a certain *je ne sais quoi* that makes for a better vine."[23] It is important here to note that Liger did not use the term *terroir* to describe this characteristic of the soil, as a modern writer might do, because even though he acknowledged the influence of the soil, he did not necessarily intend to praise wines that *tasted like* the soil.[24]

The intangible characteristic to which Liger referred contributed to variability among plots of agricultural land. His contemporaries conceived of such differences as a direct result of variations in the land's inherent qualities, with certain places divinely endowed with greater fertility or other beneficial traits. These distinctions had important social

[19] Wear, "Place, Health, and Disease" (cit. n. 16), 459–61.
[20] Lémery, *Traité universel* (cit. n. 13), 11.
[21] Ibid., 808.
[22] Liger, *Oeconomie générale de la campagne* (cit. n. 1), 272–3.
[23] Ibid., 272.
[24] For more on the connotations of the word terroir in the seventeenth and eighteenth centuries, see Parker, *Tasting French Terroir* (cit. n. 8), 93–113.

and economic implications, because at this time, regional variability was seen as necessary in order to extract wealth and value from the land and its products.[25] This vision of a patchwork of resources was at the forefront of Colbert's economic policies, as he strove to make the best possible use of regional differences in production quality to improve the overall portfolio of French products in the marketplace.[26] Regional heterogeneity formed the basis for the contrast between generic wines—produced in large quantities and sold mainly in the taverns and cabarets of Paris—and high-quality wines, which were grown in specific, well-reputed areas such as Champagne and Burgundy, and that were available only to the wealthy.[27] However, it was not only political borders that were employed to separate wines from these regions from all the rest. Inherent differences in land quality were also used to argue the relative merits of the wines produced in the well-known provinces of Champagne and Burgundy.

WINE AND CHANGING IDEAS ABOUT FRENCHNESS

In addition to a medical and chemical context that influenced the understanding of how a wine's provenance and virtues were linked, Louis XIV's reign provided a unique social and political environment that would shape French understandings of wine's relationship to spatiality. The economic unification of France was an important goal under Colbert's mercantilist policies, and the minister worked to eliminate internal tariffs and tolls so that production from all parts of France might be rallied for the economic benefit of the entire state. Thus, his vision was that the concept of the "local" be supplanted by that of the "French," as can also be seen in his reorganization of the tax collection system to provide the monarchy with greater centralized fiscal control.[28] However, this mercantilist dream was not unanimously shared by the inhabitants of France, who, in the prerevolutionary period, did not possess a defined concept of national identity. Even as late as 1764, Voltaire expressed this public reticence toward the national in his *Dictionnaire Philosophique*: "The larger this fatherland becomes, the less we love it, because shared love weakens. It is impossible to tenderly love a family that is too numerous, that we barely know."[29] This complicated Colbert's goals of consolidation, because a widespread appreciation for "Frenchness" had yet to come into being.

---

[25] Simon Schaffer has argued that in eighteenth-century Britain, the characterization of a soil's fertility was used to create social order, because practices of cataloguing and taxonomizing soil characteristics were closely tied to normative directives dictating how such land should be used. The bounty of the earth must be managed, as must social relations, in order to ensure a productive society. Similarly, in her analysis of exotic medicinal substances in colonial France around the time of Louis XIV's reign, E. C. Spary similarly argues that a nation's acquisition of a diversity of colonial lands, from which derived a diversity of botanical resources, was presented as a form of wealth acquisition. Therefore, classification of resources derived from the land became inextricably tied to ambitions for economic success. See Schaffer, "The Earth's Fertility as a Social Fact in Early Modern England," in *Nature and Society in Historical Context*, ed. Mikuláš Teich, Roy Porter, and Bo Gustafsson (Cambridge: Cambridge Univ. Press, 1997), 124–47, on 128–9; and Spary, "'Peaches Which the Patriarchs Lacked'" (cit. n. 17), 22.

[26] Minard, *La fortune du Colbertisme* (cit. n. 11), 18–20.

[27] Marcel Lachiver, *Vins, vignes et vignerons: Histoire du vignoble français* (Paris: Fayard, 1988), 275.

[28] James B. Collins, *The State in Early Modern France*, 2nd ed. (1995; repr., Cambridge: Cambridge Univ. Press, 2009), 227–38.

[29] Voltaire, *Dictionnaire Philosophique* (1754; repr., Paris: Imprimerie de Cosse et Gaultier-Laguionie, 1838).

France's viticultural landscape was as diverse as the state itself. In an era when accepted medical thought suggested that the health effects of wines depended on "the diverse climates from which they come," and that an individual should "correctly utilize those from his Country," how was one to determine the scope of a person's "Country"?[30] Should the climate factor be interpreted on a local, regional, or national scale? Historian Alix Cooper has described a similar tension for early modern naturalists wishing to classify plants as "indigenous" or "domestic," because they had to choose the scale on which to define these terms, particularly in the context of shifting borders brought on by colonialism.[31] In the provinces, for all but the wealthiest families, the question of locality was unaffected by Colbert's policies, because they drank wine grown on their own or neighboring land, if they drank wine at all (leading up to the Revolution, most peasants drank only water).[32] Families who owned vineyards would drink wine that they were unable to sell, usually in diluted form as *piquette*; this was a fermented beverage made from adding water to the pomace remaining after draining the wine from the fermentation vats.[33] Servants and many rural workers also only had access to this diluted beverage. Only wealthy families could afford to buy wines from outside their region, because transport costs would have continued to make wines from farther afield inaccessible for middle- and lower-class families in the provinces.[34] These wines may have come from regions that were gaining more prestigious reputations, such as Burgundy, Champagne, or Bordeaux.

In Paris, the effects of Colbert's mercantilist policies were more evident, as wine served in drinking establishments was not generally identified by its region of origin, but simply as "French." Thanks to policies freeing up internal trade routes, wines were shipped to Paris from across the country, even though the region around the city produced large quantities of wine at this time.[35] The legal and fiscal organization of Parisian drinking establishments affected contemporary conceptions of the wines served in them. High taxes were levied on wine sold in taverns and cabarets within the city limits of Paris, especially from 1675 to 1680; these spurred the installation of a new type of casual drinking establishment, the *guinguette*, on the outskirts of Paris where such tariffs could be avoided.[36] Another key economic factor accounting for this sudden blossoming of the guinguette as an institution was a law prohibiting Parisian wine merchants from selling wines produced within twenty leagues of the city. This law further increased prices of wines sold inside the city limits due to the elevated transportation costs required to bring in wines from beyond this embargo zone. In this way, Parisians were

---

[30] Jean-Baptiste de Salins, *Defense du vin de Bourgogne contre le vin de Champagne, par la refutation de ce qui a été avancé par l'auteur de la these soûtenuë aux Ecole de Médecine de Reims le cinq de May 1700. dans la cinquiéme partie ou corrollaire que l'on raporte ici tout entier* (Paris: Parisiss, 1702), 11, http://gallica.bnf.fr/ark:/12148/bpt6k10251456/f4.vertical.

[31] Cooper, *Inventing the Indigenous* (cit. n. 17), 14.

[32] Daniel Roche, "Le temps de l'eau rare du Moyen Âge à l'époque moderne," *Annales: Economies, Sociétés, Civilisations* 39 (1984): 383–99, on 395; Jean-Charles Sournia, *Histoire de l'alcoolisme* (Paris: Flammarion, 1986); trans. Nick Hindley and Gareth Stanton as *A History of Alcoholism* (Oxford: Basil Blackwell, 1990), 14.

[33] Liger, *Oeconomie générale de la campagne* (cit. n. 1), 337.

[34] Sournia, *History of Alcoholism* (cit. n. 32), 14.

[35] Thomas Brennan, *Public Drinking and Popular Culture in Eighteenth-Century Paris* (Princeton, N.J.: Princeton Univ. Press, 1988), 77.

[36] Roger Dion, *Histoire de la vigne et du vin en France: Des origines au XIXe siècle* (1959; repr., Paris: CNRS Editions, 2010), 505.

forced to expand their definition of local wines—unless they frequented the guinguettes just beyond the city limits—because they only had access to wines produced beyond a certain distance from their home "climate." This economic prodding pushed Parisians toward consuming "French" products rather than "Parisian" ones, and thus may have encouraged them to embody this French identity themselves. But the increasing prestige of wines from certain regions such as Champagne and Burgundy, which were lauded for their unique merits integrating healthiness and gustatory appeal resulting from the perceived quality of their soils and climates, represented an obstacle to Colbert's attempts to define a concept of a French identity economically.

TYING HEALTH TO THE LAND: THE CHAMPAGNE-BURGUNDY DEBATES

Medical practitioners participated in debates over the relative primacy of Champagne and Burgundy wines, beginning in the middle of the seventeenth century. The arguments employed in these debates illuminate the interconnectedness between knowledge about place and knowledge about health. The disputes began with the publication of Daniel Arbinet's thesis written at the Faculty of Medicine in Paris in 1652. A Burgundian himself, Arbinet devoted his medical studies to demonstrating that the wines of Burgundy were more delicious and healthful than any other wines. His work launched a series of rebuttals that lasted through the end of Louis XIV's reign, with physicians alternately showing support for the wines of Champagne and the wines of Burgundy.[37] By 1693, the debate had filtered all the way to the king's court, where Louis XIV's physician Fagon forbade the ailing king from consuming Champagne, blaming his fevers and gout on his heretofore preferred beverage. He instead advised the king to consume exclusively Burgundian wines.

Throughout the debate, proponents on both sides argued that their favored wine was the healthier choice, emphasizing the close relationship between the quality of the land and the traits of its wine. The terms of this debate clearly demonstrate the underlying motives of both parties to defend (and define) their respective provincial identities—represented by the unique characteristics of the earth and climate in each region—against each other and also against the growing power of the metropolis. In 1702, the Burgundian Jean Baptiste de Salins published his rebuttal to a thesis published two years earlier by the Faculty of Medicine at Reims University that had defended the wines of Champagne. In his pamphlet, Salins drew heavily upon arguments made in Arbinet's dissertation from half a century earlier to support his claim that the wines of Burgundy were both more appealing and more healthful than those of Champagne. Invoking the properties of the soil, the direction of the vineyard slope, and the climate, Salins accounted for Burgundy's propensity to produce outstanding wine:

> Not every soil gives every type of fruit, grain grows better in the strongest and most solid soils, & the lightest soils are the best suited for good wines. The *bonté* (quality) of a wine is not only due to the proper cultivation of the vines, but principally to the good nature of the air and the soil of the Country: the wines of the *climats* (wine-producing areas) of Beaune are located in a foundation of light soil; they are situated on hillsides dear to

---

[37] Henry Vizetelly, *A History of Champagne with Notes on the Other Sparkling Wines of France* (New York, N.Y.: Scribner & Welford, 1862), 47–56, https://archive.org/details/historyofchampag 00vize.

the god Bacchus: this foundation is neither too dry nor too humid, it receives rain from time to time, & the influences of the sky are favorable for it; the vineyards do not face the setting sun, most of them face the rising & midday sun, & are only at the forty-sixth degree of latitude between the equator and the Arctic pole.[38]

Salins made the connection between wine quality and salubrity explicit, asserting that "the Wine of Beaune, just as it is the most pleasant, is also the healthiest of all beverages."[39] Thus, in this text that is as much a medical document as a piece of publicity for Burgundian wines, Salins's argument draws together health, quality, and the nature of the vineyard's particular *climat*. This last is a term that in the seventeenth century carried strong spatial connotations and is perhaps better translated as "region" than as today's concept of "climate."

Another reading of this medical debate reveals that beyond the relative merits of the wines of Burgundy and Champagne, the framework of analysis itself was also up for debate. The Reims thesis from 1700 that supported Champagne wines drew on arguments with strong Galenic underpinnings, as seen in the following analogy between a healthy wine and a healthy humoral body, a state achieved when their respective components were in equilibrium:

> Just as the proper constitution of the blood & the humors, results particularly from the harmony of their qualities, among which there should not be any that predominate, since wine has a strong relationship with the blood, so it can be maintained that it is never more pleasant nor more healthy than when there is union among its principles.[40]

By contrast, Salins defended the superiority of wines from Burgundy by emphasizing differences in their relative composition of phlegm, salts, spirits, and tartrate, which were included in the expanded set of iatrochemical elements.[41] This particular debate thus clearly illustrates the interplay among ideas about wine, medicine, and place. The interaction was so strong that arguments about the merits of substances from different places were inextricable from contemporary debate about the underlying medical philosophies used to interpret them. Thus, a Galenic framework might be used to support a vastly different conclusion from an iatrochemical interpretation, such that each of these medical philosophies was used to support an opposing claim.

By the following decade, the debate had escaped beyond the confines of the medical academy, with the publication of a pair of antagonistic Latin panegyric poems by Parisian Professor Bénigne Grenan and writer Charles Coffin. Each of the opposing authors supported wines from their preferred region on the grounds of healthfulness, taste, and the merit of the agricultural lands from which they issued. Emphasizing the health benefits of Burgundy wines, Grenan lauded: "From you flows a precious juice, / Gentle on the nerves, on the head, friend of the chest, / And, especially rare marvel in Medicine / Remedy both sure, & delicious."[42] Also arguing in medical terms, Coffin

---

[38] Salins, *Defense du vin* (cit. n. 30), 8.
[39] Ibid., 7.
[40] "These de Reims," cited in Salins, *Defense du vin* (cit. n. 30), 4.
[41] Salins, *Defense du vin* (cit. n. 30), 13.
[42] Bénigne Grenan and Charles Coffin wrote, respectively, *Vinum Burgundum ode* and *Campania Vindicata, sive laus vini Remensis*. These were translated by Bernard de La Monnoye as *Eloges des vins de Bourgogne et de Champagne ou deux odes latines L'une pour le vin de Bourgogne, l'autre pour le vin de Champagne* (Paris: J. Estienne, 1712), 9, http://gallica.bnf.fr/ark:/12148/bpt6k10251241/f4.vertical.

proclaimed that the wines of Champagne were gentle on the head and stomach, and promoted healthy kidney function.[43] The two authors treated taste and aesthetic appeal of their favored wines with equal importance, praising both qualities vehemently. In equating taste and beauty, Coffin went as far as to fully interweave descriptions of the clarity, color, and flavor of the wine. He added that the wines of Champagne should be respected for the way they are the "faithful image of the mores of the climate [of Champagne]," in a direct appeal to provenance as the underlying cause of the wines' greatness.[44] In these plaudits, health, taste, and provenance were not viewed as discrete attributes of a wine, but as working in concert to construct a wine's merit. The publication of the French version of Coffin's poems in the exclusive monthly periodical *Mercure Galant* suggests that such a conception of wine was promoted among the wealthy.[45]

Involving both the wines themselves and the medical theories employed in understanding their health properties, the contest between the wines of Champagne and Burgundy ultimately enhanced the prestige of wines from both regions, as evidenced by their increasing prices over the same period.[46] The reputations of these regions for producing high quality, luxury wines proved to be long lasting, as both were listed among the greatest French wines in the physician Alexis-François Aulagnier's *Dictionnaire des aliments* (Dictionary of foods) published in 1859,[47] and they remain among the most distinguished wine-producing areas in the world today. The historian Marcel Lachiver has described the Champagne-Burgundy debates as propaganda, which begs the reader to consider what might be the limits of the medical thesis as a historical source. The use of commercial competition as a tactic to elevate the esteem of products was by no means unique to the wine industry in this period.[48] Encouraging competition between producers was a key strategy in Colbert's arsenal to improve the quality of French products in general, and he aimed to play regions with unequal but complementary economic strengths off one another to maximize the economic potential of the state as a whole.[49] In a similar manner, the debate—fueled by medical arguments—over the potential of the land of Champagne and Burgundy to produce high-quality wines acted as a proxy for the pursuit of the economic success of France more generally. Other authors have argued—a century after Colbert—that scientific concepts were used as the basis on which a French national identity was established,[50] but the case of the Champagne-Burgundy debates suggests that medicine and identity were linked well before the French revolution.

---

[43] Grenan and Coffin, *Eloges des vins* (cit. n. 42), 13.

[44] Ibid., 13.

[45] Charles Coffin, "La Champagne vengée, ou loüange du vin de Reims, qu'un Poëte Bourguignon a blâmé," *Mercure Galant* 1712, 67–77, https://gallica.bnf.fr/ark:/12148/bpt6k6309733h/f71.item.r =champagne.

[46] Lachiver, *Vins, vignes et vignerons* (cit. n. 27), 275.

[47] Alexis-François Aulagnier, *Dictionnaire des alimens et des boissons en usage dans les divers climats et chez les différens peuples* (Paris: Cosson, 1839), 708, http://wellcomelibrary.org/player /b21039264#?asi = 0&ai = 805&z = -0.2255%2C-0.1087%2C1%2C0.506.

[48] Lachiver, *Vins, vignes et vignerons* (cit. n. 27), 275.

[49] Minard, *La fortune du Colbertisme* (cit. n. 11), 18–20.

[50] See Carol E. Harrison, "Projections of the Revolutionary Nation: French Expeditions in the Pacific, 1791–1803," *Osiris* 24 (2009): 33–52.

THE *ORDRE DES CÔTEAUX* AND THE PRESTIGE OF CHAMPAGNE
AS A LOCALLY IDENTIFIED WINE

Beginning in the fifteenth century, wines from Champagne had begun to be termed *vins de Champagne* to differentiate them from generic French wines.[51] In the seventeenth century, the reputation of vins de Champagne began to escalate, while wines once popular in Paris, including formerly esteemed wines such as those from near Orléans, suffered a fall from grace; these moved from the tables of the court into the taverns where they were absorbed into the category of "French" wines, "mixing with the riff-raff on the borders of Paris."[52] Thus, regional labels such as Champagne were an effective cachet, elevating their product above those associated with a generic national identity.

Charles de Saint-Évremond, libertine author, French soldier, and insatiable gourmand, was one of the most important advocates of Champagne during the seventeenth century. He and his two confidants the Marquis de Bois-Dauphin and the Comte d'Olonne became known as the *Ordre des côteaux* (Order of the Slopes), because of their obsession with food and drink from specific places, including Champagnes from what they considered to be the three best hillsides of the region—Ay, Haut-Villiers, and Avenay.[53] Saint-Évremond and his pleasure-seeking companions were responsible for introducing the wines of Champagne, which Saint-Évremond had discovered during a military stint in the region, to the nobility of the Marais neighborhood in Paris. In the early seventeenth century, Champagne had a good reputation, but it was not famous or widely known.[54] Saint-Évremond's personal preference was for the wines of Ay, as he wrote in a letter to the Comte d'Olonne:

> If you ask me which I prefer of all these wines . . . I would tell you that the good wine of Ay is the most natural of all wines, the most healthy, the most purified of any odor of *terroir*, and of most exquisite pleasure for the taste of peach that is peculiar to it and, in my opinion, the most excellent of all flavors.[55]

As Thomas Parker has analyzed in great detail, at this time for Saint-Évremond, an authentic expression of a wine's origin was characterized by the *absence* of any *goût de terroir* (taste of terroir), as this term was still associated with impurity and thus to be avoided by sophisticated palates such as his.[56] Here, Saint-Évremond demonstrated the inseparability of the concepts of health, spatiality, and gustatory appeal by lauding the wines of Ay because they authentically expressed the typical style of the region; in so doing, he provided the drinker with a salutory and enjoyable experience.

Both health and authenticity were key selling points used to promote luxury products more generally in the period.[57] Moreover, by the principles of both iatrochemistry

---

[51] Lachiver, *Vins, vignes et vignerons* (cit. n. 27), 272.

[52] Ibid., 275.

[53] Florent Quellier, "Le discours sur la richesse des terroirs au XVIIe siècle et les prémiçes de la gastronomie française," *Dix-septième siècle* 254 (2012): 141–54, on 142; Claude Taittinger, *Saint-Évremond ou le bon usage des plaisirs* (Paris: Perrin, 1990), 114–15.

[54] Taittinger, *Saint-Évremond* (cit. n. 53), 111.

[55] René Ternois, ed., *Lettres de Saint-Évremond*, vol. 1 (Paris: Librairie Marcel Didier, 1967), 256–7.

[56] Parker, *Tasting French Terroir* (cit. n. 8), 96–7.

[57] Natacha Coquery, *L'Hôtel Aristocratique: Le marché du luxe à Paris au XVIIIe siècle* (Paris: Publications de la Sorbonne, 1998), 97.

and Galenic medicine, salubrity and gustatory appeal were inextricably linked, because an individual could judge the suitability of a given food or beverage to their unique temperament, as based on their taste for it.[58] Moyse Charas, an apothecary to the king, explicitly blended iatrochemistry and Galenic medicine in giving his book the title of *Pharmacopée royale Galenique et chymique* (Royal Galenic and chemical pharmacopeia). He listed taste as an important criterion in the process of "election," or the choice of appropriate medicines for a patient: "Taste is a sense that is just as useful as the rest, & even more so, because of the diversity of flavors that can be found in Mixtures, resulting from the diverse nature of Salts that make up the composition of their substances, & because the flavors are relatively easy to distinguish and describe."[59] So it is natural that the most celebrated wines of the period, those from Champagne and from Burgundy, were exalted on the grounds of being both healthful and delicious.

The involvement of Saint-Évremond and his libertine cohort in the discussion about wine and locality sheds light on the social importance of these issues, which were as contentious as public opinion about the libertines themselves. The Order of the Slope's fundamental tenet was to celebrate life's pleasures in such a manner that pleasure seeking would become an intellectual act, not one of pure gluttony. For this reason, the order preferred delicate Champagne over heavier wines, which had a tendency to cut conversations short after only a few glasses. Saint-Évremond and his companions saw fine food and drink as the "means of accessing a state of grace where spirits are liberated, tongues are untied, and where one becomes drunk not upon drink but upon one's own discourse, in the hope of making others drunk upon it."[60] However, this discrimination was criticized by many, including the satirists Nicolas Boileau-Despréaux (known simply as Boileau); Roger de Rabutin, comte de Bussy (Bussy-Rabutin); and François Le Métel de Boisrobert (the Abbé de Boisrobert), as excessive choosiness.[61] Cuisine was indeed a contested subject during the reign of Louis XIV, who was infamous for his insatiable appetite. Some used critique of culinary indulgence of any sort as a means of political expression. Archbishop Fénelon satirically communicated his disgust with the unbridled military spending and sumptuous extravagance of the king in the face of famine and a starving populace.[62] But it was significant, as the French historian Florent Quellier emphasizes, that the Order of the Slopes so strongly associated what they considered ideal foodstuffs, those that fulfilled both gustative and social expectations, with a particular geographical provenance.[63] In this way the order made a political statement of its own; by championing wines from a particular region, and even a specific village, Saint-Évremond and the order associated greatness with a local—rather than French—identity, resisting the goal of Colbertism to equate quality and eminence with a centralized, homogenized idea of Frenchness.

[58] Steven Shapin, "The Tastes of Wine: Towards a Cultural History," *Rivista di Estetica* 51 (2012): 49–94, on 56–63.
[59] Moyse Charas, *Pharmacopée royale Galenique et chymique* (Paris: Chez l'Auteur [self-published], 1676), 16.
[60] Taittinger, *Saint-Évremond* (cit. n. 53), 111.
[61] Ibid., 118–20.
[62] Olivier Leplâtre, "'Un doux repas': Politique de l'alimentation chez Fénelon," *Food and History* 5 (2007): 71–93.
[63] Quellier, "Le discours sur la richesse" (cit. n. 53), 141–3.

Colbert's efforts do seem to have been more effective for lower-quality wines that flooded Parisian taverns and guinguettes under a "French" designation. But Liger's *je ne sais quoi*—the unknowable quality linking locality, taste, and the health-giving virtues of a wine—helped the finest wines escape Colbert's fiscal lasso. Though iatrochemistry and Galenic medicine provided conceptual frameworks relating the locale of a vineyard to a wine's particular traits, the mysterious connection between them eluded complete explanation. The enigmatic nature of taste was arrogated by provincial elites such as those in the order, who utilized the inscrutability of the relationship between provenance and taste to limit access to the virtues associated with gustatory pleasure. The unknowability of taste was therefore employed in the fortification of a noble identity, and to differentiate the provincial from the metropolitan. Furthermore, because the relationship between provenance and taste was not fully knowable, it was also not fully controllable, and thus the most distinguished wines remained defined by their place of origin, rather than being associated with France as a whole, in spite of Colbert's efforts to establish a "French" economic identity. This particular form of resistance, being predicated on such an unknowable and intangible feature of the land, served to undermine the effects of measurement and discipline used by the monarchy in its attempt to define and centralize the French state.

CONCLUSION: THE ELUSIVENESS OF *TERROIR*

> We praise the ordinary in wine, the color, the odor, the flavor, the consistency, the length of the finish, & finally, the Terroir where it grows; because nature has its favorite places for this production.[64]

In examining the tight-knit relationship between conceptions of land, health, prestige, and the particular characteristics of a wine in Louis XIV's France, this article has revealed an underlying, complex tension between the quality associated with specific viticultural areas and the monarchy's ambitions to establish a single French identity. As understood through a medical framework that integrated elements of both chemical philosophy and Galenic medicine, wines expressed qualities taken up by grapevines from the soil; and thus exceptional wines, those that were simultaneously healthful and delectable, must come from extraordinary land. Though a concept of "Frenchness" as a category and identity was taking root through Colbert's economic reforms, the most renowned wines, such as those from Champagne and Burgundy, continued to flaunt the names of their locales of origin, which served to maintain their elevated status above generic "French" wines. This distinction stemmed in part from the mysterious aspect of provenance that early modern medical theories were unable to fully explain, and allowed certain wines to escape encapsulation into a budding concept of national identity by instead maintaining their ties to the land.

This enigmatic character of the relationship between locale and wine persists today, as efforts to understand wine flavor through chemical, geological, and meteorological analysis consistently fail to satisfy the wine aficionado's impression that some tangible element of the vineyard site is present in their glass. In an article that sparked strong reactions in the wine world, popular food science writer Harold McGee and chef Daniel

---

[64] Salins, *Defense du vin* (cit. n. 30), 3.

Patterson debunked a straightforward connection between soil and wine: "If rocks were the key to the flavor of 'somewhereness,' then it would be simple to counterfeit *terroir* with a few mineral saltshakers. But the essence of wine is more elusive than that, and far richer."[65] As I have shown, not only is the taste of wine more nuanced than what can be explained by a simple chemical reconstruction, but the relationship between place and wine flavor has historically encompassed a complex array of interlinked human elements—medical, cultural, economic, and political. Some modern writers have argued that today's enthusiasm for wines that express their *terroir* is driven by a desire to cling to the uniqueness of individual places in the face of globalization,[66] mirroring the way that the uniqueness of vineyards in Champagne and Burgundy enabled these wines to maintain their local identities during Louis XIV's reign. Though there may be an as yet undiscovered mechanism to transfer the taste of granite through a vine and into your glistening glass of riesling, it is more likely that the elusiveness McGee and Patterson refer to results from the complexity of the relationship between wine and place, as well as people.

[65] Harold McGee and Daniel Patterson, "Talk Dirt to Me," *T: The New York Times Style Magazine*, 6 May 2007, http://www.nytimes.com/2007/05/06/style/tmagazine/06tdirt.html?pagewanted=all&_r=0.

[66] Daniel W. Gade, "Tradition, Territory, and Terroir in French Viniculture: Cassis, France, and Appellation Contrôlée," *Ann. Assoc. Amer. Geogr.* 94 (2004): 848–67, on 848.

Why Drink Water?
Diet, Materialisms, and British Imperialism

by Joyce E. Chaplin*

ABSTRACT

In 2017, New Zealand's Whanganui River was designated as having the same rights as a human person. The decision drew upon Maori belief in the animate status of non-human beings and depended on the legal power of a Western state. This article examines those two factors in relation to the history of drinking water as an essential part of human diet, focusing on early modern England/Britain. In the early modern period, water was stripped of a life-giving force with which earlier European authorities (not unlike the Maori) had endowed it, even as water was becoming a generic component of a recommended diet—recommended, not least, by state authorities. Medical interpreters who published their works in English distanced themselves from definitions of matter that had considered water as itself vital, and instead defined the material components of a healthy diet, including water, in terms that avoided any hint of vitalism. Encounter with the dietetic advice of other cultures did not revive belief in water's vitalist properties; rather, that advice was assimilated to new expectations that beverages, especially water, should maintain a cool body and temperament. These transformations took place in an imperial context. It was the Royal Navy that declared the minimum units of drinking water necessary for humans (meaning its sailors), which was a historically novel development. To uncover these trends is to explore how change occurs, and therefore how it might occur in the future, as state power may more frequently need to align with beliefs in animate nature that today are mostly non-Western beliefs, in order to protect natural features and resources, not least for human health.

Water is Life. Under that slogan, the Standing Rock Sioux Tribe protested construction of the Dakota Access Pipeline (DAPL) that was projected to cut across land the tribe claimed, carrying crude oil while imperiling stocks of freshwater. The 2016–17 protest manifesto was taken as an Indigenous rebuke of Western belief that nature is a stock of commodities, to be ranked according to saleable value, and making oil (for the moment) more valuable than water. Global medical authorities side with the Sioux, recommending water as an integral part of a healthy diet, usually in varying quantities (depending on the drinker and where they live), and frequently describing it with two modifiers: clean and plain. Today the word "clean" specifies water that is not polluted by organic or inorganic materials known to induce ill health, and "plain" recommends water in place of drinks that contain sugar, alcohol, caffeine, artificial sweeteners, and

* Department of History, Harvard University, Robinson Hall, 35 Quincy St., Cambridge, MA 02138, USA; chaplin@fas.harvard.edu.

other additives. But it was only in the eighteenth century that water became a generic component of a recommended Western diet, stripped of a life-giving force with which earlier European authorities—not unlike the Sioux—had endowed it. That is, medical interpreters who published their works in English distanced themselves from definitions of matter and bodily health that considered water (and other nonhuman materials) as itself vital, and instead defined the material components of healthy diet, including water, in terms that avoided any hint of vitalism. This redefinition of water occurred within the global context of imperial expansion—its history is neither plain nor entirely clean, but infused with the rise of Western state power.[1]

This article, a short history of that transformation, contributes to scholarship on food as part of the histories of science and medicine, and it responds to emerging literatures on new materialisms, particularly to correct their (so far) highly selective definitional focus on recent history. The scholars who have recently proposed a "new" materialism have usefully suggested attention to nonhuman beings and phenomena as active forces—"actants"—rather than as mere passive background for human history. These calls have sometimes intersected with established literatures in the histories of science and medicine, plus environmental and animal studies (among other subfields) that had similar motivation. But they were mostly aimed at late-modern theories of language that, the new materialists have asserted, had reached dead-end contemplation of human-generated texts as unidimensional (because immaterial) reflections of human experience. As such, new materialisms have exemplified an ongoing shift in comprehension about the human place within a larger world of material entities and forces. This transformation is a major concern even beyond the academy, perhaps especially in formerly colonized places; the national constitution of Ecuador was the first (in 2008) to declare that nature itself had rights, and the Whanganui River in New Zealand was in 2017 likewise granted the legal rights of a human being.[2]

As these extra-European examples make clear, new materialist studies have so far been specific to the modern West, therefore failing as critical understandings of how nonhuman nature may have been understood in the past and in cross-cultural circumstances, especially in formerly (or presently) colonized zones. That is, many of the

---

[1] "Nutrition. Get the Facts: Drinking Water and Intake," Centers for Disease Control and Prevention (CDC), US Dept. Health & Human Services, https://www.cdc.gov/nutrition/data-statistics/plain-water -the-healthier-choice.html; "Drinking-water," World Health Organization (WHO), http://www.who.int /en/news-room/fact-sheets/detail/drinking-water. On definitions of vitalist principles within the natural world, concurrent (though not intersecting) with new expectations about drinking water, see Catherine Packham, *Eighteenth-Century Vitalism: Bodies, Culture, Politics* (New York, N.Y.: Palgrave Macmillan, 2012).

[2] The two most widely recognized foundational texts for new materialism are by Gilles Deleuze and Félix Guattari, trans. Brian Massumi, 2nd ed., *A Thousand Plateaus: Capitalism and Schizophrenia* (Paris: Les Éditions de Minuit, 1980; Minneapolis: Univ. of Minnesota Press, 1987); and Bruno Latour, *Reassembling the Social: An Introduction to Actor-Network-Theory* (New York, N.Y.: Oxford Univ. Press, 2005). Texts currently central to new materialism are Jane Bennett, *Vibrant Matter: A Political Ecology of Things* (Durham, N.C.: Duke Univ. Press, 2010); and Timothy Morton, *Hyperobjects: Philosophy and Ecology after the End of the World* (Minneapolis: Univ. of Minnesota Press, 2013); see also Diana Coole and Samantha Frost, "Introducing the New Materialisms," in *New Materialisms: Ontology, Agency, and Politics*, ed. Coole and Frost (Durham, N.C.: Duke Univ. Press, 2010), 1–46. On the personhood of things, see Christopher D. Stone, "Should Trees Have Standing? Toward Legal Rights for Natural Objects," *Southern California Law Review* 450 (1972): 450–501; and Mihnea Tanasescu, "When a River is a Person: From Ecuador to New Zealand, Nature Gets its Day in Court," *The Conversation*, 19 June 2017, http://theconversation.com/when-a-river-is-a-person-from-ecuador-to-new-zealand -nature-gets-its-day-in-court-79278.

main new materialist analyses that consider the force of nonhuman things have focused on manufactured objects from the nineteenth century onward, and particularly those from the last century, if not from the very recent past—or the present; the result is a series of important intellectual contributions, but primarily (and primarily intended) as cultural studies of the present.[3] An association between modern people of the global West and objects of human manufacture (above all products of industrial manufacture) has been critiqued and yet reified by scholarship on Indigenous communities and non-Western cultures, which focuses on those communities' comprehension of natural, nonhuman, and nonmanufactured things and forces. While not specifically intended, the net result has been to accept distinctions between Western and non-Western beliefs about the character and power of nonhuman nature as if these have a timeless validity.[4]

To confront and challenge that division, this article follows criticisms of the new materialisms for their neglect of premodern historical periods, when Western definitions of nonhuman things still sometimes assumed that they had active qualities. That assumption receded. Meanwhile, state power to define which materials were necessary to human life emerged. To uncover these trends is to explore how change occurs, and how it might occur in the future, as state power may more frequently need to align with surviving beliefs in animate nature, which now are predominantly non-Western beliefs, to protect natural features and resources.

The history of European vitalism—of theories of a super-added quality to material things, making them more than inert objects—is for this reason relevant. Once assumed to have a vital principle, water became a generic part of the Western diet by having its vital qualities stripped away. The shift was not, however, a result of the (better-studied) neovitalist revolt against mechanistic science—in fact, it preceded and may have informed that rejection. And as reflected in medical works, the transformation was more inertial than doctrinal, reflecting commonsensical understanding of matter as enlivened, rather than adherence to any school of thought. Inertia played a role, as well, in isolating water's distinctive dietetic virtues as wetting and cooling. These principles referred to ancient definitions of matter—in this case humoral ones—even as they identified a new kind of regimen known as the temperate or cooling diet, in which water had a central place quite different from the older humoral ideal of balance among qualities. Those shifts questioned earlier ideas of water's nutritive or vital power and accepted that it instead functioned as a universal moistening solvent and cooling agent for the body.[5]

[3] See, for examples, Bill Brown, "Reification, Reanimation, and the American Uncanny," *Crit. Inq.* 32 (2006): 175–207; and Robin Bernstein, *Racial Innocence: Performing American Childhood from Slavery to Civil Rights* (New York: New York Univ. Press, 2011), 194–244.

[4] Jasbir Puar, "'I Would Rather Be a Cyborg Than a Goddess': Becoming Intersectional in Assemblage Theory," *philoSOPHIA* 2 (2012): 49–66; Marcy Norton, "The Chicken or the Iegue: Human-Animal Relationships and the Columbian Exchange," *Amer. Hist. Rev.* 120 (2015): 28–60; Kim TallBear, "Beyond the Life/Not-Life Binary: A Feminist-Indigenous Reading of Cryopreservation, Interspecies Thinking, and New Materialisms," in *Cryopolitics: Frozen Life in a Melting World*, ed. Joanna Radin and Emma Kowal (Cambridge, Mass.: MIT Press, 2017), 179–202.

[5] On vitalism generally, see Hans Driesch, *History and Theory of Vitalism*, trans. C. K. Ogden (London: Macmillan, 1914); L. Richmond Wheeler, *Vitalism: Its History and Validity* (London: H. F. and G. Witherby, 1939); and Georges Canguilhem, "Aspects of Vitalism," chap. 3 in *Knowledge of Life*, trans. Stefanos Geroulanos and Daniela Ginsburg (New York, N.Y.: Fordham Univ. Press, 2008), 59–74. On modern vitalist theories as reactions to, or rejections of, the mechanical philosophy, see Elizabeth A. Williams, *A Cultural History of Medical Vitalism in Enlightenment Montpellier* (Burlington,

The new valorization of water's physical virtues was also crucial evidence of West-ern understanding of non-Western dietary practices, though not (disappointingly) with a deeper recognition of cultural differences in knowledge of body and health; rather, other cultural practices, as revealed in European expansion and particularly in English/British colonization, were assimilated to the new definition of water's cooling and hy-drating capacities. That this assessment took place in an imperial context was even more obvious given the Royal Navy's key role in declaring minimum units of drinking water necessary for its sailors, a historically novel development that indicates how this branch of the military, while consulting men of science and using eighteenth-century scientific terms of analysis, used its authority to draw such a conclusion.

Water's centrality to premodern dietetics is belied by lack of sustained scholarly at-tention to it as a topic within the historiographies on science and medicine. Some of the literature on early modern food and dietetics does include discussion of water, though never as something of significance equal to any solid food item; only recently has there been a book-length study of water, as a transhistorical and transcultural global com-modity, written for a general audience.[6] In part, this rather sparse treatment is likely due to water's commonness if not ubiquity, and hence its relative invisibility within historical records. The trend by which water became a generic dietary element has therefore tended (so far) to be interpreted, sparingly, in three ways: as the result of rec-ommendations for dietary abstemiousness (sometimes modeled on Christian asceti-cism, as with the teetotaling Methodists), according to modern chemistry's redefinitions of material substances in new elementary terms, and as part of urban history in the nine-teenth and twentieth centuries, when clean water supplies for cities became a priority.[7]

But these explanations privilege intra-European developments, even as the period in question—circa 1500 to circa 1800—was one in which Europe (never exactly sealed

---

Vt.: Ashgate, 2003); Peter Hans Reill, *Vitalizing Nature in the Enlightenment* (Berkeley and Los An-geles: Univ. of California Press, 2005); and Reill, "Eighteenth-Century Uses of Vitalism in Construct-ing the Human Sciences," in *Biology and Ideology from Descartes to Dawkins*, ed. Denis R. Alexan-der and Ronald L. Numbers (Chicago: Univ. of Chicago Press, 2010), 61–87. If Williams and Reill emphasize a shift from the mid-eighteenth century onward, Sebastian Normandin looks even later, to the late nineteenth century, for another attempt to rout vitalism in his "Claude Bernard and *An Intro-duction to the Study of Experimental Medicine*: 'Physical Vitalism,' Dialectic, and Epistemology," *J. Hist. Med. Allied Sci.* 62 (2007): 495–528. On temperance and the cooling regimen, see Ginnie Smith, "Prescribing the Rules of Health: Self-help and Advice in the Late Eighteenth Century," in *Patients and Practitioners: Lay Perceptions of Medicine in Pre-Industrial Society*, ed. Roy Porter (Cambridge: Cambridge Univ. Press, 1985), 259–65.

[6] E. C. Spary, *Feeding France: New Sciences of Food, 1760–1815* (Cambridge: Cambridge Univ. Press, 2014), 31. Robert Appelbaum, in *Aguecheek's Beef, Belch's Hiccup, and Other Gastronomic In-terjections: Literature, Culture, and Food among the Early Moderns* (Chicago: Univ. of Chicago Press, 2006), discusses water as part of an abstemious or virtuous diet (155–200), but water itself does not ap-pear in the book's index (375). David Gentilcore, *Food and Health in Early Modern Europe: Diet, Med-icine and Society, 1450–1800* (London: Bloomsbury, 2016), 159–65; Ian Miller, *Water: A Global His-tory* (London: Reaktion, 2015).

[7] Roy Porter, ed., *The Medical History of Waters and Spas* (London: Wellcome Institute for the His-tory of Medicine, 1990); Jan Golinski, *Science as Public Culture: Chemistry and Enlightenment in Britain, 1760–1820* (Cambridge: Cambridge Univ. Press, 1992), 61–2, 77–8, 272; Ursula Klein and Wolfgang Lefèvre, *Materials in Eighteenth-Century Science: A Historical Ontology* (Cambridge, Mass.: MIT Press, 2007); Danielle Fauque, *Lavoisier et la naissance de la chimie moderne* (Paris: Vuibert, 2003); Harriet Ritvo, *The Dawn of Green: Manchester, Thirlmere, and Modern Environmen-talism* (Chicago: Univ. of Chicago Press, 2009); John Broich, *London: Water and the Making of the Modern City* (Pittsburgh, Penn.: Univ. of Pittsburgh Press, 2013); Carl Smith, *City Water, City Life: Water and the Infrastructure of Ideas in Urbanizing Philadelphia, Boston, and Chicago* (Chicago: Univ. of Chi-cago Press, 2013).

off from the rest of the world in the first place) was in ever more frequent contact with other parts of the world. Equally important to new comprehensions of water were three circumstances that were bound up with the creation of extra-European empires: new attention to the provisioning of ships embarked on long-distance voyages, ethnographic accounts of non-Europeans who favored water over other beverages, and the adoption of that behavior by English-speaking settlers in American colonies.

This article focuses on sources in English. It does so primarily because of the strong English-language tradition of medical self-help that generated multiple and frequently reprinted texts on dietetics, including the significance of different kinds of water, and then of elementary water in and for itself. These texts, as well as other records in English, also reflect an imperial context, as authorities in England (later Britain) accumulated and circulated statements about medical practices in other parts of the world, but, as they did so, reflected on the need to keep the sailors who traveled the globe on behalf of the empire well watered at sea. The British Admiralty, the government bureaucracy in charge of the Royal Navy, was therefore instrumental in defining water generically as a daily necessity—meaning units of plain water rationed out per man per day—as recommended and supervised by governmental authorities at sea. Indeed, the Royal Navy was among the first governmental authorities at the level of the nation to make recommendations as to how much water each person might need to consume. While these state and imperial dimensions of dietetics were of course present in other nations (with Portugal and the Dutch Republic perhaps the most likely pioneers of long-distance maritime regimens), the English self-help medical literature was distinctive in refracting them into popularly accessible recommendations.

Water, a material substance that exists within the human body, yet has its own properties beyond corporeal humans, is therefore an excellent thing to think with, given its place within the longer histories of diet, health, and materialism. Drinking water's premodern history, within the early modern contexts of cultural encounter and empire, is especially effective in revealing the role of the state in supporting new dietetic regimens.

THE VARIED VIRTUES OF VARIOUS WATERS

Ancient and medieval dietetic theories had recommended drinking water, though in terms distinct from those of modern nutritional practice. In premodern medicine, water had been one of the four elementary substances deemed constituent to life on Earth, along with fire, air, and earth, but water was never rated as more essential than the other three. Nor had its specific contribution to human life achieved the kind of stable prescriptive status implied by today's near-universal recommendation of certain units of only certain kinds of water. Moreover, water's everyday use in ancient and medieval times would be better described as the selective intake of different kinds of waters, in the plural, not of drinking a single elemental water, which could not, in any case, really exist in sublunary form. But even various earthly waters were considered as having qualities of the superior element, especially in their ability to impart cleanliness and purity.

Water's importance to a good diet was signified by its ability to cleanse the body on the outside; faith in this had a long history. Christian baptism rituals continue to use water to mark the total purification of a newborn or converted soul. Within Britain and Ireland, ancient and medieval landscapes of holy springs and wells survived into modernity, and pious folk continued to believe that certain waters might purify and heal the body, either through bathing or drinking. Consider the word "thirst." Out of the compendium of English books published from 1473 to 1700 that are listed in Early English Books Online (EEBO), 10,214 texts contain at least one use of the word "thirst," but only 480 of them, a mere 4.7 percent, are works that could be defined as medical (including veterinary medicine, cookery, anatomical, herbal, and alchemical works). The overwhelming majority of the rest are religious works, including some with evocative titles like *Water of Life* and *Living Water* that are nevertheless devoid of dietetic content. The general idea of purification did, however, follow through into medicine, including dietetics, with water designated as the universal solvent and dilutant, able to diffuse beneficial substances within but also wash away pollutants that had somehow entered the body.[8]

Water was also basic, quotidian, known from daily experience (within what were still primarily agricultural economies) to be necessary to the life of all animals—and plants. Thomas Tryon, the important food reformer, had recommended drinking water in his *Country-Man's Companion* (1684) even before he turned to human dietetics. He presented, syllogistically, the rationale that animals drank water; humans were animals; ergo humans also needed to drink water, with all creatures needing bodily moisture. Certainly, even when people did not drink water straight, they were constantly imbibing it as an omnipresent substrate in their beer, soup, porridge, even their bread, and as the obvious element of many drinkable medications.[9]

WATER AS A SINGULAR ELEMENT OF HEALTH

To some extent, the virtues of drinking water were negatively defined. Water may not necessarily have had intrinsic merit, but if it were being drunk, the drinker was less likely to be drinking something bad, which usually meant alcohol to excess. This was the central assumption of temperance, in its original sense of moderation in all things. Only later did it exclusively mean the avoidance of drunkenness. Moderation had of course long been a guideline for health, including dietetic recommendations. Balance, a central goal of moderation, was to keep the body in an optimal state between humoral extremes of hot, cold, dry, and wet; water was supposed to restore cool or damp qualities to unhealthily overheated and dry individuals.

But the first modern promoters of temperance extracted and embellished this principle as the central tenet of health, associating it with a reasoned, industrious, virtuous life, as well as a state of coolness thought to support those characteristics. The most important early modern temperance text in English (as judged by multiple reprints and later citations, but first published in 1683) was Tryon's *The Way to Health, Long Life, and Happiness: or, a Discourse of Temperance [. . .]*, the basis for Tryon's several other publications on health. While now celebrated for its early endorsement of vegetarianism, and distinct from continental European texts on dietetics for this reason,

---

[8] Janet Bord and Colin Bord, *Sacred Waters: Holy Wells and Water Lore in Britain and Ireland* (London: Granada, 1986), esp. 1–3; Alexandra Walsham, *The Reformation of the Landscape: Religion, Identity and Memory in Early Modern Britain and Ireland* (Oxford: Oxford Univ. Press, 2011), 395–470; ProQuest, Early English Books Online (EEBO), accessed 18–20 July 2019, https://search.proquest .com/eebo, results for keyword search _thirst_.

[9] Thomas Tryon, *The Country-Man's Companion [. . .]* (London: Andrew Sowle, 1684), 27–9.

Tryon's *Way to Health* in fact recommended avoidance of meat as part of regimen adjustments that generally favored simplicity in diet, including renunciations of anything elaborate, expensive, or too transformed from its natural condition, or too likely to excite the body into a violently heated state.[10]

Following on from his conclusion about all animal life needing water, Tryon recommended that people drink water because it was elemental, and the more elemental—stripped of other things and qualities—the better. An "Internal Super-essential Water sustaineth every being," he wrote, and needed constant replenishment from the ordinary waters of the Earth to maintain life itself. The best water had "neither . . . Colour, Smell, nor Taste," because that kind of water most resembled the super-essential fluid inside a living being. As such, drinking water was, along with solid food, the thing most essential to life and health. A bit too optimistically, Tryon thought drinking water could cure even scurvy. Tryon said he had known people who lived to old age "only with Water, Bread, Pulse, Herbs or Fruits, about sixteen ounces of water, and sixteen ounces of Bread and Herbs." Those precise measurements were intriguing because extremely rare—numerically defined units of any food were then uncommon. (Padua physician Santorio Santorio's early seventeenth-century metabolic experiment on himself, in which he measured his weight, intake, and excretions, was likewise highly unusual.) While many of his justifications for drinking water may seem assimilable to present-day ideas about hydration, Tryon's description of water was definitely premodern. Water was elemental because it had a vital quality: "It is the Seminary Vertue of all things, especially of *Animals*, whose Seed [semen] is manifestly Watrish . . . so various the Uses [of water] thereof, both in the Generation, Nourishment and Increase of things, that some of the Wise Men have concluded, that *Water* was the beginning of all things, and first of all Elements." This claim challenged belief that animal flesh, because like human flesh, was the optimal nourishment. Tryon instead insisted that an invisible, generative principle existed within a deceptively clear fluid, but was somehow absent in blood-suffused hunks of meat.[11]

That intrinsically vitalist statement, however underexplained (and therefore possibly reflexive for Tryon), matched surviving ideas about water's spiritual significance, its ability to wash away the evils that assailed human life on Earth. A religious text by John Hacket, roughly contemporary with Tryon's, emphasized that drinking only water, however extreme a behavior, was nevertheless conducive to moral life, as scripture had shown about the Rechabites, the ancient tribe exhorted to live in the wilderness and drink no wine. Thus, Hacket wrote: "The *Rechabites* are contented to be more sober than any, and lap the water of the Brook, like *Gideons* Souldiers. Which moderation of diet (though, as I said in the beginning, as it is an extreme, and as it is a Vow for ever to drink no Wine, I do not urge it to your imitation) yet it did enable them to avoid

[10] Thomas Tryon, *The Way to Health, Long Life, and Happiness: or, a Discourse of Temperance [. . .]* (London: Andrew Sowle, 1683); Tristram Stuart, *The Bloodless Revolution: A Cultural History of Vegetarianism from 1600 to Modern Times*, 1st US ed. (New York, N.Y.: W. W. Norton, 2006), 60–77; Smith, "Prescribing the Rules of Health" (cit. n. 5), 259–62, 264–5.

[11] Tryon, *Way to Health* (cit. no. 10), 145, 147, 191; [Thomas Tryon], *The Way to Save Wealth* (London: G. Conyers, [1695?]), 71; Ralph H. Major, "Santorio Santorio," *Annals of Medical History* 10 (1938): 373–5; Tryon, *A Discourse of Waters [. . .]* (London: T. Sowle, 1696), 2–3; Smith, "Prescribing the Rules of Health" (cit. n. 5), esp. 259; all emphasis in the original.

Luxury, and swinish drunkenness, into which sin whosoever falls makes himself subject to a four-fold punishment."[12]

And so, a simple cup (or palmful) of plain water was a primal draft from nature, the basic stuff any thirsty creature might drink, while also representing human abstemiousness, cool-headedness, sacrifice, fasting, mortification, or even punishment—whether of oneself or by others. Saints and wandering Jews subsisted on water, as did felons and paupers denied more elaborate drink. Although water had long connoted these virtues and conditions, it was only in the late seventeenth century that it was labelled "Adam's ale," as in the puritan William Prynne's 1643 complaint that prisoners of conscience were reduced to that beverage, plus a penny's worth of bread.[13]

For every recommendation to drink water, however, there was almost sure to be an angry rebuttal. There was dissent, above all, over two points: whether water's main virtues remained its humorally based wetting and cooling qualities, and whether water had a vital or nutritive capacity. Richard Short mocked the "novelists" who prescribed a new practice of drinking water to English people: "the whole world runs a madding in novelties, and our *English* men will not be left behind." Short agreed that water was itself inoffensive. It was an elemental substance; after all, it was the "Mother" (with fire the father) of all life, which again was a statement of nondoctrinal, humoral vitalism. Like Tryon, Short regarded water as an "aliment," one that in fact could itself nourish, to the point that it had actually made some deep-drinking princes fat. It was also a vehicle for carrying nourishment, and a medicine, especially for fevers. But its benefits were not universal—they varied by climate and from one human body to the next. Short advised that people in hot regions needed cool diets, "therefore tis better to drink water in *Africa* and *Lybia*, then in Northern Countries." England's cold climate should be a warning against drinking water there, "because cold distempered Countries require hot drink." By hot, Short meant a chymical attribute (as within the humoral oppositions of ancient medical theories), whether heated by fire or not. Wine and beer were for this reason better than drinking water, and English beer seemed particularly well matched to English bodies. Finally, Short rejected arguments in favor of a thing's benefits according to its naturalness and simplicity. "I Marvaile that some new light of this doting age, doe not bring upon the stage the eating of Acorns, as well as drinking of water: for in the infancy of the world, men and beasts had their meat and drinke in common. They both eat acorns, and both drank water." But that was no excuse for modern folk to continue doing either.[14]

Whatever their disagreements, dietetic commentators tended to accept another traditional view of water—that the best type to drink was the one that matched one's healthy body or had qualities that might restore health. The ancient Hippocratic "airs, waters, places" traditions had warned against exposure to the water (and other natural phenomena) of foreign places, implying that one's native waters were best, without

---

[12] John Hacket, *A Century of Sermons upon Several Remarkable Subjects Preached by the Right Reverend Father in God, John Hacket, Late Lord Bishop of Lichfield and Coventry* (London: printed by Andrew Clark for Robert Scott, 1675), 880.

[13] Spary, *Feeding France* (cit. n. 6), 31; Appelbaum, *Aguecheek's Beef, Belch's Hiccup* (cit. n. 6), 155–200; *Oxford English Dictionary* (*OED*), 2nd ed. (1989); online, 3rd ed., 2000–, s.v. "Adam's ale," accessed 30 January 2020, https://www.oed.com/.

[14] Richard Short, *Peri psychroposias, of Drinking Water against our Novelists, that Prescribed it in England: Whereunto is Added, peri thermoposias, of Warm Drink* [. . .] (London: printed for John Crook, 1656), preface (unpaginated), 1, 2–5, 7–8, 17, 27, 36, 62.

therefore urging any need to drink them beyond slaking thirst or preparing foods and medicines. And an excess of water could be as bad as excessive intoxicating liquor. For individuals with bodies already regarded as having too much of a cool and watery character (women, the phlegmatic, and so on), too much water was not as recommended as it might be for those whose bodies were hot and dry; these included primarily adult men but also other choleric types. Denizens of cooler climates, like people of Britain and Ireland, were less likely to benefit from water than people closer to the Equator.[15]

Reflecting a modern revival in Hippocratic thought about place-specific medical phenomena, commentators explained the natural varieties of waters according to their sources. English medical works therefore distinguished among waters from rivers, lakes, ponds, springs, rain, or snow. Each type was thought to have different levels of purity and healthfulness, with pond water generally rated lowest and spring water highest. Tryon concluded, elaborately, and with another statement of water's potentially vital character, that river water was best for horses, it being made "friendly and fat by its running through various sorts of Earth," which thus rendered it "far wholsomer for all Creatures to drink, and performing all uses in *Housewifery* better." Spring water was next best, then pump water, which was least good for household tasks. Pond water was the worst for drinking; because of its lack of motion, it had the opposite of the desired qualities, meaning the least smell and taste, the absence of those qualities making it best for "Man or Beast to drink." Rain and melted snow were frequently suspect, the latter because it was associated with people in mountainous regions who suffered from goiter, though its popularity for cooling food and drink was slowly overriding that concern, with the Italian taste for ices spreading through other parts of Europe from the sixteenth century onward. Whatever the ranking and preference of these types of water, the attention to their multiple natures, and particularly their interaction with earth, suggests a continuing belief that water was not inert, but itself capable of holding and conveying either friendly virtues, or else stinkingly unwholesome attributes.[16]

Classification of water based on its source also expressed contemporary uncertainty as to whether water contained nourishment. Tryon's and Short's statements about water having and inducing fatness indicated belief that it did more than supply neutral fluid. But they and others also thought water had significant utility in conveying things that were obviously nutriment or else medicine. Everyone stated a preference for light, clear, tasteless, odorless water, hence the almost universal suspicion of pond water. The clearest water was hardest to find, except in the cases of populations that lived near large or multiple springs. But the desired clarity of the generally preferred water reflected a suspicion that drinking water should not contain anything additional. Again, however,

[15] Joyce E. Chaplin, *Subject Matter: Technology, the Body, and Science on the Anglo-American Frontier, 1500–1676* (Cambridge, Mass.: Harvard Univ. Press, 2001), 52, 149–53; Steven Shapin, "How to Eat Like a Gentleman: Dietetics and Ethics in Early Modern England," in *Right Living: An Anglo-American Tradition of Self-Help Medicine and Hygiene*, ed. Charles E. Rosenberg (Baltimore: Johns Hopkins Univ. Press, 2003), 21–58; Chaplin, "Earthsickness: Circumnavigation and the Terrestrial Human Body, 1520–1800," *Bull. Hist. Med.* 86 (2012): 515–42; Rosenberg, "Epilogue: Airs, Waters, Places. A Status Report," *Bull. Hist. Med.* 86 (2012): 661–70.

[16] Tryon, *Country-Man's Companion* (cit. n. 9), 27–30; Gentilcore, *Food and Health* (cit. n. 6), 160; James C. Riley, *The Eighteenth-Century Campaign to Avoid Disease* (Houndmills, UK: Macmillan, 1987), 1–30; Melissa Calaresu, "Making and Eating Ice Cream in Naples: Rethinking Consumption and Sociability in the Eighteenth Century," *Past & Present* 220 (2013): 35–78.

any references to water's living and revivifying principles were offhandedly vitalist, without sustained critique or analysis (or references to other scientific authorities), at least in literatures that discussed water in terms of dietetics.[17]

Fine-tuning of that background set of beliefs came from another field: chemistry. This has been a well-examined topic, particularly so in histories of the main figures in modern chemistry (Cavendish, Priestley, Lavoisier), and in terms of the benefits of mineral spas. Although certain springs on the European continent (and a lesser number in Britain and Ireland) had been renowned for their health-giving waters, and sometimes continued a religious history of spiritually powerful waters, their popularity and commodification were a largely late eighteenth-century story. Chemical experimenters tended to conclude that spa waters had no special (let alone miraculous) properties of vivification. Such waters simply contained greater proportions of certain minerals than water found elsewhere. From this knowledge, entrepreneurs concocted artificial mineral waters, available to those who might never travel to Spa or Bath or Buxton. Formally trained medical men were willing to support the claims about spa waters so long as patients consulted them and deferred to their expertise. Meanwhile, chemical experimenters were beginning to define water (along with air) as a composite of certain elementary substances, not as an element itself; air contained oxygen, and water consisted of hydrogen and oxygen.[18]

Put together, these developments constituted a trend to understand water as one kind of substance—water, not waters—though the substance was capable of being permeated with others, whether mineral or possibly organic. Its nutritive or poisonous capacity was incidental, not intrinsic. Water might be vital to life, but descriptions of its own vitality were being adjusted if not erased. To a very great extent, the range of advice about water—and the nature of the disagreements about it—remained stable over centuries, certainly surviving into the late eighteenth century, however embellished these more recent iterations may have been with the most recent chemical definitions.

Thus, two best-selling medical advisors, George Cheyne (early eighteenth century) and William Buchan (late), helpfully reminded readers that water was highly beneficial, except when it wasn't. Cheyne veered toward the optimistic end of the spectrum in his early *Essay on the Gout* (1720), rehearsing the established opinion that water's main benefit was as "our best and most universal Dilutent." In that function it conveyed many effectual remedies and could itself succeed "when other more pompous Medicines have fail'd." Cheyne was even more confident of water in his subsequent and much-reprinted *Essay of Health and Long Life* (1724). He nodded toward chemistry by recommending water's substantial nature, as compared to other substances. Though air was compressible, water was not. While mercury, light, and air were three other examples of a "Simple *Fluid*," only water could be regarded as "fit for human *Drink*," suited, as it was, for "*diluting, moistening* and *cooling*, the Ends of *Drink* appointed by *Nature*." Cheyne again extolled water as the only "universal Dissolvent,"

---

[17] Gentilcore, *Food and Health* (cit. n. 6), 160; George Cheyne, *An Essay of Health and Long Life* (London: G. Strahan, 1724), 17.
[18] David Harley, "A Sword in a Madman's Hand: Professional Opposition to Popular Consumption in the Waters Literature of Southern England and the Midlands, 1570–1870," in Porter, *Medical History of Waters* (cit. n. 7), 148–55; Noel G. Coley, "Physicians, Chemists, and the Analysis of Mineral Waters," in Porter, *Medical History of Waters* (cit. n. 7), 156–68; Golinski, *Science as Public Culture* (cit. n. 7), 61–2, 77–8, 272; Gentilcore, *Food and Health* (cit. n. 6), 161–2.

conveyor of whatever might be medically beneficial or else inert: "*Tea* is but an *Infusion* in Water of an *innocent* Plant." Background diluter it might be, but water was essential. The early "*eastern Christians*" had survived on minimal food and "mere Element for Drink"; St. Anthony Abbot lived to his 105th year on "mere Bread and Water, adding only a few Herbs at [the] last."[19]

William Buchan's *Domestic Medicin[e] or, The Family Physician*, originally published in London in 1769 (and eventually the most reprinted of all self-help medical texts in English), likewise recommended water: "Water is not only the basis of most liquors, but also composes a great part of our solid food. Good water must therefore be of the greatest importance in diet." Good water was free of visible foreign bodies, and was thus light, colorless, odorless, tasteless. When that kind of water was available, it could do no harm, notwithstanding attempts to blame it for something or other. Buchan, for instance, disputed that "snow-water" in the Alps imparted goiter; more likely, the water had picked up something unhealthful that pure melted snow lacked. Nor did he think water itself transmitted or prolonged malaria, a disease associated with watery places, characterized by its intermittent fever. Rather, water was "the greatest febrifuge in nature," a tonic cooler whenever the body overheated; something else about marshy places and drinking "bad water" must make people ill. When in doubt, the well-known practices of filtration and exposing water to sun and air would remove any impurities, leaving behind the purer water associated with health. (That Buchan did not also recommend boiling and distillation may have been because those required the greater expense of buying either fuel or special equipment.)[20]

Finally, there were, by the early eighteenth century, medical works that focused exclusively on water, calling out its new centrality to dietetics. John Smith's *Curiosities of Common Water* (1723) claims it to be a universal panacea, washing away all ills and every kind of dietary indiscretion. With greater balance, Thomas Short's *Rational Discourse of the Inward Uses of Water* (1725) both deplores the recommendation of water as "universal Drink and Remedy" and denounces the claim that it has no benefit. Water was placed everywhere on "the terraqueous Globe" because God had made it "the most universal Use and Benefit" for all creatures. Short agreed with those who categorized water as a dilutant and vehicle; he extolled it as a virtuous drink. Noah had escaped water only to succumb to wine—and thence a history of humans making a medicine (alcohol, when taken medicinally) into a source of sin. Short regarded water and blood as roughly homogeneous in their vital agencies; for that reason, water increased "seed" in men and milk in wet nurses. Still, water required knowledge for its optimal use. Individual drinkers needed to know themselves well; the hot, dry, and choleric needed more water than those who were phlegmatic or cold. The knowledge of physicians and natural philosophers (Short dedicated his text to Hans Sloan, physician, savant, founder of the British Museum) was also needed. Men of science had used microscopes to verify that water had content, though not (Short said) of the living kind. Rather, "Water consists of *imperceptibly small* Particles," inert material things, to which anything alive was foreign. The best water had the smoothest such particles, which permitted its fluid nature, making it good for the human body. Other elements within it generated various

---

[19] George Cheyne, *Essay on the Gout* (London: G. Strahan, 1720), 23, 26, 33, 40–1; Cheyne, *Health and Long Life* (cit. n. 17), 12, 17, 30, 42, 48, 63. The emphasis in all quotations is in the original.

[20] William Buchan, *Domestic Medicin[e] or, The Family Physician* (Philadelphia: John Dunlap, 1772), 40, 41, 74, 98.

effects on health, whether good or bad, even making certain kinds of water specific to certain diseases.[21]

While overtly vitalist definitions of water were receding, residual ideas about its distinctive cooling power within living bodies nevertheless lingered. It is significant that the new understanding of water, derived from dietetics, found a place in the work of formally trained, vitalist medical authorities who questioned the ability of mechanistic philosophy to explain life and health. This was certainly the case with William Cullen, the Scottish doctor (David Hume's physician) who taught William Black, the chemical experimenter who likewise denied the universal claims of mechanistic theories. In his *First Lines of the Practice of Physic* (1777), Cullen recommended water as the most beneficial dilutant and the best vehicle of excretion. Especially when cold, it was particularly effective in increasing heat at the surface of the body, therefore cooling the interior; ipso facto, it was the ultimate remedy for fevers. Cullen noted that Spanish and Italian physicians prescribed six to eight pounds of water each day—the *Diaeta acquea*—to cure a fever. Given Cullen's and Black's emphasis on how heat seemed impossible to explain merely mechanistically, the former's explanation of water's effects on the human body is intriguing evidence that vitalistic ideas were never wholly extinguished, but instead transformed. The ideas survived, moreover, in a conduit that ran from early modern dietetics into modern chemistry, from food to science.[22]

As is always the case with prescriptive literature, it is impossible to determine how closely people in the past followed advice to drink water and to seek good water to drink. Certainly, they were told to do so, if they could read or have someone read to them. But better evidence on the actual practice of drinking water comes from different kinds of sources, including, and maybe especially, those about Britain's empire during the eighteenth century.

WATER AND COLONIALISM

Eighteenth-century advice literature does at the least indicate the emergence and acceptance of water-drinking "cool" regimens as healthy and virtuous. If anything, this advice was initially more powerful in imperial and colonial settings than within England/Britain itself. While the British Admiralty would, ultimately, become the most powerful champion of drinking water, its actions followed upon earlier colonial preference for cold water and bodily simplicity. This pattern of influence complicates the existing literature on modern discussion of drinking water as an urban concern—and therefore terrestrial—to the neglect of its extra-European and maritime histories, which imply more firmly the eventual hand of the nation-state in regulating diet. The imperial context that would generate this official policy, and its concomitant medical statements about water, was paralleled by the valorization of non-European and creole people who, outside Europe, drank water or water-based infusions instead

---

[21] John Smith, *The Curiosities of Common Water: Or the Advantages Thereof in Preventing and Curing Many Distempers* (London: printed for J. Billingsley, 1723)—this had at least six reprintings on both sides of the Atlantic; Thomas Short, *A Rational Discourse of the Inward Uses of Water* (London: printed for Samuel Chandler, 1725), iii, vii, 2, 7, 8–9, 17, 14–16, 18, 21, 41.

[22] Reill, *Vitalizing Nature* (cit. n. 5), 96–7, 122; William Cullen, *First Lines of the Practice of Physic* [. . .] (Edinburgh: printed for William Creech, 1777), 35, 121–3.

of alcoholic beverages. Some non-European drinks that were both hot and nonalcoholic, however, forced a more precise designation of how water—even when hot—might be a cooling substance.[23]

Drinking water in the colonies at first seems to have been explained according to traditions of dietary simplicity and privation, if not saintliness. Adjustment to colonial places, eventually called "seasoning," included bodily assimilation to local waters, whether brewed or cooked into other foods, or drunk straight when nothing else was available. The very first printed account of an English colony in the new world, Thomas Hariot's 1588 description of the fleeting Roanoke settlement, chided any would-be colonist for expecting "dainte food." Such food was neither available nor necessary: "excepting for twentie daies, wee liued only by drinking water and by the victuall of the countrey." Similarly, in the religiously radical colony at Plimoth Plantation, which lacked adequate food in its first two years (1620–21), let alone beer or liquors, the colony's first governor, William Bradford, had insisted that the local water was "wholesome enough to us that can be content therewith."[24]

Travel and colonization also presented multiple examples of non-European people whose most common drink was not alcoholic. In Thomas Tryon's imagined dialogue between a Frenchman and an Indian Brahmin, for example, the Brahmin says, "nor do we think it lawful for us to heat our Veins, and distemper our Blood with Wine, since Water more kindly quencheth our Thirst." But such total denial astonished most Europeans, who, whatever the recommendations of moderation, considered outright refusal of beer, watered wine, or even stronger spirits to be oddly abstemious, a temporary medical concession, or a perhaps inadvisable shunning of a possible daily assistance to health. Ottoman Turks, for example, whose adherence to Islam meant they drank no alcohol, were nevertheless regarded within Europe as culturally sophisticated—and militarily formidable. It was principally from Arab culture in the Near East that the English, like other Europeans, learned how to drink coffee—hot coffee—as an alternative to alcoholic beverages that were at room temperature or cold. One account of the Near East described meals given even by the ruling sheikhs where one could expect to "drink only Water," then smoke tobacco and drink coffee afterward, leaving the dinners without the singing and laughter afforded by wine, but conversely rich in "serious Conversation." Positive assessment of Islamic prohibition of alcohol was by no means universal. Another account claimed that Muslim men would "drink only water, sherbet, and coffee" when in company, but sneak wine in private and intoxicate themselves on opium whenever they liked. Still, even this criticism did not dispute that if water and coffee and sherbet had been the main beverages, people would be soberer.[25]

---

[23] John Brewer, *The Sinews of Power: War, Money and the English State, 1688–1783* (Cambridge, Mass.: Harvard Univ. Press, 1990); Mark Harrison, "Science and the British Empire," *Isis* 96 (2005): 56–63.

[24] Thomas Hariot, *A Brief and True Report of the New Found Land of Virginia* (London: R. Robinson, 1588), 8, 45; William Bradford, *Of Plymouth Plantation, 1620–1647*, ed. Samuel Eliot Morison (New York, N.Y.: Knopf, 1952), 26, 143; Chaplin, *Subject Matter* (cit. n. 15), 52, 149–53.

[25] Thomas Tryon, *A Dialogue between an East-Indian Brackmanny or Heathen-Philosopher, and a French Gentleman Concerning the Present Affairs of Europe* (London: Andrew Sowle, 1683), 17; Laurent d'Arvieux, *The Chevalier D'Arvieux's Travels in Arabia the Desart; Written by Himself* [. . .] (London: printed for D. Browne, 1723), 134–5; Evan Jones, *A New and Universal Geographical Grammar: Or, a Complete System of Geography* (London: printed for G. Robinson and T. Evans, 1772), 2:89; Gentilcore, *Food and Health* (cit. n. 6), 160.

Discovery of yet other beverages taken hot incited commentary not only on their natures but on whether anything made with hot water—or just the hot water itself—was beneficial. Was water more than a solvent, as it had been thought to be when made into various teas or porridges that carried herbal medicaments? Was it the hot water itself that carried a benefit? Or would heat destroy water's virtues, thus destroying the health of the drinker? A 1726 advice manual, which included pieces from both Thomas Tryon and George Cheyne, reported ambiguously about dietetic practices among people in South Asia: "the *Indians* advise Fasting, and eating Water and Rice boiled together, which Liquor they drink." These were customs that fit Tryon's and Cheyne's recommendations of temperance while also bestowing some value on hot water. Henry Stubbe, commenting on American chocolate prepared as a hot beverage, said that it had a good effect on kidney diseases: "Thus *hot water* drunk daily before Diner cures the Stone, and Gravel in the Reins, as *Zecchius* affirms, and *Trallianus*: and the benefit others have found by it doth manifest."[26]

These suggestions seemed not to have been convincing. Multiple authors rejected the use of hot beverages, whether simple water or other heated drinks with water in them. One commentator said these practices had represented corruption of diet among ancient Europeans, implying a deficiency of knowledge that should have been overcome. "The Custome of the Ancients to drink Cold or Hot water alone, as some, or mixt with their wines, as others, is generally known," he wrote.

> Yet this difference may be observ'd, that the mixture of hot water was counted the worse wantonness, as *Philo* (*de vitâ Theoretica*) and Others note. The use of this was cheifly in Winter . . . Yet concerning their *aqua frigida* they were likewise grown to singular wantonness, keeping snow under ground, for such uses, till Summer and as *Seneca* implies, (*epist.* 78.) they did mix snow with their wine.[27]

Richard Short likewise warned that warm or hot drinks were categorically ridiculous; drink was meant to cool, to provide moisture, to quench, which artificially heated liquids could not do. He noted that animals by instinct drank cold water, which must therefore be the most natural practice. And to any eager suggestion that people elsewhere might have other practices, he rejected their efficacy for European bodies. True, the Chinese and Japanese were said to drink hot water (and presumably hot tea), but they had been accustomed from infancy to do so, unlike Europeans. Short felt so strongly on this point that he concluded his book with it: "Thus I end, wishing not any of my Friends to drink warm drink."[28]

The pure draft of cold water, on the other hand, remained a signature colonial icon, and appeared as such in the memoirs of a famous colonial American, Boston-born Benjamin Franklin. Relating his youth in London in the 1720s, when he was an apprentice at two print shops, Franklin said he had refused to follow the English workers' example in drinking beer, let alone their rationale for doing so:

---

[26] *The Way to Health and Long Life: or, a Discourse of Temperance* [. . .] (London: printed for G. Conyers, 1726), 65; Henry Stubbe, *The Indian Nectar, or, A Discourse Concerning Chocolata* (London: printed by J. C. for Andrew Crook, 1662), 122. See also Marcy Norton, *Sacred Gifts, Profane Pleasures: A History of Tobacco and Chocolate in the Atlantic World* (Ithaca, N.Y.: Cornell Univ. Press, 2008).

[27] Juvenal, *Decimus Junius Juvenalis, and Aulus Persius Flaccus*, trans. Barten Holyday (Oxford: printed by W. Downing for F. Oxlad Sr., J. Adams, and F. Oxlad Jr., 1673), 85.

[28] Short, *Of Drinking Water* (cit. n. 14), 132–6, 154–6, 157.

I drank only Water; the other Workmen, near 50 in Number, were great Guzzlers of Beer. On occasion I carried up and down Stairs a large Form of Types in each hand, when others carried but one in both Hands. They wonder'd to see from this and several Instances that the Water-American as they call'd me was *stronger* than themselves who drunk *strong* Beer.[29]

Franklin tried to convince a fellow printer, who drank a pint of beer at each meal every day, "that the Bodily Strength afforded by Beer could only be in proportion to the Grain or Flour of the Barley dissolved in the Water of which it was made; that there was more Flour in a Penny-worth of Bread, and therefore if he would eat that with a Pint of Water, it would give him more Strength than a Quart of Beer." His companion evidently heard him out, but kept drinking beer. Small wonder. At the time of Franklin's residence in London, the 1720s, water was not yet the standard, recommended beverage, and all manner of "strong" drink was assumed to make its drinker strong. It is significant that Franklin was at this point, in his late teens, already a semivegetarian and disciple of Tryon, an early (and unusual) known adherent of temperance in its original broad sense of keeping the body in a state of cool moderation. Franklin also resented the money beer would cost; his unconvinced companion kept paying this out, four or five shilling he might have saved—"and thus these poor Devils keep themselves always under." Whatever else his fellow apprentices thought about his moral censoriousness, they associated Franklin's difference from them with where he came from. The "Water-American's" simplicity of diet matched the presumed Otherness of his place of birth.[30]

This account, from Franklin's famous "autobiography," was written at the end of his life, in the 1770s and 1780s, reflecting back on events that took place decades earlier, when the cool regimen was a radical proposal and not yet mainstreamed. It is impossible to tell, therefore, how much of the Water American's story was inflected by the creole patriotism that had emerged in the meantime, which likewise stressed simplicity of diet, as in the earliest accounts of the American settlements. In the 1770s, again based in London, this time as a colonial lobbyist, Franklin wrote angry defenses of Americans' nonimportation of British goods, noting them as part of their protest against taxes on British exports. He championed the American ability to breakfast on a simple gruel or porridge made of corn meal and water, rather than drink British transshipped, hatefully taxed tea first thing in the morning. The American patriot's blend of corn and water reappeared in Joel Barlow's mock-epic *The Hasty-Pudding* (1796), in which the poet/narrator suffers, while visiting Europe, from the excesses of its cities:

> . . . Paris, that corrupted town,
> How long in vain I wander'd up and down,
> Where shameless Bacchus, with his drenching hoard,
> Cold from his cave, usurps the morning board.
> London is lost in smoke and steep'd in tea.

[29] *Benjamin Franklin's Autobiography: A Norton Critical Edition*, ed. Joyce E. Chaplin (New York, N.Y.: W. W. Norton, 2012), 45–6.

[30] Joyce E. Chaplin, *The First Scientific American: Benjamin Franklin and the Pursuit of Genius* (New York, N.Y.: Basic, 2006), 76; Sarah Hand Meacham, *Every Home a Distillery: Alcohol, Gender, and Technology in the Colonial Chesapeake* (Baltimore: Johns Hopkins Univ. Press, 2008).

But in rural Savoy, Barlow found food that was true, pure, and American. To start the day right, "in boiling water stir the yellow flour," milk on the side, if you like.[31]

These references to extra-European practices were to suffer the fate of assimilation, absorption into an existing set of culturally external customs—in this case, the long-standing European tradition of virtuous abstention. It is notable that "thirst" continued, into the eighteenth century, to connote spiritual longing, if in an increasingly secularized sense. In Eighteenth-Century Collections Online (ECCO), the compendium of more than 180,000 English-language titles published from 1701 to 1800, the word "thirst" appears in 29,286 texts. The majority of these—10,903 (37.22 percent)—are in the genre of literature, with the second largest group—9,323 (31.83 percent)—in religion and philosophy. A total of 69 percent of the references, therefore, are to poetic or spiritual thirsts. In contrast, only 2,756 texts that mention thirst (9.41 percent) are classifiable under medicine, science, and technology. This percentage represents an increase over the earlier EEBO average of 4.7 percent for such titles in medicine, but not enough of an increase, obviously, to erode the overwhelming sense, in English, that the human longing to drink expressed the needs of their souls much more than those of their bodies.[32]

Franklin and Barlow had conformed to the new European plan for a cool diet by insisting on a sobriety not afforded by London beer and Paris wine. And the medical reasoning that the Chinese or Muslim Arabs might have about coffee or tea versus water, as substitutes for alcoholic beverages, did not become part of British or Anglo-American commentary. Whatever the possibility of assessing non-Western definitions of body and diet, even to appropriate them, this opportunity did not form an overt part of the debate over the virtues of drinking plain water. That inherent imperialism was even more apparent in the Royal Navy's eventual stocking of fresh water for daily drink on its vessels.

NAVAL RATION

As both the medical advice literature and colonial accounts show, it was during the eighteenth century that drinking water became a more central part of dietetic practice, though this did not take place without debate over which kinds of water and how much could be recommended. By the end of the century, the British Admiralty would, through its officers and sponsored publications, have formed answers to both of those questions—first, by recommending whatever water was clean and available, and second, by beginning to state daily requirements of drinking water per sailor when provisioning ships. The establishment of uniform naval policies regarding provision of drinking water therefore positions the history of dietetics within military history,

---

[31] [Benjamin Franklin], "Homespun," Second Reply to "Vindex Patriae," *Gazetteer and New Daily Advertiser*, 2 January 1766, in Benjamin Franklin, *The Papers of Benjamin Franklin*, eds. Leonard W. Labaree et al., 43 vols. to date (New Haven, Conn.: Yale Univ. Press, 1959–), 13:7–8; Joel Barlow, *The Hasty-Pudding: A Poem, in Three Cantos* (New York, N.Y.: [John Bull?] for Fellows & Adam, 1796), 1.2.11, 1.4.10, 1.7.1–5.

[32] Gale Cengage, Eighteenth Century Collections Online, https://www.gale.com/primary-sources /eighteenth-century-collections-online, accessed 18 July 2019, results for keyword search *thirst*: Literature and Language (10903), Religion and Philosophy (9323), History and Geography (3084), Medicine, Science and Technology (2756), Social Sciences (2109), General Reference (630), Fine Arts (317), Law (164).

and, in the cases of official voyages of "discovery," also within imperial history, extending the history of the modern military-industrial-science complex into the realm of quotidian drink.

The victualling of vessels in the Royal Navy, which after 1628 was under the governance of the Board of Admiralty, was an official part of the defense of England; after 1707, this entailed the defense of Great Britain, and increasingly, of the British Empire. At first, the work of provisioning ships was subcontracted to victuallers who were paid to supply various units of food and drink, the latter mostly composed of beer, quite obviously most sailors' preference (not to be succeeded by hard-alcohol "grog" until the late eighteenth century); water was an emergency choice or a substrate for food and medicaments. Victuallers were paid set rates—so much per man aboard per month—for whatever they supplied, such as beef, beer, biscuit, water, and so on. Naval vessels were also given orders to procure fresh supplies, including water, once they were underway and beyond the home supply line. The victualling procedure regularly failed in times of war, when sudden preparations at a larger scale overstretched the capacity of private suppliers. Attempts to improve the situation recurred through the seventeenth century, with a Board of Victualling Commissioners established in 1657. A new Victualling Board was created in 1701, and underwent renewed reforms after 1711 as Great Britain, later the United Kingdom, continually refined its expanding bureaucratic structure.[33]

To drink water at sea became a priority in the eighteenth century, when the system of contracting provisions routinely included supplying fresh water at contract depots along with other drink, principally alcohol. Water remained for the moment a precious commodity under sail, reserved almost entirely for drinking (by humans and by livestock, if the ship carried the latter), with sea water used for all washing and even some cooking. Although longer-distance voyages prompted better planning for carrying water, attention to this provision rose even before routine British forays into the Pacific and Indian Oceans made it absolutely necessary for these much more extended voyages, where resupply would be difficult if not impossible.[34]

But water was not supplied as a freestanding commodity—it couldn't be. Before 1815, it was ordinarily supplied in wooden casks, meaning that the water depended on a successful symbiosis with its wooden vessel. The surviving Admiralty correspondence about victualling that discusses water quite often reflects dissatisfaction with casks that were too new (imparting bad flavor), too old (therefore falling apart), or too few for a particular ship's complement or mission (going into battle or across the Atlantic with insufficient casks was considered foolhardy). When he headed out on a voyage he estimated would last two years, privateer William Dampier requested eighteen barrels of water. This was to give him an appropriate number of refillable barrels for such a voyage, not water to last throughout. It was all the more important that the wood of the containers be fit for service month after month. Thus, Captain Thomas Lake complained, in 1743, that the water for his ship was black and "stinking" from being supplied in new, inadequately seasoned casks. Supplying staves for the casks

---

[33] N. A. M. Rodger, *The Wooden World: An Anatomy of the Georgian Navy* (London: Collins, 1986), 91–2; Rodger, *The Command of the Ocean: A Naval History of Britain, 1649–1815* (New York, N.Y.: W. W. Norton, 2005), 33–4, 42–3, 192–4, 304–7.

[34] Janet Macdonald, *The British Navy's Victualling Board, 1793–1815: Management Competence and Incompetence* (Woodbridge, UK: Boydell, 2010), 42.

Table 1. *National Archives, Records of the Admiralty (ADM), Keyword Search for Cask AND Water.*

1500–1599: 7
1600–1699: 113
1700–1799: 70
1800–1899: 68
1900–1924: 8

was usually more difficult than finding the water to fill them, even though the water could not be transported without them.[35]

Naval water casks had, however, a historically specific reign. For the period 1500–1889 (table 1), the volume of references to "water" and "cask" in Admiralty records held at the National Archives shows a dramatic rise in the 1600s, with a steady if lower rate continuing into the next two centuries.

As will be explained below, shipping water in casks became much rarer after 1815. And it was invisible as an official process before 1600. Between those two points, however, from roughly 1600 to roughly 1815, casks were water, and water was cask, and when sailors resorted to drinking water, they also drank a woody residue.

In contrast to the prevailing concern over the wooden casks, only occasionally was there a preference expressed for a type of water, as specified by location. In 1675, one captain stated that the water from Gravesend was described as unfit for a voyage. Another captain's letter to a contractor, dated 1701, requested "Thames water" specifically—though it was not clear whether this meant straight from the river, or just from the greater London area. But most often, the desire was simply that sufficient water be supplied in seasoned but not-too-old casks—just to have enough to drink at sea.[36]

Keeping enough water aboard naval vessels was of mounting concern in the long eighteenth century for two reasons: longer distances were being traversed more routinely; and there was a steadier rate of warfare, including naval action, during the period of the Second Hundred Years' War (ca. 1689–ca. 1815). Indeed, these two background conditions interacted, with British ships unwelcome in many foreign territories throughout the world at times either of mere rivalry or open conflict. It was still the policy for naval vessels to carry more food than water, the edibles to last twice as long as the water would, simply because of the greater weight of, and difficulty in handling, water. A 1692 letter from the Victualling Office had specified the authorities' observance of an "ancient" rule that naval vessels must carry four hogsheads worth of water for every hundred men per month, if serving in the English Channel, and three times that amount for ships bound for Africa and India. Estimating a late seventeenth-century hogshead at about 60 US gallons, this would mean roughly 240 gallons of fresh water for 100 men each month, or 2.4 gallons per man per month. That was far too little for everyday drink; it was obviously for much more incidental if not emergency use, not unlike medicine.

---

[35] William Dampier to Navy Board, 3 October 1698, Records of the Admiralty, Naval Forces, Royal Marines, Coastguard, and Related Bodies (**hereafter** ADM) 106/516/232, National Archives, Kew, Richmond, UK; Thomas Lake to Navy Board, 16 December 1743, ADM 106/977/238, National Archives; Macdonald, *Victualling Board* (cit. n. 34), 21.

[36] Captain H. Killigrew to Navy Board, 14 June 1675, ADM 106/311, National Archives, Kew, Richmond, UK; Captain G. Hicks to Navy Board, 16 March 1701, ADM 106/546/32, National Archives.

Refilling casks on land was feasible for some routes, though hard to guarantee for trans-oceanic routes, especially in times of war when British ships were not welcome in a variety of places.[37]

It was precisely the question of whether drinking water had some medical benefit that prompted new thinking about its availability on ships. Access to water was a concern of the famous eighteenth-century attempts to combat scurvy, for which there was no prevailing theory, prophylactic, or cure. At least since the sixteenth century, physicians and mariners had generated a rather long list of possible antiscorbutics, including spruce beer, citrus, regular exercise, fresh bread, cheerfulness, procedures to leach at least some salt out of the salted provisions, and on and on. A sense that the disease was rarer on land than at sea generated statements that earthly conditions engendered the proper habitat for humanity. Land itself was considered a cure—contact with earth, proximity to growing plants, fresh water to drink were regarded as restorative after extended time away from a terrestrial environment. Whatever the merits of one's natal land, in cases of scurvy, any landed area worked just fine—and so any water from land was beneficial, a striking rejection of the Hippocratic airs, waters, places tradition that had argued the opposite. At this point, ongoing exasperation with the faulty wood of casks was augmented by suspicion that any water preserved for sea might lose its earthly, antiscorbutic benefit. Efforts to collect water from awnings rigged to capture rainfall were interesting in this regard too, as when Captain Philip Carteret ordered the floor cloth of his cabin taken up to catch rain in the middle of the Pacific Ocean. Such measures produced water untainted by long storage in wood, yet ignored suspicion of rainwater as inferior to any taken from terrestrial sources. (And in any case, rainwater rarely supplied an entire ship beyond a few days.)[38]

It was at this point in time that chemical experimenters and physicians helped the Admiralty determine what water could and should be supplied at sea. These men of science focused mostly on two tasks: isolating fresh water from salt water, and verifying water's essential qualities for human consumption. The tasks were related, with attempts to make salt water fresh through the addition of alkali substances giving way, gradually, to attempts at distillation; comprehension of water was therefore shifting from adding something to it to balance any saline quality, to taking something out of it (the salt itself), leaving a purer form of water behind. One experiment of the 1750s, published in the Royal Society's *Philosophical Transactions*, used both an added alkali and then distillation to produce something said to be comparable to fresh water obtained on land, earthly nature reconstituted at sea.[39]

Subsequent attempts, at least from the 1770s, just used distillation, reflecting a new understanding of water as a substance that might contain others, but that could itself be isolated as a singular fluid. An Arctic voyage commanded by Constantine John Phipps Mulgrave, in 1773, tested a boiler still designed by physician Charles Irving,

[37] Janet Macdonald, *Feeding Nelson's Navy: The True Story of Food at Sea in the Georgian Era* (London: Chatham Publishing, 2004), 78; Victualling Office to Navy Board, 27 January 1692, ADM 106/424/9, National Archives, Kew, Richmond, UK.

[38] Helen Wallis, ed., *Carteret's Voyage round the World, 1766–1769*, 2 vols. (Cambridge: Cambridge Univ. Press, 1965), 1:151; Thomas Pasley, *Private Sea Journals, 1778–1782* (London: J. M. Dent and Sons, 1931), 229; Chaplin, "Earthsickness" (cit. n. 15), 540.

[39] Mr. Appleby and W. Watson, "An Account of Mr. Appleby's Process to Make Sea-Water Fresh; With Some Experiments Therewith; Communicated to the Royal Society," *Philosophical Transactions* 48 (1753–54): 69–71.

with which Irving claimed to have introduced distillation to the Royal Navy—a boast, not a fact. Irving accompanied Mulgrave at sea, testing his apparatus, investigating sea ice for its potential drinkability, and measuring the salt content of seawater from different depths. Irving's friend and protégé, former slave and sailor Olaudah Equiano, who had participated in Irving's experiments, also joined the expedition. Both Irving and Equiano pronounced the device a success. It could make up to forty gallons a day, with the advantage that it could be rigged up in the galley and kept on the boil without much more fuel than needed for everyday cooking. But Irving's most interesting statement reflected on the minimum capacity of his apparatus. Twice, it produced only twenty-three gallons a day, just over a quart per man, "which, though not a plentiful allowance, is much more than is necessary for subsistence." This was considerably more than the "ancient" practice the Victualling Board had described a century earlier, the estimated 2.4 gallons per man per month. And from this point forward, a more lavish supply would be stocked on some ships, expressly for drinking— and nothing else. John Jervis, as an admiral during the Napoleonic Wars, had given his men free access to water butts on deck, as noted in correspondence expressing his irritation that female passengers were depleting the precious supply by taking it away for washing. Rather than ration the water, Jervis simply ordered "the ladies" ashore.[40]

This investment in the lives of sailors was significant. Through the Napoleonic Wars, men with maritime experience were regarded as the property of the state, as evidenced in the continuing practice of impressment, when licensed press gangs would apprehend even the most lightly qualified maritime workers to serve aboard naval vessels at times of war, using force if necessary. Sailors themselves were regarded as wards of the state, no more able than women, children, or the insane to govern their bodies or protect their personal rights. To counter ever-louder protests over what seemed an abridgement of the rights of this category of freeborn Englishmen, the British state was busy making amends, if not exactly the ones sailors themselves might have wanted. The more careful provisioning of naval vessels, and the statements about investment in the medical well-being of sailors, were important modifications of state power, particularly of state investment in the health of its citizens.[41]

Naval authorities continued to consult physicians and men of science to verify the quality of portable drinking water. For a time, distillation was the dominant way of obtaining fresh water once under sail. James Lind, famous recommender of citrus (among other foodstuffs) for preventing scurvy, was also an advocate for distilling water while at sea. Others advised this only as a last resort. In his guidelines for preserving health at sea, Gilbert Blane declared fresh water "one of the articles most essential to the health of a ship's company," recommending spring water as less likely to carry noxious effluvia than any source of surface water and stating that a ship should carry equipment for a

[40] Macdonald, *Victualling Board* (cit. n. 34), 101; Constantine John Phipps Mulgrave, *A Voyage towards the North Pole Undertaken by His Majesty's Command, 1773* (London: printed by W. Bowyer and J. Nichols, for J. Nourse, 1774), 11, 28 (quart/man/day), 60, 141–7, 205–9, 214, 215, 219; Olaudah Equiano, *The Interesting Narrative of the Life of Olaudah Equiano, or Gustavus Vassa, the African, Written by Himself*, ed. Werner Sollors (1789; repr., New York, N.Y.: W. W. Norton, 2001), 125–6, 130–4; Jedediah Stephens Tucker, ed., *Memoirs of Admiral the Right Hon[orabl]e the Earl of St. Vincent*, 2 vols. (London: R. Bentley, 1844), 1:193, 414.

[41] Nicholas Rogers, *The Press Gang: Naval Impressment and its Opponents in Georgian Britain* (London: Bloomsbury, 2008); Denver Brunsman, "Subjects vs. Citizens: Impressment and Identity in the Anglo-American Atlantic," *J. Early Repub.* 30 (2010): 557–86.

still "in case of distress." Even the water put into casks at official supply depots underwent new scrutiny. In the early nineteenth century, for example, the Admiralty ordered chemical analyses of two of the wells it used to supply ships at Chatham.[42]

Distillation of seawater into drinking water would be replaced, however, by the adoption of large storage tanks, filled while in port, and sloshingly abundant (it was hoped) even for long-distance travel. Toward the end of the Napoleonic Wars, the Admiralty began discussing this solution, from circa 1809 until the conclusion of hostilities in 1812. An order from April 1815, from John Wilson Croker, secretary to the Admiralty, stated that iron water tanks were to be introduced generally on His Majesty's ships, in place of casks (or tanks containing lead, which were recognized as conducive to poisoning). Croker's letter listed tanks already issued, by class of warship. Thereafter, over the nineteenth century, metal storage tanks would become the way to carry water at sea by default, for both military and commercial vessels. The tanks would end the long conversations about the interface between wood and water since their main feature was a materially undifferentiated capacity; they provided so much water for so many people, wherever the water happened to be brought, increasingly by hoses, from land to ship.[43]

GOOD FOR EVERYBODY

The new practice stated not only that people (sailors) needed to drink water, but that they would or even should do so in certain quantities, which amounted to recommendations that displaced any lingering interest in different types of water. The timing of this shift is worth noting, because it does not map neatly over the previously identified trends in vitalist thought. Tryon's early works, for example, which used vitalist imagery, nevertheless predated the vitalism of the late eighteenth century, which questioned the instantiation of mechanistic theories of nature. By the time vitalism was being drained from medical recommendations about water, the neovitalists were in full protest against the mechanical philosophy. The succeeding links from Tryon, to Cullen, to Black indicate that dietetics were indebted to older medical theories, ones conducive to vitalist ideas, yet ultimately remade into nonvitalist statements about the material benefits of drinking water, as championed by a branch of the British state that, at the turn of the eighteenth century and height of the Napoleonic Wars, was highly visible if not feared.

The new norm for maritime drinking water then seeped back into English medical advice literature. Former sailor Peter Crosthwaite placed the tenets of the cool regimen within maritime and imperial contexts in his *Ensign of Peace* (1775), summarizing his experiences at sea and in India, the better to extol water as "the greatest dissolvent that we have." He commended the robust health of the natives of India who, like the more cautious of his sailor companions, washed frequently in water and took

---

[42] Johannes Haarhoff's "The Distillation of Seawater on Ships in the 17th and 18th Centuries," *Heat Transfer Engineering* 30 (2009): 237–50, is a useful chronology of events, if an internalist analysis of them; Macdonald, *Victualling Board* (cit. n. 34), 87; Gilbert Blane, *A Short Account of the Most Effectual Means of Preserving the Health of Seamen* ([London? publisher unknown, 1780?]), 12–14.
[43] Macdonald, *Feeding Nelson's Navy* (cit. n. 37), 84–5; J. W. Croker order to his command, 21 April 1815, ADM 359/35A/90, formerly National Archives, now Caird Library and Archive, National Maritime Museum, London.

it as their only beverage. "Water is their common drink," he said of Indians, whose hospitality included giving visitors access to good wells; "washing, and using good water for diet," likewise maintained health on ships. Crosthwaite referred to formal medical authorities to support his recommendations. He cited Bucan, above all, on the evils of "artificial drink," the beer or grog sailors often preferred.[44]

For his own part, William Buchan, in his *Domestic Medicine*, described sailors simply as another category of "laborious" persons and said that they (and by implication others) were wise to drink water in place of alcohol. He particularly praised Captain James Cook for his attention to "good water" for his men. Given that these were recommendations for working men who (whether willingly or not) constituted some of the vigorous sinews of nation and empire, they differed from those for water rationed to the unwillingly incarcerated, for whom water still carried a punitive aspect, as when prisoners, convicts, orphans, or workhouse inhabitants were reduced to such a regimen. The children in London's Foundling Hospital, for example, subsisted on "plain and simple" fare, including broths and porridges, "their Bread coarse, and their Drink Water." A report on English prisons, from 1777, explained that "those who drink only water, and have no nutritious liquor, ought to have at least a pound and a half of bread every day," thereby conveying the new conclusion that water had no nutritive capacity, but was mere fluid.[45]

The shifts in thinking about water were summarized in the third edition of the *Encyclopaedia Britannica* (1797). Under the topic of "DRINK," the entry declared that "the general use of drink is, to supply fluid," not nutrition; it dilutes, thus is "opposed to nourishment." And anything humanly drinkable was a version of water, the universal, necessary fluid. A separate entry for "WATER" described it as "a well known fluid, diffused through the atmosphere . . . and abounding in a certain proportion in animals, vegetables, and minerals." Though once thought to be one of the four elements, chemical experimenters had found it to be a "compound substance" with further "substances combined with it." Those impurities could now be isolated and identified, to "select the purest water for the purposes of life, and to avoid water which might be improper and hurtful; or, when good water cannot be had, to separate those substances from it which render it impure." Any once-living element within it, from animal or vegetal traces, was likely to be an impurity that caused a telltale bad smell and taste. Older traditions of vitalism were thus ignored, except for the concept of "holy water," which this encyclopedic reference work (from a Protestant nation) firmly identified as Catholic. It is by no means the case that popular opinions about the nutritive capacity of water vanished, nor that belief in multiple kinds of waters faded. But a new, official comprehension of drinking water as a nonliving fluid essential to life had emerged.[46]

---

[44] Peter Crosthwaite, *The Ensign of Peace* (London: printed for J. Wilkie, 1775), 5, 25–6, 106–18, 140–213.

[45] William Buchan, *Domestic Medicine* [. . .] (Boston, Mass.: printed by Joseph Bumstead, for James White, Court-Street, and Ebenezer Larkin, Jr., Cornhill, 1793), 31–2 (this edition included the reference to Cook, missing from earlier editions [cit. n. 20]); *An Account of the Hospital for the Maintenance and Education of Exposed and Deserted Young Children* [. . .] (London: Foundling Hospital, 1759), 50; John Howard, *The State of the Prisons in England and Wales* [. . .] (Warrington, UK: William Eyres, 1777), 60–1.

[46] *Encyclopaedia Britannica*, 3rd ed. (1797), s.v. "drink," 125; s.v. "water," 806, 807–10, 816.

WATER IS LIFE

At the 2016–17 Standing Rock protest, Native Americans asserted an Indigenous right to a substance that itself was vital. Their claim referenced, in two equal parts, a non-Western belief about nonhuman materials as themselves animate, and an expectation of state intervention when drinking water was threatened. The "No DAPL" protest overlapped with a scandal in Flint, Michigan, which had begun in 2014, when improperly treated public drinking water had become severely contaminated with lead. The pair of US controversies emphasized how access to uncontaminated water for human consumption is now regarded as a universal right, and therefore as the responsibility of government authorities. This is a modern development. To have pure water on tap is a very recent expectation, dating, at the earliest, from the nineteenth century.[47]

That expectation is also the result of earlier state definitions of the human need for certain units of water. This was different from the preference for various kinds of waters, as expressed in earlier English/British descriptions of optimal drinkable fluids. While many governmental authorities were no doubt involved in this shift of expectation, the British Admiralty was one extremely visible and highly powerful state actor that endorsed the drinking of water, generically defined, as a universal need, and as specified in standard units of measurement per thirsty drinker. This itself was a twinned redefinition. First, it rejected earlier belief that there was no such thing (on Earth) as one kind of water, but instead that multiple versions of it existed, with different kinds of vital principles within them, some of them peculiarly beneficial for humans; instead, water had become a universal fluid universally good for human bodies. Second, the policy accepted new, publicly recognized authorities as having the power to define what water was drinkable and how much was needed—and indeed, *that* it was needed. Military authorities did this at a critical high point of British imperial activity and at a concurrent moment of selective assimilation of non-Western and colonial beliefs to European definitions of water and health.

Whatever the medical rationale that would subsequently support this new set of beliefs, it was the power of the Western nation-state, as expressed through a branch of the British military, that was an early proponent of its truth. Today, it is notable that Indigenous concepts of nature have been instrumental in designating nature as comparable to humanity in its capacity to have legally protected rights. Given earlier Western conceptions of animate nature, this is also a redesignation, a restoration of what had been removed within European culture. But the other necessary component of this current shift in law is state power. Within our current condition of planet-wide environmental emergency, this new alliance is critical, uniting something deleted from global Western culture, but preserved elsewhere, along with something else that had developed out of the modern history of the West.

The full history of that alliance's expression and contestation, especially in relation to multiple imperial and colonizing arenas, would be valuable in understanding how body and diet were redefined, not least as political subjects, in the early modern period. This small history of water affirms the claims of new materialism—that materiality

---

[47] Jean-Pierre Goubert, *The Conquest of Water: The Advent of Health in the Industrial Age*, intro. Emmanuel Le Roy Ladurie, trans. Andrew Wilson (Princeton, N.J.: Princeton Univ. Press, 1989), 21.

matters, and that human comprehension of material things matters—while urging attention to all places and times, beyond the new materialism's current focus on the modern if not contemporary West. A broader focus in this article has identified the significance of the Western state in defining water as essential to human life, but it has also identified the historical suppression of European ideas about living water, preventing (for some time) any amalgamation with similar, non-European ideas. For the moment, histories of demands for subsistence, of definitions of minimal nutritional standards, and of rationing have focused on food and other substances that supply bodily energy. As the world's stocks of freshwater become scarcer, and if conflicts such as those at Standing Rock and in Flint become more common, our knowledge of how water has become defined as essential to life—and on whose authority—is still a glass only half full.[48]

[48] Nick Cullather, "The Foreign Policy of the Calorie," *Amer. Hist. Rev.* 112 (2007): 337–64; Lizzie Collingham, *The Taste of War: World War II and the Battle for Food* (London: Penguin, 2011); Dana Simmons, *Vital Minimum: Needs, Science, and Politics in Modern France* (Chicago: Univ. of Chicago Press, 2015); Michael Klare, *The Race for What's Left: The Global Scramble for the World's Last Resources* (New York, N.Y.: Metropolitan, 2012).

The Shape of Meat:
Preserving Animal Flesh in Victorian Britain

by Rebecca J. H. Woods*

ABSTRACT

In the mid-nineteenth century, animal flesh was subject to a range of treatments in an effort to preserve meat grown on the fringes of the British Empire (in Australia and New Zealand, South and North America) for consumption in urban centers in Britain. Focusing on the publications of the British Society for the Encouragement of Arts, Commerce and Manufacture, and allied sources such as the *Lancet*, this article demonstrates that the more a preservative technique transformed animal flesh, the more likely consumers—often presumed to hail from the poor and working classes—were to resist it. This resulted in frustration among elite "men of science and industry," who held that tinned, canned, dried, or chemically treated meats were a "great boon" to precisely these classes. By refusing to consume industrial charqui, which was salted and dried, or by purchasing imported tinned Australian beef or mutton only unwillingly, the lower classes frustrated the ambitions of would-be tastemakers in the Society of Arts, who interpreted consumer resistance in their articles and published reports as the lower orders' refusal to act in their own best interest. Importantly, it was the very changeability of meat—its figurative malleability as well as its material inconstancy—that enabled industrial transformations, consumer resistance, and its cultural symbolisms, making it a particularly rich object of study for historians of science.

In the 1860s, political economists, statisticians, journalists, industrialists, and technologists in Britain's imperial metropole looked out beyond their shores and saw vast herds of cattle and flocks of sheep languishing for want of a market. Some of these animals cropped the grass—newly seeded and fiercely maintained—of British colonies proper.[1] Many of them browsed the pampas and prairies of South and North America, part of the sphere of British financial and sometimes cultural influence, but politically if not economically independent of Rule Britannia.[2] These herds and flocks were often at

* Department of History, University of Toronto, Room 2086 Sidney Smith Hall, 100 St. George St., Toronto, ON M5S 3G3, Canada; rebecca.woods@utoronto.ca.

I am grateful to Emma Spary, Anya Zilberstein, Suman Seth, W. Patrick McCray, Lisa Messeri, and Emily Wanderer for their comments on drafts of this article.

[1] Recent scholarship emphasizes the constructed nature of New World grasslands. See Tom Brooking and Eric Pawson, eds., *Seeds of Empire: The Environmental Transformation of New Zealand* (New York, N.Y.: I. B. Tauris, 2011); and Maura Capps, "Fleets of Fodder: The Ecological Orchestration of Agrarian Improvement in New South Wales and the Cape of Good Hope, 1780–1830," *J. Brit. Stud.* 56 (2017): 532–56.

[2] Richard Perren documents British investment in the North and South American livestock industries in "Capital and Markets," chap. 3 in *Taste, Trade and Technology: The Development of the International Meat Industry Since 1840* (Aldershot, UK: Ashgate, 2006).

least partly composed of descendants of the British animals sent out as ovine and bovine counterparts to human colonizers a hundred or more years before.[3] They had been deliberately encouraged to reproduce, both in order to solidify European claims to indigenous lands and in the service of imperial markets such as that for wool. Now, having reproduced at a much faster rate than human populations, there were far more flesh-bearing domesticates than local populations could possibly ingest.[4] And yet, when these observers turned their gaze homeward, they saw hordes of people—primarily the poor and working classes, they claimed—clamoring for more meat.

Merchants, distributors, and victualers in Britain, on the one hand, and Australasian stockmen and ranchers on the other, were keen to "promote equalisation of supply and demand," but meat was a notoriously difficult article to redistribute.[5] On the hoof, transport over such great distances was uneconomical, while on the hook, meat was too vulnerable to putrefaction to remain edible for the duration of a voyage from the Americas or Australasia.[6] Thus, finding a way to forestall natural processes of decay for long enough to transport meat across the ocean (or oceans) and distribute it to consumers in Britain seemed the likeliest solution to what appeared to these interested parties as a problem of misplaced supply and mismatched demand. Expressing a faith in progress typical of their time, contemporaries were certain that "science and mechanical skill will ere long master the difficulty" of so doing.[7] And indeed, the technically inclined and profit minded applied themselves to this project with great zeal. Subscriptions were taken and companies formed.[8] The Society for the Encouragement of Arts, Manufactures and Commerce (hereafter the Society of Arts)—a more practical analogue to the Royal Society—offered a prize to anyone who could devise "a process for preserving fresh meat better than by any method hitherto employed, applicable to the preservation of meat in countries where it is now almost valueless, so as to render it an article of commerce."[9] Over a twenty-year period, meat was salted, tinned, enveloped in paraffin,

---

[3] On breeds of livestock in the British Empire, see Rebecca J. H. Woods, *The Herds Shot Round the World: Native Breeds and the British Empire* (Chapel Hill: Univ. of North Carolina Press, 2017), especially chap. 5, "A Universal Type." On Spanish antecedents to British sheep and cattle, see Elinor Melville, *A Plague of Sheep: Environmental Consequences of the Conquest of Mexico* (New York, N.Y.: Cambridge Univ. Press, 1994).

[4] Alfred Crosby, *Ecological Imperialism: The Biological Expansion of Europe, 900–1900*, 2nd ed. (Cambridge: Cambridge Univ. Press, 2004). For a characterization of this issue with respect to New Zealand specifically, see Rebecca J. H. Woods, "Breed, Culture, and Economy: The New Zealand Frozen Meat Trade," *Agricultural History Review* 2 (2012): 288–308. For the role of domesticated livestock in colonial territorial acquisition and the establishment of markets, see Virginia De John Anderson, *Creatures of Empire: How Domestic Animals Transformed Early America* (New York, N.Y.: Oxford Univ. Press, 2004); Melville, *Plague of Sheep* (cit. n. 3); and William Cronon, *Changes in the Land: Indians, Colonists, and the Ecology of New England* (New York, N.Y.: Hill and Wang, 1983).

[5] "Animal Food Supplies," *Lancet* 102 (1867): 94–7, on 94.

[6] C. Knick Harley, "Steers Afloat: The North Atlantic Meat Trade, Liner Predominance, and Freight Rates, 1870–1913," *J. Econ. Hist.* 68 (2008): 1028–58; Richard Perren, "The North American Beef and Cattle Trade with Great Britain, 1879–1914," *Econ. Hist. Rev.* 24 (1971): 430–44.

[7] "Australian Meat," *Lancet* 93 (1869): 239.

[8] One agricultural historian writes that the mid-1860s "saw a rash of canning factories" established overseas: E. J. T. Collins, "Food Supplies and Food Policy," in *The Agrarian History of England and Wales*, vol. 7, *1850–1914*, pt. 1, ed. Collins (Cambridge: Cambridge Univ. Press, 2000), 33–71, on 36.

[9] "Proceedings of the Society: Food Committee," *Journal of the Society of Arts* 15 (4 January 1867): 99–102, on 100. (**Hereafter** referred to as *J. Soc. Arts.*) See also Richard Perren, *Taste, Trade and Technology* (cit. n. 2), 8.

soaked in chemicals, vacuum packed, frozen, refrigerated, and treated in a host of other ways in a great collective effort to hold putrefaction at bay.[10]

As E. C. Spary and Anya Zilberstein argue in the introduction to this volume, food is profoundly relational and transformational—"a site of direct encounter between individuals and larger social structures or transformations over which they may have little power" and that can emerge as a flashpoint between conflicting interests, commitments, and contested forms of expertise.[11] Meat in mid-nineteenth-century Britain was no exception. Not only a "very perishable material," in the words of the *Lancet*, vulnerable to putrefaction and decay, animal flesh was, and is, a highly changeable substance, materially and metaphorically. Conceptually, meat held a wide array of significations and values, ranging from those implicated in national identity to those made in service of scientific authority. Materially, meat was subject to industrial processes capable of transforming it from a familiar article of diet into highly debated gustatory novelties.[12] In 1860s Britain, industrialists, entrepreneurs, and "men of science" assumed that innovation would solve what they perceived as Britain's problem of supply and demand, but the products they proffered met with resistance from consumers—more specifically, poor and working-class consumers, the "teeming masses" who constituted the stated beneficiaries of these efforts to "increas[e] and cheapen . . . the supply of animal food" in Britain, and who exercised their power by choosing whether or not to purchase imported preserved meats.[13] In effect, the more a process transformed that which it sought to preserve—by cooking, chemical application, desiccation, or by some other process—the less appeal it held for consumers. The more consumer resistance a tinned or dried product generated, the more the frustrations of would-be tastemakers grew.

This dialectic emerges from the published record of the Society of Arts' Food Committee, established in 1867, and more specifically, from its subcommittee on meat, and the broader discussion of meat preservation carried on in the *Journal of the Society of Arts* and other specialist literature largely between the years 1860 and 1880. Reading these records against the grain, this article demonstrates how the very changeability of meat itself became an opportunity for poor and working-class consumers to resist and refuse both the claims and the products of scientific and industrial expertise. Members of the Society of Arts and their contemporaries writing in the *Lancet* and more broadly understood "science" to encompass both the technical ingenuity requisite for the preservation of meat as well as expert evaluations of its value as food, which was usually expressed in terms of "wholesomeness," "nutritive power," or "nutritive value."

---

[10] For the history of food preservation generally, see Sue Shephard, *Pickled, Potted and Canned: The Story of Food Preserving* (London: Headline Book Publishing, 2000); Stuart Thorne, *The History of Food Preservation* (Totawa, N.J.: Barnes & Noble Books, 1986); and C. Anne Wilson, ed., *"Waste Not, Want Not": Food Preservation from Early Times to the Present Day* (Edinburgh: Edinburgh Univ. Press, 1991).

[11] E. C. Spary and Anya Zilberstein, "On the Virtues of Historical Entomophagy," in this volume, 12, 7.

[12] Mark R. Finlay, "Quackery and Cookery: Justus von Liebig's Extract of Meat and the Theory of Nutrition in the Victorian Age," *Bull. Hist. Med.* 66 (1992): 404–18; Lesley Steinitz, "Making Muscular Machines with Nitrogenous Nutrition: Bovril, Plasmon and Cadbury's Cocoa," in *Food and Material Culture: Proceedings of the Oxford Symposium on Food and Cookery 2013*, ed. Mark McWilliams (Totnes, UK: Prospect, 2014), 289–303; Steinitz, "The Language of Advertising: Fashioning Health Food Consumers at the *Fin de Siècle*," in *Food, Drink, and the Written Word in Britain, 1820–1945*, ed. Mary Addyman, Laura Wood, and Christopher Yiannitsaros (London: Taylor & Francis, 2017), 135–63.

[13] "Animal Food Supplies" (cit. n. 5), 94.

They expected science to solve Britain's "meat deficit."[14] But these self-appointed experts in the Society of the Arts and elsewhere were forced to admit that consumer preference would also determine the success or failure of a given product. Charqui—an industrially dried and salted nineteenth-century South American precursor to beef jerky—offers a particularly stark example of the way this scientistic push met with resistance, but tinned and canned meat were subject to similar contestation. Ultimately, freezing and refrigeration—processes which seemed to transform dead meat the least—carried the day, becoming by the close of the century a common article of diet.

AN EMPIRE OF MEAT EATERS

The way in which Great Britain developed foodstuffs, and consequently British diets, underwent profound transformation over the course of the nineteenth century. This transformation was part of a much larger contemporaneous structural shift in British culture and economy that redefined both the substance of diet and the culture of consumption from the macroeconomic level down to the plates of working people across Britain.[15] Early in the century, a regional system of procurement and distribution reigned, where grains and livestock raised in various districts of the British Isles were exchanged for consumption in population centers, and local market gardens and urban dairies provided fresh vegetables and milk for cities across the industrializing north and in London.[16] Diets were seasonally varied and locally determined, with a significant portion of foodstuffs (such as bread and preserves) made in the home.[17] As Britain's population grew, and became increasingly concentrated in London, Manchester, Birmingham, and other newly industrialized conurbations, this arrangement came under strain.[18] By midcentury, urban centers, especially London, absorbed more and more livestock from Scotland, Wales, Ireland, and eastern Europe, while reliance upon imported grains grew. In addition to sugar and tea, Britons increasingly came to subsist on purchased foods fabricated from imported grain, and eventually, imported meat as well.[19] As procurement networks industrialized, so too did the diet

---

[14] See also Benjamin Aldes Wurgaft, "Meat Mimesis: Laboratory-Grown Meat as a Study in Copying," in this volume.

[15] John Burnett documents this transition in *Plenty and Want: A Social History of Diet in England from 1815 to the Present Day* (London: Thomas Nelson & Sons, 1966).

[16] See Jack Cecil Drummond, *The Englishman's Food: A History of Five Centuries of English Diet* (London: J. Cape, 1940), esp. pt. 3; and Craig Muldrew, *Food, Energy and the Creation of Industriousness: Work and Material Culture in Agrarian England, 1550–1780* (Cambridge: Cambridge Univ. Press, 2011).

[17] Andrea Broomfield, "Rushing Dinner to the Table: The 'Englishwoman's Domestic Magazine' and Industrialization's Effects on Middle-Class Food and Cooking, 1852–1860," *Victorian Periodicals Review* 41 (2008): 101–23, on 102.

[18] Thorne, *History of Food Preservation* (cit. n. 10), 17; Richard Perren, "Changes in Town Markets, 1840–64," chap. 3 in *The Meat Trade in Britain 1840–1914* (London: Routledge and Kegan Paul, 1978), esp. 32; Robyn S. Metcalfe, *Meat, Commerce and the City: The London Food Market, 1800–1855* (London: Pickering & Chatto, 2012). Although other cities in Europe experienced similar transformations, London's size and rapid growth meant that the strain to its existing systems of procurement was particularly acute. See Hans Jürgen Teuteberg, "Urbanization and Nutrition: Historical Research Reconsidered," in *Food and the City in Europe since 1800*, ed. Peter J. Atkins, Peter Lummel, and Derek Oddy (Burlington, Vt.: Ashgate, 2007), 13–24, on 18.

[19] Derek J. Oddy, "Food Quality in London and the Rise of the Public Analyst, 1870–1939," in Atkins, Lummel, and Oddy, *Food and the City* (cit. n. 18), 91–104, on 99. See also Sidney Mintz's classic account of the British industrial diet in *Sweetness and Power: The Place of Sugar in Modern History* (London: Penguin, 1985).

of most Britons (with the possible exception of society's highest echelons), inaugurating what Chris Otter has termed "the British nutrition transition," the first instance of a truly globalized food chain, and the origins of twentieth-century diets across the developed world.[20]

The significance of this broad shift was felt deeply with respect to meat. The association between national identity and meat eating has deep roots in English thought and culture, but it took on a new charge as a mark of Britishness in the eighteenth century with the intersection of contemporary dietetics and anti-French rhetoric, as Anita Guerrini has argued.[21] In material terms as well, meat— understood at the time to refer to beef, sheep meat, and pork (to the exclusion of fowl, fish, and game, the last of which was primarily a delicacy reserved for the gentry)—was central to the British diet, and Britons consumed quantities of meat far in excess of their continental counterparts.[22] This material enthusiasm for meat reinforced cultural preference so that even as the components and qualities of what constituted an adequate diet were debated throughout Europe at this time, few contested that meat was requisite.[23] When Wentworth Lascelles Scott, a statistician and expert on food adulteration, proclaimed it "a primary necessity of our national existence" in a paper read before the Society of Arts, he expressed a majority position.[24]

So, when a complex of factors converged to put the squeeze on Britons' access to meat in the 1860s, the issue of the nation's meat supply ranked high for policy makers, agronomists, and other interested parties. By midcentury, rising wages for industrial workers, which translated to greater purchasing power, contributed to a burgeoning demand for meat, while repeated zoonotic outbreaks in Europe undercut regular sources of imported foreign livestock, as did the transition to grain growing for export in traditional cattle districts of central Europe.[25] As the cost of butcher's meat rose accordingly, many learned commentators feared that animal protein was

---

[20] Chris Otter notes that the diversity of meats consumed rose with social standing, and that the characteristic high-fat, high-carbohydrate diet that emerged from the "British nutrition transition" was a phenomenon of the working class. See Otter, "The British Nutrition Transition and its Histories," *Hist. Comp.* 10 (2012): 812–25, on 813. See also Andrea Broomfield, who notes the persistence of traditional estate cooking after the early nineteenth century, but only among the landed gentry: Broomfield, "Rushing Dinner to the Table" (cit. n. 17), 102. For the impact of industrialized transportation networks on Britain's meat trade specifically, see Perren, "Changes in the Domestic Livestock Trade, 1840–64," chap. 2 in *Meat Trade in Britain* (cit. n. 18).

[21] Anita Guerrini, "Health, National Character and the English Diet in 1700," *Stud. Hist. Phil. Biol. Biom. Sci.* 43 (212): 349–56.

[22] Peter J. Atkins, "'A Tale of Two Cities': A Comparison of Food Supply in London and Paris in the 1850s," in Atkins, Lummel, and Oddy, *Food and the City* (cit. n. 18), 25–38, esp. 35; Perren, *Meat Trade in Britain* (cit. n. 18), 3.

[23] Ulrike Thoms, "The Technopolitics of Food: The Case of German Prison Food from the Late Eighteenth to the Early Twentieth Centuries," and Corinna Treitel, "Nutritional Modernity: The German Case," both in this volume.

[24] Wentworth Lascelles Scott, "On the Supply of Animal Food to Britain, and the Means Proposed for Increasing It," *J. Soc. Arts* 14 (21 February 1868): 255–68, on 256. In a paper read before the Society of Arts in 1875, Scott was described as "Public Analyst to the Counties of Durham and North Stanford"; Scott, "Food Adulteration and the Legislative Enactments Relating Thereto," *J. Soc. Arts* 23 (2 April 1875): 427–37, on 427. See also Scott, "On Food; Its Adulterations, and the Methods of Detecting Them," *J. Soc. Arts* 9 (1 February 1861): 153–62.

[25] The "Great Cattle Plague" of 1865 was particularly notable, and had a palpable effect on the availability of meat in Britain. See Arvel B. Erickson, "The Cattle Plague in England, 1865–67," *Agricultural History Review* 2 (1961): 94–103; Collins, "Food Supplies" (cit. n. 8), 36.

beyond the reach of "our teeming and poorer population."[26] As the 1860s drew to a close, Scott claimed, "the entire country is in a state of mitigated starvation."[27] Although such dramatic terms distorted the actual availability of meat in Britain, the rhetoric of scarcity persisted into the 1870s, especially in relation to the nation's meat supply.[28] Periodic hungers gripped Britain—notably during the Napoleonic wars, and again in the hungry 1830s and 1840s, the memory of which probably contributed to anxiety surrounding what an eminent agricultural historian called the "mid-Victorian meat famine"—yet economic historians largely agree that all but the very poorest in Britain were relatively well supplied with animal protein during the latter half of the nineteenth century, even if they paid dearly for it.[29] Per capita consumption was in fact considerably higher than in other European countries at 90 pounds per head in the 1860s.[30] Yet the cultural and rhetorical emphasis placed on meat eating as a marker of Britishness meant that the midcentury "meat-deficiency" was felt acutely, apparently by those who commented on it, and presumably also by those who experienced it directly.[31]

Anxiety about the presumed inadequacy of the nation's meat supply, moreover, was an expression of worry over Britain's standing as an industrial leader and was deeply tied to Britain's imperial identity. "It is more than likely," according to George Carrick Steet, a fellow of the Royal College of Surgeons who presented before the Society of Arts in 1865, "that our position among the nations is not a little due to [our] national taste for good, strong food and plenty of it."[32] Without sufficient meat to fuel labor, Steet worried, the working class might not be able to bear the mantle of Britain's industrial primacy. "If . . . our energies of body and mind are to be kept going it is absolutely necessary that proper supplies of aliment should be forthcoming," Steet continued, "and if that is not to be had at home we must go to other countries to seek for it."[33] If only, an anonymous contributor to the *Lancet* editorialized in 1867, "meat could be as easily transported from one country to another as tea, sugar, and

[26] "Australian Meat" (cit. n. 7), 239. Between 1850 and 1870, domestic production of meat rose by less than 3 percent; see Perren, "Foreign Imports and the Domestic Supply, 1840–64," chap. 5 in *Meat Trade in Britain* (cit. n. 18), 69.

[27] Scott, "Supply of Animal Food" (cit. n. 24), 256.

[28] Collins, "Food Supplies" (cit. n. 8), 33.

[29] E. J. T. Collins, "Rural and Agricultural Change," in Collins, *Agrarian History* (cit. n. 8), 7:107–16; quoted in Perren, *Taste, Trade and Technology* (cit. n. 2), 8. Michael Nelson notes that nineteenth-century diets in Britain were "radically different" across the classes, with the very poor eating very little meat, while "well-off families" were well supplied. See Nelson, "Social-Class Trends in British Diet, 1860–1980," in *Food, Diet and Economic Change Past and Present*, ed. Catherine Geissler and Derek J. Oddy (Leicester: Leicester Univ. Press, 1993), 101–20, on 102 and 103. Nonetheless, per capita meat consumption rose by approximately 50 percent between 1840 and 1914, according to Richard Perren; see Perren, *Taste, Trade and Technology* (cit. n. 2), 3. See also Perren, *Meat Trade in Britain* (cit. n. 18), 3; and Forrest Capie and Perren, "The British Market for Meat, 1850–1914," *Agr. Hist.* 50 (1980): 502–15. For hunger and the British state more generally, see James Vernon, *Hunger: A Modern History* (Cambridge, Mass.: Belknap Press of Harvard Univ. Press, 2007).

[30] Perren, *Meat Trade in Britain* (cit. n. 18), 3.

[31] For meat eating as a mark of Britishness, see Steven Shapin, "'You are What You Eat': Historical Changes in Ideas about Food and Identity," *Hist. Res.* 87 (2014): 377–92. For the dietary perspectives of the poor in late nineteenth-century Britain, see Anna Davin, "Loaves and Fishes: Food in Poor Households in Late Nineteenth-Century London," *Hist. Workshop J.* 41 (1996): 167–92. Quotation from Scott, "Supply of Animal Food" (cit. n. 24), 255.

[32] G. C. Steet, "On the Preservation of Food, especially Fresh Meat and Fish, and the Best Form for Import and Provisioning Armies, Ships, and Expeditions," *J. Soc. Arts* 13 (1865): 309–15, on 315.

[33] Ibid.

grain can, much benefit would result both to the owners of land and stock on one side of the Atlantic, and to the imperfectly fed populations on the other side."[34] Meat, though, is very little like sugar, tea, or grain, all of which are bulky yet relatively light commodities, making them cheap to transport, and easy to store for long periods without degradation. Instead, both the weight of live animals and the tendency toward decay of dead meat kept the profit margins of the live trade industry slim, and hampered the feasibility of a long-distance trade in dead meat.

THE FOOD OF THE PEOPLE

In 1866, in the midst of this perceived crisis of supply and demand, the Society of Arts established a committee on "the food of the people," commonly called the Food Committee.[35] Motivated in part by recent governmental inquiry into the "defective amount of nutritious food available for the population at large," the Food Committee's principal charge was to "inquire and report respecting the food of the people," and to bring scientific thought and method to bear on what was, according to the members of parliament, aristocrats, and the occasional medical man who staffed the committee, no mere "question of humanity and charity," but "a grave national question, vitally affecting 'arts, manufactures, and commerce,' and the very sources of national strength."[36]

Inquiries into the national diet, food supply, and "production, importation, preservation, and preparation of articles suitable for food" were precisely the kind of reasons for which the Society of Arts had been founded in 1754.[37] An organization devoted to the "encouragement of Arts, Manufactures and Commerce in Great Britain," it drew members from throughout Britain's educated strata.[38] Improvement-minded worthies like Lord Romney, elected the society's second president in 1761, and other "men of great property" came together with men of lesser property but great reputation for applied practical or technical knowledge, to offer premiums "for such Productions, Inventions, or Improvements as shall tend to the employing of the Poor, to the Increase of Trade, and to the Riches and Honour of this Kingdom, by Promoting Industry and Emulation."[39] At regular Wednesday meetings, where members were invited to read papers on their relevant expertise, in the pages of its journal, and among the specialist committees, members of the society sought scientific and technical solutions to problems of national and imperial significance.[40]

As a central article of diet, rhetorically and materially, meat and the questions of where to get it and how to preserve it, distilled precisely the issue of imperial order

[34] "Animal Food Supplies" (cit. n. 5), 94.

[35] "Food Committee," *J. Soc. Arts* 14 (16 November 1866): 781.

[36] Ibid.; "Food Committee" (cit. n. 9), 99. On the broader nutritional and public health fallout from the Privy Council's 1863 report on the diet of the poor, see Edwin Chadwick, "'Mutton Medicine,' and the Fever Question," *Bull. Hist. Med.* 70 (1996): 233–65. For a list of inaugural members, see "Proceedings of the Society: Food Committee," *J. Soc. Arts* 15 (21 December 1866): 69.

[37] "Food Committee" (cit. n. 9), 99.

[38] The statement was made by William Shipley in 1754, and was quoted in James Harrison, *Encouraging Innovation in the Eighteenth and Nineteenth Centuries: The Society of Arts and Patents, 1754–1904* (Gunnislake, UK: High View, 2006), vii.

[39] Ibid.; D. G. C. Allan and John L. Abbott, "General Introduction," in *The Virtuosi Tribe of Arts and Sciences: Studies in the Eighteenth-Century Work and Membership of the London Society of Arts*, ed. Allan and Abbott (Athens: Univ. of Georgia Press, 1992), xv–xxii, on xvii–iii.

[40] Harrison, *Encouraging Innovation* (cit. n. 38), xxii, 37–8.

and political economy at the heart of the Society of Arts' mandate. Food, as E. C. Spary has shown in the context of eighteenth-century France, played a constitutive role in building, maintaining, and transforming bodies in the service of European imperial expansion.[41] Because food is and has been such an important marker of cultural identity, the ability of cosmopolitan French imperialists to bring the gustatory trappings of their nation with them on board oceanic expeditions and to Caribbean plantation communities was of utmost importance. Fulfilling this desire required devising new techniques for making characteristic French cuisine portable, such as the creation of the stock cube.[42] Preserved foods, including meat, were thus a crucial way in which European bodies could remain European in the colonies, but they were also an opportunity for colonial matter to reformulate metropolitan bodies. Harry Chester, who gave the opening address at the Food Committee's inaugural meeting on 21 December 1866, declared that "science was required to devise means" for dealing with the "millions of tons of beef and mutton [that] were wasting in distant quarters of the earth . . . [so] that commercial enterprise might be enabled to bring it to this country in a condition suitable for food."[43] Like the stock cubes of eighteenth-century French imperialists, but in reverse, preserved meat in the nineteenth century was to be the lynchpin with which to orchestrate bodies, nation, and empire. The herds and flocks of colonial places were now seen as the raw material through which to reconstitute the industrial human bodies of the imperial metropole in stronger, more efficient laborers.[44] Within this logic, the "resources" of global-imperial places—not only dead meat, but grain, sugar, and other foodstuffs as well—would be redirected toward the maintenance and constitution of the very same colonizing bodies that were intended to supplant indigenous populations throughout the empire.

THE QUESTION OF MEAT

So significant was this matter to the well-being of Britain that by the third meeting of the Food Committee, a subcommittee devoted to "the question of meat" was appointed; it was hoped the subcommittee would oversee the distribution of the Trevelyan Prize, on offer since 1864 to reward a recipient for any superior method of preserving fresh meat.[45] But the "question of meat" and its nutritive value was a complex one in

[41] E. C. Spary, "Self Preservation: French Travels Between *Cuisine* and *Industrie*," in *The Brokered World: Go-Betweens and Global Intelligence, 1770–1820*, ed. Simon Schaffer, Lissa Roberts, Kapil Raj, and James Delbourgo (Sagamore Beach, Mass.: Science History Publications, 2009), 355–86.

[42] Ibid., 364–9.

[43] "Food Committee" (cit. n. 9), 100.

[44] M. Norton Wise and Crosbie Smith chart the wide-ranging shift toward thermodynamical models of work and efficiency in nineteenth-century Britain in a three-part series published in the journal *History of Science*: Wise and Smith, "Work and Waste: Political Economy and Natural Philosophy in Nineteenth Century Britain (I)," *Hist. Sci.* 27 (1989): 263–301; "Work and Waste: Political Economy and Natural Philosophy in Nineteenth Century Britain (II)," *Hist. Sci.* 27 (1989): 391–449; and "Work and Waste: Political Economy and Natural Philosophy in Nineteenth Century Britain (III)," *Hist. Sci.* 28 (1990): 221–61.

[45] "Proceedings of the Society: Food Committee," *J. Soc. Arts* 15 (15 February 1867): 189–91, on 191. Two further subcommittees on milk and fish were appointed at the same meeting. W. C. Trevelyan had first "placed in the hands" of the Society £70 to be offered "for any subject the Council [of the Society] thinks fit" in June 1863. It was designated for "Preserved Fresh Meat" in November of the same year. See "Annual Report," *J. Soc. Arts* 11 (26 June 1863): 546–50, on 548; and "Subjects for Premiums for the Sessions 1863–4 and 1864–5," *J. Soc. Arts* 11 (13 November 1863): 1–8, on 5.

the 1860s. Contemporaries understood meat to be a changeable substance, and they recognized the challenges this posed for its preservation. Meat consisted of "organic bodies of highly complex constitution" subject to "constant mutation," whether the animal "yielding it" was alive or dead, explained the *Lancet*.[46] If the former, the set of changes to which meat was subject—growth, regulation, "constant renewal and repair"—were "controlled by the vital force" of the creature in question, but upon death, "other changes immediately commence, resulting in putrefactive decomposition."[47] Although microorganisms were identified in the late eighteenth century, their precise role in putrefaction remained obscure until almost the close of the nineteenth. That bacteria were associated with decayed flesh was well known thanks to the prevalence of microscopy as a method of analysis, but their presence in decaying organic matter was assumed to be effect, rather than cause.[48] Steet explained that "as soon . . . as life ceases," the "constituents of flesh and other structures" composing the animal body "have a tendency to resolve themselves into new compounds by the union of their elements with atmospheric air and with one another."[49] Existing methods of preservation that excluded oxygen or precluded oxidation, such as canning or pickling, offered good support for this theory, and so interpretations of putrefaction remained rooted in oxidation.[50]

Novel methods developed for preserving meat around midcentury likewise focused on excluding oxygen from contact with meat. Many tried to accomplish this by providing a "protecting shield or bulwark" such as a tin or a can between the meat and the "oxidizing influences of the atmosphere," as Scott explained.[51] But preservation could also be accomplished by "deoxidating . . . chemical substance[s]" that "rapidly absorb[ed]" oxygen, or by "the addition of some substance which . . . prevents or arrests oxidation or putrefaction by its mere presence."[52] Preservative additives ranged from the familiar (salt, smoke) to the novel "chemical antiseptics," of which bisulphite of lime constituted Scott's preferred method.[53] Over the course of the Subcommittee on Meat's existence—alongside its parent committee, it met regularly until 1879—it regularly sampled specimens of meat preserved by various methods and interviewed expert "witnesses" from the medical and practical professions whose work was related to meat.[54]

---

[46] "Animal Food Supplies" (cit. n. 5), 94.

[47] Steet, "Preservation of Food" (cit. n. 32), 312; Analytical Sanitary Commission, "Records of the Results of Microscopical and Chemical Analyses of the Solids and Fluids Consumed by All Classes of the Public," *Lancet* 59 (1852): 294–7, on 294; "Animal Food Supplies" (cit. n. 5), 94.

[48] Thorne, *Food Preservation* (cit. n. 10), 13.

[49] Steet, "Preservation of Food" (cit. n. 32), 312.

[50] Thorne, *Food Preservation* (cit. n. 10), 13, 14. For the reception of germ theory in Britain, see Michael Worboys, *Spreading Germs: Disease Theories and Medical Practice in Britain, 1865–1900* (Cambridge: Cambridge Univ. Press, 2000); Peter J. Atkins, "The Pasteurisation of England: The Science, Culture and Health Implications of Milk Processing, 1900–1950," in *Food, Science, Policy, and Regulation in the Twentieth Century: International and Comparative Perspectives*, ed. David F. Smith and Jim Phillips (New York, N.Y.: Routledge, 2000): 37–52.

[51] Scott, "Supply of Animal Food" (cit. n. 24), 266.

[52] Ibid.

[53] Ibid., 267.

[54] Quotation from "Food Committee" (cit. n. 9), 99. One of the first experts to appear before the subcommittee was Johann Thudichum, the noted German biochemist and former student of Justus von Liebig: "Proceedings of the Society: Food Committee," *J. Soc. Arts* 15 (8 March 1867): 237–41.

But other, simpler ways of preserving meat, from sun drying to smoking or salting, existed and were also widely employed. Despite their taste for smoked herring and bacon, the British had a tendency to associate nonindustrial methods of preservation—especially those drawing on the evaporative power of the sun—with "less civilized peoples" due to their antiquity and presumed simplicity.[55] "The process of drying or desiccating is . . . hardly a scientific process at all," a miscellaneous note on food preservation stated in the *Journal of the Society of Arts* in 1875.[56] However, despite the high imperial chauvinism at work in relegating atmospheric preservation to the primitive, Victorians displayed a great interest in other cultures' preservative techniques, particularly if they seemed useful for the "very important object of increasing and cheapening the supply of animal food" in Britain by facilitating importation from distant lands.[57] Some of the earliest efforts to exploit the "enormous meat stores of Australia and South America" were based on indigenous methods of preservation that harnessed the power of the sun and the atmosphere (the "desiccating class," in Scott's typology, "which include all methods for robbing food products of their natural moisture").[58] In this way, the British Empire provided not only a source of meat for preserving, but the method by which to do so.

THE RAW AND THE DRIED

Just as imperialists mined the globe for resources, including the "enormous meat stores" of distant lands, so too they mined indigenous cultures for methods of preservation that might provide the raw material, so to speak, for novel industrial methods of preserving meat. In scouring indigenous cultures at the fringes of Britain's imperial expansion for methods of preservation, British industrialists and innovators hoped to subject desiccation to industrialization. Pemmican, a kind of dried animal flesh mixed with berries and fat and used extensively by indigenous peoples of the North American Plains and the Pacific Northwest, came under consideration. Though it came to provide the basis of a regional Great Plains energy regime, at least until the destruction of the bison in the late nineteenth century, for metropolitan entrepreneurs, pemmican was little more than a curiosity.[59]

However, British industrialists went much further in adapting, extending, and industrializing the production of charqui.[60] Charqui originated as sun-dried llama meat among the Quechua people of present-day Peru, but by the nineteenth century it applied to horseflesh and beef as well as to that of native ungulates, and was used more broadly to sustain the laboring bodies of gauchos and enslaved laborers throughout South America. By applying copious amounts of salt, exposing it to the open air,

---

[55] Samuel Rideal, "The Use and Abuse of Food Preservatives," *J. Soc. Arts* 48 (1908): 384–93.

[56] "Miscellaneous: Food Preservation," *J. Soc. Arts* 23 (1 October 1875): 917–20, on 917.

[57] "Animal Food Supplies" (cit. n. 5), 97.

[58] Scott, "Supply of Animal Food" (cit. n. 24), 267, 262.

[59] George Colpitts, *Pemmican Empire: Food, Trade, and the Last Bison Hunts in the North American Plains, 1780–1882* (New York, N.Y.: Cambridge Univ. Press, 2015).

[60] "Food Preservation" (cit. n. 56), 917. Biltong, the dried meat of southeast Africa, which would have been familiar to British colonists in South Africa, did not seem to have been considered as a possible method for cheapening meat for the masses and only became a subject of medical/dietary notice during the Second Boer War (1899–1902). See, for example, "Emergency Rations," *British Medical Journal* 1 (1900): 1243–4; W. D. Haliberton, "The Composition and Nutritive Value of Biltong," *Brit. Med. J.* 1 (1902): 880–2; and "The Preservation of Foods," *Brit. Med. J.* 1 (1908): 936–8.

and then packing it in barrels, firms funded by British investment transformed cattle flesh into industrial charqui and introduced it to British palates as a solution to the mid-nineteenth-century "meat-deficiency."[61] Very quickly, though, "an almost universal opinion against it" took root, first among the lower classes who were its target consumers and subsequently among experts and promoters who conceded that charqui was neither toothsome nor nutritious.[62]

Charqui's ill-fated introduction to Britain illuminates the extent to which the relative success or failure of any preservation technique depended on the degree to which any given type of preserved meat represented the "genuine article"—that is, raw, uncooked flesh.[63] To varying degrees, all methods of preservation induced change in the tissues and fibers that compose meat, but the more limited the extent of that change, the better. Conversely, the more profoundly a given method changed the texture or appearance of meat, or sensory response to it, the more likely it was to fail. According to Steet, the British public's dislike of charqui was rooted in "its appearance and unsavoury smell in the raw state." Its smell, he explained, "resemble[ed] the odour of a small country chandler's store," where the rendering of animal carcasses and the byproducts of butchery for candle making would have taken place on site.[64] And its form and texture were most unlike those of fresh-killed animal flesh. In an exposé of the lives of the working poor aimed at middle-class readers, one contemporary journalist claimed to have overheard a pair of shoppers mistake a roll of rubber roofing felt for imported charqui, suggesting both the extent of transformation meat underwent in this particular preservative process and consumers' disdain for it.[65]

It is therefore hardly surprising that charqui failed to find a ready reception among its intended consumers. The further preserved meat was from resembling the fresh variety, the less likely it was to be considered tasty. And taste was so closely associated with evaluations of "nutritive value" that the less tasty a product was, the more chance there was of its "wholesomeness" or "nutritive power" coming into question. Steet undertook a chemical analysis of charqui, the results of which suggested that it "undoubtedly lost a large proportion of its best constituents" compared to fresh "lean English" beef. This led him to declare that charqui "would never commend itself to the stomachs and appetites of our people."[66] Indeed, in their published disquisitions on the question of preserving meat for importation to Great Britain, experts toggled directly back and forth between matters of taste and matters of nutrition, revealing just how closely the two were aligned. Mr. Warriner, a "practical cook," "did not decry charqui because of its untempting appearance." He claimed this during the discussion that followed Steet's paper at the Society of Arts, adding, "the great question was whether it contained the necessary nutritive qualities which all food should possess."[67]

---

[61] Scott, "Supply of Animal Food" (cit. n. 24), 256.

[62] Steet, "Preservation of Food" (cit. n. 32), 313.

[63] Wurgaft (cit. n. 14).

[64] Steet, "Preservation of Food" (cit. n. 32), 313, 314. On chandlery, see Joan Tighe, "An Early Dublin Candle Maker," *Dublin Historical Record* 41 (1988): 115–22. On industrial candle making more generally, see Jeremy Zallen, *American Lucifers: The Dark History of Artificial Light, 1750–1865* (Chapel Hill: Univ. of North Carolina Press, 2019), esp. chaps. 2 and 4 on the smell associated with this work.

[65] [James Greenwood], "The Depths of Poverty: A London Exploration. III—Poverty's Larder," *Englishwoman's Domestic Magazine* 15 (1866): 86.

[66] Steet, "Preservation of Food" (cit. n. 32), 313, 314. Nor was Steet alone in his opinion. The Society of Arts revisited charqui in its 1875 miscellany on food preservation, declaring that, "though largely used in South America, it will not 'go down' among ourselves"; see "Food Preservation" (cit. n. 56), 917.

[67] Steet, "Preservation of Food" (cit. n. 32), 316.

Edward Smith, a fellow of the Royal Society and also among Steet's audience, connected "the hardness and saltness" of charqui—"both bad qualities, the former showing the absence of the nutritious elements of meat . . . and the latter exercising a prejudicial effect upon the human frame"—with "the questionable flavour of this South American beef as at presently cured."[68] Steet offered a scientific rationale for this conclusion: salting "abstracted . . . the greater part of the nutritive element of meat," rendering the fiber "drier, harder, and less digestible."[69] Unpalatability thus called into question the "nutritive value" of preserved meats, and experts found themselves conceding to popular opinion.

Nor were other preserved meats immune to these challenges. A promotional notice for Australian preserved meat—mutton and beef, that is, cooked and sealed in tins—linked taste and nutritive value closely together, and claimed that "it is wholesome . . . it contains all the nourishment of meat of good quality, and . . . it is tender, sweet, and sound."[70] But an earlier evaluation of a similar product by the *Lancet*'s Sanitary Commission on Australian Boiled Beef concluded, "as might be supposed," that the tinned meat "was deficient in true meat flavour, and contained but little of the constituents of extract of meat."[71] Generally, the *Lancet* explained in 1867, a successful preservative method "must be neither costly nor complicated, nor injurious to the flavour, nor destructive to the nutritive qualities of the substance with reference to its use as food."[72] Taste and "nutritive value" went hand in glove in the mid-nineteenth century, the former often trumping claims to the latter.

A GREAT BOON REBUFFED

The relationship between class and consumption in the case of meat is a complex one. The palatability and suitability of a particular cut of meat, or of a particular animal breed, had long been a factor in British dietary habits, which is to say that different kinds of meat were believed to be suitable for different classes of people. In the early nineteenth century, for example, the meaty, marbled Leicester sheep—whose status in Britain approached that of an ovine national hero—was held to be the most appropriate breed for working-class consumption, while the more refined merino breed was touted as better suited for the plates and palates of the wealthy, on account of its more "gamey" flavor and texture.[73] Predictably, the starkest expression of the connection between animal protein and social status was made with respect to charqui. According to the *Lancet*, charqui was "a bad sort of animal food, which has been amply dried and definitively condemned and rejected as unsuited for European palates." In fact, it was fitting only for the lowest of the low, such as "the slave populations of the West Indies," among whom it was extensively employed "as an article of food."[74] Other social castoffs closer to home might also profit from charqui, such as Britain's prison population; Warriner, the practical cook in Steet's audience, suggested, "they should feed convicts on charqui, and leave English beef and mutton for the honest labouring population."[75]

[68] Ibid., 317.
[69] Ibid., 312. See also Scott, "Supply of Animal Food" (cit. n. 24), 267.
[70] "Australian Meat," *Lancet* 97 (1871): 681.
[71] "Report of the Lancet Sanitary Commission on Australian Boiled Beef," *Lancet* 89 (1867): 550.
[72] "Animal Food Supplies" (cit. n. 5), 95.
[73] Woods, *Herds Shot Round the World* (cit. n. 3), 63–4.
[74] "Animal Food Supplies" (cit. n. 5), 95.
[75] Steet, "Preservation of Food" (cit. n. 32), 316.

Although experts yielded to popular opinion in the case of charqui, they clung to what they claimed were the benefits of other preserved meats more persistently, and none more so than tinned boiled beef. Those involved in procuring, preserving, and marketing imported meat in Britain insisted that industrialized meat products represented a "great boon" to the lower orders. Time and again, contemporaries described their work in these terms. The *Lancet* pronounced in 1869: "Cheap meat, of good quality, would be an immense boon to the poorer classes, many of whom must experience extreme difficulty in procuring animal food at all."[76] The medical superintendent of the Sussex Lunatic Asylum noted in 1872 that preserved meat from Australia was in extensive use "in many of our public institutions," and yet, he lamented, "there is still a great prejudice against it . . . in the minds of many." If only "it could be proved to them that preserved meat as nutritious as fresh meat can be put within their reach" at a reasonable price, "it would . . . be a great boon to the lower classes."[77]

The poor, however, were unwilling to accept preserved meat, whether in tins or barrels, or gassed or dipped in wax, on the terms dictated by doctors, importers, or other would-be adjudicators of the issue. And this refusal generated resentment. Ever attuned to the national interest, Steet commended his fellow citizens "for the wonderful appetites we possess, and the appreciation we have for the best quality of food."[78] But the poor stood in the way of truly efficient use of scarce nutritional resources. Their "ignorance" with regard "to the proper use of food . . . was very great," according to the Right Honourable Henry Austin Bruce, member of British Parliament. He held that "probably in this matter they were among the most backward in Europe, certainly they were the most wasteful."[79] With hardly more sensitivity to the demands upon the time of working-class women, many of whom were employed outside the home and therefore had little time to devote to the kitchen, Scott concurred that "much might be done . . . towards husbanding the supplies we already have if a better and more common-sense plan of cookery were adopted by all classes, but especially the poor."[80]

Experts admitted that preparation for the table was of the utmost significance when it came to handling tinned meat, because unlike many other preserved varieties, this article was in fact cooked during the preservative process. Canning methods varied somewhat, but nearly all involved the application of extreme heat to effectively sterilize the contents of a given tin. Containers of meat were "immersed . . . in a bath of boiling brine," and held there until their contents reached a given temperature.[81] Steam produced during this process was allowed to escape through a small hole left in the lid, which was soldered closed once "air has been expelled and was entirely excluded," thereby "preserving the contents in vacuo."[82] The high temperature necessary for preservation, and the length of time tins tended to remain in their "bath" thoroughly cooked the meat.[83] Not

---

[76] "The Australian Meat Question," *Lancet* 93 (1869): 71.

[77] S. W. D. Williams, "The Nutritive Value of Australian Preserved Meat," *Lancet* 98 (1872): 287–8, on 287.

[78] Steet, "Preservation of Food" (cit. n. 32), 315.

[79] "Food Committee" (cit. n. 45), 190.

[80] Steet, "Preservation of Food" (cit. n. 32), 315. Broomfield notes that for middle-class as well as working-class women, industrialization meant that they "lost their ability—and arguably the liberty— to raise and preserve food themselves, let alone cook it expertly." Broomfield, "Rushing Dinner to the Table" (cit. n. 17), 108.

[81] "Proceedings of the Society: Food Committee," *J. Soc. Arts* 15 (3 May 1867): 375–80, on 375.

[82] Ibid.; "Animal Food Supplies" (cit. n. 5), quotation on 96.

[83] Analytical Sanitary Commission, "Records of the Results" (cit. n. 47), 296. See also Thorne, *Food Preservation* (cit. n. 10); and Shephard, *Pickled, Potted, and Canned* (cit. n. 10).

only did this detract from its desirability as a substitute for fresh meat, it meant that anyone heating canned meat for a meal risked "reduc[ing it] to shreds."[84] Even when the contents of a tin of meat were "of first-rate quality," as the Society of Arts' Food Committee found upon examining two tins of Australian beef, "[it] is so much cooked in its preparation that any further application of heat deteriorates it and diminishes its usefulness as an article of food." Although tinned meat would be "quite fit to eat as it is found in the tins, cold," they concluded that "few persons would like it in that condition."[85] Experts like Scott suspected that the process of canning "overheat[ed its contents] to the detriment of its nutritive power."[86] Consequently, it was "very desirable that further attention should be given to the method of preservation, in order to provide an article which could be converted into stews, curries, &c., without disintegration, which is now inevitable if it is further cooked."[87]

Despite widespread recognition that the process of putting meat into cans profoundly transformed it, experts insisted on holding consumers to blame for its poor reputation, along with its producers. What consumers did with canned meat in the privacy of their kitchens detracted from its alimentary value, members of the Society of Arts believed. It was well known, claimed Harry Chester to the Food Committee, "how deplorable was the cooking among the lower, indeed among all but the highest classes in this community."[88] Yet diagnosis alone was no cure for the disease. "Great as the evil" of wasteful cookery among the poor was, Chairman Bruce of the Food Committee believed that "greater still was the difficulty of dealing with it, because we had to do with the settled habits, and often fixed prejudices of the people."[89] The well-known German physiologist Johann Thudichum spoke to the Subcommittee on Meat in February of 1867 and proclaimed that "as the common people do not know how properly to cook the simplest thing, they would not succeed in imparting appetizing qualities to preserved food materials."[90] In a milder tone, the editors of the *Lancet* in 1872 commended recent "public attention" to "the subject of supplementing our native supply of fresh meat by importations from Australia," but noted that until "improved methods of curing . . . come into vogue," great care was needed "to do little more than heat without cooking them afresh."[91] Recognizing how critical preparation for the table was to the success of canned meat, one importer went so far as to hand out "receipts for preparing and cooking the food" to buyers of preserved Australian meat "on behalf of the shippers of the meat . . . and, in order to clear away difficulties," by which he no doubt meant danger to the product's reputation that might result from improper preparation.[92]

---

[84] "Australian Boiled Beef" (cit. n. 71), 550.

[85] "Proceedings of the Society: Food Committee," *J. Soc. Arts* 16 (3 January 1868): 103–5, on 104.

[86] Scott, "Supply of Animal Food" (cit. n. 24), 266.

[87] "Food Committee" (cit. n. 85), 104.

[88] "Food Committee" (cit. n. 9), 101.

[89] "Food Committee" (cit. n. 45), 190. In fact, as Joanna Bourke has documented, the poor and working classes were eager to improve their culinary skills: Bourke, "Housewifery in Working-Class England 1860–1914," *Past & Present* 143 (1994): 167–97.

[90] Thudichum proceeded to note the lack of suitable cooking apparatus among the poor, suggesting (in an echo of the wider rhetoric surrounding meat preservation) that "the introduction of a really practical and economical cooking apparatus for the poor would be a great boon." See "Proceedings of the Society: Food Committee," *J. Soc. Arts* 15 (8 March 1867): 240.

[91] "Australian Meat," *Lancet* 99 (6 Jan 1872): 32.

[92] "Australian Meat Question" (cit. n. 76), 71.

Others accused the lower orders of stymying the supposed public mindedness of preserved meat purveyors, of putting the great "boon" of imported preserved meat in jeopardy by refusing to consume it. In 1872, the *Lancet*'s editors were "glad to find the subject of supplementing our native supply of fresh meat by importations from Australia is attracting an amount of public attention which promises well for the introduction of a valuable article of diet, and a reduction in the present outrageous price of meat." That attention, though, was due to its "very general use . . . in well-to-do families; for, of course, the poor are the last to take up with anything that is wholesome and cheap but a little out of the common way."[93]

Although tinned meat, along with other commercially preserved foods, began as specialty items for the wealthy classes in the early nineteenth century, by the 1850s and 1860s it had become more affordable and widely accessible, and consumption had become more broadly popular and stratified according to class. With meat, as with other tinned foods like salmon and imported fruit, the finest-quality importations were marketed to better-off consumers, while cheaper, lower-quality products were "bought as small luxuries by the working class."[94] And though the market for canned meat expanded in this period, precise measures of consumption are difficult to obtain. Tinned meat was a relatively small proportion—never more than an estimated 5 percent—of Britain's live animal, and later frozen and refrigerated meat, imports.[95] Together with widespread and repeated claims of consumer resistance within the published record, this suggests that the lower orders never adopted tinned meat with the enthusiasm (or gratitude) that its promoters expected.

When lower-class consumption of canned meat appeared to lag, the self-proclaimed experts operating out of the Society of Arts took umbrage at what they considered their overly discriminating palates. "Of all classes of the community," averred Smith, and despite their "deplorable" skills in the kitchen, "the poor were, perhaps, the most dainty." This had sounded the death knell for charqui—"the higher classes did not want it, and the lower classes would not eat it"—and threatened other potential sources of foreign meat like pork and bacon.[96] Henry Grainger, an expert in the bacon trade whom the Society of Arts' Food Committee interviewed in 1867, did not think American bacon "would answer" on the British market, "owing to the extreme fastidiousness of the people in respect of food."[97] (It was "over-greas[y]" and "want[ed] firmness in the meat.")[98] The "very finest quality of salt beef" was similarly defeated. Though "offered in establishments in various parts of the country" on favorable terms, "it was a total failure." Simply put, Grainger explained, "the miners and colliers would not eat it."[99] To the great frustration of these "men of science," the poor refused to act in their own supposed self-interest by taking advantage of the novel sources of meat protein, suggesting not only that the assumed centrality of meat to the British diet was up for debate, but that what counted as meat was itself subject to negotiation. How it was preserved, and the extent of transformation caused by a given preservative process, mattered.

---

[93] "Australian Meat" (cit. n. 91), 32.
[94] Richard Perren, "Food Processing Industries C: Food Manufacturing," in Collins, *Agrarian History* (cit. n. 8), 7:1085–1100, on 1096.
[95] Perren, *Meat Trade* (cit. n. 18), 124.
[96] Steet, "Preservation of Food" (cit. n. 32), 317.
[97] "Proceedings of the Society: Food Committee," *J. Soc. Arts* 15 (17 May 1867): 414–17, on 414.
[98] Ibid., 415.
[99] Ibid., 414.

The upper orders, moreover, liked to think that they could adjudicate these matters. Experts writing in the medical press and presenting before the Society of Arts were quick to emphasize that preserved meat was never meant for the middle classes, much less the aristocracy. "This description of meat was never intended to come into competition with the sirloin of beef and the leg of mutton, which were the food of the rich," but rather "with the inferior parts, which for the most part fell to the lot of the poorer classes as being within their means."[100] It was an offering to the poor and working classes that was intended to provide them with the protein inputs necessary to fuel their labor during a time when the cost of butchers' meat put the fresh article out of reach. Recognizing the significance of meat as a marker of social distinction, the *Lancet* "strongly suspect[ed] that those who can afford to give 6d. per lb. for meat, would prefer rather to strain a point and give something more in order to secure a more satisfactory article in the shape of raw and uncooked meat."[101]

CONCLUSION

Indeed, the "shape of raw and uncooked meat" proved to be the sticking point in mid-nineteenth-century preservation efforts. Meat's changeability afforded both opportunities for transformation, and the grounds upon which to reject those efforts. Meat could be made more like sugar, tea, or grain by salting, drying, smoking, or tinning—processes that rendered it relatively stable and far more portable than in its untreated state. But opinion was universal among experts and consumers, as evidenced by the latter's hesitancy to purchase preserved meats, and the former's stated views that dried or precooked meat would never take the place of the fresh variety. Tinned meat was little more than the best of bad options, promoted in one breath, while in the next, experts acknowledged that "up to the present time, science has failed to show how meat can be popularly, as well as permanently, cheapened to the masses, inasmuch as all methods hitherto adopted have rendered fish, flesh, and fowl alike unpalatable."[102] Very few methods the Food Committee reviewed during the course of its existence were candidates for the Society's Trevelyan Prize, which by the 1870s had risen from £70 to £100. Tinned meats were "excluded by the terms of the offer," precisely because they came to the consumer cooked.[103]

The "great desideratum" was "beef and mutton preserved in a fresh state . . . and in such a manner as to be thoroughly palatable to the consumer."[104] Chemical treatments, which were part of a general enthusiasm for antisepsis in the 1880s, promised to do so, but the public was dubious.[105] Scott maintained that the "neutral sulphite of lime . . . when oxydised, is merely converted into sulphate of lime—a substance perfectly harmless and inert," but others were not so sanguine.[106] The editors of the *Lancet* dismissed

---

[100] Steet, "Preservation of Food" (cit. n. 32), 316.

[101] "Australian Meat" (cit. n. 7), 239.

[102] "Preserved Meat," *Lancet* 97 (1871): 133.

[103] "Proceedings of the Society: Address," *J. Soc. Arts* 25 (17 November 1876): 5–13, on 8. The address was given by Chairman Alfred S. Churchill. A claimant to the prize was mentioned in 1877, but neither the method of preservation nor the name of the claimant was mentioned; "Proceedings of the Society: Food Committee," *J. Soc. Arts* 25 (29 June 1877): 782–90, on 788.

[104] "Food Preservation" (cit. n. 56), 917.

[105] Atkins, "Pasteurisation of England" (cit. n. 50); Oddy, "Food Quality in London" (cit. n. 19).

[106] Scott, "Supply of Animal Food" (cit. n. 24), 267.

chemical antisepsis as "obviously worthless," for in "effecting the preservation of the meat, they would render it unfit for use as food."[107] Here again, the method of preservation itself "became . . . a ground of offense, and one which no subsequent treatment could wholly or satisfactorily remove."[108] When Professor Frederick Settle Barff presented his own chemical preservative—boroglycerine, a combination of boric acid and glycerine—to the Society of Arts in 1882, a hospital physician in the audience noted "how suspicious the public were of any changes with regard to their food," and emphasized Barff's need to make "absolutely clear, that the ingredients he used" to prepare "this practically fresh meat . . . were not in the slightest degree deleterious."[109] Evidently, the aptly named Barff was unable to do so, as he was pilloried for his signature method in *Punch* magazine shortly thereafter (fig. 1).

Barff's chemically preserved offerings, though, were late to the table regardless of their questionable salubrity. A year earlier, the *Lancet* had declared, "there can be no longer any doubt as to the possibility of bringing supplies of meat . . . to this country in a frozen state and landing them in a condition fit for food."[110] The first shipments of Australian frozen meat reached London in early 1880, and by the 1890s, frozen meat had fast eclipsed its tinned and desiccated brethren in popularity.[111] Pinpointing the quantity of tinned, salted, or dried meat imported relative to frozen and refrigerated meat is difficult due to the imperfect statistical records for the middle to late nineteenth century, but existing quantifications suggest that they were never more than a small fraction of the dead meat that Britain imported from abroad.[112] By 1894, they constituted only 15 percent of the total yearly importation of frozen beef and mutton.[113] Meat from the freezer was so "[in]distinguishable from ordinary fresh-killed butchers' meat," according the Society of Arts' miscellany on food preservation, that it could not even "strictly be called preserved meat."[114] It preserved the "meat-juice,"[115] which Steet and others "looked upon as the active principle of meat."[116] It interfered relatively little with the flavor of the flesh, and based on the rapid expansion of the trade—from 1,095 cwts of chilled beef imported in 1874 to 839,748 cwts beef and mutton in 1883, and 4,117,337 in 1894—British consumers bought it far more readily than tinned or salted meat.[117] The Trevelyan Prize went unclaimed (none "engaged in the importation of meat preserved by means of cold" could demonstrate "any such precise claim to the credit of the invention as would warrant the committee

[107] "Animal Food Supplies" (cit. n. 5), 96.

[108] "The Frozen Meat Supply," *Lancet* 130 (1887): 433.

[109] F. Barff, "A New Antiseptic Compound, and its Application to the Preservation of Food," *J. Soc. Arts* 30 (1882): 516–21, on 524. Derek J. Oddy notes the "numerous complaints" made against the use of such chemical preservatives in the 1890s; Oddy, "Food Quality in London" (cit. n. 19), 99.

[110] "Frozen Meat and Fish," *Lancet* 118 (1881): 816.

[111] Perren, *Meat Trade in Britain* (cit. n. 18); Rebecca J. H. Woods, "From Colonial Animal to Imperial Edible: Building an Empire of Sheep in New Zealand, c. 1880–1900," *Comparative Studies of South Asia, Africa, and the Middle East* 35 (2015): 117–36.

[112] Perren notes that "unenumerated meats" were only 5 percent of the total of dead meat imported from 1882 to 1890; Perren, *Meat Trade in Britain* (cit. n. 18), 124.

[113] Perren, "Food Processing" (cit. n. 94), 1097.

[114] "Food Preservation" (cit. n. 56), 918.

[115] "Australian Boiled Beef" (cit. n. 71), 550.

[116] Steet, "Preservation of Food" (cit. n. 32), 312.

[117] James Trowbridge Critchell and Joseph Raymond, *A History of the Frozen Meat Trade: An Account of the Development and Present Day Methods of Preparation, Transport, and Marketing of Frozen and Chilled Meats* (London: Constable, 1912), 423.

PUNCH'S FANCY PORTRAITS.—No. 84.

PROFESSOR BARFF,

MEMBER FOR BORO-GLYCERIDE.   OUR PRESERVER!

Figure 1. *Frederick Settle Barff presenting his chemical preservative, boroglycerine (*Punch *239, 20 May 1882).*

in thus awarding the prize"), and some familiar complaints were heard—of unfounded prejudice "among the working class," and of "freemasonry between butchers and cooks" who "united against anything that was cheap."[118] But frozen meat found its

[118] "Proceedings of the Society: Annual General Meeting: Report of the Council, Article 27: Food Committee," *J. Soc. Arts* 29 (1 July 1881): 645–57, on 654; "Food Preservation" (cit. n. 56), 917.

place within the diets of Britons such that by the early twentieth century, "Canterbury lamb" from New Zealand was presumed to be that of the Canterbury district of Kent in England.[119] By rendering a changeable substance inert, artificial refrigeration had achieved that "great desideratum": the preservation of raw meat in such a way as to satisfy the tastes of the poor, "popularly, as well as permanently, cheapen[ing meat] to the masses."[120]

[119] This comment was made by H. Moncriff Paul in the discussion following E. Montague Nelson's paper, "The Meat Supply of the United Kingdom," *J. Soc. Arts* 43 (15 March 1895): 420–9, on 427–8. R. Ramsay, "The World's Frozen and Chilled Meat Trade," in *The Frozen and Chilled Meat Trade: A Practical Treatise by Specialists in the Trade*, 2 vols. (London: Gresham, 1929), 1:3–30, on 5.
[120] "Preserved Meat" (cit. n. 102), 133.

The Introduction of Chemical Dyes into Food in the Nineteenth Century

by Carolyn Cobbold*

ABSTRACT

This article examines the introduction of chemical dyes into food in the nineteenth century in four different countries: the United States, Britain, Germany, and France. From the early 1860s, chemists produced aniline and azo dyes from coal tar on an industrial scale for the burgeoning European textile industry. However, by the end of the century, hundreds of the new dyes were also being added to food, a use for which they were not designed. This article examines the disagreements among chemists over whether these new chemical substances should be seen as legitimate food ingredients or as food adulterants. This was a period when chemists were establishing themselves as professionals, with chemistry being promoted as a science capable of transforming everyday commodities and solving public health issues. However, chemists' attempts to mediate the use of chemical dyes as food coloring were complicated by a lack of consensus within the chemical community about how to detect the use of such dyes in food and how to test their toxicity. Chemists also were conflicted in their response to the debate depending on whether they were employed by food or dye manufacturers, or working as food inspectors for the state and local authorities. In their efforts to gain authority as food experts, chemists found themselves in a crowded market of interested parties, including food manufacturers, consumers, and politicians. The article describes the diverse opinions of chemists, manufacturers, consumers, and regulators in Britain, France, Germany, and the United States, and the varied regulatory responses in these countries to the use of new chemical dyes in food.

INTRODUCTION

When chemists discovered how to synthesize new dyes from coal-tar waste in the second half of the nineteenth century, they had no idea that their bright new colors would become a key ingredient for the food industry. Manufactured for the booming European textile industry, the aniline and azo dyes were heralded in the press as "wonders" of chemistry.[1] Yet within a few years of their discovery, the new synthetic dyes were being

* Clare Hall, University of Cambridge, Herschel Road, Cambridge, CB3 9AL, UK; cac85@cam
.ac.uk.

With many thanks to Emma Spary and Anna Zilberstein for asking me to contribute to this volume and for their excellent editing skills. Thanks also to Emma, Hasok Chang, and Lesley Steinitz for all their guidance, wise words, and friendship during my researches into dyes, and to all my colleagues at Clare Hall, Cambridge.

[1] Anon., "A Ramble into the Eastern Annexe of the International Exhibition," *Ladies' Treasury*, 1 November 1862, 342.

widely used in food production, becoming an early example of laboratory-created chemicals, manufactured on a mass scale, and used in many different and unintended ways. The widespread, and largely undisclosed, use of the new dyes in food production created a dilemma for chemists, who, for decades, had been warning about food producers' unscrupulous and fraudulent use of artificial food coloring to deceive the consumer.[2] Indeed, throughout the nineteenth century, chemists had used the fear of food adulteration to help boost their fortunes and status, creating new roles for themselves as public analysts or municipal chemists, and protecting the consumer from dishonest food producers and retailers.[3] By the end of the century, however, it was additives being manufactured by chemists themselves that became the target of adulteration debates. Chemists struggled to address the complex and contradictory questions surrounding whether, and which, chemical additives were adulterants or legitimate ingredients. Attempting to balance their roles as creators of useful new chemical substances and self-professed guardians of food quality, the chemists found themselves just one group among many vested parties, including food manufacturers, consumers, and politicians, in a battle to control industrializing food practices.

As Stanziani and other historians have shown, in situations such as these, distinct interest groups contend with one another to shape government policy.[4] However, in responding to the use of chemical dyes in food, chemists struggled to speak with a unified voice and were unable to reach agreement through experiment as to the harmlessness of consuming the dyes. This was a situation where new chemical substances were being introduced into food in different countries and traded internationally, with no official oversight and no agreed-on national or international regulation. Both consumers and businesses turned to chemists to protect their interests and arbitrate on the issue, and chemists saw this dilemma as an opportunity to boost their authority. But reaching consensus proved a dilemma.

Today, more than a century and a half after the first chemical dyes began to be used to color food, there is still no global agreement as to which dyes are safe and which are harmful to use. Large food companies market some of their products as being free from artificial coloring, yet continue to use chemical dyes in other products. Examining how nineteenth-century chemists and regulators in four different countries reacted to the unexpected use in food of new chemical substances provides important insights into the acceptance and mediation of novel scientific products and food ingredients in the marketplace and demonstrates how issues surrounding food safety are culturally specific.

---

[2] Friedrich Christian Accum, *A Treatise on Adulterations of Food and Culinary Poisons: Exhibiting the Fraudulent Sophistications of Bread, Beer, Wine, Spirituous Liquors, Tea, Coffee, Cream, Confectionery, Vinegar, Mustard, Pepper, Cheese, Olive Oil, Pickles and Other Articles Employed in Domestic Economy and Methods of Detecting Them* (London: J. Mallett, 1820); Arthur Hill Hassall, *Adulterations Detected, Or, Plain Instructions for the Discovery of Frauds in Food and Medicine* (London: Longmans, 1857); Hassall, *Food: Its Adulterations, and the Methods for Their Detection* (London: Longmans, 1876).

[3] Peter Atkins, Peter Lummel, and Derek J. Oddy, *Food and the City in Europe since 1800* (Farnham, UK: Ashgate, 2007); Atkins, *Liquid Materialities: A History of Milk, Science and the Law* (Farnham, UK: Ashgate, 2010); Karl-Peter Ellerbrock, "Lebensmittelqualität vor dem ersten Weltkrieg: Industrielle Produktion und staatliche Gesundheitspolitik," in *Durchbruch zum modernen Massenkonsum*, ed. Hans-Jürgen Teuteberg (Stuttgart: Franz Steiner Verlag, 1987), 127–88; Michael French and Jim Phillips, *Cheated Not Poisoned?: Food Regulation in the United Kingdom, 1875–1938* (Manchester: Manchester Univ. Press, 2000); Vera Hierholzer, *Nahrung nach Norm: Regulierung von Nahrungsmittelqualität in der Industrialisierung 1871–1914* (Göttingen: Vandenhoeck & Ruprecht, 2010); Alessandro Stanziani, "Information, Quality and Legal Rules: Wine Adulteration in Nineteenth Century France," *Bus. Hist.* 51 (2009): 268–91.

[4] Stanziani, "Information, Quality, and Legal Rules" (cit. n. 3), 268.

France's centralized state and Germany's federalized system both adopted a system of norms and standards that varied regionally. The United States and Britain, meanwhile, took radically different approaches. The United States introduced strict centralized legislation, permitting only federally approved chemical "food colors," while the British government ruled out any specific legislation on the chemical coloring of food, leaving it to the courts to rule on their use in different applications.

Artificially coloring food was not a new phenomenon in the nineteenth century. Dyes extracted from plants such as saffron, turmeric, or beetroot; animals such as cochineal beetles; or metals and minerals including arsenic, copper, and lead had been used in food production for centuries. Debates over deception and harm associated with the use of colors added to food also was nothing new, as historians such as Madeleine Ferrières have shown.[5] Colors have been introduced into food for centuries for a variety of reasons, from meeting consumer expectations and making food products look fresher, more consistent, and more appetizing; to disguising poor food quality and increasing profits by facilitating the watering down or bulking out of food and drink. What makes the late nineteenth century so interesting for historians of food and science, however, was the emergence of new industrially manufactured chemical substances being used in food at a time when politicians were increasingly turning to legislation and chemists to protect the public from food adulteration.

The burgeoning populations of European and North American cities, living in crowded and unsanitary conditions and increasingly removed from the sites and sources of their food, meant that the safety of food and drink, and other public health issues, were of growing social and scientific concern by the middle of the nineteenth century.[6] For many decades, chemists had been warning about the use of toxic metal colorings, along with other harmful and fraudulent food practices, which they claimed were becoming more prolific. This had resulted in several countries, including Britain, France, and Germany, adopting food legislation and appointing consultant chemists, known in Britain as public analysts, to help monitor food and drink.[7]

However, while chemists had successfully devised tests to detect the presence and toxicity of arsenic and other metallic dyes, testing for the new chemical dyes proved extremely challenging. Part of the problem in the early years of aniline dyes was that these new chemical substances were being introduced into the food supply without the knowledge of consumers, dye manufacturers, municipal chemists, or, in many cases, food manufacturers themselves. These were new substances extracted by chemists from coal-tar waste and manufactured industrially as dyes for the textile industry.[8] The dyes were sold to wholesale chemists and marketed under innocuous names, such as "butter yellow," that masked their chemical origins. Even as late as 1901, owners and directors of food manufacturers, appearing before a British Parliamentary inquiry into the use of colorings in food, demonstrated a lack of knowledge or concern about the constitution of the

---

[5] Madeleine Ferrières, *Sacred Cow, Mad Cow* (New York, N.Y.: Columbia Univ. Press, 2006).

[6] Patrick Zylberman, "Making Food Safety an Issue: Internationalized Food Politics and French Public Health from the 1870s to the Present," *Med. Hist.* 48 (2004): 1–28.

[7] Accum, *A Treatise on Adulterations*, (cit. n. 2); Hassall, *Adulterations Detected* (cit. n. 2); Atkins, Lummel, and Oddy, *Food and the City* (cit. n. 3); Ellerbrock, "Lebensmittelqualität" (cit. n. 3), 127–88: French and Phillips, *Cheated Not Poisoned?* (cit. n. 3); Hierholzer, *Nahrung nach Norm* (cit. n. 3).

[8] Andrew Pickering, "Decentering Sociology: Synthetic Dyes and Social Theory," *Perspect. Sci.* 13 (2005): 352–405; Anthony Travis, *The Rainbow Makers: Origins of the Synthetic Dyestuffs Industry in Western Europe* (Bethlehem, Penn.: Lehigh Univ. Press, 1993).

"colourings" they purchased from wholesalers.[9] Meanwhile, food manufacturers who were aware of the origin of the colors supplied to them by wholesalers were reluctant to reveal the use of artificial food coloring in their products, and, in many cases, were marketing their products as "pure" and "unadulterated."

This lack of transparency, and the widespread use of the new dyes in tiny amounts that made them almost impossible to detect and that tended to produce a risk of accumulative poisoning rather than making individual products toxic, was a challenge for regulators and analysts. To ban all coloring of food would be viewed as restricting consumer choice and almost impossible to enforce or monitor. However, attempting to determine safe and acceptable thresholds for the multitude of different dyes was beyond the technical capability of analysts and threatened their credibility and authority.

A crucial issue for chemists was their inability to accurately test whether, and which, dyes were being used in food items and whether or not their long-term use was problematic. Chemists struggled to agree on standardized tests to detect and assess the aniline and azo dyes, with dozens of new dyes appearing on the market monthly, whose properties changed when combined with other chemicals.

One of the first chemists to devise tests for detecting aniline dyes was the French chemist Armand Gautier, who advised that "aniline should be sought for in all wines found to be adulterated with other substances."[10] Gautier suggested that the mixing of previously recognizable, and often acceptable, dyes, such as blueberries, with these new, undetectable, coal-tar dyes was widespread in the French wine industry by the mid-1870s, and was not being noticed by either consumers or municipal chemists. Gautier's tests were based on classical analytical experiments usually involving titration, a technique that involved inducing different color reactions to eliminate the presence of certain dyes. However, these techniques, used widely in the textile industry, proved unmanageable when trying to detect any one of hundreds of known and unknown chemicals being covertly added to food. The lack of standardized ways of measuring color also made the task impossible for chemists.[11]

The German chemist Theodor Weyl was another pioneer of devising tests for synthetic dyes. Weyl was one of a fast-growing fraternity of German chemists, including manufacturing chemists producing thousands of new, commercially valuable chemical substances; consulting chemists advising government and industry on public health issues; and hygienic and physiological chemists working in the medical and nutritional fields.[12] All these chemists had a stake in safeguarding the reputation of chemistry and its products and understood the importance of the growing chemical industry to the economy. They recognized the advantage of extending the range of chemical products, while acknowledging both the pervasive manner in which the new chemical substances

---

[9] *Report of the Departmental Committee Appointed to Inquire into the Use of Preservatives and Colouring Matters in the Preservation and Colouring of Food* (London: HMSO, 1901), 99–103.

[10] A. Gautier, "The Fraudulent Colouration of Wines," *Analyst* 1 (1876): 109–12; Gautier, "Continuation of the Fraudulent Coloration of Wines," *Analyst* 1 (1876): 130–5.

[11] Colorimetry, the measurement of color, proved an intractable issue for much of the twentieth century. Sean F. Johnston, *A History of Light & Colour Measurement: Science in the Shadows* (Boca Raton, Fla.: CRC, 2001).

[12] Jeffrey A. Johnson, "Germany: Discipline-Industry-Profession: German Chemical Organisations 1867–1914," in *Creating Networks in Chemistry: The Founding and Early History of Chemical Societies in Europe*, ed. Anita Kildebæk Nielsen and Soňa Štrbáňová (Cambridge: Royal Society of Chemistry, 2008), 113–38.

were infiltrating everyday life and the need to reassure the public. However, Weyl and his fellow chemists were unable to agree on either the best tests to use or on how to interpret the many different tests devised. Devising tests to detect and assess hundreds of new, and constantly appearing, substances, most of which were new to the testers and whose properties changed when combined with other chemicals, proved to be a daunting undertaking.[13]

With no consensus as to how to detect the new dyes in food or assess their safety, social, economic, and political aspects played as large a part as the science in determining the validity of using chemicals as food ingredients. Municipal chemists across Europe and in the United States themselves were divided as to whether the new dyes should be used as alternatives to the toxic metallic dyes. Some chemists claimed the new chemical colors were a safer and scientific alternative to the toxic metallic and mineral dyes, while others worried that their harmlessness was not proved and that they could still be used to deceive the public.

INCREASE IN FOOD CHEMISTS AND PUBLICLY APPOINTED ANALYSTS FROM THE 1870s

From the 1870s, consulting chemists were appointed by local municipalities in Britain, France, and Germany to represent the interest of the food consumer. At the same time, food producers and retailers began recruiting consultant chemists to help them find ways to improve food production, shelf life, and food reliability and quality.[14] The second half of the nineteenth century was a period when chemists, consumers, social reformers, free- and fair-trade advocates, manufacturers, and retailers all had a stake in forming economic and cultural food knowledge.[15] Municipal chemists, or public analysts, were caught in the middle of debates between different interest groups. Some of these groups, particularly advocates of the growing sanitarian movements in Europe and the United States, raised the issue of public health and economic fraud to further their interests, while others, such as manufacturers and retailers, promoted the benefits of competition and consumer choice. Food manufacturers argued that banning the use of artificial coloring in food would reduce consumer choice, while food reformers argued that consumers needed protection from an increasingly opaque and competitive food chain. The chemists themselves were similarly divided.

French chemists Paul Cazeneuve, Saturnin Arloing, and Alfred Riche argued that experiments showed that many coal-tar dyes were no more toxic than vegetable dyes and

---

[13] Theodor Weyl, *Handbuch der Hygiene*, vol. 3 (Jena: Fischer, 1895); Weyl, "Ueber eine neue Reaction auf Kreatinin und Kreatin," *Berichte der Deutschen Chemischen Gesellschaft* 11 (1878): 2175–7; Weyl and Henry Leffman, *The Coal-Tar Colors: With Especial Reference to Their Injurious Qualities and the Restriction of Their Use* (Philadelphia: P. Blakiston, 1892).

[14] Atkins, Lummel, and Oddy, *Food and the City*; Ellerbrock, "Lebensmittelqualität," 127–88; French and Phillips, *Cheated Not Poisoned?*; Hierholzer, *Nahrung nach Norm* (all cit. n. 3).

[15] For more on consumerism and the political economy in this period, see Frank Trentmann and Martin Daunton, eds., *Worlds of Political Economy: Knowledge and Power in the Nineteenth and Twentieth Centuries* (New York, N.Y.: Palgrave Macmillan, 2004); Matthew Hilton, *Consumerism in Twentieth-Century Britain: The Search for a Historical Movement* (Cambridge: Cambridge Univ. Press, 2003); Mitchell Okun, *Fair Play in the Marketplace: The First Battle for Pure Food and Drinks* (Chicago: Northern Illinois Univ. Press, 1986); Jean Baudrillard, *The Consumer Society: Myths and Structures* (Thousand Oaks, Calif.: Sage, 1998); John Benson, *The Rise of Consumer Society in Britain, 1880–1980* (London: Longman, 1994); Alain Chatriot, Marie-Emmanuelle Chessel, and Matthew Hilton, *The Expert Consumer: Associations and Professionals in Consumer Society* (Burlington, Vt.: Ashgate, 2006).

other long-standing and legitimized food additives, such as salt, particularly as the quantity of chemical dyes used was so small.[16] The British consulting chemist Benjamin Newlands announced proudly that he had himself introduced the use of aniline dyes to color sugar, following rumors of poisoning by the use of chloride of tin to darken sugar.[17] Chemists argued that any drawbacks associated with the use of coal-tar dyes in food should be judged against consumer choice and the practicalities of production.

August Dupré, public analyst for the borough of Westminster in London, claimed that the public had become used to prepared food being colored and often preferred it.[18] Analysts were principally worried about consumers being conned; an example was the purchase of margarine thinking it was butter. Analysts argued that the public should be able to buy and eat yellow margarine, if they preferred it to white margarine, so long as it was sold and labeled as margarine and not butter. Otto Hehner, the public analyst for Nottinghamshire, West Sussex, the Isle of Wight, and Derby in Britain, however, complained that food producers were using the arguments of consumer choice as an excuse to artificially color their food, for reasons of convenience and cost. Public analysts, he declared, should "disregard the popular wish and raise the standard of purity of food," arguing that "the popular wish had been used by sophisticators for many years past as an excuse for every abomination" and that "no kind of adulteration was ever carried out without its being alleged to be in obedience to the public wish."[19] By portraying themselves as the guardians of national food supply, public analysts and municipal chemists argued that they, rather than food manufacturers, should determine the regulations surrounding food additives and what represented genuine or illegitimate food ingredients.

Hehner was one of several chemists working in Britain who had been born or trained in Germany. He became president of the Society of Public Analysts, a professional body set up in 1874 to represent the interests of British municipal chemists, and vice-president of the London-based Institute of Chemistry. In private and in anonymously written articles in the sanitary press, Hehner argued passionately that chemists should protect consumers and honest food producers, as well as the reputation of a nation's food supply, from the fraudulent use of chemical dyes and other undisclosed additives. In a meeting with fellow chemists in 1890, Hehner contended that public analysts had a responsibility to check the use of poisonous substances in food and therefore should attempt to identify individual dyes used. He recognized that the debate over the new chemicals being used in food would advance the public analysts' attempts to distinguish themselves from other chemists. To succeed, they needed to portray themselves as impartial public figures with a wide and competent knowledge of both chemistry and physiology. Hehner argued that several aniline colors, including Victoria yellow or Dinitrocresol, and Martius yellow or Dinitronaphthol, were "of a decidedly poisonous nature" and that it was the expectation and duty of public analysts to "know the exact nature of the colouring matters

---

[16] Paul Cazeneuve, *Les colorants de la houille au point de vue toxicologique et hygiénique* (Lyon: Affaire de la succursale de la B. Anilin & Soda Fabrik à Neuville-sur-Saône, 1887).

[17] For more about the use of chloride in tin, see the sugar debate reported in Charles Cassal, "On Dyed Sugar," *Analyst* 15 (1890): 141–60.

[18] August Dupré, Evidence to the Select Committee, 22 January 1900, *Report of the Departmental Committee Appointed to Inquire into the Use of Preservatives and Colouring Matters in the Preservation and Colouring of Food: Together with Minutes of Evidence, Appendices and Index* (London: HMSO, 1901), 201–3.

[19] Otto Hehner, Evidence to the Select Committee, 19 January 1900, *Report of the Departmental Committee* (cit. n. 18), 191–4.

employed." Positioning the public analysts as disinterested mediators in debates surrounding healthy food additives, Hehner stated: "just as a public analyst might not need to know all details of technical and manufactory processes, so manufacturing chemists might possibly be ignorant of the physiological action of the colours which they added to food materials."[20] August Dupré also stressed the important role of public analysts in determining the legitimacy of food additives, stating that the Society of Personal Analysts should not take at face value the claims of food manufacturers "that the colouring matter they added was innocent" and not "injurious to health."[21]

However, Hehner and other consulting chemists also recognized the practical benefits of artificial food coloring in the struggle to feed rapidly urbanizing populations. As the food supply chain lengthened, chemical dyes, many of which also acted as preservatives, helped to ensure that food stayed healthy and looked fresh. In an age of Malthusian politics, where feeding the population was a primary concern of governments, the benefits of chemical additives were not trivial. It is also important to reflect that consulting chemists were a relatively new breed of professionals who depended financially on both government and commercially sourced work.

British consulting chemists especially found themselves conflicted because the adversarial common law court system in Britain resulted in chemists being employed as expert witnesses and having to make public statements about the use of the new chemical dyes as both public analysts working for local authorities and consultants working for food manufacturers and retailers. Interestingly, the fervent views expressed by the analysts in private were often different from those they expressed in court or in other public settings, when they were being paid to represent manufacturers or other clients. Historians such as Graeme Gooday have highlighted the conflicts faced by chemists and other consulting scientists in the nineteenth century who were increasingly paid as "expert witnesses" in British court cases.[22]

One of several commercial disputes involving the use of artificial coloring in food was a case brought in Birmingham's Victoria Courts against Thomas Davis, a grocer charged with selling dyed sugar beet crystals as Demerara sugar.[23] Hehner, acting as an expert witness for the defense, claimed that dyeing West Indian sugar was a common practice and that the aniline dye used was harmless to health. Dye was used in the preparation of nearly every article of food, Hehner added; in the case of sugar, the purpose of dyeing was to give the sugar a consistent and uniform color, making it more attractive and in keeping with the public taste. Hehner's trial arguments provide no clue as to his aversion to artificial colorings. Indeed, they conflict with observations he frequently made in the press and in private discussions among other analysts about the harm and dishonesty of using artificial preservatives and colorings in food. Hehner's disillusionment with food industrialization and the use of chemical preservatives and colorings continued throughout his life.[24] Eventually he became an outspoken critic of the use of chemical colorings along with fellow chemist Charles Cassal, who founded *The British Food*

---

[20] Charles Cassal, "On Dyed Sugar" (cit. n. 17).

[21] Ibid.

[22] Graeme Gooday, "Liars, Experts and Authorities," *Hist. Sci.* 46 (2008): 431–56; Christopher Hamlin, "Scientific Method and Expert Witnessing: Victorian Perspectives on a Modern Problem," *Soc. Stud. Sci.* 16 (1986): 485–513.

[23] "Birmingham Court Case," *British Food Journal & Analytical Review*, May 1900, 143.

[24] *Times* (London), 20 September 1923; cited in French and Phillips, *Cheated Not Poisoned?* (cit. n. 3), 30.

Journal and Analytical Review in 1899 to promote food purity and increased food regulations and edited it until 1914.[25] Hehner's comments in court reveal how consulting chemists were able to adjust their opinions according to the needs of their clients, whatever their personal views.

However, because of the analysts' inability to accurately test for the dyes, courts were a disquieting setting for them. Cassal argued that it was more important to establish whether an artificial dye had been used than to determine which precise dye it was. Indeed, making statements that were too precise about the specific dyes used was a dangerous tactic in court, he pointed out; if an analyst specified any ingredient, a manufacturer or retailer could swear in court that this specific compound or chemical had not been used if it was not exactly identical to the one that the public analyst claimed to have been the adulterant. Cassal explained that it was the duty of the analyst only to prove adulteration, rather than detail the exact chemical composition of the adulterant.[26] Cassal defended this tactic by suggesting that the intricacy of identifying individual dyes was beyond the requirements of the court procedure or the comprehension of legal officers. However, analysts like Cassal were acutely aware of their limited ability to detect and identify individual chemical dyes.

Analysts were also divided as to whether it was the dyes' ability to deceive the consumer or the lack of proof surrounding their safety that was of the greatest public concern. Hehner's comments related to concerns surrounding the possible toxicity of the coal-tar dyes, rather than the issue of fraud. Although Cassal agreed with Hehner that many of the chemical dyes might well be toxic if used in large quantities, he claimed that the principal worry of most analysts was the use of the colorings to camouflage food or defraud the consumer. Dupré claimed that he had artificially colored sugars frequently submitted to him. But he said he would be wary about accusing manufacturers of adulterating sugars unless they were being passed off as something different, such as beet sugar sold as Demerara. He also pointed out the difficulty of obtaining coloring convictions under Britain's 1875 Sale of Food and Drugs Act, if the coloring matter was harmless, or could not be proven to be harmful.[27] These comments reveal the uncertain role of dyes in food and the problems associated with defining them as either legitimate additives or as illicit "adulterants."

Pierre-Antoine Dessaux has described the complicated process involved in mediating the recognition of chemical ingredients in food, with analytical chemists just one group in a marketplace that included food producers, retailers, other chemists, and consumers. Chemists realized that they needed to work closely with the food and drinks industry, as well as with public authorities, in order to secure any scientific authority.[28] With a lack of accurate and standardized tests, chemists had to prove their usefulness to both industry and consumers to obtain any credibility as arbiters of food quality and mediators on the use of chemical food additives. However, the relationship between the chemists and the food industry proved to be a challenging one.

---

[25] Derek Oddy, "Food Quality in London and the Rise of the Public Analyst, 1870–1939," in Atkins, Lummel, and Oddy, *Food and the City* (cit. n. 3), 91–104, on 95–6; Edward Collins and Oddy, "The Centenary of the British Food Journal, 1899–1999—Changing Issues in Food Safety Regulation and Nutrition," *British Food Journal* 100 (1998): 433–550.

[26] Cassal, "On Dyed Sugar" (cit. n. 17).

[27] Ibid.

[28] Pierre-Antoine Dessaux, "Chemical Expertise and Food Market Regulation in Belle-Epoque France," *Hist. & Tech.* 23 (2007): 351–68.

FRENCH CHEMISTS MEDIATE THE USE OF CHEMICAL DYES
IN FOOD AND DRINK, 1883–1905

In 1883, Charles Girard, head of the Paris Municipal Laboratory of Chemistry, published a report showing the prevalence of adulteration in French food and drink products. The Paris Préfecture de Police, who were responsible for monitoring food, had set up the Paris Laboratory in 1878. Its establishment was prompted by the French organic chemist Jean-Baptiste Dumas, who was a municipal council member in Paris, following increasing concerns about the use of artificial colorings in wine.[29] Businessmen, who were concerned as to the impact the report would have on trade, were furious. They argued that preventing the artificial coloring of food and beverages would be an encroachment on freedom, as people should be able to drink aniline-colored water if they wished to.[30] In a second report, published in 1885, Girard claimed that using chemistry to defeat adulteration would be advantageous to French trade: "Commerce today is quick to change the discoveries of science into instruments of fraud. Falsification, which was formerly based on a few clumsy formulae, has become scientific—and we cannot be successful against it if we don't attack it with weapons equal to its own."[31]

Girard and other chemists claimed that they were supporting the interests of honest businessmen by thwarting the attempts of fraudulent producers and traders to boost their profits by deceiving the public. Girard and Paul Brouardel, chemist and chair of the Consultative Committee of Hygiene (CCH), argued that chemists should be the final arbiters of food safety.

Businesses, however, argued that chemists alone should not be allowed to define the quality of wine, as consumers had their own predilections regarding taste, smell, and color, while wines differed from province to province according to the variety of grape grown, and other conditions such as soil, topography, production method, and climate. Natural variation in food therefore made the creation and application of uniform food regulations problematic.[32] In 1905, a Paris-based representative of the food and beverage retail industry, Georges Berry, warned about giving scientists and bureaucrats the power of defining what constituted food, claiming that it would give "the gift of infallibility to chemists whose theories are challenged every day by their own colleagues," and these new experts "will be given the right to decree formulae with which nature and consumer tastes will have to comply."[33]

Both industry executives and politicians pointed out that chemists were often unable to agree with one another in cases of adulteration. Chemists such as Paul Cazeneuve

[29] Henry Huet-Desaunay, *Le Laboratoire Municipal et les falsifications ou recueil des lois et circulaires concernant la vente des produits alimentaires et hygiène publique* (Paris: Richou, 1890); Harry W. Paul, *From Knowledge to Power: The Rise of the Science Empire in France, 1860–1939* (Cambridge: Cambridge Univ. Press, 2003), 211–20; Dessaux, "Chemical Expertise" (cit. n. 28).

[30] Denys Cochin, *Revue des Deux Mondes*, 15 June 1883; cited in Anon., "Adulteration in Paris," *Analyst* 8 (1883): 239–40.

[31] Charles Girard, *Documents sur les falsifications des matières alimentaires et sur les travaux du Laboratoire Municipal: Rapport À Monsieur Le Préfet de Police: Deuxième rapport* (Paris: Imp. Municipale, 1885); cited in Paul, *From Knowledge to Power* (cit. n. 29), 215.

[32] Vera Hierholzer, "Searching for the Best Standard: Different Strategies of Food Regulation during German Industrialization," *Food and History* 5 (2007): 295–318; Dessaux, "Chemical Expertise" (cit. n. 28); Uwe Spiekermann, "Redefining Food: The Standardization of Products and Production in Europe and the United States, 1880–1914," *Hist. & Tech.* 27 (2011): 11–36.

[33] Testimony of George Berry in *Journal Officiel*, Assemblée Nationale 1904, 2356, 2358; cited in Dessaux, "Chemical Expertise" (cit. n. 28).

acknowledged the lack of consensus within the chemical community. Cazeneuve believed that scientists should not make laws, but should instead support government and industry to ensure that food and drink was fit for consumption. Cazeneuve, a Senator of the Rhône District, was a politician as well as a consulting chemist who worked for several different businesses, including manufacturers of dyes. As a result, Cazeneuve was able to see the mutual benefits of close cooperation between chemists and industry.

The growing concern surrounding the use of artificial dyes in wine, and contradictory opinions as to their impact, prompted the CCH to commission a report into the matter. This report determined that most dyes probably were harmless in the small amounts used, but that more experiments were required to ensure the safe use of chemical dyes in consumables.[34] Cazeneuve suggested that insufficient experiments had been performed on chemical dyes to accurately evaluate their influence on digestion. He recommended establishing a register of harmful and harmless colors, and that dyes sold for food and drink should be traded under a manufacturer's seal of quality using the manufacturers' authorized names, and not names made up by retailers. Most of Cazeneuve's own tests indicated that chemical dyes were not harmful when consumed in small quantities. However, he argued that all dyes sold for use in food should be pure or only combined or treated with substances known to be nonpoisonous.[35] Indeed, many of the initial debates about the safety of consuming aniline dyes centered on their possible contamination with arsenic, a substance often used in their manufacture. In 1885, an ordinance of the Paris police commissioner had barred the use of fuchsine and other coal-tar dyes in food and drink. But Cazeneuve wanted a food law in France that was closer to the German July 1887 legislation, which separated colors containing known poisons such as arsenic, antimony, mercury, and lead, from pure aniline and azo colors. As a result of Cazeneuve's mediation and protestations of food and drink manufacturers, an 1890 Paris police ordinance included both a list of banned coal-tar dyes and a list of coal-tar dyes acceptable for use, in small quantities, in food and drink.[36]

The situation in France, as in Britain, highlighted both the uncertainty surrounding what constituted food improvement and food adulteration, as well as the lack of consensus between chemists at the time. While chemists such as Girard, who were appointed to police the nation's food supply, urged greater measures be taken against the new dyes, others, including physiological and organic chemists, believed most of the new dyes were harmless in small amounts, and, in some cases, provided benefits to food production.

MONITORING AND REGULATING CHEMICAL DYES IN FOOD IN GERMANY, 1871–1911

Chemists in Germany, where most of the aniline and azo dyes were being manufactured by the latter decades of the nineteenth century, were similarly divided as to whether,

---

[34] G. Bergeron, *Rapport sur les propriétés toxiques de la fuchsine non arsenicale* (Recueil des travaux du Comité consultative d'hygiène), 7:321; cited in Cazeneuve, *Les colorants de la houille* (cit. n. 16).

[35] Paul Cazeneuve, "Les couleurs de la houille et la revision des listes légales des colorants nuisibles et non nuisibles," *Annales d'hygiene pub.* 18 (5 July 1887): 5–8, cited in *Revue des sciences médicales en France et l'étranger* 31 (26 July 1888), on 517; Paul Cazeneuve and R. Lépine, "Sur les effets produits par l'ingestion et l'infusion intraveineuse de trois colorants, dérivés de la houille," *Compt. Rend. Académie de Sciences* 101 (1885): 1167; "Présentations d'ouvrages manuscrits et imprimés," *Bulletin de l'Académie de Médecine*, 2nd ser., 16 (1886): 303–11, on 310. It is possible the food items singled out were those with the most persuasive producers, as Stanziani has argued in the case of French wine.

[36] Hugo Lieber, *The Use of Coal Tar Colors in Food Products* (New York, N.Y.: H. Lieber, 1904), 32.

and how, chemical dyes should be legitimized as a food ingredient. Germany was a young nation comprised of provinces with different histories, geographies, and cultures, and tastes and food products differed from region to region. At the same time, new production techniques and novel food ingredients, as well as a changing understanding of physiology and of the importance of different food groups, created confusion and anxiety among consumers, and led to calls for more "natural" food products and a desire to stamp out adulteration. However, as in France and Britain, what counted as illicit versus legitimate food intervention and ingredients was contested, and became the subject of ongoing debate between food producers, retailers, consumers, chemists, and government officials.[37]

In 1871, the year of Germany's unification, the German Reich set up the Imperial Health Office (IHO) in Berlin, employing chemists as important observers and decision makers in food negotiation.[38] Chemists became crucial constituents in the state's new food regulatory and monitoring network by 1907, with up to 180 private, state, and university food laboratories all forming part of the regime.[39] However, with different food cultures across the new country and during a period when communities of chemists centered in particular cities were growing fast, there were significant regional differences in interpretation and execution of the national food laws. In 1879, the Munich-based Institute of Hygiene took 80,000 food samples compared with a total of 7 samples in Dortmund, 15 in Köln, and 3 in Münster in 1878.[40]

In 1899, an investigation by the Imperial Health Office into the use of chemical colorings in sausages confirmed extensive disparity in food practice and monitoring across Germany. Replying to the IHO, Berlin's Chemical Laboratory Institute for Microscopy and Bacterial Research confirmed that most sausages tested contained coal-tar dyes. Offenbach's Chemical Research Department, however, reported only one artificially colored sausage from the previous three years, which had been dyed with cochineal, a dye extracted from beetles. The Medicinal Bureau in Hamburg, on the other hand, sent the IHO comprehensive lists of a wide range of dyes used in sausages, including eosin, rosaline, and other coal-tar dyes, while Leipzig University's Institute of Hygiene noted the regular use of azo dyes. After gathering evidence from public and private laboratories across Germany, the IHO concluded that coal-tar dyes were regularly used in sausages and other meat products; it suggested that the substitution of previously used dyes, such as cochineal with chemical colors, had become normalized with the "opportunity provided and incentive created to find alternative colors" as a result of the German food law of 5 July 1887. This law had banned several, mainly metallic, dyes known to be harmful, thus driving producers to search for replacement dyes and, effectively, legitimizing any dye not banned by the act.[41]

[37] Hans Jürgen Teuteberg, "Adulteration of Food and Luxuries and the Origins of Uniform State Food Legislation in Germany," *Zeitschrift für Ernährungswissenschaft* 34 (1995): 95–112; Hierholzer, "Searching for the Best" (cit. n. 32); Ellerbrock, "Lebensmittelqualität" (cit. n. 3); Hierholzer, *Nahrung nach Norm* (cit. n. 3); Jutta Grüne, *Anfänge staatlicher Lebensmittelüberwachung in Deutschland: Der Vater der Lebensmittelchemie Joseph König*, ed. Hans Jürgen Teuteberg (Stuttgart: Franz Steiner, 1994).

[38] *Denkschrift über die Aufgaben und Ziele, die das kaiserliche Gesundheitsamt sich gestellt hat* (Berlin, 1878); cited in Hierholzer, "Searching for the Best" (cit. n. 32), 81.

[39] Joseph König and Adolf Juckenack, "Preußen," in *Die Anstalten zur technischen Untersuchung von Nahrungs- und Genußmitteln sowie Gebrauchsgegenständen, die im Deutschen Reiche*, ed. König and Juckenack (Heidelberg: Springer, 1907), 3–159.

[40] Ellerbrock, "Lebensmittelqualität" (cit. n. 3).

[41] "Various Papers Relating to an Investigation into the Artificial Colouring of Sausages in Germany," Imperial Health Office R86/2255, Bundesarchiv, Berlin.

Food producers, retailers, and consumers across Germany were divided as to what constituted legitimate or illicit ingredients, while chemists, nutritionists, regulators, and the food industry had differing opinions on the permissible limits of chemical additives. Consumer habits and conceptions about what colors different foods should be, as well as the commercial constraints of industrial food production and a longer supply chain, the lack of consensus among chemists, and the political difficulties surrounding the need to feed a growing urban population economically and safely, all fed into an uncertain and heated debate as to whether, and how many, chemical additives should be used in food production.[42]

Germany's broad legislative and regulatory framework did not include any prescriptive definition of what constituted food adulteration. As a result, the growing numbers of chemists working in different areas of food chemistry, technology, and nutritional science began to produce their own opinions and guidelines to monitor and regulate food.[43] During the 1890s, the Free Association of German Food Chemists compiled a series of uniform norms and guidelines for food content and preparation. These standards were adopted by the government and published by the IHO.[44] However, food producers and retailers as well as consumers also set up their own organizations, such as the Union of German Food Producers and Retailers, which published their own food norms and guidelines in the *Deutsches Nahrungsmittelbuch* in 1905.[45]

To appease continuing complaints from food businesses that chemical analysts were being too purist and insufficiently pragmatic about the issue of food adulteration, the German Justice Minister ruled in 1883 that Germany's courts had to consult experts from the food industry as well as independent and state food chemists in disputes over food adulteration. Food producers' arguments in court that food was not adulterated if the processing it underwent was economic, safe, and "normal" infuriated chemists, who advocated stricter rules about food norms and ingredients, particularly with regard to chemical additives.[46] Professor Carl Neufeld, head of the Munich Research Laboratory, objected to the expertise of food analysts being challenged in such a way, arguing that "the food chemist is not, as the food industry would like to call him, a mere analyser, he is a key contributor to any judgment."[47] Josef König, head of the Münster Agricultural Research Station, argued that "when the habits of industry can be considered with impunity then hardly any tampering (or adulteration) or bad habits can

---

[42] Ellerbrock, "Lebensmittelqualität" (cit. n. 3); Hierholzer, *Nahrung nach Norm* (cit. n. 3); Grüne, *Anfänge staatlicher Lebensmittelüberwachung* (cit. n. 37).

[43] Alfons Bujard and E. Baier, *Hilfsbuch für Nahrungsmittelchemiker* (Berlin: Salzwasser Verlag, 1894); Alfred Hasterlik, *Die praktische Lebensmittelkontrolle. Ein Leitfaden für die Nahrungs- und Genußmittelpolizei und für das Lebensmittelgewerbe* (Stuttgart: Eugen Ulmer, 1906); Carl A. Neufeld, *Der Nahrungsmittelchemiker als Sachverständiger: Anleitung zur Begutachtung der Nahrungsmittel, Genussmittel und Gebrauchsgegenstände nach den gesetzlichen Bestimmung, mit praktischen Beispielen* (Berlin: J. Springer, 1907); J. Gustav Rupp, *Die Untersuchung von Nahrungsmitteln, Genussmitteln und Gebrauchsgegenständen. Praktisches Handbuch für Chemiker* (Heidelberg: C. Winter, 1900).

[44] *Vereinbarungen zur einheitlichen Untersuchung und Beurtheilung von Nahrungs- und Genußmitteln sowie Gebrauchständen für das Deutsche Reich. Ein Entwurf festgestellt nach den Beschlüssen der auf Anregung des Kaiserlichen Gesundheitsamtes einberufenen Kommission deutscher Nahrungsmittelchemiker* (Berlin: Julius Springer, 1897–1902); Hierholzer, "Searching for the Best" (cit. n. 32).

[45] Specialist journals included *Zeitschrift für Untersuchung der Nahrungs- und Genussmittel sowie Gebrauchsgegenstände (ZUNG), Deutsche Nahrungsmittel-Rundschau, Konserven-Zeitung*, and *Monatsschrift für die volkswirtscaflichen, gesetzgeberischen und kommerziellen Interessen der Margarine-Industrie*.

[46] Ellerbrock, "Lebensmittelqualität" (cit. n. 3), 141.

[47] Neufeld, *Der Nahrungsmittelchemiker* (cit. n. 43), 4.

be punished, such grounds may be invoked for any falsification."[48] Neufeld and König acknowledged that chemical colorings were used extensively in the food industry and that most were safe in small quantities. However, both argued that additives such as chemical colorings could be used to deceive consumers and if used in many different food products could be consumed in larger quantities than recognized.[49] Like their peers in other countries, the German food chemists as a group neither advocated for nor objected to the use of chemical dyes in food, but were anxious to ensure that professional chemists played a role in the decision-making process. They pressed for more independent analysis and reproached the food industry for using professional advisors and skilful advocacy to thwart regulations. These calls were successful; in 1910, the Prussian government passed a new law that emphasized the importance of the food chemists' expertise, while allowing commercial expert witnesses representing the food industry to appear in contentious cases.[50]

The independent food chemists and the German food industry had each set up their own versions of food norms, and often disagreed when it came to the use of colorings. While municipal food chemists sought to ban the use of chemical colorings, or at least limit their use, *Das Deutsche Nahrungsmittelbuch*, the food industry's handbook, declared that "the coloration of a food stuff for nutritional physiological reasons has to be seen as a material improvement and purely from a physiological perspective, be treated with equal importance to seasoning." The food industry claimed that Ivan Pavlov's experiments had shown that food color was part of the psychology of food, and argued that as the color of food encouraged people to eat it, so the dyeing of food to improve its attractiveness should be considered an improvement upon uncolored food.[51] Indeed, the visual perception and psychology of color is an area that is attracting new scholarship in the history of science.[52]

Lack of agreement within the chemical community in Germany, as in France, together with a reluctance to condemn the use of chemicals as food additives without substantial proof of harm, led to regulation that banned a limited number of specified aniline and azo dyes. Ironically, the introduction of the legislation allowed German and French chemical and food ingredient wholesalers to market all dyes that had not been banned by legislation as "harmless," thereby further legitimizing the use of chemical additives.

HARVEY W. WILEY AND THE REGULATION OF CHEMICAL DYES AS FOOD ADDITIVES IN THE UNITED STATES, 1883–1907

On the continent of North America, meanwhile, a very different approach to food legislation developed, particularly with regard to chemical coloring. Historians often depict

---

[48] For König's statement, see *Bericht uber den XIV. Internationalen Kongress fur Hygiene und Demographie Berlin, 23–29 September 1907*, vol. 2 (Berlin, 1908), 321; Ellerbrock, "Lebensmittelqualität" (cit. n. 3), 141.

[49] Neufeld, *Der Nahrungsmittelchemiker* (cit. n. 43), 462.

[50] Ellerbrock, "Lebensmittelqualität" (cit. n. 3), 141.

[51] Wilhelm Kerp, *Nahrungsmittelchemische Tagesfragen: Über die durch die gewerbliche Herstellung der Lebensmittel an diesen hervorgebrachten Erscheinungen* (Berlin: Julius Springer, 1914), 191; "Wird die Farbung eines Nahrungsmittels aus Gründen der Ernährungsphysiologie als seine stoffliche Verbesserung und als eine in physiologischer Hinsicht dem Würzen analoge Behandlung angesehen werden müssen," from *Deutsches Nahrungsmittelbuch* (Heidelberg: C. Winters, 1905), 12.

[52] Regina Lee Blaszczyk and Uwe Spiekermann, eds., *Bright Modernity: Color, Commerce and Consumer Culture* (Washington, DC: Palgrave Macmillan, 2017).

the 1906 US Pure Food Act as a model for modern food regulation, and its chief drafts-man, the state chemist Harvey W. Wiley, as a champion for the consumer, who tried to put a brake on the unrelenting industrialization of food.[53] Wiley, like chemists in Europe, identified chemical additives, including coal-tar colors, as an area where analytical chemists could exert their authority. Also in common with his European counterparts, Wiley soon realized that he needed to cooperate with the food and chemical manufacturers. Historians have demonstrated how large US food producers successfully used regulation to their commercial advantage, and have showed how Wiley himself garnered big-business support to help get his measures passed.[54]

As in Europe, concern over the adulteration and industrialization of food was widespread in nineteenth-century America, with food chemists, large food manufacturers, politicians, rural communities, and the middle classes all demanding reform as part of a national Pure Food Movement. As the head of the Department of Agriculture's Bureau of Chemistry between 1883 and 1912, Wiley was in an excellent position to help raise the status of analytical chemists, and saw food analysis and the Pure Food Movement as an opportunity to achieve this.[55]

As "pure food" activists targeted the new industrial food practices and chemical ingredients, Wiley claimed that chemistry itself was pivotal to policing food and enforcing food standards. In 1903, the Bureau of Chemistry started publishing purity standards for food products compiled by the National Committee of Food Standards, a group of representatives from the Department of Agriculture, the Association of Official Agricultural Chemists, and the Association of State Dairy and Food Departments, as well as other experts.[56] The Bureau of Chemistry then began work on regulating food additives, particularly preservatives and dyestuffs. States across the country had conflicting regulations concerning the use of coal-tar dyes, and food producers and retailers wanted a uniform approach.

The original 1906 Pure Food Act contained no regulations specifically on coal tar derived dyestuffs, and Wiley realized that it would take time to determine which of the many hundreds of dyes being distributed in the market were safe to use. To oversee the analysis, he appointed Bernard Hesse, a chemist who had worked for the German chemical company BASF (Badische Anilin und Soda Fabrik) in Germany for several years before becoming an independent consultant in the United States.[57]

---

[53] Wiley, in his autobiography, and his official biographer, Oscar Anderson, were instrumental in creating this early image; Harvey Washington Wiley, *The History of a Crime against the Food Law: The Amazing Story of the National Food and Drugs Law Intended to Protect the Health of the People, Perverted to Protect Adulteration of Foods and Drugs* (Washington, DC: printed by the author, 1929); Oscar E. Anderson, *The Health of a Nation: Harvey W. Wiley and the Fight for Pure Food* (Chicago: Univ. of Chicago Press, 1958).

[54] Harvey Levenstein, *Revolution at the Table: The Transformation of the American Diet* (New York, N.Y.: Oxford Univ. Press, 1988; Berkeley and Los Angeles: Univ. of California Press, 2003); Levenstein, *Fear of Food: A History of Why We Worry about What We Eat* (Chicago: Univ. of Chicago Press, 2012); Clayton C. Coppin and Jack C. High, *The Politics of Purity: Harvey Washington Wiley and the Origins of Federal Food Policy* (Ann Arbor: Univ. of Michigan Press, 1999); Donna J. Wood, "The Strategic Use of Public Policy: Business Support for the 1906 Food and Drug Act," *Bus. Hist. Rev.* 59 (1985): 403–32; James Harvey Young, *Pure Food: Securing the Federal Food and Drugs Act of 1906* (Princeton, N.J.: Princeton Univ. Press, 1989).

[55] Coppin and High, *Politics of Purity* (cit. n. 54), 3, 35–50.

[56] Spiekermann, "Redefining Food" (cit. n. 32).

[57] Williams Haynes, *American Chemical Industry: A History*, 6 vols. (New York, N.Y.: Van Nostrand, 1945), 2:61; "Bernard Hesse," *Oil, Paint and Drug Reporter* 91 (1917), 9.

By September 1906, Hesse had collected about ninety samples of dyes recommended by dye distributors for use in food, several of which he believed were harmful. But by February 1907, he had only managed to test the red samples, warning Wiley that testing might have to continue after regulations had been developed, because each test was so time consuming. As a compromise, Hesse suggested producing a shortlist of dyes that could be "considered harmless and permitted for use." He observed that legislators in other countries had banned certain colors judged to be harmful, but noted that these laws had simply encouraged the unrestricted use of artificial colors not banned, including all coal-tar dyes not yet examined and all newly discovered colors.[58] However, the more precautionary approach of the United States, which licensed only colors shown to be safe, was problematic because, by1906, there were already too many dyes on the market to police effectively. Hesse estimated that there were 695 different chemical dyes produced by 37 companies on the world market, with more constantly being discovered, making it impossible to test every dye.[59]

This led to a situation wherein the US government decided to ban all dyes, whether they were toxic or not, except for a few that could be certified as safe to use. To determine which dyes should be selected as "permitted dyes," Hesse consulted various toxicological studies, including those undertaken by Weyl and Sigmund Fraenkel. He also visited dye manufacturers, including Hartford-based Schoelkopf; Hanna of Buffalo; Newark-based Heller; Merz and Company, Hartford; and the US subsidiaries of BASF and Hoechst; as well as the specialist dye importers and suppliers to the food industry, H. Kohnstamm and Company, and H. Lieber and Company.[60] He surveyed 13 manufacturers worldwide and four importers, noting that "5, or 29 per cent have not found it in their interest to contribute either specimens or information" and adding that there was "very little unanimity among the different concerns furnishing coal-tar colors for use in food products as to which of their products are desirable, necessary, or suitable for such use."[61]

From more than two hundred samples, Hesse chose seven colors he decided could be safely used in food, if produced in a pure form, based on the toxicological tests of Weyl and others. The seven dyes were two yellows (napthol yellow S and tartrazine); four reds (carmoisine, rhodamine B, amaranth, and erythosine); and light green SF bluish. Hesse, however, altered his opinion less than a week afterward, and following another review, swapped four of his original choices of dyes for a different selection. The final list was eventually approved by the federal government in Washington and was included in Food Inspection Decision 76, "Dyes, Chemicals and Preservatives," issued on 18 June 1907; it was comprised of red (amaranth GT, Green Number 1107); scarlet (Ponceau 3R, 56); bluish red (Erythrosine 517); orange (Orange 1, 85); yellow (Napthol Yellow S, 4); green (Light Green SF Yellowish, 435); and blue (Indigo Disulphonic Acid, 692).[62]

---

[58] Bernhard Conrad Hesse, *Coal-Tar Colors Used in Food Products* (Washington, DC: Government Printing Office, 1912); Sheldon Hochheiser, "Synthetic Food Colors in the United States: A History under Regulation" (PhD diss., Univ. of Wisconsin, 1982).

[59] Hesse, *Coal-Tar Colors* (cit. n. 58), 15.

[60] Ibid.; Bernhard C. Hesse, "The Industry of the Coal-Tar Dyes. An Outline Sketch," *Journal of Industrial & Engineering Chemistry* 6 (1914): 1013–27.

[61] Hesse, *Coal-Tar Colors* (cit. n. 58), 16–18.

[62] Hochheiser, "Synthetic Food Colors" (cit. n. 58); US Department of Agriculture, Office of the Secretary, FID (Federal Inspection Decision) 76, "Dyes, chemicals and preservatives in food," 13 July 1907; cited in Hesse, *Coal-Tar Colors* (cit. n. 58), 9.

By creating the concept of "permitted" chemical dyes, the US legislation effectively opened up a new legitimized and specialized market for food dyes, which created an opening for the US chemical companies to differentiate themselves from their larger European competitors. Most overseas dye manufacturers, claiming that it was impossible to produce dyes of the purity required by the US government, chose to stop marketing food dyes in the United States. In retrospect, commentators have argued that the seven colors recommended by Hesse as permitted colors in 1907 "later turned out to be comparatively poor choices."[63] The seven permissible dyes chosen from the hundreds of different dyes being indiscriminately and covertly used in the food and drink industry were a subjective and convenient choice of Hesse's and remained the only officially permitted coal tar derived dyes for food sold in the United States for most of the twentieth century.

The legislation adopted in the United States that instigated the concept of specified "permitted food dyes" contrasted with the approach adopted in both France and Germany. These two countries effectively permitted the use of any dye not specifically banned by government. In Britain, the law forbade food producers and retailers from poisoning or cheating the consumer without banning or permitting any specific chemical dyes.

BRITISH APPROACH TO REGULATIONS ON CHEMICAL DYES

While the US government produced a list of recommended food dyes in 1907, and many European countries had prohibited certain coal-tar dyes for food use by 1900, Britain produced no legislation specifically addressing the new dyes until 1925, and then merely banned a few named coal-tar dyes. A list of permitted coal-tar dyes was not introduced into British law until 1957, fifty years after similar US legislation.

Britain's lack of legislation on chemical dyes in food production was not due to less anxiety about the dyes. In 1899, public and political concern surrounding chemical additives in food, including synthetic colorings, prompted the British Parliament to establish a governmental committee to "inquire into the use of preservatives and colouring matters" and determine whether their use was "injurious." The use of a select committee, which comprised a group of cross-party parliamentarians questioning a cross-section of industry executives, health and scientific advisors, and consumer group representatives, was one way in which the government sought to mediate between conflicting interest groups.

For twenty-six days the committee listened to evidence given by seventy-eight witnesses, including prominent food manufacturers and public analysts.[64] While there were a few outspoken critics of chemical food colorings among the analysts, most prominently Cassal, Hehner, and Alfred Hill, public analyst and medical officer for Birmingham, their inability to detect coal-tar dyes in food and assess the harmfulness of these substances led to a lack of consensus among analysts and hampered attempts to regulate their use. Indeed, even the analysts who usually were more critical of the use of

---

[63] Melanie Miller, "Food Colours: A Study of the Effects of Regulation" (PhD diss., Aston Univ., Birmingham, 1987), 411–12.

[64] This evidence culminated in the *Report of the Departmental Committee Appointed to Inquire into the Use of Preservatives and Colouring Matters in the Preservation and Colouring of Food: Together with Minutes of Evidence, Appendices and Index* (London: HMSO, 1901).

coal-tar dyes in food were more muted in their protests before the select committee. Analysts seemed to be making different arguments in different spaces, changing their approach according to their various roles as commercial consultants, public and political advisors, food custodians, or scientific researchers. Once again, Hehner's testimony before the select committee, like his comments in the Birmingham Court Case, appeared to contradict the more critical opinions he had conveyed in private among colleagues and written in the sanitarian press. Hehner told the select committee that he had observed an increasing use of aniline dyes in butter, but was not concerned because the quantities used were so small. When asked about the use of Martius yellow, he noted that he had never come across it but suggested that the small size of samples would mean that its presence would be difficult to ascertain.[65] These comments contrast with his observations at the SPA (Society of Public Analysts), where he described Martius yellow as being "eminently poisonous."[66] For someone who campaigned vociferously in the press about the increased use of chemical ingredients in food, Hehner took a distinctly conciliatory approach at the parliamentary hearing, telling the committee that coloring matter was "continually present—almost invariably in the London milk." While critical of the practice, he did not advocate banning it: "I think it is a deceptive practice, and undoubtedly was due originally to the milkman being anxious to hide the blueness of his milk. At the same time, it is such a universal practice now that the consumer would probably refuse the natural milk if he got it."[67] In a setting where influential food manufacturers were also present and where the press was reporting witnesses' testimony, it could be argued that Hehner, in presenting a less critical view of the novel food ingredients, was mindful of his continuing need to earn income as a commercial consultant to the food industry.

According to the government laboratory (first called the Board of Inland Revenue Chemical Laboratory from 1842 to 1894), coal-tar colors were increasingly used in food and drinks, especially in margarine (found in 100 of 133 samples tested by the laboratory), sauces and ketchups (present in 5 of 10 samples), fruit syrups (12 of 23), and cordials (12 of 24), but they were also widely used in sausages (72 of 226), butter (40 of 364), fruit jellies (9 of 28), sugars (24 of 149), potted meats (27 of 165), and temperance drinks (56 of 769). The government select committee noted that these figures probably significantly understated the actual use of coal-tar dyes, because of the difficulty of identifying the dyes in food and drink, given the minute quantities and varying strengths in which they were used.[68]

Analysts were divided as to how to respond to the new substances, created by chemists but manipulated outside of the laboratory in ways that made them hard to identify and even harder to manage. Walter William Fisher, the public analyst for Oxford, Berkshire, and Buckinghamshire, noted that the quantity of chemical dyes "used is so small that it is as a rule impossible to absolutely identify any particular dye." He added: "It is say, one in 100,000, or 200,000 or 300,000 only."[69] Fisher observed that dyes such as tropoelin were used in butter, and eosin in jellies, as well as other nitrocompounds

[65] Hehner, Evidence to the Select Committee (cit. n. 19).

[66] Hehner is quoted in Charles Cassal, "On Dyed Sugar" (cit. n. 17).

[67] Hehner, Evidence to the Select Committee (cit. n. 19).

[68] *Report of the Departmental Committee* (cit. n. 64), paragraphs 63, 65.

[69] Walter Fisher, Evidence to the Select Committee, 17 January 1900, *Report of the Departmental Committee* (cit. n. 64), 163–9.

known to be toxic, but doubted that such tiny quantities could be harmful.[70] The British analyst Alexander Wynter Blyth stressed that the use of such small quantities of coal-tar dyes and the fact that analysts were usually given only small samples of food for testing made it problematic to identify the presence of artificial colorings. While he agreed that the small amount of dye used in any one item of food meant that it was unlikely to have an injurious effect, Wynter Blyth expressed concern about the cumulative effect of the widespread use of chemical dyes, particularly on children, as a cheap substitute for dyes like saffron and cochineal. He noted that the work of German and French chemists had shown some aniline dyes to be poisonous, but stated that it was impossible in Britain at that time to produce a list of injurious and harmless colors. Other analysts called for more spending by the British government on chemical research, which might enable legislators to identify which toxic chemicals to ban. Hill recommended that Britain follow the example of France and Belgium in prohibiting certain colorings, invoking the analysts' public health responsibilities: "I think it is desirable to protect the public from having such drugs administered to them against their will, or against their knowledge."

Cassal claimed that "it should be the business of the legislature to ensure that the necessary information should be obtained at the expense and under the direction of the Government," pointing out that the pronouncement of whether specific dyes were poisonous or not in the quantities used should not be put upon the "shoulders of individual public analysts" to fight in court.[71] According to Cassal, there was no definition of a poison or a drug under the Sale of Food and Drugs Act, and that proving a substance or item of food to be injurious was a costly and difficult matter.[72] By 1901, several European countries already had banned certain coal-tar dyes understood to be toxic, and the United States would soon produce a list of permitted dyes, reducing the onus on analysts to prove toxicity in the law courts.

The inability of chemists to determine, and agree on, the harmlessness or otherwise of the hundreds of new dyes on the market and being used indiscriminately in food made it very difficult for chemists to assert themselves as guardians of food quality, particularly in Britain where food legislation did not specify which dyes could or could not be used in food preparation. Issues of consumer choice and economic fraud played a prominent role in the hearings, with analysts arguing that the British public had the right to consume what they wished, so long as they were not deceived. Even Cassal, normally an ardent campaigner against artificial dyes, accepted the use of color in butter, on account of "long custom" or "trade continuity."[73]

After listening to the testimony, the select committee concluded: "In regard to the colouring matters of modern origin, while we are of opinion that articles of food are very much preferable in their natural colours, we are unable to deduce from the evidence received that any injurious results have been traced to their consumption. Undoubtedly some of the substances used to colour confectionery and sweetmeats are highly poisonous in themselves; but they are used in infinitesimal proportions, and

---

[70] Alfred Hill, Evidence to the Select Committee, 28 November 1899, *Report of the Departmental Committee* (cit. n. 64), 78–84.

[71] Charles Cassal, Evidence to the Select Committee, 15 January 1900, *Report of the Departmental Committee* (cit. n. 64), 129–36.

[72] Ibid.

[73] Ibid.

before any individual had taken enough of colouring matter to injure him, his digestion would probably have been seriously disturbed by the substance which they were
employed to adorn."

The only item for which the committee recommended banning the use of coloring
was milk, "because of large quantities consumed and expectation among consumers
that it is a 'natural' product." As such, the response by the committee represented a negotiation between freedom of trade, consumer choice, and the support of commercial
practices, versus consumer protection and greater transparency.[74]

CONCLUSION

As the industrialization of food production developed apace in the later decades of the
nineteenth century, problems emerged in the food supply thanks to new technologies
and chemical additives. Across Europe and North America, chemists promoted chemistry as the main weapon in fighting perceived new forms of food adulteration. However, in
their claim to offer rational and scientific ways of managing the increasing complexities
of the food supply, they had to negotiate and compromise with producers, retailers, and
politicians. Libertarian arguments surrounding consumer choice and free trade competed with debates surrounding public health and risk management. By focusing on
synthetic chemical dyes, we can consider how chemists and other stakeholders in different countries attempted to arbitrate the wider use of a new scientific product and technology in society. Food and drink industries turned to legislation and analysts to strengthen
their position, while analysts themselves used the adulteration debate to boost their status and that of the chemical industry. The more muted response of the British analysts
arose partly as a result of the laissez-faire economic culture in Britain, where food producers and politicians were reluctant to impose stringent food regulations. Chemists
also had less political influence in Britain compared with other countries. In France,
for example, chemists such as Cazeneuve ran for political office; in the United States,
analytical chemist Wiley became a powerful bureaucrat within the US state administration; and in Germany, politicians, business leaders, and academics worked closely
together to strengthen the German chemical industry.

A lack of political consensus and contestations surrounding the use of coal-tar dyes
also arose partly because of the inability of chemists to generate reliable tests for detecting the presence of synthetic dyes in food, as well as the fact that these new substances
being used as food ingredients proved so difficult to stabilize and standardize. The chemists involved in arbitrating the use of chemical dyes in food differed in their own understanding of the new substances. In many ways, the approach taken, and consensus
reached, in each country reflected the interests, allegiances, and status of all the stakeholders involved and their cultural, political, and institutional settings. The degree of cultural influence was even reflected in the foods that became central to the debates in each
country. The use of chemical dyes in wine became of greatest concern in France, compared with the adulteration of sausages in Germany, and dairy products in Britain and
the United States.

As a result, approaches to regulation differed from country to country. Determining
limits for certain food additives, or banning certain ingredients and approving others,

---

[74] *Report of the Departmental Committee* (cit. n. 9), paragraphs 126, 128, 130, 131.

became the approach for food legislation favored in the United States. Continental European countries, while banning some dyes, tended instead to opt for the creation of "norms" for the constitution of certain foodstuffs, standards which often reflected regional differences in production and consumer expectations. Britain's food legislation, however, did not specify the use of coal-tar dyes at all, whether as permissible or prohibited. Its laws, enacted during the second half of the nineteenth century, merely stated that food producers and retailers should not harm or deceive the public, without specifying any substances or additives that could or should not be used. Legislation appointed public analysts to monitor the system with any disputes being resolved in the law courts.

The debates surrounding the use of chemical coloring in food were not just about the healthfulness or otherwise of artificially colored foods; they formed the backdrop against which consumers were able to trust, or not, the novelties being produced by science as part of the industrialization of the food supply. However, it was not the scientists alone who determined the validity of novel substances as legitimate food ingredients. The validation of certain aniline and azo dyes as food dyes was the result of mediation and negotiation between a wide range of stakeholders, with regulations differing from country to country according to diverse cultural and institutional practices.

The Technopolitics of Food:
The Case of German Prison Food from the Late Eighteenth to the Early Twentieth Centuries

by Ulrike Thoms*

ABSTRACT

Since the *Annales* School started to investigate the history of food in the 1960s, institutional diets have been an important field of research. The history of food encountered the general source problem of the history of everyday life of the lower, often illiterate, classes, as they have left hardly any written sources. Even more, food is highly perishable, so that it leaves nothing as a source itself. In contrast to this, institutions kept books registering the expenses and consumption of food, and can be used to access the history of food. This article presents a different view of institutional food. It shows that prison food does not just mirror the general developments of the food of the people, as the Annalistes had hoped for. Instead, it shows that the dietetic order in prison described and expressed the social, political, and judicial concepts of the day, as well as concepts of the body. In addition, it shows that the dietetic order resulted from multiple negotiations of the various actors involved, including those in law, administration, science, and medicine, and last but not least, the public. Using the example of Prussia, this article investigates changes in prison food standards and cross-checks this with quantitative developments as well as with the personal experiences of prisoners from 1700 to 1914.

In 1777, a report concerning the "State of the Prisons in England and Wales" was published in the small English town of Warrington. It would be the starting point of a movement of international dimensions. Its author was John Howard, a tradesman's son, who, thanks to a fortune inherited from his father, was a man of independent means, occupying himself with multiple philanthropic projects.[1] Among other things, he held the post of high sheriff of Bedfordshire, an honorary position that involved inspecting prisons; in

---

* Baseler Str. 165, 12005 Berlin, Germany; ulrike.thoms.1@gmx.de.

[1] John Howard, *The State of the Prisons in England and Wales with Preliminary Observations, and an Account of Some Foreign Prisons* (Warrington: William Eyres, 1777); Roy Porter, "Howard's Beginning: Prisons, Disease, Hygiene," in *The Health of Prisoners*, ed. R. Creese, W. F. Bynum, and J. Bearn (Amsterdam: Rodopi, 1995), 5–26; Jan-Carl Janssen, "Entwicklung, Praxis und kriminalpolitische Hintergründe des Strafvollzugs in England, Wales und Schottland im nationalen und internationalen Vergleich" (PhD diss., Universität Greifswald, 2018); Maximiliane Friedrich, "John Howard und die Strafvollzugsreformen in Süddeutschland in der ersten Hälfte des 19. Jahrhunderts" (PhD diss., Univ. of Würzburg, published in Frankfurt am Main: Peter Lang, 2013); Thomas Nutz, *Strafanstalt als Besserungsmaschine: Reformdiskurs und Gefängniswissenschaft 1775–1848* (München: Oldenbourg, 2001).

the course of this task, he witnessed intolerable conditions. Yet, Howard was not alone in his critique. By the end of the eighteenth century, a discursive community of doctors, jurists, clergymen, and prison officials spanning Europe and the United States set itself the task of reforming the existing condition of prisons.[2] Diet was a major aspect of these programs. Hardly one of their countless reports on life in prison lacked a portrayal of the inadequate diet of prisoners, while the many prisoner autobiographies penned into the twentieth century would turn this perspective into a stereotype and establish it still more completely. Subsisting on bread and water alone remains synonymous with prison diet to this day.

To move beyond this static image, this article examines the relationship between broader changes in the diet of the lower classes and prison fare in Europe. It builds in part on a research program begun by *Annales* school historians such as Maurice Aymard, who relied heavily on statistics from institutions like hospitals and prisons as sources for writing the history of common food in Europe, a methodology that emphasized structural factors and long-term developments.[3] Historians of everyday life were particularly inspired by the *Annalistes* and were glad of these hints as to sources on the food of non-elite people, something otherwise not documented prior to the development of household statistics.[4] As the *Annalistes* rightly recognized, prison food is relatively well documented, and the available sources offer quantitative information regarding the acquisition, preparation, and consumption of food, as well as a variety of different perspectives on eating habits. Prison regulations, moreover, mirror the norms of other nascent institutions, while scientific papers concerned with prison administration and improvement allow historians to trace the establishment, questioning, and revision of such norms. In addition to such sources, this article also encompasses popular books about prison life, which reflected changing public perceptions of prison diets, as well as inmates' autobiographical writings, which, especially in the case of political prisoners, afford insight into their unique personal experiences.[5]

One reason for the existence of this wealth of sources is the lively discourse on the penal system, especially during the eighteenth and nineteenth centuries, when the modern prison was born. It formed part of the phenomenon Norbert Elias has named the "civilizing process."[6] From the perspective of law, death sentences and corporal punishment were increasingly criticized as inhumane. They were replaced by the idea

---

[2] Thomas Nutz, "Global Networks and Local Prison Reforms: Monarchs, Bureaucrats and Penological Experts in Early Nineteenth-Century Prussia," *Germ. Hist.* 23 (2005): 431–59; C. J. A. Mittermaier and Lars Hendrik Riemer, *Das Netzwerk der "Gefängnisfreunde" (1830–1872): Karl Josef Anton Mittermaiers Briefwechsel mit europäischen Strafvollzugsexperten* (Frankfurt am Main: Klostermann, 2005).

[3] For a general appraisal of the *Annales* school, see Peter Burke, *The French Historical Revolution: The Annales School 1929–2014* (Cambridge, UK: Polity, 2015). Specifically on its impact on the history of food, see Eva Barlösius, "The History of Diet as a Part of the *Vie Matérielle* in France," in *European Food History*, ed. Hans Jürgen Teuteberg (New York, N.Y.: Leicester Univ. Press, 1992), 90–108.

[4] Maurice Aymard, "Pour l'histoire de l'alimentation: Quelques remarques de méthodes," *Annales: Économies, Sociétés, Civilisations* 30 (1975): 431–44, on 433; Uwe Spiekermann, "Haushaltsrechnungen als Schlüssel zum Familienleben? Ein historischer Rückblick," *Hauswirtschaft und Wissenschaft* 42 (1994): 154–60.

[5] For a detailed appraisal of sources on prison food, see Ulrike Thoms, *Anstaltskost im Rationalisierungsprozess: Die Ernährung in Krankenhäusern und Gefängnissen im 18. und 19. Jahrhundert* (Stuttgart: Franz Steiner, 2005), 30–7.

[6] Norbert Elias, *Über den Prozeß der Zivilisation: Soziogenetische und psychogenetische Untersuchungen*, 2 vols. (Frankfurt am Main: Suhrkamp, 1976).

of changing the offender through education. This new attempt was mainly conducted by prison societies, with visiting services and priests working toward the betterment of the soul. But it became increasingly clear that soul and body were tied together, and that the needs of the latter, as well as desires, had to also be taken into account.

The European community of law experts and prison doctors, who developed modern prison theory and practice, long maintained the "less-eligibility principle," which dictated that the food of a prisoner should be a punishment. On the other hand, statisticians with a distinct interest in food, such as Édouard Ducpétiaux (1804–68) in France and Ernst Engel (1821–96) in Germany, both pointed to insufficient food provisions as one factor in the high mortality and morbidity rates of prisoners. However, it is important to note that both approaches emerged from an international community of statisticians involved with the development of national statistics and household budgets. They derived and applied their statistical knowledge from, and also applied it to, private households and public institutions during the epochs of nationalism and imperialism, when a healthy *Volkskörper* or "body of the people" came to be considered an important, positive entity for the state. The experts who employed statistical tools and data worked closely with medical professionals. This kind of collaboration was particularly true for Abraham Adolf Baer (1834–1908), a German who had started his professional career as a prison doctor in 1872 and later became an important figure in the development of social hygiene. Drawing upon the results of the German physiologist Carl von Voit (1831–1908), Baer became the European expert in prison hygiene and food.[7] In 1887, he presented his findings on the close relations among food, mortality, and morbidity to the *Congrès Penitentiaire International* in Rome.

What may be concluded from the story of Engel, Ducpetiaux, Voit, and Baer is, first, that prison reform was a truly European enterprise at this time, and, second, that it was scientists who produced knowledge about the nutritional and health status of the prisoners, against the backdrop of a concept of social reform based on the standard of living of the working poor. In other words, we see that prison food reform was located at the crossroads of science, social reform, and state administration. Because their reforms were focused on comparison of the living conditions of prisoners, workers, and soldiers, their project of improving prison food allows us to link the history of science to the history of material culture, consumption, and everyday history, studying whether, how, and where such knowledge was intended to become practical, and whether or not it was ultimately successful. At the same time, this approach shows that we need to be cautious in using statistical data from institutions as a substitute for data on the food of the people that is unavailable from other sources.[8]

Certainly, prison food did reflect the way in which overall standards of living changed, but we must keep in mind first, that consumers had no choice over its content, and second, that these changes stemmed from a rationale which was closely tied to the transformation of penal reform and forms of punishment. The making of prison diet involved multiple actors: administrators, doctors, legislators, society as a whole, clergymen and, not least, the prisoners themselves, all of whom had a range of different

---

[7] Abraham Adolf Baer, *Die Gefängnisse, Strafanstalten und Strafsysteme, ihre Einrichtung und Wirkung in hygienischer Beziehung* (Berlin: Enslin, 1871). An extensive list of his publications is in Thoms, *Anstaltskost* (cit. n. 5), 835.

[8] For an attempt along these lines, see Andreas Kühne, *Essen und Trinken in Süddeutschland: Das Regensburger St. Katharinenspital in der Frühen Neuzeit* (Regensburg: F. Pustet, 2006), 23.

perspectives and agendas. In the negotiations that ensued, they all subjected various aspects of diet to standardization. Such processes responded to the transformation of norms, the sociocultural and socioeconomic changes undergone by the penal regime, and also the emergence of the food sciences and their imposition as a bureaucratic-managerial technique. The dietetic order in prison at once described and expressed the social order, political and juridical concepts of order, and concepts of the body and character. In these negotiations, the agency of the various actors involved from the medical domain, their collaboration with state institutions, the participating public domain, and institutional bureaucracy is made clear. The end point of the transformation was a rationalized diet, which became a self-evident part of modern penal regimes, in tandem with an increasing emphasis on education and psychology in the penal process. Just how this arose out of the complex discursive exchanges between specialist administrators, prison reformers, physicians, and the emerging nutrition sciences, is the subject of this article.[9]

In particular, the article considers changes in prison food standards from the eighteenth to the early twentieth centuries, illustrated by examples selected from Prussian penal establishments, both because these set the tone for German developments as a whole, and because of the availability of relevant source materials. A large quantity of dietary regulations, archival documents, and publications illuminate the circumstances under which food standards were produced and the discussions that surrounded them. Autobiographies and, last but not least, detailed statistical data attest to longer-term developments and allow standards to be investigated from microcosmic and macrocosmic perspectives. Although discourse about diet was rooted in a community that spanned Europe, diet was regionally specific throughout this period, and it is for this reason that the case studies are taken from Germany, and particularly Prussia. This approach has the added advantage that both structures, and the changes made to them, can be cross-checked against the fairly detailed records of food use that were included in the statistics produced by Prussian penal institutions, which are differentiated by region. My study follows these developments chronologically across four broad historical phases, respectively covering the periods from 1700 to 1770, 1770 to 1800, 1800 to the mid-nineteenth century, and from 1850 up to the start of World War I. But to begin with, we should briefly characterize the development of the modern prison system and penal regime.

THE DEVELOPMENT OF THE PRISON SYSTEM

Recent research has shown that the modern penal regime developed independently of the transformations of penal law, and that discourses arising out of education, theology, and medicine were more significant for the development of that regime than the discussions of jurists.[10] The educational concerns of the Enlightenment and the concomitant flourishing of the sciences were particularly prominent in this process. Enlightened authors like Christian Wolff (1679–1754) and Gottfried Wilhelm Leibniz (1646–1716) decisively influenced Frederick the Great of Prussia (1712–86), leading him to place a higher priority upon imprisonment at the expense of corporal

---

[9] Thoms, *Anstaltskost* (cit. n. 5).
[10] See Desiree Schauz, *Strafen als moralische Besserung: Eine Geschichte der Straffälligenfürsorge 1777–1933* (München: Oldenbourg, 2012), with reference to Nutz, "Global Networks" (cit. n. 2).

or capital punishment. These authors, for reasons of both state and welfare policy, opposed corporal and financial punishments on the grounds that these also penalized families; corporal punishment, in particular, rendered criminals permanently unable to work. They also opposed despotic judges and their harsh sentences, insisting that the penalty must fit the crime. Such views were taken up and implemented in the Prussian general law, and in one law in particular—the revolutionary *Allgemeines Landrecht* from 1794 onward—in the process of classifying crimes and developing a system that hierarchically distinguished confinement, punishment, imprisonment, jails, and workhouses according to severity of offence.

Yet prisons, penitentiaries, workhouses, and houses of correction or improvement formed a confusing muddle, being administered by a combination of private and public jurisdictions, city magistrates, and regional poor administrations, among others. These institutions, very variable in size, not only served the purpose of providing penal services, but also cared for the poor and provided social welfare. A significant hindrance to reform was what was called dualism in the correctional system, whereby the administration of houses of correction and of larger central prisons was subordinated to the ministry of the interior. This meant that the ministry of justice struggled to exert any authority over these establishments, despite the fact that they served the justice system. The responsibility of the latter was limited to those prisons and houses of correction designated for prisoners undergoing ordinary investigation, or else for the execution of sentences of imprisonment imposed within the law courts. As if the situation in Prussia were not already complicated enough, state territorial legislation in the newly acquired provinces presented a further problem. There was a mishmash of over sixty different regulations relating to service and enforcement, stemming from a wide variety of perspectives on the aims and form of the prison sentence. Little changed in consequence of the formation of the German Empire in 1871 and the penal code that resulted from it. The differences between the many regulations that the individual *Länder* (states) and provinces had to follow remained, because the law only established a framework for the Reich as a whole.[11] In light of the persistence of these long-established penal structures, it is impossible to speak of an ordered "system." Instead, very different institutions existed side by side in Germany, while institutions with the same name were organized very differently. Even the question of funding was differentially resolved: in some cases, inmates were informed that officials would provide their meals; in others, they were forced to buy food from wardens; while in yet other cases, there was an in-house innkeeper. Some institutions only supplied a fixed quantity of bread, while others provided full meals, and sometimes several times a day.

Doctors, theologians, and, from the mid-nineteenth century onward, also prison officials, were involved in debates about these circumstances. The British penal

---

[11] Dirk Blasius, *Kriminalität und Alltag: Zur Konfliktgeschichte des Alltagslebens im 19. Jahrhundert* (Göttingen: Vandenhoeck & Ruprecht, 1978), 73–4; Alexander Klein, *Vereinheitlichung, Vereinfachung und Verbilligung des dualistischen Teiles des Gefangenwesens, der Gefängnisverwaltung und des Vollzugs a) der Zellhaft, b) der Untersuchungshaft und c) der gerichtlich verhängten Freiheitsstrafe (Zuchthaus-, Gefängnis-, Haft- und Festungsstrafe) in Preußen* (Berlin: n.p., 1917); Peter Jacobs, *Der Dualismus im preußischen Gefängniswesen: Darlegung der Gründe für die Umstellung des gesamten Strafvollzuges unter das Ministerium des Innern* (Bonn: Hanstein, 1906); Thomas Berger, *Untersuchungen zur Geschichte des Strafvollzuges an Zuchthausgefangenen in Preussen 1850–1881: Praxis und Reformversuche* (PhD diss., Univ. of Freiburg, 1974), 292–4.

reformer John Howard produced publications that awakened official interest in Germany; this interest was directed in particular at institutional deprivation and the state of prisoners' health, which were being reported with thrilling sensationalism. Tales of poor diet and rampant prison epidemics developed into a stereotype that was constantly reaffirmed in a rich literature generated by the growing genre of prisoner autobiographies.[12]

Howard's publications were particularly influential in spurring prison inspections across Germany. The comprehensive reports of Heinrich Balthasar Wagnitz (1755–1838), a clergyman, and the physicians Justus Gruner (1777–1820) and Nikolaus Heinrich Julius (1783–1862), gave a final boost to the development of the so-called prison sciences, and in so doing had a wide-ranging influence upon the development of the penal system.[13]

Beginning in 1864, knowledge about prisoners and prisons came to be systematically collected and evaluated through a rapidly expanding literature on the topic found in specialist journals on prison science, annual publications containing detailed statistics on prisons and penal institutions, and reports from prison societies.[14] With the aid of this knowledge, a new technopolitics of punishment emerged.[15] The same doctors, priests, and officials engaged in prison reform and active in the Society of German Prison Officers, who were founding their own periodicals and forming international networks, were also those who introduced changes to the diet on offer in prisons.[16] While the theologians who involved themselves in the prison sciences mostly concerned themselves with prisoners' spiritual well-being and moral improvement, the doctors particularly addressed their material circumstances, including bodily needs such as nutrition, and continually fell out with prison officers over questions of food standards and dietary adequacy.[17]

All of these efforts—especially the reform of diet—formed part of what Michel Foucault named "the birth of the prison," during which new principles of regulation, refinement, and generalization of the art of punishing were established. This process

[12] Thoms, *Anstaltskost* (cit. n. 5), 738–9.

[13] Balthasar Wagnitz, *Historische Nachrichten und Bemerkungen über die merkwürdigsten Zuchthäuser in Deutschland* (Halle: Gebauer, 1791); Julius Gruner, *Versuch über die recht- und zweckmäßigste Einrichtung öffentlicher Sicherungsinstitute deren jetzigen Mängel und Verbesserungen nebst einer Darstellung der Gefangen- Zucht- und Besserungshäuser Westphalens* (Frankfurt am Main: Eßlinger, 1802). On Gruner, see K. Zeisler, "Justus von Gruner: Eine biographische Skizze," *Berlin in Geschichte und Gegenwart* (1994): 81–105; and Justus Karl Gruner, Gerd Dethlefs, and Jürgen Kloosterhuis, *Auf kritischer Wallfahrt zwischen Rhein und Weser: Justus Gruners Schriften in den Umbruchsjahren 1801–1803* (Cologne: Böhlau, 2009); Monika Schidorowitz, *H. B. Wagnitz und die Reform des Vollzugs der Freiheitsstrafe an der Wende vom 18. zum 19. Jahrhundert* (St. Augustin, France: Gardez!, 2000); Nikolaus Heinrich Julius, *Vorlesungen ueber Gefaengnis-Kunde, oder ueber die Verbesserung der Gefaengnisse und sittliche Besserung der Gefangenen, erweitert herausgegeben, nebst einer Einleitung ueber die Zahlen, Arten und Ursachen der Verbrechen in verschiedenen europaeischen und amerikanischen Staaten u.s.w.* (1797; repr., Münster: LIT, 1996).

[14] See the journals *Blätter für Gefängniskunde* 1 (1864/65)–75 (1944); and *Jahrbücher für Gefängniskunde und Besserungsanstalten* 1 (1842)–11 (1847). See also the detailed figures in *Statistik der zum preußischen Ministerium des Innern gehörenden Straf- und Gefangen-Anstalten für das Jahr 1869–1878/79* (Berlin: Decker, 1869–95); and *Statistik der zum preußischen Ministerium des Innern gehörenden Strafanstalten und Gefängnisse 1879/80–1907* (Berlin: Decker, 1880–1909).

[15] This is the term used by Michel Foucault in *Überwachen und Strafen: Die Geburt des Gefängnisses*, 8th ed. (Frankfurt am Main: Suhrkamp, 1989), 116ff.

[16] Nutz, "Global Networks" (cit. n. 2); Mittermaier and Riemer, *Das Netzwerk* (cit. n. 2).

[17] *Verhandlungen der Versammlung der Deutschen Strafanstaltsbeamten* (Heidelberg: Weiss, 1867 and 1889).

aimed at standardizing punishment in order to reduce its political and economic costs, while at the same time multiplying its effects; in other words, the process produced a new economy of punishment.[18] One important element of this process was the standardization of food, which this article will first examine in terms of the guidelines for food that were developed in German prisons.

1700–1770: GUIDELINES ON WORK AND NEEDS

The impossibility of overseeing the prison system in Germany stemmed in part from the decentralized character of the state system. Each German state organized its penal system differently, and there were town jails and state jails, as well as those run by the police and the magistracy.[19] Early on, corporal, financial, and capital punishment predominated, serving a view of punishment as retribution, so that all prisons served as places where criminals could be held before trial, or else they were a location for carrying out corporal or capital punishment. The earliest act of law under the *Constitutio Criminalis Carolina*, in 1532, introduced imprisonment as a form of punishment, although its effect was limited to individual cases.[20]

The growth of sensibility was accompanied by increasing opposition to the death penalty or corporal punishment; new views on improvement led judicial practice also to employ imprisonment, which became the dominant penal practice from the sixteenth century onward. With this change came new institutions, including the workhouses and poorhouses. Indebted to Protestant ethics, these institutions treated teaching inmates to work as the most effective way to socialize all those who, whether because of their feckless lifestyle or for some other reason, were misfits within the social order; the aim was to turn them into productive citizens through education. Just like Bridewell Prison in London, founded in 1555, or Amsterdam's *Tuchthuis*, founded in 1696, the inmates of German workhouses were supposed to be brought under a regime of strict discipline and order, bound into a minutely regulated daily schedule, and habituated to hard labor in order to convert them into useful members of society.[21] Those who worked and fulfilled their allocated tasks were entitled to sufficient food, unlike those who disobeyed the institution's rules or were lazy. The content of the workhouse diet varied significantly between institutions. Many had no regulations at all concerning the care of inmates, who were the responsibility of friends and family,

---

[18] Foucault, *Überwachen* (cit. n. 15), 113ff.

[19] The number of publications on prison history is overwhelming; see, for example, the extensive bibliography in Falk Bretschneider, "Europe," History of Prisons—A Selected Bibliography," http://www.falk-bretschneider.eu/biblio/biblio-index.htm.

[20] Gerd Kleinheyer, "Wandlungen des Delinquentenbildes in den Strafrechtsordnungen des 18. Jahrhunderts," in *Deutschlands Entfaltung: Die Neubestimmung des Menschen*, ed. Bernhard Fabian, Wilhelm Schmidt-Biggemann, and Rudolf Vierhaus (Munich: Kraus, 1980), 208–30; Erwin Bumke, *Deutsches Gefängniswesen* (Berlin: Vahlen, 1928), 1–15; Eberhard Schmidt, *Entwicklung und Vollzug der Freiheitsstrafe in Brandenburg-Preußen bis zum Ausgang des 19. Jahrhunderts: Ein Beitrag zur Geschichte der Freiheitsstrafe* (Berlin: De Gruyter, 1915), 125; Schmidt, "Die Carolina," in *Die Carolina: Die Peinliche Gerichtsordnung Kaiser Karl II. von 1532*, ed. Friedrich Christian Schroeder (Darmstadt: Wissenschaftliche Buchgesellschaft, 1986), 51–73.

[21] Hubert Treiber and Heinz Steinert, *Die Fabrikation des zuverlässigen Menschen: Über die "Wahlverwandtschaft" von Kloster- und Fabrikdisziplin* (1980; repr., Münster: Westfälisches Dampfboot, 2005); Dario Melossi and Massimo Pavarini, *The Prison and the Factory*, 40th anniversary edition (1977; repr., London: Palgrave Macmillan, 2018). This idea of social discipline prevails in Foucault's work; see Foucault, *Überwachen* (cit. n. 15).

or who had to make an arrangement with wardens. Often, the wardens themselves received no payment for their duties, but were expected to make their living from the sale of food and meals to prison inmates. This practice was conducive to fraud and extortion, such that some prisoners who were already in debt left prison even more heavily burdened. In many institutions, prisoners only received a bread ration, but in others they might receive hot meals, sometimes even several times a day. In such cases, the institutions often had detailed meal regulations—far from offering a few general principles, they might have detailed menus for each day of the week.[22] These rules were part of the institution's constitution, and so were an integral component of its internal and external order, which at once depicted and produced the social hierarchy of those housed within it.[23] Those who worked and fulfilled the requirement to be productive were entitled to food, but not those who failed to obey the regulations or were lazy.[24]

These rules were legitimated by religion, in the first instance. A letter from Paul to the Thessalonians stated, "The one who is unwilling to work shall not eat."[25] But a second guideline were inmates' wants or needs, determined in accordance with custom and convention. In general, specifications remained fairly loose. Although precise amounts were given in the case of beer, bread, and meat, the quantities of all other foodstuffs were not provided as exact measures or numbers. On the one hand, this was a way of accounting for individual variation in needs and amount of work done; but on the other hand, it also made allowances for the economic circumstances of institutions and the changing state of the markets. Plagued in preindustrial times by variable supplies, failed harvests, and hunger crises, markets posed significant problems for systematic attempts at regulation. Indeed, we find that in bad times, portion sizes were reduced, while institutions laid up stores after good harvests and bought generously, as well as serving the odd treat that might count as a luxury—for example, fresh vegetables.[26]

Yet, it was in keeping with the religious origins of such institutions that those who could *not* work, because they were too young, old, or sick, should be cared for by the institution, as was customary in convents or monasteries. For example, all inmates of poorhouses, orphanages, and workhouses in Waldheim received the same basic care,

---

[22] For examples, see the regulations of the penitentiaries at Stargard (1720), Stettin (1723), and St. Georgen bei Bayreuth (1732), discussed in Herbert Lieberknecht, "Das altpreussische Zuchthauswesen bis zum Ausgang des 18. Jahrhunderts, insbesondere in den Provinzen Pommern und Ostpreussen" (PhD diss., Univ. of Berlin, published in Charlottenburg: Klambt, 1921), 113, 116; and Rolf Endres, "Das Straf-Arbeitshaus St. Georgen bei Bayreuth," *Jahrbuch der Sozialarbeit* 4 (1981): 80–105. On food in hospitals, see Barbara Krug-Richter, *Zwischen Fasten und Festmahl: Hospitalverpflegung in Münster 1540 bis 1650* (Stuttgart: Franz Steiner, 1994); and Marcel Mayer, *Hilfsbedürftige und Delinquenten: Die Anstaltsinsassen der Stadt St. Gallen 1750–1798* (St. Gallen, Switz.: Staatsarchiv, 1987).

[23] However, these rules did not include fixed figures, so as to allow an institution to react to rising food prices, such as during a hunger crisis. At such moments, the quantities were reduced. Sometimes administrations fixed allowances in terms of cost rather than size of food portion.

[24] Petrus Cornelis Spierenburg, *The Prison Experience: Disciplinary Institutions and Their Inmates in Early Modern Europe* (New Brunswick, N.J.: Rutgers Univ. Press, 1991); Thorsten Sellin, *Pioneering in Penology: The Amsterdam Houses of Correction in the Sixteenth and Seventeenth Centuries* (Philadelphia: Univ. of Pennsylvania Press, 1930). Mercantilist views were of the utmost importance here. Within the mercantilist model, workhouses were conceptualized as institutions of economic development. New branches were established, and youngsters were educated for these new jobs. On the impact of mercantilism on the founding of workhouses, cf. Thoms, *Anstaltskost* (cit. n. 5), 50–3.

[25] 2 Thess. 3:10.

[26] Geheimes Staatsarchiv Preußischer Kulturbesitz, Berlin (**hereafter** cited as GStA PK), Rep. 96, Nr. 246C, fol. 6.

but the sick received more meat and beer.[27] In the workhouse at Lüneburg, it was stipulated in 1750 that the needs of the poor should be thoroughly met, while prisoners received only bread and had to purchase all other provisions.[28] However, this requirement did not necessarily restrict them; rather, it gave them the freedom to buy pleasures like alcohol or tobacco. Closer inspection of the meal regulations reveals striking structural similarities among institutions, alongside variations in detail; basic care consisted of bread and beer. As a rule, there was only a single hot meal at lunch, and the daily bread ration generally had to serve for breakfast and dinner too. Occasionally warm soup was served in the evening. The midday meal consisted of cereal and legume dishes, while cabbage and meat were served on Sundays.

These similarities are explained by the dietary customs of the general population for whom the institutional diet was intended, but also by a lively exchange conducted by the managers of such institutions. If a town planned a new workhouse, its representatives would ask other towns or known institutions to send details of their organization and regulations.[29] It was also common to undertake fact-gathering journeys to the institutions in question, to learn about the organization and management of poorhouses and workhouses.[30] Large workhouses—for example, an institution in Amsterdam—even conducted a tourist industry, offering open days and charging an admission fee.[31] The later travel narratives of philanthropists like Howard or Wagnitz, those published after 1800, may be understood as professionalized versions of these fact-gathering trips. Both such publications and the exchange of regulations by post contributed to the standardization of care and the process of normalization via punishment.

THE CLASH OF TRADITION AND EFFICIENCY: CRISES OF TRANSITION, 1770–1800

The period between 1770 and 1800 was colored by the aftereffects of the Seven Years' War (1756–63). Damage, plunder, and widespread poverty throughout society led to increasing use of workhouses and penitentiaries, and income from donations declined. Despite the war, the population was growing; in the case of Brandenburg, for example, the population increased from 670,488 in 1748 to 801,222 in 1766, and again to 997,903 by 1794, reaching 1,066,770 inhabitants by 1805.[32] Over the

---

[27] Report of the penitentiary commission concerning their arrangements for the poorhouses, orphanages, and penitentiaries of Torgau, Waldheim, and Leipzig, 18 August 1766, Brandenburgisches Landeshauptarchiv, Potsdam (**hereafter** cited as BLHA), Rep. 23C, Niederlausitzische Stände, Nr. 1706; Adolf Streng, *Geschichte der Gefängnisverwaltung in Hamburg* (Hamburg: Verlagsanstalt und Druckerei, 1890), 15–19.

[28] Adolph Ludolph, "Das Werk- und Zuchthaus und die Kettenstrafanstalt zu Lüneburg: Ein Beitrag zur Geschichte der Entwicklung des Strafvollzuges" (PhD diss., Univ. of Göttingen, 1930), 14.

[29] As, for example, when other institutions requested the workhouse at Bremen to send out its regulations: cf. Otto Grambow, "Das Gefängniswesen Bremens" (PhD diss., Univ. of Göttingen, published in Borna-Leipzig: Noske, 1910).

[30] Letter from Hornemann to Zuchthaus-Commission, 18 August 1766, BLHA, Rep. 23C, Niederlausitzische Stände, Nr. 1706.

[31] Christian Marzahn and Hans-Günther Ritz, *Zähmen und Bewahren: Die Anfänge bürgerlicher Sozialpolitik* (Bielefeld: AJZ, 1984), 36; Georg Wilhelm Keßler, ed., *Der alte Heim: Leben und Wirken Ernst Ludwig Heim's: Aus hinterlassenen Büchern und Tagebüchern* (Leipzig: F. A. Brockhaus, 1846).

[32] Bernd Kölling, ed., *Agrarstatistik der Provinz Brandenburg 1750–1880* (St. Katharinen: Scripta Mercaturae, 1999), 41–54, cited in GESIS Datenarchiv, Köln. Histat, Studiennummer 8464, Datenfile Version 1.0.0, https://histat.gesis.org/histat/de/table/details/8A253565BFA99A2760226062E0CC8E65#tabelle.

same period, there was a huge upsurge in the number of those convicted of criminal offences.[33]

The population suffered on account of the high price of food, while towns and states registered declining tax revenues. In Prussia, Frederick II implemented a whole package of political measures to counter this crisis, striving to rationalize the administrative apparatus. In connection with this program, he carried out extensive revisions of existing prisons, continuing a judicial reform on which he had already embarked. The *Allgemeines Landrecht*, the General State Law of 1794, brought about a general reassessment of imprisonment. Existing structures were not prepared to cope with such rapid growth, and prison crowding sometimes led to catastrophic hygienic outcomes in existing institutions, an effect exacerbated by the turbulence of the Napoleonic and post-Napoleonic periods. In the first instance, a method was sought for providing nourishment as cheaply as possible. Like medieval hospitals, prisons grew their own vegetables and potatoes, brewed their own beer, and baked their own bread, using prisoners as cheap labor.[34] In order to keep costs down, the food-purchasing policy implemented was designed to guarantee the lowest cost for foodstuffs.[35] All such initiatives came from the prison administration. Yet, scientific practitioners also engaged in similar endeavors, experimenting with innovative foods and food preparation techniques that were intended to reduce costs while still satisfying the prisoners.

Howard described the circumstances resulting from inadequate nourishment of prison inmates in strong language; according to him, individuals entered the prison in full possession of their bodily strength, but, for lack of adequate nourishment, they left it "famished, scarce able to move, and for weeks incapable of any labor."[36] Suffering from diseases contracted in prison not only meant that the idea of imprisonment—the betterment of the offender—was endangered, but also that prisoners presented a risk of infection to the area and a danger to society as a whole, as well as a burden upon their families or society.[37]

Four large volumes of travel narratives published between 1791 and 1794 by Howard's German counterpart, the prison priest Heinrich Balthasar Wagnitz, were particularly influential. In line with his goal of forming the strongest possible state, the Prussian ruler took up these initiatives because he was susceptible to arguments concerning waste in relation to both time and labor. He commissioned the Jewish physician Nikolaus Heinrich Julius (1783–1862), who had converted to Catholicism in 1809, to conduct tours of inspection of Prussian institutions, and permitted him to disseminate his findings in "Lectures on Prison Science," which King Frederick attended in person.[38] It was this personal interest that allowed a philanthropic reform movement to have concrete influence over penal policy. The new "prison science" systematically

---

[33] Statistical accounts of court decisions are only available for the nineteenth century onward. See Dirk Blasius, *Kriminalität und Alltag: Zur Konfliktgeschichte des Alltagslebens im 19. Jahrhundert* (Göttingen: Vandenhoeck & Ruprecht, 1976), 80–2; Herbert Reinke, "Die 'Liaison' des Strafrechts mit der Statistik: Zu den Anfängen kriminalstatistischer Zählungen im 18. und 19. Jahrhundert," *Zeitschrift für Neuere Rechtsgeschichte* 12 (1990): 169–79.

[34] Thoms, *Anstaltskost* (cit. n. 5), 107–66.

[35] Ibid., 166–208.

[36] Howard, *State of the Prisons* (cit. n. 1), 12.

[37] Ibid., 9.

[38] Schidorowitz, *H. B. Wagnitz* (cit. n. 13).

addressed all aspects of imprisonment, including diet.[39] The initiative should be under-stood as associated with the program of "medical police," which spread the idea of medicalization.[40] "Medical police" was a precursor to modern social hygiene, treating the provision of care as a state responsibility. In order to organize care rationally, the knowledge of the new sciences was to be integrated within it.[41] Within dietetics, nu-trition had a place.[42] But at this time, the dominant humoral theory defined health as a balance of the humors, which could be achieved through appropriate diet. As a qual-itative concept, humoral theory was flexible enough to accommodate the new discov-eries unproblematically, yet structured enough to allow these to be ordered. Such views are still to be found in the writings of Howard, who was already thinking quan-titatively, proposing that there should be a requirement to measure diet and record the regulations in a table that could be hung up in institutions for all to see, allowing in-mates to know and insist upon their rights while simultaneously preventing fraud.[43] This call to quantify can also be found in the work of German prison reformers; its concordance with humoral theory demonstrates the extent to which this was a transi-tion period.[44] Although there were still no experimentally grounded precepts of nutri-tion science that could actually be put into practice, the preconditions for their success-ful implementation were already in place. For administrators, quantification offered the advantage of allowing expenditure to be calculated, though actually they had ear-lier required institutions to possess food regulations that specified quantities precisely and were as cost effective as possible, while simultaneously allowing financial ac-counts to be kept, as well as an account of the effects of the (expensive) diet on health and illness, morbidity, and mortality.[45]

Administrations increasingly came to rely on doctors' testimony. For, after all, it was the doctors who became responsible for prisoners who sickened on poor fare and then required a special diet. The higher incidence of certain diseases in prisons and workhouses was known; a whole range of diseases reflected these institutions in their very names, particularly "jail fever," which was probably a form of typhus.[46] Such labeling was of ancient date; jail fever is even mentioned in the Hippocratic

[39] Nutz, *Strafanstalt als Besserungsmaschine* (cit. n. 1); Jürgen Blühdorn, "Beiträge zur Ent-wicklung und Pflege der Gefängniswissenschaft an den deutschen Universitäten seit Anfang des 19. Jahrhunderts" (PhD diss., Univ. of Münster, 1963).

[40] Wolfgang Eckart, "Sozialhygiene, Sozialmedizin," in *Enzyklopädie Medizingeschichte*, ed. Wer-ner E. Gerabek, Bernhard D. Haage, Gundolf Keil, and Wolfgang Wegner (New York, N.Y.: De Gruyter, 2005), 1344–6.

[41] Christian Barthel, *Medizinische Polizey und medizinische Aufklärung: Aspekte des öffentlichen Gesundheitsdiskurses im 18. Jahrhundert* (New York, N.Y.: Campus, 1989). On the development of the concept of medical police by Johann Peter Frank, see Anne Pieper, "Johann Peter Frank: Vom Arzt zum Gesundheitspolitiker," *Deutsches Ärzteblatt* 100, pt. A (2003): 1951–2.

[42] Philipp Sarasin, *Reizbare Maschinen: Eine Geschichte des Körpers 1765–1914* (Frankfurt am Main: Suhrkamp, 2001).

[43] Howard, *State of the Prisons* (cit. n. 1), 26, 27, 103; for the remark concerning the publication of the norms, see 107.

[44] Thoms, *Anstaltskost* (cit. n. 5), 296–8. On the transitional period of humoral pathology in the eighteenth century, see Thoms, "Die Kategorie Krankheit im Brennpunkt diätetischer Konzepte," in *Essen und Lebensqualität: Natur- und kulturwissenschaftliche Perspektiven*, ed. Gerhard Neu-mann, Alois Wierlacher, and Rainer Wild (Frankfurt am Main: Campus, 2001), 77–106.

[45] This was the argument strategy of the administrator of the penitentiary at Luckau in 1773; cf. Thoms, *Anstaltskost* (cit. n. 5), 292.

[46] Nikolaus Heinrich Julius pointed to the scurvy epidemic in Milbank Jail to prove the diet's de-ficiencies; see Julius, *Vorlesungen* (cit. n. 13), 100.

corpus. At the beginning of the nineteenth century, there was a literature devoted to these typical prison ailments, which included prison cachexia, prison scrofula (variants of emaciation and tuberculosis), as well as jail addiction.[47]

From this perspective, physicians were the responsible experts, something underscored by prison officials themselves in writing that prisoners should get as much food "as the doctor orders."[48] Nevertheless, this placed doctors in a highly awkward position. Recognition as dietary experts must have been welcome, since they were still striving to professionalize medicine. As doctors, they had sworn a Hippocratic oath, and were obliged to be humane and heal the sick. Yet, they were also state officials, and had in view the interests of society in general. An episode occurring in 1787 illustrates the kinds of conflict that could result. In that year, a new meal policy that sharply reduced the portion size served to sick prisoners was implemented in the workhouse in Münster, Westphalia, at the insistence of the administrator. When a fever epidemic broke out, the institutional physician P. M. Roer wrote a report that very clearly highlighted the dilemma involved in coupling food with work: "Convalescents rush to leave the sickbed because they are so tormented by hunger; so I must grant them more food, when they promise to work their full quota, and I cannot stop them. In that condition, matter is still escaping from their bodies which is harmful to others in the workshops."[49]

Nevertheless, the administration took a tough line, refusing to improve the food, with the justification that "the people in this house are being held for the purpose of correction."[50] It emphasized that in other institutions prisoners received no more than bread and water. This touched on an essential feature of prison diet up to the mid-nineteenth century. Very often comparisons were made among penal institutions. In addition, the food of free workers and soldiers was compared to the food of prisoners; in other words, the food of prisoners served as a standard for the nutritional minimum. By these means, it was hoped that the throng of vagabonds in workhouses would considerably decline if food were made as scarce as possible. Such meager diets, then, were seen to be both a just punishment for prisoners serving a sentence, and also sufficient for working people who earned a living from heavy labor. Intensively under discussion in debates surrounding the English Poor Law reform of 1835, this fundamental tenet of social reform is still being discussed today.[51]

In working out new regulations, the Prussian administration fell back on individual arrangements that were taken as exemplary, and developed standard guidelines from these. The Spandau Citadel prison, for example, was seen as a model penitentiary from 1804 onward. In Berlin, the Stadtvogtei jail became a model for magistrates' prisons where inmates waited for trial, "because it was here that the practical observations were collected that gave rise to the fundamental principles on the basis of which the . . . General Plan for the Improvement of Criminal Institutions of 16 September 1804 was drawn up."[52] In 1833, nutrition in all Prussian prisons was placed under the oversight of the ministry of the interior. This new policy was explicitly justified on the basis that a stay in

---

[47] Ibid., 326.

[48] GStA PK, Gen.-Dir. Kurmark, Tit. CCXXXIII, Justiz-Sachen Criminalia, No. 1, Bl. 26.

[49] Landesarchiv NRW, Abteilung Westfalen, Münster (**hereafter** cited as LA NRW), Fürstentum Münster Kabinettsregistratur, Nr. 2833, fols. 15–17f.

[50] LA NRW, Fürstentum Münster Kabinettsregistratur, Nr. 2833, fol. 14.

[51] Richard Sparks, "Penal 'Austerity': The Doctrine of Less Eligibility Reborn?," in *Prisons 2000*, ed. Roger Matthews and Peter Francis (London: Palgrave Macmillan, 1996); Edward W. Sieh, "Less Eligibility: The Upper Limits Of Penal Policy," *Criminal Justice Policy Review* 3 (1989): 159–83.

[52] GStA PK, Rep. 96A, Geheimes Zivilkabinett, Tit. 65E: 13.

workhouses had such a slight deterrent effect that recidivism was increasing, since inmates could "enjoy a better standard of care [there] than the poorer artisans."[53] Tellingly, the policy was based on the regulations of the Benninghausen poorhouse and workhouse, the institution with the smallest portion sizes of all. The compilation stemmed from a questionnaire circulated to all the institutions concerning their existing regulations, which had revealed huge differences between the way things were organized that were not the results of pure chance, but arose from differences between the provinces in terms of their social and economic structure; these led to variations in dietary standards. The ministry now energetically opposed the wide-ranging demands of some institutions and the excessive permissiveness shown toward individuals or groups of prisoners in others, making it plain that uniformity was desired. Likewise, excessive differences raised accusations of partisanship or preferentiality, fueling bitterness and distrust.[54] Nonetheless, substantial divergences remained between the provinces, as the following account shows.[55]

THE GROWING INFLUENCE OF PHYSIOLOGY

This development toward uniformity of treatment was driven ahead by physiology, the seat of judgment before which, in the view of nineteenth-century contemporaries, all were equal. Until the mid-nineteenth century, prisoners who were financially able could request the right to feed themselves. But since this practice could not be justified in physiological terms, but was grounded in social habits and differences, it was abandoned. It is important to note that the democratic character of physiology and the resulting uniform needs of bodies was not self-evident to all, but had to be established. Even at the end of the nineteenth century, prisoners from the social élite still argued that their bodies, habituated to bourgeois fare, were unable even to digest the coarse prison diet. On this basis, they claimed, prison fare was an unjust hardship for them, since they personally experienced the qualitative and quantitative distance between their normal diet and prison fare far more severely than a worker or a farmer.

Social democrats were still advancing this argument at the very end of the nineteenth century, claiming that they were in prison not for lowly crimes but for their political convictions. So, for example, Wilhelm Liebknecht would argue that political prisoners should have a fundamental right to feed themselves, such as they had traditionally possessed in the eighteenth century. If in their prison tales, such individuals reported themselves as starving on a bed of straw, that was mere rhetoric; the reality was that political prisoners held cheerful meals, even feasts, passing shopping lists to their relatives that not infrequently included delicacies, and they drank wine rather than humble beer. This was in marked contrast to the requirement for contemplation and reflection that, in modern penal institutions, counted as the high road to self-improvement. It is easy to imagine that this question of the right to self-nourishment gave rise to envy and ill will among the prison population. In 1876, self-nourishment was abolished at a stroke on paper; but in practice it continued, probably because the institutional administrators, who came from

---

[53] LA NRW, Letter from Vincke to Oberpräsidenten des Regierungsbezirkes Münster, Oberpräsidium Münster Nr. 2607, fol. 218.

[54] BLHA, Horn, 29 August 1833, Rep. 23 C, Niederlausitzische Stände, Nr. 1892.

[55] Ibid.

the same social élite as those particularly affected by this change, also viewed prison food as unsuited to their status.

Furthermore, these political prisoners had not yet gone to final trial but were merely in custody, so the notion of punishment did not exercise a powerful sway over them; and the fact that self-nourishment relieved the institution of the burden of their care may have contributed as well. But a binding limit was added in 1898 that, while acknowledging these realities, capped expenditure and completely banned the right to self-nourishment for workhouse inmates condemned to severe punishment.[56] But as important as this was in regard to the establishment of uniform dietary norms based on physiological laws, self-nourishment played only a minor role in the prisons of the judicial administration and the ministry of the interior. If there were still 172 prisoners feeding themselves in 1872, by 1913 this figure had declined to 13.[57] Nonetheless, it is an important indication of the changes in discourse and reality. The rise of physiology and the far-reaching hopes pinned to its implementation in institutions are apparent in the countless gelatin trials undertaken in institutions around Europe from the eighteenth century onward. These not only represented the first controlled nutrition experiments, and thus the state of progress of nutritional physiology, but they also attest to the practical orientation of research into nutrition. The starting point was experimentation upon the uses of waste bones carried out by the natural philosopher Denis Papin (1647–1714). Using a pressure cooker, he had "dissolved" bones to make a product, which, despite its origins in waste material, was claimed to have even greater nutritive value than meat. According to the chemist, one pound of jelly was the nutritive equivalent of six pounds of meat.[58] Persuaded that their products represented "concentrated stock tablets," numerous chemists undertook similar experimentation with a view to using the products in public institutions and the military. One such individual, the chemist Sigismund Friedrich Hermbstaedt (1760–1830), was commissioned by the Prussian king himself to conduct extensive trials at Berlin's Charité hospital in 1804.[59] Although there were mishaps, and the inmates rejected the product on account of its taste, officials were infected with enthusiasm for this scientific advance; that same year, a general plan for Prussian penal institutions explicitly called for prison administrations to be supplied with pressure cookers to allow the production of "bone jelly."[60] Scientific and public interest led the French Academy of Sciences to repeated practical experimentation, and in 1814 it created a scientific commission tasked with implementing trials of the gelatin administered in hospitals. The positive results of this trial at a Paris hospital led to the introduction of these soups in many Parisian poorhouses. Yet doubts as to their nutritional value endured, nourished by auto-experimentation by an assistant physician, who tried living on nothing but gelatin and bread for five days, only to experience continual hunger and significant stomach and intestinal problems. Other experimenters reported that their experimental animals refused the gelatin, preferring

---

[56] Alexander Klein, *Die Vorschriften über Verwaltung und Strafvollzug in den Preußischen Justizgefängnissen* (Berlin: Vahlen, 1905), 90, 182.

[57] See the data in *Statistik der zum preußischen Ministerium . . . für das Jahr 1869–1878/79*; and *Statistik der zum preußischen Ministerium . . . 1879/80–1907* (both cit. n. 14).

[58] Antoine Cadet de Vaux, *Die Gallerte aus Knochen ein angenehmes, wohlfeiles und kräftiges Nahrungsmittel, deren leichte Bereitung in allen Haushaltungen und Hospitälern, und deren Wichtigkeit für Kranke und Arme* (Frankfurt am Main: Varrentrapp und Wenner, 1803), 2, 4.

[59] GStA PK, Rep. 96A, Geheimes Zivilkabinett, Ti. 8m.

[60] GStA PK, Gen. Dir. Magdeburg Tit. CLII Justiz-Sachen Magdeburg, Nr. 25, S. 9 des Berichts.

to die of starvation, while one manufacturer of gelatin reported that not even the rats in his factory would eat his product.[61] The academy took these reports under advisement, but in 1831 it founded a new commission headed by François Magendie (1783–1855). Building on earlier research by other chemists, including Lavoisier, Magendie introduced new experimental protocols and gained fundamental knowledge of the role of individual foods in metabolic processes.[62] Thanks to his precise and controlled clinical trials of nutrition, a new era of metabolic research dawned. Under Magendie's leadership, the commission exploited the entire arsenal of scientific methods available to them. The existing literature on gelatin and food experiments was systematically cross-checked, and soups made from gelatin served in institutions were not just tested once, but chemically analyzed over a period of months. Month-long feeding trials on dogs ensued, in which the general nutritional condition of the experimental animals and the composition of their blood were painstakingly subjected to examination, and the role of individual foodstuffs in the nutritive process was thoroughly scrutinized. By this means, the commission achieved fundamental insights into the nutritive process. By the time its work ended in 1844, Magendie's claim that gelatin was not nourishing, but rather interfered with the uptake of foods, had become standard.[63] This spelled the end of gelatin in Parisian hospitals. What survived of the gelatin debate were insights into the significance of albumen, and also the fact that it was the first time that physicians were recognized as decisive scientific authorities.

The first German publication devoted exclusively and explicitly to prison diet referenced the gelatin trials. This article, published in 1845 in a journal devoted to prison science, focused in particular on the link between diet and mortality.[64] From the outset, it declared its central aim to be the release of prisoners who were in a healthy state and fit to work, explaining that prison was not intended to be a physical punishment, and that prisoners should be adequately fed.[65] This made the central problem one of identifying the proper quantity of food. While bodily needs varied by age, activity, and gender, in a prison, averages had to be used, with a small supplement to ensure that the inmates' health was preserved in all cases. The question was thus no longer whether inmates should receive meat, but how much of it they should receive. Since research had demonstrated the indispensability of nitrogen for survival, meat stood at the top of the hierarchy and was addressed first, while bread, potatoes, legumes, and vegetables followed in rank order, without any quantitative indices. The essay remained broadly theoretical, invoking medical arguments against poor food as the

---

[61] "Rapport des médecins, chirurgiens et pharmaciens de l'Hôtel-Dieu," *Comptes-Rendus Hébdomadaires de séances de l'Academie des Sciences* 6 (1841): 244–95.

[62] On Magendie, see Rolf Langmann, "Das Werk des französischen Physiologen François Magendie" (PhD diss., Univ. of Düsseldorf, published in Oberhausen: Wolters, 1936); and Frank Stahnisch, *Ideas in Action: Der Funktionsbegriff und seine methodologische Rolle im Forschungsprogramm des Experimentalphysiologen François Magendie (1783–1855)* (Münster: LIT, 2003).

[63] For more details, see Jörg Liesegang, "Die Gelatine in der Medizin: Geschichtliches zu der Verwendung der Gelatine in der Medizin des ausgehenden 17. bis zu dem beginnenden 20. Jahrhundert" (PhD diss., Univ. of Heidelberg, 2006), 308–13; Frank Stahnisch, "Den Hunger standardisieren: François Magendies Fütterungsversuche zur Gelatinekost 1831–1841," *Medizinhist. J.* 39 (2004): 103–34.

[64] "Ueber die Verpflegung der Gefangenen in medicinischer Hinsicht und über die Sterblichkeit zu Marienschloss," *Jahrbücher der Gefängniskunde und Besserungsanstalten* 78 (1845): 3–23. The anonymous author concluded that although the debate was not yet over, gelatin could not replace meat (10).

[65] Ibid., 7; compare the categorical imperative (20): "Therefore the grief over loss of liberty is the actual penalty that the prisoner suffers, rather than the bodily constraint or the deprivation of corporal pleasures to which he is accustomed."

cause of increased mortality and morbidity, and also invoking economic arguments; these stated that reduced mortality and morbidity, plus increased capacity to work, would be the purpose of the benefits of better food for prisoners. However, it lacked all reference to practical and usable measures of food quality.

Some ten years later, the military physician Wilhelm Hildesheim published a book of one hundred and fifteen pages, setting out normal care regulations for the army. This work, while purely limited to physiological and chemical facts and empty of moralizing debate or reflection, also put forward some comprehensive proposals from which recommendations for military rations would be derived.[66]

1850–1914: THE TECHNOPOLITICS OF DIET, FROM THEORY TO PRACTICE

Hildesheim's normal diet may be seen as the product of a particular technopolitics of diet, conceptualized in quantitative terms, that gave protein a central role. It led to a standardization of fare around bodily needs alone, instead of financial resources, custom, or source, which gave rise to the standardization within Prussia as outlined above. But it was a logical consequence of this shift that there could be no further cuts in nourishment levels established in accordance with these standards, even in the interests of discipline. It was on this basis that plans to cut food provisions for disciplinary reasons were abandoned in Prussia as early as 1866, after a remarkably brief discussion. This contrasted with the situation in other countries, where starvation punishments were still being imposed by the end of the nineteenth century; one example occurred in Belgium in 1893.[67]

Overall, an objectification of debate is evident, and doctors played an increasing role within it, appealing in particular to the developing view of mortality and morbidity as indicators of adequate care, supported by statistics—in other words, by arguments concerning physical soundness.[68] This meant that the prison regime in Prussia in a sense represented a kind of large-scale social experiment. As physiologists underlined: "To establish the diet of prisoners is of special value for the science of nutrition, since it will fix the lowest possible amount of food that can satisfy bodily needs, from which we may gain some idea of the minimum food standard for the poorer levels of society."[69] As stated above, the argument based on rights played an important role in prison diet debates: convicts in prison should not have it better than workers outside it. But that perspective changed under the influence of applied physiology. Henceforth, physiologically established, rather than socially accepted, dietary norms determined the food of the prisoners. Captivity alone constituted the true punishment, while an

[66] Wilhelm Hildesheim, *Die Normal-Diät: Physiologisch-chemischer Versuch zur Ermittelung des normalen Nahrungsbedürfnisses der Menschen, behufs Aufstellung einer Normal-Diät mit besonderer Rücksicht auf das Diät-Regulativ des neuen Reglements für die Friedens-Garnisons-Lazarethe, und die Natural-Verpflegung der Soldaten sowie auf die Verpflegung der Armen* (Berlin: Hirschwald, 1856); on prisoners' food, which is presented in the form of calculations, see 66, 80.

[67] Eric Maes, Ann-Sofie Vanhouche, Peter Scholliers, and Kristel Beyens, "A Vehicle of Punishment? Prison Diets in Belgium Circa 1900," *Food, Culture & Society* 20 (2017): 77–100.

[68] For a study very much in this vein, see L. Böhm, "Vorschläge zur Verbesserung der Speiseetats in den Gefangenanstalten," *Deutsche Vierteljahrsschrift für öffentliche Gesundheitspflege* 1 (1869): 371–9, on 371.

[69] Felix Hirschfeld, *Nahrungsmittel und Ernährung der Gesunden und Kranken* (Berlin: A. Hirschwald, 1900), 105.

undamaged body formed the basis for moral betterment and improved the chance of reintegration into society after release from prison.[70]

For these reasons, meat had been almost totally abolished from the Prussian prison diet in the early nineteenth century. However, when a cholera epidemic occurred, meat was given to prisoners to provide the greatest possible resistance to the disease. With that success, a decision was taken in 1867 to offer it to prisoners on a regular basis. The ministry gave permission to include two portions of meat per week within the normal dietary allowance, and hoped to improve the gustatory quality of the otherwise rather paltry diet.[71] In order to oversee the results of this change, the ministry began to publish detailed annual statistical reports on prisons and houses of correction, which supplied information not only on inmate numbers, but also on prisoners' standard of living. These documents contained a precise record of the quantities of food supplied, to the nearest gram, and also data on mortality and morbidity in the institution. Such reflections were articulated explicitly in preparation for the new dietary standards that came into force in 1872. But the regular publication of such statistics meant that they became more than merely an instrument for the bureaucracy to oversee its own activity. Rather, they opened the state intervention and treatment of prisoners up for public debate, and grounded that debate in a solid basis of data that could also serve for scientific studies.[72] By this means, nourishment was removed from the speculation and sensationalist portrayals that colored autobiographical accounts.

The consequence was a fundamental invigoration of scientific debate during the 1870s, driven by physiologists, who saw prison diet as not only a tempting field for scientific enquiry, but also one in which they could involve themselves as advisors and experts. This was particularly true of Carl von Voit, who had studied medicine and politics before becoming professor of physiology at Munich in 1860.[73] A student of Max Pettenkofer, Voit was involved in important ways in the development of calorimetry and of instruments for measuring metabolic exchange. With their aid, he had been investigating animal metabolism since 1858, and its human counterpart since 1866. His work made him incredibly popular, and he belonged to an extended scientific network, founding the "Munich School," which attracted students and researchers from across the globe. Voit's main theme was protein and the establishment of the "Kostmaß," a measure of the metabolic significance of individual nutrients. He was interested above all in what he termed the "middling worker," for whom he proposed a dietary standard of 500 grams carbohydrates, 118 grams protein, and 56 grams fat per day. This "Kostmaß" was to become the RDA (Recommended Daily Allowance) of the nineteenth and early twentieth centuries, with the slight but important difference that it fixed a minimum intake, but not a maximum. However, it was a very popular concept, and can be found in

---

[70] "Über die Verpflegung" (cit. n. 64), 20.

[71] LA NRW, Erlaß vom 21. Juni 1867, Oberpräsidium Münster, Nr. 2606, Bd. 1, 198.

[72] On the role of the legitimation of scientific experts in the public domain, see Eric Engstrom, Volker Hess, and Ulrike Thoms, eds., *Figurationen des Experten: Ambivalenzen der wissenschaftlichen Expertise im ausgehenden 18. und frühen 19. Jahrhundert* (New York, N.Y.: P. Lang, 2005), especially the article by Karl Hildebrandt, "Experten wider Willen: Statistische Projekte und ihre Akteure um 1800," 167–90.

[73] Edith Heischkel-Artelt, "Carl von Voit als Begründer der modernen Ernährungslehre," *Ernährungs-Umschau* 10 (1963): 232–4. On Voit's, influence on Rubner, see the essays in Theo Plesser and Hans-Ulrich Thamer, eds., *Arbeit, Leistung und Ernährung: Vom Kaiser-Wilhelm-Institut für Arbeitsphysiologie in Berlin zum Max-Planck-Institut für Molekulare Physiologie und Leibniz Institut für Arbeitsforschung in Dortmund* (Stuttgart: Franz Steiner Verlag, 2012).

almost every publication on food from the nineteenth century, from popular books for workers to scientific papers. In addition, Voit would investigate various aspects of nourishment, such as soldiers' rations, the diet in orphanages, and, inevitably, prison diet, at government behest. In 1877, Voit published a collected volume, addressed to doctors and administrators, that included a study of nutrition in two Munich penitentiaries written by his pupil Adolf Schuster, which was highly critical of the inadequate quantities of animal protein offered to inmates.[74]

However, the scientific debate was furthered above all by institutional physicians, who yearned for an improvement in prison fare in the light of high mortality and morbidity rates. Without being prompted, they presented the ministry of the interior with a set of calculations of the nutritional value of prison food, in hopes of starting a conversation about dietary improvement.[75] Adolf Baer stands out in this regard. As a young doctor, he had settled in the Naugard in 1862, becoming its prison doctor four years later. In 1871, he moved to the Plötzensee workhouse in Berlin, a model establishment of its day, and was named its sanitary advisor in 1879. Baer would become an international authority on matters of prison diet. Using the results of nutritional research, he calculated the nutrient content of the Naugard prison diet and compared this with the requirements established by nutritional physiologists. Although his perspective remained predominantly quantitative, Baer nevertheless counseled caution in light of the incomplete state of that research. As an institutional physician, he was aware that what counted was not simply the right quantity of nutrients or the correct measure of protein; form, flavor, and variety of foodstuffs also played an important role.[76] Baer gave detailed descriptions of what was known as *Abgegessensein*, a condition in which prisoners found the monotonous diet so revolting that they vomited at the very sight or smell of the food. He also viciously attacked the argument that prison fare should not be better than that available to the free worker. If this were so, he argued, the punishment would not only be captivity but also a starvation diet—that is, deprivation of liberty would transform into corporal punishment. In addition, he flagged the fact that workers had more variety in their diet, and freedom to eat what they wanted, including the occasional treat, unlike prisoners. On the basis that there was variation between prisoners, he also defended the individual tailoring of prison diet, calling for the physician to be permitted to prescribe supplements and what he called a *Mittelkost* for those prisoners unable to tolerate ordinary prison fare.

Many of Baer's views indeed penetrated the Prussian decree on diet that, after prolonged debate and preparation, came into force in 1872. Yet, the basic framework of care, as laid down in the early nineteenth century, was still identifiable within it. Bread took a central place, alongside potatoes, legumes, and vegetables, although meat, which had been removed at the start of the nineteenth century, was reintroduced. Cheese and herring for supper were added in 1887, as well as additional protein-rich foods in 1896

---

[74] Adolf Schuster, "Untersuchung der Kost in zwei Gefängnissen," in *Untersuchung der Kost in einigen öffentlichen Anstalten: Fuer Ärzte und Verwaltungsbeamte zusammengestellt*, ed. Carl von Voit (München: Oldenbourg, 1877), 142–85.

[75] See, for example, Falger, "Ueber die Verpflegung der Gefangenen in den öffentlichen Straf-Anstalten: Aus dem Jahresberichte pro 1963 über die Straf-Anstalt zu Münster," *Vierteljahrsschrift für gerichtliche und öffentliche Medicin*, new ser., 6 (1867): 342–53.

[76] Adolf Baer, "Ueber Beköstigung der Gefangenen," *Vierteljahrsschrift für gerichtliche und öffentliche Medicin*, new ser., 14 (1871): 291–324; Baer, *Die Gefängnisse und Strafanstalten und Strafsysteme: Ihre Einrichtung und Wirkung in hygienischer Beziehung* (Berlin: Mittler, 1871).

with the introduction of a portion of seafood per week, and even a luxury stimulant, in the form of coffee, in 1887.[77] The ideal subject for all of these improvements was, and remained, the working man. The nourishment and needs of women were scarcely ever separately discussed; although, interestingly, they received the same quantity of meat, and only the supplies of bread and vegetables were reduced. But in compensation, women received a ration of coffee.[78]

Improvements in one area demanded economic adjustments in other areas, if necessary by avoiding or at least reducing additional costs. If we compare the way in which allowances developed, we can see that meat, fat, and milk allowances increased significantly in the course of the nineteenth century, while those for bread, cereal products, potatoes, and legumes declined. Overall, therefore, prison fare was becoming far more comprehensive and varied. With Josef König's concept of rational public nutrition or the theory of nutritive cost, a model developed that oriented a cost-efficient fulfillment of needs around protein. The theory of nutritive cost developed by this food chemist might be summed up as follows: as much as necessary, but as little as possible and at lowest possible cost.[79] This maxim continued to inform popular lecture series up to the advent of vitamins.[80]

If we compare the actual amounts eaten according to the statistical reports, as listed in table 1, with the allowances from institutional diet sheets, significant differences are observable. These follow a characteristic pattern. There are few deviations from the prescribed meat allowances, but a reduction, sometimes quite marked, is apparent in the allocation of plant foods. Considering their comparatively low price, the amounts of plant foods were relatively generous; some prisoners could not even manage to consume the full amount they received. Meat or fish, by contrast, were highly prized foods that were also accorded an important position in popular views of prison diet and in the social rank of food. For this reason, prisoners always regarded portion sizes of these foods with a leery eye, and any curtailment led to protests.[81] It was precisely here that problems arose with the implementation of excessively schematic standards, problems that might produce waste rather than savings; and it was on that basis that fixed rations were abandoned in hospitals at the beginning of the twentieth century.

UNDAMAGED BODIES? THE CONSEQUENCES OF NUTRITIONAL DISCOURSE

The rationalization of institutional diet, which led to homogenization and schematization in the first instance, proved a success. If eighteenth-century institutions had experienced high mortality and morbidity that varied with the price of foodstuffs, the ongoing improvement in food supply from 1850 onward led to ongoing reductions in mortality and morbidity. In Prussian institutions, the number of sick days halved between 1869

---

[77] Baer, "Ueber Beköstigung" (cit n. 76), 322–4.

[78] Ulrike Thoms, "Gender Issues: Die Ernährung weiblicher und männlicher Strafgefangener im 19. und frühen 20. Jahrhundert," *Med. Gesell. Gesch.* 35 (2018): 37–66.

[79] Josef König, "Theorie der Volksernährung Berlin," *Schriften der Centralstelle für Arbeiter-Wohlfahrtseinrichtungen* 8 (1895): 179–208.

[80] Uwe Spiekermann, "Historische Entwicklung der Ernährungsziele in Deutschland: Ein Überblick," in *Ernährungsziele unserer Gesellschaft. Die Beiträge der Ernährungsverhaltenswissenschaft*, ed. Ulrich Oltersdorf and Kurt Gedrich (Karlsruhe: Bundesforschungsanstalt für Ernährung, 2001), 97–112.

[81] Thoms, *Anstaltskost* (cit. n. 5), 548–9.

Table 1. Quantities of individual foodstuffs in Prussian prison care from 1815 to 1905 (kilograms per head per annum). Compiled from Statistik der zum preußischen Ministerium . . . für das Jahr 1869–1878/79; *and* Statistik der zum preußischen Ministerium . . . 1879/80–1907 *(both cit. n. 14).*

| Year | Plant foods | | | | | | Luxury foods | |
	Bread	Cereal products	Potatoes	Legumes	Vegetables	Total	Beer	Coffee
1815	340	55	424	18	77	913	7	-
1829/31	213	35	312	29	71	660	209	-
1833	258	32	284	39	27	640	2	-
1849	258	34	218	38	27	575	2	-
1872	243	46	287	47	27	650	2	1
1887	214	27	370	28	62	701	-	3.7
1905	214	21	422	25	41	723	-	3.7

| Year | Animal foods | | | | |
	Meat	Fats	Fish	Dairy products	Total
1815	-	20	-	49	69
1829/31	2.3	6	-	28	36
1833	0.8	7	-	8	15.8
1849	1	6	-	-	7
1872	12	9	-	12	32
1887	11	15	5	67	98
1905	11	15	17	74	116

and 1914; bodies were no longer being harmed, and efforts at improvement could now be focused exclusively upon the soul.[82]

Several consequences flowed from this for administrations. If diet produced such evident improvements in health, individual special exceptions, introduced as safety measures, could be abolished in favor of greater uniformity of diet once strict regulations had been removed. Self-nourishment was no longer permitted after 1900, while the supplements that had been granted for additional work vanished, apparently unproblematically. Interestingly, quantitative statistical analyses of consumption also stopped in 1903. Apparently, these had become redundant as instruments of control once the system introduced in the nineteenth century had stabilized. This same moment also marked an end to theoretical and scientific debate over prisoners' diets, which had become largely canonical. The scientific rules of prison food were established, and were so widely acknowledged that there was no need to discuss and disseminate them further. What survived was the discourse of the countless published prison autobiographies. Notwithstanding all the improvements made to diet, food played an ever more central role within these; any reduction in quantity was registered with extreme sensitivity, and the bodies of eaters were closely monitored. Closer inspection reveals that prisoners possessed their own, subjective, theories of nourishment, and even developed fixed ideas; one example was the notion that they were

[82] Ibid., 778.

being poisoned.[83] Those with a more educated background might command a well-established body of popular knowledge, thanks to the popularization of nutritional science. These concerns also attest to the huge importance of food in the institution, which was often the only diversion to relieve the tedium of an otherwise monotonous daily routine; for inmates in solitary confinement, the warden who brought the food might offer the sole opportunity for conversation. Amid the experience of oppression and the pressure to conform, an inmate's self-expression (*Eigensinn*) might find an outlet in this subjective and individual aspect of prison fare, which, in the final analysis, was diametrically opposed to the technopolitics of nutrition that sought to preserve the integrity of the individual by means of constructs of identity.[84] But this theme of resistance to psychic harm is one that demands its own separate treatment.

[83] For an extensive discussion of this point, see Thoms, *Anstaltskost* (cit. n. 5), 736–57.
[84] Alf Lüdtke, *Eigen-Sinn: Fabrikalltag, Arbeitererfahrungen und Politik vom Kaiserreich bis in den Faschismus* (Hamburg: Ergebnisse, 1993).

Nutritional Modernity:

The German Case

by Corinna Treitel*

ABSTRACT

Germans were instrumental in shaping "nutritional modernity," an era stretching from the 1840s to the present, in which nutrition became a vibrant field of scientific study as well as a preeminent tool of social control. Nutrition emerged as a scientific discipline in German laboratories and statistical studies in the nineteenth century. Responding to recurrent bouts of food insecurity associated with key moments of national crisis, moreover, German scientists turned hunger into a major social problem whose best solution lay in their hands. The article begins by considering major nineteenth-century contributions to nutrition science, from Carl Voit's intake-output method to Max Rubner's caloric vision of the human body. Second, the article investigates the development of nutrition as a scientific discipline against the backdrop of recurrent cycles of nutritional insecurity. Fear over the political threat posed by a hungry proletariat at the fin-de-siècle turned nutrition into a labor problem and stimulated the emergence of *Volksernährung*, an applied branch of nutrition science that was nationalized when the country faced severe food shortages during the First World War. Finally, the article turns to scientists' social action in important fields such as nutritional prescription and popularization, from their frequent appearance as endorsers of reformist cookbooks in the nineteenth century to their role in developing didactic visual materials and mass scientific spectacles for the state in the twentieth. German nutritional modernity in its scientific aspects emerged from the confluence of these three trends: the scientization of nutrition in the nineteenth century, the medicalization and nationalization of hunger at the fin-de-siècle, and a vibrant tradition of scientific popularization that began in the nineteenth century and continued well into the twentieth. The German case provides a particularly useful venue for exploring how the social and scientific aspects of nutrition became entangled with the project of national governance.

Today, world food production continues to rise, yet malnutrition—whether too much food of the wrong kind or too little food of the right kind—remains a topic of concern. Eight hundred and fifteen million people globally suffer from chronic hunger, especially in sub-Saharan Africa and large parts of Asia, while the number of overweight and obese people is increasing everywhere. In *The State of Food Security and Nutrition in the World* (2017), several United Nation institutions trace the causes of chronic hunger to armed conflicts and climate change and warn that prolonged food insecurity may well lead to

* Department of History, Washington University in St. Louis, 1 Brookings Drive, St. Louis, MO 63130-4899, USA; ctreitel@wustl.edu.

All translations are mine.

more violence.[1] Meanwhile, chronic overnutrition puts severe strain both on individual bodies and on the health systems that serve them. It is no wonder, then, that promoting good nutrition and food security has become central to both national and global governance.

Scientists of all varieties, from the behavioral to the biomedical, have played leading roles over the last two centuries in making nutrition governable. Among other feats, they have unpacked the mechanisms of metabolism, created a scientific language of nutrients, prescribed nutritional standards, erected food safety regimes, and developed metrics for malnutrition. At the same time, they have found it difficult to achieve consensus over core questions, from what people should eat to how best to achieve food security. Other social groups, moreover, have consistently challenged their authority to offer definitive answers to these questions at all. Taken together, these developments belong to what might be termed "nutritional modernity," an era stretching from the 1840s to the present, in which nutrition emerged and was contested both as a vibrant field of scientific study and as a preeminent tool of social control.

German science and German scientists played leading roles in shaping nutritional modernity, both at home and abroad. In the nineteenth century, nutrition became a scientific discipline in German laboratories and statistical studies. Because of Germany's global influence in scientific research and training, its approach to nutrition spread quickly to other parts of the world. Recurrent bouts of food insecurity, moreover, connected the new discipline from its very inception to significant moments of German political instability. These included the revolutions of 1848; the last third of the nineteenth century, when conflicts unleashed by political unification, rapid industrialization, and a population explosion threatened to tear the new country apart; and the First World War, which ended in national defeat. At each juncture, German scientists seized the opportunity to turn hunger into a major social problem whose best solution lay in their hands. In short, the German case offers a fruitful vantage point from which to study the onset of nutritional modernity, not just because of Germany's leadership in establishing nutrition science as a discipline, but because Germany's particularly rocky path into the twentieth century so deeply entangled the social and scientific aspects of nutrition with the project of national governance.

This article explores those entanglements in three steps. First, it considers major nineteenth-century German contributions to nutrition science, from the intake-output method of metabolic research pioneered by Carl Voit, to Max Rubner's caloric vision of the human body. The second section investigates the development of nutrition as a scientific discipline from unification in 1871 through defeat in 1918. It traces the process by which fin-de-siècle fears over the political and economic threat posed by hungry workers turned nutrition into a labor problem and stimulated the emergence of *Volksernährung*, an applied branch of nutrition science that medicalized and eventually nationalized hunger. Finally, the article turns to nutritional prescription as a field for social action by scientific experts, from their frequent appearance as endorsers of reformist cookbooks in the nineteenth century to their role in developing didactic visual

---

[1] Food and Agriculture Organization of the United Nations, International Fund for Agricultural Development, UNICEF, World Food Programme, and World Health Organization, *The State of Food Security and Nutrition in the World in Brief 2017: Building Resilience for Peace and Food Security* (Rome: FAO, 2017), 5, 10, 13, 16–7, http://www.who.int/nutrition/publications/foodsecurity/state-food-security-nutrition -2017-inbrief-en.pdf?ua = 1.

materials for the state in the twentieth. German nutritional modernity in its scientific aspects emerged from the confluence of these three trends: the scientization of nutrition in the nineteenth century, the medicalization and nationalization of hunger at the fin-de-siècle, and a vibrant tradition of scientific popularization that began in the nineteenth century and continued well into the twentieth.

INVENTING NUTRITION SCIENCE

Nutritional modernity rested in part on a new science of nutrition invented primarily in German-speaking Europe between the mid-nineteenth and the early twentieth century. Our modern discourse of proteins, fats, carbohydrates, and calories achieved a chemical life here. German researchers, moreover, led the way in consolidating two main sites for the production of nutritional knowledge: the laboratory, where the discipline of nutritional physiology first emerged; and the field, where statistical techniques were deployed to determine what ordinary citizens were eating and why. Deep anxieties about industrialization drove this effort to invent a new science of nutrition. What were citizens eating—and even more importantly, what *should* they be eating—in an age of industrial labor, industrial conflict, and, eventually, industrial war? "Nutritional knowledge," as historian Dietrich Milles put it, "was an early attempt to analyse and combat the consequences of industrialization with scientific instruments."[2] Nowhere was this truer than in Germany, which emerged in this period as a scientific and political powerhouse while simultaneously undergoing massive conflicts over modernization. Produced at the intersection of these developments, the new science of nutrition—its methods, research questions, and social applications—generated intense power struggles that shaped the nascent discipline. This section surveys German global leadership in making nutrition scientific from the 1840s through the 1890s, explores major battle lines, and concludes with a brief look at the early twentieth century, when the science launched by German researchers took new directions in other national contexts.

German nutrition science grew out of Justus Liebig's forays into "animal chemistry" in the 1840s. Antoine Lavoisier had argued in the 1780s that respiration was a form of combustion in which the combination of carbon and hydrogen with oxygen in the lungs released animal heat. Then, in the 1820s–1830s, chemists discovered that foods consisted mainly of three basic foodstuffs: "the saccharine, the oily, and the albuminous," as William Prout put it in 1827, or in terms common by the 1840s, carbohydrates, fats, and proteins.[3] Without being quite sure how the body transformed these nutrients, Liebig (1803–73) proposed that foods could be divided into two large classes: "respiratory" foods such as vegetable oils or potatoes, composed primarily of fats and carbohydrates; and "plastic" foods such as beef or legumes, composed primarily of proteins. Respiratory foods heated the body, he hypothesized, while plastic foods replaced blood and tissue burned up when the body performed physical work. Spelled out in two important books, *Animal Chemistry* (1842) and *Researches on the Chemistry of Food* (1847), Liebig's speculations set the agenda for the "New Nutrition," which emphasized the

[2] Dietrich Milles, "Working Capacity and Calorie Consumption: The History of Rational Physical Economy," in *The Science and Culture of Nutrition, 1840–1940*, ed. Harmke Kamminga and Andrew Cunningham (Amsterdam: Rodopi, 1995), 77. For the French case, see Dana Simmons, *Vital Minimum Need, Science, and Politics in Modern France* (Chicago: Univ. of Chicago Press, 2015), 93–101.
[3] Kamminga and Cunningham, *Science and Culture*, ed. Kamminga and Cunningham (cit. n. 2), 3.

centrality of proteins, fats, and carbohydrates (as well as water and a few mineral salts) to animal nutrition.[4] First, although he turned out to be wrong on many of the details, Liebig's foray into animal chemistry made the case forcefully that nutrition followed chemical laws and could be studied by chemical methods.[5] Second, Liebig's valorization of plastic food as the sole source of fuel for the working body turned protein, especially from meat, into a research focal point. In the popular phrase "meat makes meat," indeed, protein also became the core of nutritional prescription for the rest of the century as well as an object of intense debate, both inside Germany and out.[6]

But how did the body metabolize the basic building blocks of food, especially protein? Inspired by Liebig's call to unravel the chemical laws of animal nutrition, the young chemist Carl Voit (1831–1908) went to work as an assistant to the physiologist Theodor Bischoff (1807–82) at the Physiological Institute in Munich. In a series of controlled feeding experiments, Voit and Bischoff measured how much urea dogs fed on different diets produced. They focused on urea, a nitrogenous waste product found in urine and feces, because it allowed them to gauge indirectly how much protein (the only nitrogen-containing foodstuff) had been metabolized.[7] Convinced that they had developed a new method for studying protein metabolism, Bischoff and Voit nonetheless knew that measuring "nitrogen balance" was only a first step toward a complete chemical view of metabolism. Seeking greater precision, they teamed up next with chemist Max Pettenkofer (1818–1901) in 1859 to build a large respiration apparatus that could house a living research subject for several days. By combining feeding and respiration experiments, they hoped to measure precisely all the carbon, hydrogen, nitrogen, and oxygen consumed and excreted by their research subjects. So expensive that it required a private infusion of money from the Bavarian king, the apparatus allowed the trio to do just that. Indeed, although Voit remained strangely reticent to say so publicly, these experiments suggested that Liebig's division between respiratory and plastic foods was too simple. Even more importantly, the "intake-output method of quantification" in physiology, as historian Frederic Holmes dubbed it, demonstrated that living bodies behaved in a regular, quantitative fashion and could be studied by the same methods as inorganic nature. That method became the foundation of the Munich School of metabolism and dominated the New Nutrition into the twentieth century.[8]

[4] Justus von Liebig, *Die organische Chemie in ihrer Anwendung auf Physiologie und Pathologie* (Braunschweig: Friedrich Vieweg und Sohn, 1842); Liebig, *Chemische Untersuchung über das Fleisch und seine Zubereitung zum Nahrungsmittel* (Heidelberg: C. F. Winter, 1847).

[5] William H. Brock, *Justus von Liebig: The Chemical Gatekeeper* (Cambridge: Cambridge Univ. Press, 1997), 184–86, 213; Mark Finlay, "Early Marketing of the Theory of Nutrition: The Science and Culture of Liebig's Extract of Meat," in Kamminga and Cunningham, *Science and Culture* (cit. n. 2), 48–74; Kenneth J. Carpenter, *Protein and Energy: A Study of Changing Ideas in Nutrition* (Cambridge: Cambridge Univ. Press, 1994), 48–52.

[6] Carpenter, *Protein* (cit. n. 5), 50. On changes in German meat consumption, see Hans-Jürgen Teuteberg and Günter Wiegelmann, "Ein Grundnahrungsmittel in der Einzelberatung: Die Entwicklung des Fleischverbrauchs," chap. 4 in *Der Wandel der Nahrungsgewohnheiten unter dem Einfluß der Industrialisierung* (Göttingen: Vandenhoeck & Ruprecht, 1972). For a pan-European analysis, see Peter Koolmees, "Meat in the Past: A Bird's Eye View on Meat Consumption, Production and Research," in *Meat Past and Present: Research, Production, Consumption*, ed. W. Sybesma, P. A. Koolmees, and D. G. van der Heij (Zeist, Neth.: TNO Nutrition and Food Research Institute, 1994), 5–32. The British medico-politics of meat is explored in Anita Guerrini, "Health, National Character and the English Diet in 1700," *Stud. Hist. Phil. Biol. Biom. Sci.* 43 (2013), 349–56.

[7] Th. L. W. Bischoff and Carl Voit, *Die Gesetze der Ernährung des Fleischfressers durch neue Untersuchungen festgestellt* (Leipzig: C. F. Winter, 1860).

[8] Frederic L. Holmes, "The Intake-Output Method of Quantification in Physiology," *Historical Studies in the Physical and Biological Sciences* 17 (1987): 235–70, on 256–68, 270; Holmes, "The Formation

This enthusiasm for quantitative chemical methods also moved outside the strict confines of the laboratory and emboldened other scientific challengers to Liebig's views. In an 1866 experiment that is now a classic of nineteenth-century science, Adolf Fick (1829–1901) and Johannes Wislicenus (1835–1902), respectively a physiologist and a chemist at the University of Zürich, decided to test Liebig's assertion that only plastic (protein-rich) foods were metabolized during muscular work. Liebig and his followers, they pointed out, had fallen into a fallacy:

> The action of the muscle [Liebig reasoned] is connected with the destruction of its substance, by far the greater part of which is of an albuminoid nature; therefore the destruction by oxidation of albuminoid bodies is the essential condition of the muscle's mechanical action. The fallacy of this line of argument will be immediately apparent if, for instance, we apply it to a locomotive. "This machine consists essentially of iron, steel, brass, etc.; it contains but little coal; therefore its action must depend on the burning of iron and steel, not on the burning of coal." In like manner it is by no means self-evident that it is specially the oxidation of albuminous compounds which produces muscular force.[9]

To determine whether, in fact, fats and carbohydrates were involved in muscular work, Fick and Wislicenus performed an elegant experiment on their own bodies. After excluding protein from their diet the night before, they hiked up a Swiss mountain while feeding themselves only fat and starch, then enjoyed a meaty (and protein-filled) dinner at the summit hotel. Having carefully collected their own waste products throughout the experiment, they analyzed the amount of urea their bodies had produced during the climb, concluding that "the muscle-machine can undoubtedly be heated by means of the non-nitrogenous fuel."[10]

Results such as these suggesting the interchangeability of nutrients during metabolism helped precipitate another major challenge to Liebig's protein-centered views in the 1870s, when Max Rubner (1854–1932) came to study with Voit at the Munich Physiological Institute. Much to his mentor's displeasure, Rubner conducted a series of experiments that took direct aim at the division between plastic and respiratory foods. In the process, he pioneered a bioenergetic approach to metabolism that is still with us today. Between 1878 and 1883, Rubner developed the experimental foundation for his isodynamic law, which cast proteins, fats, and carbohydrates as interchangeable according to strict caloric equivalents. Further work set "standard values" for these caloric equivalents: 4.1 calories/gram of protein, 9.3 calories/gram of fat, and 4.1 calories/gram of carbohydrates, as laid out in *The Laws of Energy Consumption in Nutrition* (1902). In yet another series of experiments, Rubner used a self-registering calorimeter to show that living organisms obeyed the first law of thermodynamics. All of this helped turn the body into a "heat machine" that took proteins, fats, and carbohydrates—mutually interchangeable through caloric equivalents—into mechanical work.[11]

---

of the Munich School of Metabolism," in *The Investigative Enterprise: Experimental Physiology in Nineteenth-Century Medicine*, ed. William Coleman and Holmes (Berkeley and Los Angeles: Univ. of California Press, 1988), 179–210.

[9] A. Fick and J. Wislicenus, "On the Origin of Muscular Power," *Philosophical Magazine* 31 (1866): 485–503, on 485–6.

[10] Ibid., 485–9, 502.

[11] Corinna Treitel, "Max Rubner and the Biopolitics of Rational Nutrition," *Central European History* 41 (2008), 3; Max Rubner, *Die Gesetze des Energieverbrauchs bei der Ernährung* (Leipzig: F. Deuticke, 1902); Anson Rabinbach, *The Human Motor: Energy, Fatigue, and the Origins of Modernity* (Berkeley and Los Angeles: Univ. of California Press, 1992), 124–9.

But just as Voit had challenged Liebig, and Voit's students had challenged Voit, by the early twentieth century, experimentalists were becoming increasingly frustrated with the limits of the New Nutrition. Conflicts over protein and meat had generated dozens of studies that suggested, unexpectedly, that there were still other essential foodstuffs beyond the protein/fat/carbohydrate nexus. Russell Chittenden (1856–1943), an American physiologist trained in Germany, for instance, did a series of well-regarded experiments at Yale in which human subjects were fed a low-protein, meat-free diet. The results showed that the working body needed much less protein than previously thought and that meat was not necessary at all.[12] Even more interesting was that follow-up experiments in which animals were fed exclusively on proteins, fats, carbohydrates, and mineral salts—the foodstuffs of the New Nutrition—eventually sickened and died. This led researchers to hypothesize that health might depend on additional "accessory factors." Just what those other substances might be started to come into focus in 1912, when the chemist Casimir Funk (1884–1967), working in London on diseases such as beriberi, scurvy, pellagra, and rickets, suggested that a new class of organic compounds found in food were necessary to prevent these diseases. He called them "vit-amines": "vit" because they were vital to life and "amines" because he believed that they all contained an amine group ($-NH_2$). When that turned out to be wrong, the name was shortened to vitamins and the "Newer Nutrition," a field devoted to the study of vitamins and other micronutrients, was born. In Germany, vitamins were taken up as part of a research tradition in *Wirkstoffe* or biocatalysts, a category that also included hormones and enzymes. By the 1920s, *Wirkstoffe* had become the "organism's home pharmacy," and the *Deutsche Forschungsgemeinschaft*, the German equivalent of the National Science Foundation, poured in financial support.[13] A younger generation of nutritional scientists that included Wilhelm Stepp (1882–1964), Arthur Scheunert (1879–1957), Adolf Bickel (1875–1946), Emil Abderhalden (1877–1950), Werner Kollath (1892–1970), and many others quickly moved into vitamin research looking for new questions, new funding, and new power.[14]

---

[12] Russell H. Chittenden, *Physiological Economy in Nutrition: An Experimental Study* (New York, N.Y.: Frederick A. Stokes, 1904); Carpenter, *Protein* (cit. n. 5), 119–21. For German challengers to the protein dogma, see Corinna Treitel, *Eating Nature in Modern Germany: Food, Agriculture, and Environment, c. 1870–2000* (Cambridge: Cambridge Univ. Press, 2017), 95–110.

[13] Heiko Stoff, "Enzyme, Hormone, Vitamine von der Deutschen Forschungsgemeinschaft geförderte Wirkstoffforschung, 1920–1970," in *Die Deutsche Forschungsgemeinschaft 1920–1970: Forschungsförderung im Spannungsfeld von Wissenschaft und Politik*, ed. Karin Orth and Willi Oberkrome (Stuttgart: Franz Steiner Verlag, 2010), 325–6. For vitamins elsewhere, see Sally Horrocks, "The Business of Vitamins: Nutrition Science and the Food Industry in Interwar Britain," in Kamminga and Cunningham, *Science and Culture* (cit. n. 2), 235–58; Rima Apple, *Vitamania: Vitamins in American Culture* (Rutgers, N.J.: Rutgers Univ. Press, 1996); Mark W. Weatherall and Harmke Kamminga, "The Making of a Biochemist I: Frederick Gowland Hopkins' Construction of Biodynamic Chemistry" and "The Making of a Biochemist II: The Construction of Frederick Gowland Hopkins' Reputation," *Med. Hist.* 40 (1996): 269–92 and 415–36; Kamminga, "Vitamins and the Dynamics of Molecularization: Biochemistry, Policy and Industry in Britain, 1914–1939," in *Molecularizing Biology and Medicine: New Practices and Alliances, 1910s–1970s*, ed. Soraya de Chadarevian and Kamminga (Amsterdam: Harwood, 1998), 83–105; and Adel den Hartog, "The Discovery of Vitamins and its Impact on the Food Industry: The Issue of Tinned Sweetened Condensed Skim Milk 1890–1940," in *Food and the City in Europe since 1800*, ed. Peter J. Atkins, Peter Lummel, and Derek J. Oddy (Burlington, Vt.: Ashgate, 2007), 131–42.

[14] Heiko Stoff, *Wirkstoffe: Eine Wissenschaftsgeschichte der Hormone, Vitamine und Enzyme 1920–1970* (Stuttgart: Franz Steiner, 2012), 56; Ulrike Thoms, "'Vitaminfragen—kein Vitaminrummel?' Die deutsche Vitaminforschung in der ersten Hälfte des 20. Jahrhunderts und ihr Verhältnis zur Öffentlichkeit," in *Wissenschaft und Öffentlichkeit als Ressourcen füreinander: Studien zur Wissenschaftsgeschichte im 20. Jahrhundert*, ed. Sybilla Nikolow and Arne Schirrmacher (Frankfurt am Main: Campus, 2007), 75–96.

Alongside this effort to make nutrition scientific by bringing it into the laboratory, researchers also went out to the field to study not just what Germans were eating but, perhaps more importantly, prescribe what they should be eating. On both counts much work had to be done. Institutional feeding studies of soldiers, prison populations, and hospital patients had been an object of interest for decades, but little was known about how Germans in other social groups ate.[15] In the 1850s, Ernst Engel, a statistician who eventually headed the Prussian Statistical Bureaus, pioneered working-class household budget studies as part of a social reform program. Engel's Law, still used by economists today, held that as families became poorer, more and more of their household budget went to buy food.[16] Inspired by this approach, researchers began in the 1870s and 1880s to survey the consumption of proteins, fats, and carbohydrates for different groups: male workers involved in heavy or light work, women at various life stages, growing children, the elderly, and so on.[17] One important milestone came in the 1870s, when Carl Voit published a study of what the average German worker was eating and, based on these results, laid out general guidelines on what average German workers should be eating. That turned out to be 118 grams protein, 500 grams carbohydrates, and 56 grams fat every day, with at least half of the protein coming from animal sources, preferably meat. Enthusiastically taken up by civil servants, doctors, and social planners under the banner of *Normalkost* (standard diet), Voit's meat-centered standard quickly became the yardstick for evaluating diets across the industrializing world.[18] In one well-known example, Wilbur Atwater (1844–1907), a chemist who studied with Voit and Rubner before introducing German nutrition science to the United States, combined household surveys with calorimetric work to reveal the connection between purchasing power and the nutritional state of American workers. Atwater revised Voit's 118-gram protein benchmark slightly upward for American workers to 125 grams, thus casting doubt on the universality of the German standard. More significant was his analysis of why so many American workers were so poorly nourished. Echoing Rubner and many other experts of the period, Atwater insisted that the problem was not workers' purchasing power but rather their ignorance about how to combine the laws of physiology and economics to eat healthfully on a limited budget.[19]

Analyses such as these, which combined laboratory work, field study, and social commentary disguised as science, eventually tipped a younger generation of German experts with left-leaning political commitments into open revolt. Alfred Grotjahn (1869–1931), a physician associated with Gustav Schmoller's *Verein für Sozialpolitik* (Association for Social Policy), was a case in point. In 1902, he attacked Voit's *Normalkost* as useless,

[15] Ulrike Thoms, *Anstaltskost im Rationalisierungsprozeß: Die Ernährung in Krankenhäusern und Gefängnissen im 18. und 19. Jahrhundert* (Stuttgart: Franz Steiner, 2005); and Thoms, "The Soldier's Food in Germany, 1850–1960," in *Setting Nutritional Standards: Theory, Policies, Practices*, ed. Elizabeth Neswald, David. F. Smith, and Thoms (Rochester, N.Y.: Univ. of Rochester Press, 2017), 97–118.

[16] Ernst Engel, *Das Rechnungsbuch der Hausfrau und seine Bedeutung im Wirthschaftsleben der Nation* (Berlin: L. Simon, 1882).

[17] Elizabeth Neswald, "Nutritional Knowledge between the Lab and the Field," in Neswald, Smith, and Thoms, *Setting Nutritional Standards* (cit. n. 15), 38–41.

[18] The average worker was a 70-kilogram male doing moderate physical labor: Carl Voit, "Ueber die Kost in öffentlichen Anstalten," *Zeitschrift für Biologie* 12 (1876), 1–59. Voit returned to these numbers in Voit, *Untersuchung der Kost in einigen öffentlichen Anstalten* (Munich: R. Oldenbourg, 1877), 20; and Voit, "Physiologie des allgemeinen Stoffwechsels und der Ernährung," in *Handbuch der Physiologie*, vol. 6, ed. L. Hermann (Leipzig: F. C. W. Vogel, 1881), 518–28.

[19] Neswald, "Nutritional Knowledge," (cit. n. 17), 40–1; Treitel, "Max Rubner" (cit. n. 11), 18.

pointing out that there was a difference between what Germans were being told they should eat and what they were in fact eating. Using budget studies, Grotjahn showed that what the average worker by his behavior seemed to *want* to eat was what the middle classes ate: a meaty, sweet, and fatty diet, containing on average 274 grams meat, 68 grams fat, and 68 grams sugar per day. Most workers, however, could not afford to purchase enough of this diet to make it nutritious. The result, Grotjahn charged, was that the vast majority of German workers lived wretchedly in a chronic state of "undernutrition" (*Unterernährung*).[20]

Grotjahn's critique of nutrition science was not, of course, a rejection of the new discipline. Since its inception in the 1840s, German scientists had played a leading role in making nutrition scientific. The chemical-quantitative approach of the New Nutritionists had yielded new knowledge about how the body used macronutrients (proteins, fats, and carbohydrates). The intake-output method, moreover, gave the field a tool of lasting importance that was exported around the world. And even the persistent debates—over the division between plastic and respiratory foods, the centrality of protein, and the desirability of meat—had yielded new insights (e.g., Rubner's isodynamic law) and produced the Newer Nutrition, with its focus on unpacking the role of micronutrients in animal health. Field studies conducted from the mid-nineteenth century onward finally plunged the new discipline into the emerging realm of social policy. Far from being a rejection, Grotjahn's critique of nutrition science was thus an affirmation of the discipline's immense social and political importance.

MAKING HUNGER SCIENTIFIC

From the beginning, the science of nutrition had close ties to the social politics of hunger. Crop failures and resulting food scarcities in the 1840s helped precipitate political revolution across Europe and left a lasting mark on the generation of scientists who witnessed it. Profoundly shocked by what he had seen, for instance, Liebig's turn to animal chemistry grew from his conviction that chemistry should serve the cause of dietary reform, agricultural reform, and social improvement.[21] By the late nineteenth century, the "social question" of how to address urban poverty before it became politically toxic drove the new field of social policy. The hunger of industrial laborers now became associated with the new discourse of *Volksernährung* (public nutrition), defined by Max Rubner in 1908 as "nutrition for the great mass of the people."[22] When hunger became a military as well as a political problem in the Great War, nutritional scientists finally assumed important positions of leadership in managing *Volksernährung*, which now became a national category. At all of these political moments, in short, a diverse group of often warring players wielded nutrition science to construct and manage hunger as a social-epistemic object of modern governance.

[20] Alfred Grotjahn, *Über Wandlungen in der Volksernährung* (Leipzig: Duncker & Humblot, 1902), 4–5, 15, 62–3. For more on the nutrition transition described by Grotjahn, see Hans J. Teuteberg, "Der Verzehr von Nahrungsmitteln in Deutschland pro Kopf und Jahr seit Beginn der Industrialisierung (1850–1975): Versuch einer quantitativen Langzeitanalyse," in *Unsere tägliche Kost: Geschichte und regionale Prägung*, ed. Teuteberg and Günther Wiegelmann (Münster: F. Coppenrath, 1986), 225–79. On industrial food systems generally, see Tim Lang, "Food Industrialisation and Food Power: Implications for Food Governance," in *Taking Food Public: Redefining Foodways in a Changing World*, ed. Psyche Williams-Forson and Carole Counihan (New York, N.Y.: Routledge, 2011), 11–22.
[21] Brock, *Justus von Liebig* (cit. n. 5), xi.
[22] Max Rubner, *Volksernährungsfragen* (Leipzig: Akademische Verlagsgesellschaft, 1908), 45.

With memories of the hungry forties still fresh and disappointment over the failed revolutions of 1848 still deep, scientific liberals turned in the 1850s to making the new science of nutrition a science for the people.[23] Jacob Moleschott (1822–93), a physiologist, framed nutritional deficits as a basic driver of sociopolitical inequality. In his telling, protein, especially from meat, was the foodstuff of freedom, and his best-selling *The Science of Food, For the People* (1850) presented the newest scientific discoveries on metabolism, with the express aim of guiding readers toward the most meat-rich diet they could afford.[24] *The Food Question in Germany* (1855) by physician Hermann Klencke (1853–1904) also concerned itself with food and freedom, but refused to valorize meat. Bristling with nutritional tables and cost calculations, the book argued that a diet based on pulses—"the meat and milk of the plant world"—could be just as nutritious as, but much less expensive than, a meaty one.[25] In the service of German liberalism, in short, both Moleschott and Klencke embraced the New Nutrition, but drew different lessons from it. Their disagreement about whether plant or animal food should take pride of place on the German table, indeed, remained a persistent fault line among experts well into the twentieth century.

That fault line continued alongside new ones that opened in the 1880s, when *Volksernährung* emerged as a key discursive category for thinking about German hunger. An older literature on food provisioning for institutions such as the armed forces, prisons, hospitals, and public kitchens had approached mass feeding as an administrative task of stilling the pangs of hunger. In *The Army and Public Nutrition* (1880), in contrast, physician C. A. Meinert sought to apply Voit's theories to recast mass feeding as a chemical enterprise with physio-political consequences. Underneath the sober language of proteins, fats, and carbohydrates, Meinert's analysis held the warning that nutritionally inadequate diets, especially in protein, were producing Germans unfit to work hard, either on the factory floor or on the battlefield. New and cheaper forms of preserved meat (meat extracts, tinned beef, and so on), he advised, should be produced for these groups if the nation was to achieve its full potential as a world power.[26]

Meinert's attention to a jumble of disparate feeding populations under the banner of *Volksernährung* disappeared by the end of the 1880s, when reformers repurposed the term for proletarian hunger. Here, new fault lines developed between those advocating inadequate income or inadequate knowledge as the problem, and state help or self-help as the solution. Consider in this regard *Public Nutrition, How It Is and How It Should Be* (1888) by physician Emanuel Wurm (1857–1920). Part of a socialist pamphlet series, the book devoted more than a hundred pages to the discoveries made by Voit, Pettenkofer, and others. It also delivered scientific recommendations on how to compose a cheap and

---

[23] On liberalism and scientific popularization, see Andreas Daum, *Wissenschaftspopularisierung im 19. Jahrhundert: Bürgerliche Kultur, naturwissenschaftliche Bildung und die deutsche Öffentlichkeit 1848–1914* (Munich: R. Oldenbourg, 2002).

[24] Jacob Moleschott, *Lehre der Nahrungsmittel: Für das Volk* (Erlangen: Ferdinand Enke, 1850), 185; Moleschott, *Der Kreislauf des Lebens: Physiologische Antwort auf Liebig's Chemische Briefe* (Mainz: Victor von Zabern, 1852), 459–60. See also Harmke Kamminga, "Nutrition for the People, or the Fate of Jacob Moleschott's Contest for a Humanist Science," in Kamminga and Cunningham, *Science and Culture* (cit. n. 2), 15–47.

[25] Hermann Klencke, *Die Nahrungsmittelfrage in Deutschland*, vol. 1 (Leipzig: Eduard Kummer, 1855), 160, 168. Klencke was well known for his muckraking journalism, which helped precipitate Germany's first food adulteration law in 1879; see Treitel, *Eating Nature* (cit. n. 12), 46.

[26] C. A. Meinert, *Armee- und Volksernährung: Ein Versuch Professor C. von Voit's Ernährungstheorie für die Praxis verwerthen*, 2 vols. (Berlin: Ernst Siegfried Mittler & Sohn, 1880), 1:10–11, 22.

nutritious diet, while warning that acquisition of scientific knowledge could never compensate for lack of income. In a final chapter on *Volksernährung*, indeed, Wurm deployed statistics gathered by the city of Berlin to show that working-class families spent 50 percent or more of the household budget on food. Most Berlin laborers, he concluded, simply could not afford to eat the diet that experts told them they should be eating. He put it this way: "Self help has a limit! Even equipped with the best knowledge about a nutritious and affordable diet, the proletariat is simply not able to buy 250 grams of meat each day, as Professor Voit and the other physiologists unanimously recommend!" To address the problem of *Volksernährung*, Wurm concluded, the state had to improve wages for industrial laborers.[27] A decade and a half later, Grotjahn's *On Changes in Public Nutrition* (1902) also used budget studies to attack nutritional physiologists for being out of touch with the social reality of workers. Perhaps more importantly, Grotjahn managed to avoid specific political prescriptions while still making the case powerfully that malnutrition among the urban proletariat now posed for the German state a social policy challenge of the first order.[28]

But not all experts saw *Volksernährung* in this way. Consider the case of Max Rubner, who by this point was the world's most famous nutritional physiologist. In *Questions of Public Nutrition* (1908), he followed Wurm and Grotjahn in treating *Volksernährung* as a problem of the urban working poor. No industrial state, he agreed, could compete globally without a biologically fit population. Still, he had nothing but scathing criticism for socialist physicians claiming the mantle of authority on nutritional matters. Wurm and Grotjahn, he charged, uncritically accepted Voit's *Normalkost* in order to locate the true cause of proletarian malnutrition in low wages. Their political commitments, he implied, clouded their scientific judgment. Rather than physicians armed with statistics, physiologists wielding laboratory methods were those best placed to establish the specific nutritional needs of different bodies laboring in different circumstances. Establishing not a one-size-fits-all *Normalkost*, but a range of sufficient diets, should be the goal. Indeed, the political need for such knowledge was so pressing that Rubner closed by calling for the establishment of a state-funded Nutrition Office to lead research on *Volksernährung* in the new century.[29] In his follow-up, *Changes in Public Nutrition* (1913), Rubner again dismissed the view that insufficient income was the primary cause of malnutrition. Instead, he blamed a national fixation on meat exacerbated by widespread nutritional ignorance, both of which inclined the poor to eat an inadequate meat-centered diet rather than a nutritious plant-centered one. Establishing cheap workers' restaurants near factories, ensuring that tenement houses had kitchens for preparing home-cooked meals, initiating a campaign of nutritional enlightenment targeted at working mothers, and expanding infant feeding and school lunch programs—coupled with laboratory research—were the only measures that could adequately address the *Volksernährung* problem. Rubner, in other words, spoke the language of bourgeois progressivism.[30]

At the same time that debates over *Volksernährung* broke out among experts of different professional and political persuasions, they were also erupting between experts

---

[27] Emanuel Wurm, *Die Volksernährung, wie sie ist und wie sie sein sollte* (Dresden: Schnabel, 1888), 233–4.

[28] Grotjahn, *Über Wandlungen* (cit. n. 20).

[29] Rubner, *Volksernährungsfragen* (cit. n. 22), 46, 48, 51–63, 67, 86–7, 103–6, 141–2.

[30] Max Rubner, *Wandlungen in der Volksernährung* (Leipzig: Akademische Verlagsgesellschaft m.b.H., 1913), 118–33; Treitel, "Max Rubner" (cit. n. 11), 10–17.

and lay people. Who, after all, owned *Volksernährung*? Scientists such as Rubner enjoyed immense cultural capital but did not dominate the terrain. Instead, they had to share it with many other groups that had a stake in the social politics of hunger. Life reformers in particular played an important role here in contesting scientific authority and expanding the compass of *Volksernährung*.[31] Take Gustav Simons (1861–1914), a prominent bread reformer. In 1907, Simons argued that the nation's experts restricted their thinking about *Volksernährung* to the poor at great peril to the country. *Volksernährung*, he insisted, encompassed all Germans. Bitterly criticizing "liberal professors" for their myopia, Simons pointed out that in the event of another war, a nation of schnapps-swilling, meat-habituated, white-bread-and-refined-sugar-consuming Germans faced terrible odds. To make his point, he marshaled evidence to show that Germany's sixty million people had grown dangerously dependent on a globalized food economy. What would happen, he wondered, if access to this market ceased?[32]

Simons's warnings were prescient. When war did break out in 1914, Germany's nutritional situation declined rapidly as an Allied blockade cut the country off from global trade.[33] As the nation's experts turned their collective attention to the food war, *Volksernährung* took the national turn that Simons and others had been urging. Perhaps the best indicator of this shift was *German Public Nutrition and the English Starvation Plan* (1914). Produced by an expert panel of nutritional physiologists, agronomists, statisticians, and economists, plus one home economist, the report tackled the question of whether Germany's food economy could be managed to survive a long war.[34] The answer was a resounding yes. Indebted to the quantitative chemical tradition that had produced the New Nutrition and had most recently been trained on the poor, the panel expanded a massive calculating effort to specify how the nation's nutritional economy should be remade. A typical passage read as follows:

> In 1912–13, 89,000 tons of rye were used in grain distilleries. In the event of a prohibition of this rye, if it were ground to 78 per cent rye meal and 20 per cent bran . . . [rye] would provide 69,400 tons of rye meal and 17,800 tons of bran. The rye meal (3,220 calories and 6.7 percent protein to the kilogram) would furnish for human food 224 milliard calories and 4,700 tons of protein. The bran (2.575 calories and 12.5 percent of protein to the kilogram) would contain 45.8 milliard calories and 2,200 tons of protein, and if used in the production of milk (23.6 percent of the calories and 36.3 percent of the protein) would furnish, in the form of milk, a further 10.8 milliard calories and 800 tons of protein. Altogether

---

[31] For a review of the vast literature on the German life reform movement, including its dietary prescriptions, see Treitel, *Eating Nature* (cit. n. 12), 11–12. Particularly relevant is Sabine Merta, *Wege und Irrwege zum modernen Schlankheitskult: Diätkost und Körperkultur als Suche nach neuen Lebenstilformen 1880–1930* (Wiesbaden: Franz Steiner Verlag, 2003).

[32] Gustav Simons, *Die deutsche Volksernährung* (Soest, Neth.: Selbstverlag, 1907), 9, 97, 100; Treitel, *Eating Nature* (cit. n. 12), 164–5.

[33] The literature on German food and the First World War is synthesized and expanded in Alice Autumn Weinreb, "The Geopolitics of Total War," chap. 1 in *Modern Hungers: Food and Power in Twentieth-Century Germany* (New York, N.Y.: Oxford Univ. Press, 2017). The European experience is treated in Ina Zweiniger-Bargielowska, Rachel Duffett, and Alain Drouard, eds., *Food and War in Twentieth-Century Europe* (Burlington, Vt.: Ashgate, 2011).

[34] Members of the panel included Wilhelm Caspari, Karl Oppenheimer, Max Rubner, and Nathan Zuntz (nutritional physiologists); Karl Ballod and Robert Kuczynski (statisticians); Friedrich Aereboe, Otto Lemmermann, Kurt von Rümker, and Hermann Warmbold (agronomists); and Franz Beyschlag, Paul Krusch, and Bruno Tacke (geologists). The sole female member was Hedwig Heyl, a home economist. Paul Eltzbacher, ed., *Die Deutsche Volksernährung und der englische Aushungerungsplan* (Braunschweig: Friedrich Vieweg & Sohn, 1914), 5. The report was translated into English as *Germany's Food: Can It Last? Germany's Food and England's Plan to Starve Her Out. A Study by German Experts* (London: Univ. of London Press, 1915).

this would be a profit of 224 + 10.8 = 234.8 milliard calories and 4,700 + 800 = 5,500 tons of protein.[35]

Admitting that German consumption habits would have to change dramatically—away from a meat-centered diet and toward a more lacto-vegetarian one—the commission concluded nonetheless that careful use of domestic resources would ensure that physiological requirements could be met and even exceeded.[36] A remarkable expression of technocratic optimism, and widely praised during the war, the document was later criticized bitterly for being overly protein- and calorie-centered and for assuming that calculations about macronutrient volumes at the national level could translate into effective food policy on the ground.[37] Despite what the experts had calculated, after all, German civilians suffered terrible hunger during the war. But if nothing else, the war definitively nationalized *Volksernährung*.

Fear of repeating the hunger years of the Great War kept *Volksernährung* central to scientific and political discussions throughout the years of Weimar democracy. In the 1920s, the status of *Volksernährung* as a national problem was epitomized at the federal level by the *Reichsausschuß für Ernährungsforschung* (Imperial Committee for Nutrition Research). Established in 1921 by the federal government, the committee brought together a now familiar set of nutritional experts—Max Rubner, Emil Abderhalden, Arthur Scheunert, and others—to help guide national nutritional policy. Betraying its debt both to general scientific trends and very specific national traumas, the group identified several key areas for research. These included *Wirkstoffe* and their impact on human health, the development of synthetic foods (e.g., fat from coal, sugar from wood), and nutritional recycling (e.g., extracting nutrients from the horns and claws of slaughtered animals for animal feed). At the same time, the group pressed forward with the project of educating lay people about the *Volksernährung* problem and enlisting their support in solving it. The nine-volume series *Die Volksernährung*, sponsored by the Ministry for Agriculture and Nutrition and written by experts in the field, such as Abderhalden, Rubner, and Scheunert, was one important result.[38]

As all of this suggests, nutritional experts played a key role in making German hunger a social-epistemic object. Moleschott, Klencke, and other liberal popularizers at mid-century had already connected the New Nutrition to the nation's political struggles. By the 1880s, *Volksernährung* had emerged as a powerful new discursive category for documenting proletarian hunger as well as debating its causes and devising appropriate solutions. The Great War nationalized *Volksernährung*, which remained a key category of analysis for German nutritional leaders well into the mid-twentieth century. Throughout these years, numerous conflicts animated the *Volksernährung* discussion: high-protein versus low-protein diets, or meat- versus plant-centered ones; laboratory researchers versus physicians, or experts versus lay people; state help versus self-help, or bourgeois progressivism versus socialism. The breadth and depth of these debates, only hinted at here, suggested just how important it was to contemporaries to get *Volksernährung* right.

---

[35] Eltzbacher, *Die Deutsche Volksernährung* (cit. n. 34), 187.

[36] Ibid., 194–6, 221.

[37] "The Food Supply of the German People," *Lancet* (1915), 389–99; "The Food Supply of the German People," *Lancet* (1915), 399–400; August Skalweit, *Die Deutsche Kriegsernährungswirtschaft* (Stuttgart: Deutsche Verlagsanstalt, 1927), 8–10.

[38] R. O. Neumann, *Das Brot* (Berlin: Springer, 1922), 4–7. The pamphlet series was edited by the *Reichsministerium für Ernährung und Landwirtschaft*, and titled *Die Volksernährung* (Berlin: Julius Springer, 1922–30).

PRESCRIBING SCIENCE

Given the stakes in getting Germans to change their eating habits, it should come as no surprise that considerable attention went to nutritional education. Scientists, physicians, life reformers, home economists, and many other players sought to communicate nutrition science in such a way as to persuade lay audiences to eat differently. Figuring out how to do so in a practical and engaging manner, however, proved difficult. Getting Germans to move from acquiring nutritional knowledge to aligning their behavior with nutritional prescription proved even harder.

One problem that surfaced early on was the gendered division of labor in writing about food; women authored cookbooks, men penned scientific texts. Despite a common target audience—German housewives—male and female authors worked in mutually sealed-off discursive chambers. Consider in this regard two of the century's best-selling authors: the scientific popularizer Hermann Klencke, and Henriette Davidis (1800–76), the country's leading cookbook author. Klencke's *Chemical Cook- and Housekeeping Book* (1857) was an early attempt to educate housewives about kitchen chemistry so that they could exercise frugality in time, money, and ingredients, while still preparing tasty, nutritious meals for their families. Declaring that kitchens were "the chemical workshop of the housewife," Klencke urged women to abandon the "art of cooking" in favor of the "science of cooking." Early chapters lectured women about food's chemical elements, the chemistry of metabolism, and Liebig's division between respiratory and plastic foods. Later chapters prescribed how much carbon and nitrogen German men should consume each day. A "Chemical Food Table" listing the percentage of nitrogen and carbon in various common foods, and including sample menus, sought to guide housewives in meeting these consumption targets cheaply.[39] *The Chemical Cook- and Housekeeping Book* certainly presented new information to German housewives. Remarkably, however, it did not include recipes. Meanwhile, female cookbook authors such as Davidis attended carefully to kitchen practice, but largely ignored scientific developments. Her *Practical Cookbook* (1844) contained hundreds of recipes, but no discussion of chemistry. Given the date of the first edition, that was unsurprising; more surprising was that later editions barely gave a nod to scientific developments. The thirty-sixth edition of 1897, for instance, came in at a whopping seven hundred pages, of which only three discussed the New Nutrition.[40] On the topic of food, in short, male authors emphasized food science while female authors focused on food practice.[41]

Given this disconnect between nutritional prescription and cookery guides, Hedwig Heyl's *Kitchen ABC* (1888) was something of a breakthrough. Heyl (1850–1934) was

---

[39] Hermann Klencke, *Chemisches Koch- und Wirthschaftsbuch oder die Naturwissenschaft im weiblichen Berufe: Ein Lehrbuch für den denkende Frauen und zum Gebrauche in weiblichen Erziehungsanstalten* (Leipzig: Eduard Kummer, 1867), 33, 41, 43, 62, 70–1, 90.

[40] Henriette Davidis, ed. Luise Holle, *Praktisches Kochbuch für die gewöhnliche und feinere Küche* (Bielefeld: Velhagen & Klasing, 1897), 709–11.

[41] German cookbook literature is discussed in Walter Artelt, "Die deutsche Kochbuchliteratur des 19. Jahrhundert," in *Ernährung und Ernährungslehre im 19. Jahrhundert*, ed. Walter Artelt and Edith Heischkel (Göttingen: Vandenhoeck & Ruprecht, 1976), 350–85; and Ulrike Thoms, "Kochbücher und Haushaltslehren als ernährungshistorische Quellen," in *Neue Wege zur Ernährungsgeschichte: Kochbücher, Haushaltsrechnungen, Konsumvereinsberichte und Autobiographien in der Diskussion*, ed. Dirk Reinhardt, Uwe Spiekermann, and Thoms (Frankfurt am Mein: Peter Lang, 1993), 9–50. For a differently gendered division of labor in France, see E. C. Spary, *Feeding France: New Sciences of Food, 1760–1815* (Cambridge: Cambridge Univ. Press, 2017), 45–50.

Table 1. *Hedwig Heyl constructed this chart to guide housewives in composing daily nutritional intake for the "average male worker" in a way that fulfilled Voit's nutritional standards cheaply. Reprinted from Hedwig Heyl,* Das ABC der Küche *(cit. n. 42), 8.*

WEIGHT in grams (g)	FOOD	PROTEIN (g)	FAT (g)	STARCH (g)
212	meat*	42	21	–
35	fat	–	35	–
76	dietary protein from vegetables	76	–	–
500	starch from condiments and vegetables	–	–	500
TOTAL		118	56	500

Note. Heyl did not include water, which is approximately 75% of the meat's weight.

a social reformer who developed home economics as a field. Awarded several prizes and continuously reprinted through the late 1930s, *Kitchen ABC* presented scientific knowledge to women in a language they could understand and in a format they could easily use. After an opening chapter on budgeting, the book discussed chemical elements, macronutrients, metabolism, and digestive anatomy. It also provided extensive quantitative guidance in the form of multiple tables and charts that

- listed common foods with their percentage of protein, fats, and carbohydrates;
- summarized nutritional recommendations from Voit and others by social group;
- showed how a housewife might compose her husband's daily food intake so as to meet Voit's nutritional standards cheaply (table 1);
- recorded how long, to the minute, it took the body to digest various foods; and
- detailed how much gas common foods produced (nuts produced the most, while eggs, pickles, fish, salad, and cheese produced the least).

The rest of the book contained recipes, but, unlike Davidis's, encased them in a calculating framework. Each recipe, for example, came with information about cost and preparation time as well as advice on how to use up leftovers.[42] Economics, chemistry, and cookery merged here into a seamless whole.

When the state decided to get into the business of nutritional enlightenment, however, it initially ignored Heyl's synthetic approach and instead continued in the tradition of male scientific popularizers. Consider in this regard the *Health Booklet* (1895) that was published by the newly formed Imperial Health Office to educate citizens about the basics of disease prevention: "The food we enjoy consists of a series of food-stuffs (nutrients) and these in turn are made up of basic chemical elements. . . . Food-stuffs can be divided chemically into a nitrogenous group, the proteins, and two nitrogen-free groups, carbohydrates and fats." Invoking Voit, it prescribed the proper mix of nutrients for a healthy adult male living in a German climate doing moderate work as 120 grams protein (18.8 grams nitrogen), 475 grams carbohydrates, and 280 grams fat. A color-coded chart, originally developed by food chemist Josef König and widely reprinted into the twentieth century, made it easier than ever for readers to grasp the chemical composition of the foods they were likely to eat (fig. 1). The book also provided a

[42] Hedwig Heyl, *Das ABC der Küche* (Berlin: Carl Habel, 1897), 7–8, 10, 12, 87.

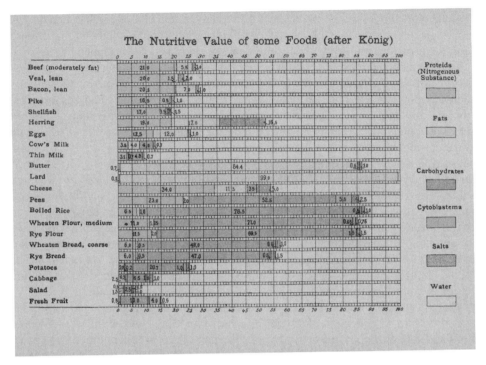

Figure 1. *An English version of the color-coded table of nutritive values assembled for the* Health Booklet *(1895), the official health guide of Imperial Germany. It privileges animal foods by putting them first and coding protein red. Fresh fruits and vegetables at the bottom, in contrast, come off looking nutritionally poor. Reprinted from* The Imperial Health Manual *(Dublin: Bailliere, Tindall & Cox, 1896), 64.*

sample set of menus by which the average male worker might reach these chemical goals, and an accompanying table showed that he could do so for just fifty-eight pfennig per day. But there were no recipes.[43]

In addition to this persistent gendered divide between nutritional prescription and cookery advice, a second challenge for the project of nutritional enlightenment was that popularizers could not agree among themselves whether Voit's standards were the right standards for the country. And in this, they largely reproduced fault lines mentioned previously. The socialist book series *Worker's Health Library* was a case in point, as two of its titles carried competing recommendations. Benno Chajes (1880–1938), a physician, contributed a pamphlet in 1907 that explained the basics of metabolism in sixteen pages, extolled meat, and reproduced the *Health Booklet*'s color-coded chart.[44] In 1912, in contrast, Julian Marcuse (1862–1942), a psychiatrist with close ties to the life reform movement, used the discoveries of Voit and Rubner to recommend a life reform (mostly vegetarian and teetotaling) diet as the cheapest and most nutritious for

[43] Kaiserliches Gesundheitsamt, *Gesundheitsbüchlein: Gemeinfaßliche Anleitung zur Gesundheitspflege* (Berlin: Julius Springer, 1895), 54, 56, 60–1.
[44] Benno Chajes, *Nahrung und Ernährung*, Arbeiter-Gesundheits-Bibliothek, vol. 8 (Berlin: Buchhandlung Vorwärts, 1907).

German workers.[45] None of the pamphlets, moreover, touched on kitchen practice. A working-class housewife could be forgiven for wondering not just how to make sense of these disparities in expert advice, but what she should cook for dinner.

The war helped remake the terrain of nutritional enlightenment in at least two ways: first, by forcing male and female authors to work together on reorganizing the nation's food economy; and second, by getting them to rethink nutritional standards inherited from the previous century. Take the *Pamphlets on Public Nutrition*. A state-sponsored series, it contained several titles on wartime cookery that were coauthored by female home economists, who wrote the recipes, and male physiologists, who explained how eating scientifically would help Germans win the war. In 1915, Hedwig Heyl teamed up with Nathan Zuntz (1847–1920), an animal physiologist, on *The Low-Fat Kitchen*. Like other volumes in the series, it guided housewives in cooking low-fat, low-meat meals that decisively broke with Voit's standards. Those revisions, in turn, were a sign of wartime exigencies; fat was needed to make glycerin, while meat inefficiently took up nitrogen, which as a key ingredient in both fertilizer and explosives was the war's most sought-after chemical element.[46] As this example suggests, the war functioned as a crucible of exchange between different players in the nation's food economy and forced them to alter their message about what Germans should be eating.

That forced collaboration paid big dividends after the war. Weimar ushered in a golden age of nutritional education and, with it, a shift to a more fluid set of recommendations and heterogeneous group of players. One sign of a new collaborative spirit came with the publication of *The Foundations of a Proper Diet* (1925), a best-selling book by Ragnar Berg (1873–1943) and Martin Vogel (1887–1935). Berg, a food chemist, had worked for years at the fabled naturopathic sanatorium of Heinrich Lahmann (1860–1905) in Dresden. Vogel, a physician, was at the time involved in organizing the GESOLEI exhibit, perhaps Weimar's greatest health spectacle, and eventually became director of the German Hygiene Museum, the country's premier institution of health education.[47] Before the war, a physician as prominent as Vogel would have been unlikely to work with someone, such as Berg, tainted by naturopathic quackery. But the war had upended old epistemological hierarchies. Now, shunning the protein-, meat-, and calorie-centered focus of yesteryear and embracing the Newer Nutrition of vitamins and trace minerals instead, Berg and Vogel co-opted life reform dietary prescriptions to craft a syncretic message. They criticized the transformation of German foodways away from self-provisioning with fresh foods, and toward a convenience economy in which preserved foods—conserved meats, canned fruits and vegetables, powdered milk—dominated. Rejecting the protein-and-meat obsession of their predecessors, moreover, Berg and Vogel recommended that Germans convert to a modified lacto-vegetarian diet first pioneered by life reformers. Combining a new interest in vitamins

[45] Julian Marcuse, *Volksernährung*, Arbeiter-Gesundheits-Bibliothek, vol. 29 (Berlin: Buchhandlung Vorwärts, 1912).

[46] Hedwig Heyl and Nathan Zuntz, *Die fettarme Küche*, Flugschriften zur Volksernährung, vol. 9 (Berlin: Zentral-Einkaufsgesellschaft, 1915). Other volumes included Karl Weinhausen, Wilhelmine Tschernoglasow, and Max Rubner, *Die Wintergemüse als Volksnahrung* (1914), vol. 11; Josephine Nagel and Emil Abderhalden, *Die Kartoffelküche in der Kriegszeit* (1915), vol. 12; Eduard Kallert and Johanna Martin, *Der Klippfisch als Volksnahrungsmittel* (1915), vol. 15; and Josephine Nagel and Carl Oppenheimer, *Die Kriegsküche im Sommer 1916* (1916), vol. 25.

[47] Ragnar Berg and Martin Vogel, *Die Grundlagen einer richtigen Ernährung* (Dresden: Deutscher Verlag für Volkswohlfahrt GmbH, 1925); "Vogel, Martin," World Biographical System Online (WBIS), De Gruyter, D623-604-3, https://wbis.degruyter.com/.

with an old interest in cost and nutritional values, they recommended that Germans eat fresh vegetables and fruits every day—in weight, five to seven times as much as all other foods combined—and drink at least a half liter of milk.[48] There were no recipes, but the nutritional advice had changed dramatically, as had the social location of the people giving it.

Meanwhile, other nutritional enlighteners turned to new forms of mass media to get their message out, both visually and aurally. With state support from the Interior Ministry and the *Reichsausschuß für hygienische Volksbelehrung* (Committee for Hygienic Education), Weimar's main body for health education, *Reichsgesundheitswoche* (Imperial Health Week), in 1926 included hundreds of films, talks, and exhibits held at thousands of locations around the country. Scientifically informed nutritional prescription leavened with references to wartime hunger were a persistent theme, while clever visuals simplified the message to a child's level. In the take-home pamphlet *Health is Happiness: Remarks for Everyone*, for instance, a smiling and rosy-cheeked mother and her children hoisted lacto-vegetarian items upward while moving actively across the page. On the next page, a man, balding and overweight, sat at a table being served a large joint of meat with a side of lobster and champagne; the waiter was a tuxedo-dressed skeleton (fig. 2).[49] Here, fitness, youth, fecundity, and the future were tied visually to the diet of life reformers, while death was posited as the end result of precisely those gluttonous meat- and alcohol-saturated dietary habits that had exacerbated wartime hunger.

Perhaps the most innovative among Weimar nutritional educators was Max Winckel (b. 1875), a food chemist. Bridging old divides—between male food scientists and female home economists, as well as between academic scientists and life reformers—Winckel worked with a heterogeneous set of actors to remake nutritional education for a high modernist age. In 1926, for instance, he worked with H. Walther, a high school home economics teacher, on a series of nutritional programs for the *Hour of the Housewife and Mother*, a national radio show for women.[50] Informing housewives that thirty to forty billion marks passed through their hands each year, Winckel and Walther instructed them on meeting their citizenly duties by mobilizing the family table around a mostly lacto-vegetarian diet with German-sourced ingredients.[51] Similarly, in his journal *Die Volksernährung*, Winckel began from the premise that "to be healthy is not a private matter but a citizenly duty," yet declined to offer a one-size-fits-all *Normalkost* for nutritional health. Instead, he assembled a pastiche of voices and approaches. On the journal's pages, advocates of nutritional autarky and free marketeers, life reformers and nutritional physiologists, Jewish Germans and Nazi Germans, men and women, all voiced their views freely and often in competition with each other. That cacophony, moreover, was produced by a virtual who's who of the country's major nutritional players, from the late imperial through the Nazi eras. Hedwig Heyl, Max Rubner, Arthur Scheunert, and Wilhelm Stepp all published there, as did well-known advocates for a life reform diet such as Ragnar Berg, Max Bircher-Benner

---

[48] Ragnar Berg and Martin Vogel, *Die Grundlagen einer richtigen Ernährung* (Dresden: Deutscher Verlag für Volkswohlfahrt GmbH, 1930), 6–8, 185–6, 239–40, 247–8.

[49] *Ein Merkbüchlein für Jedermann: Gesundheit ist Lebensglück* (Berlin: Reichsausschuß für hygienische Volksbelehrung, 1926), 10–11.

[50] Treitel, *Eating Nature* (cit. n. 12), 142.

[51] Max Winckel, ed., *Nahrung und Ernährung* (Berlin: Richard Schoetz, 1927), 58.

Figure 2. *Text and image worked together to encourage more eating of plant foods and less of animal foods in a pamphlet distributed at Weimar Germany's Imperial Health Week in 1926. Reprinted from* Ein Merkbüchlein für Jedermann *(cit. n. 49), 10–1.*

(1867–1939, inventor of muesli), and Rudolf Just (a naturopath). Prominent Jewish physiologists such as Carl Oppenheimer (1874–1941) and Wilhelm Caspari (born in 1872 and died in the Lodz ghetto in 1944) also contributed, as did Nazi-era leaders, who after 1933 remade the country's food economy for war and genocide. These included Richard Walther Darré (1895–1953) and Herbert Backe (1896–1947), both of whom led the Ministry of Agriculture, as well as Adolf Bickel, Sigwald Bommer

(1893–1963), Karl Kötschau (1892–1982), Werner Kollath, and Wilhelm Ziegelmayer (1898–1951).[52]

In addition to gathering a heterogeneous set of voices, Winckel also took full advantage of the new visual language of signs to amplify his message. In 1926, he began working with DATSCH, an educational design firm, to create images that could communicate nutritional concepts in lay terms. What, for instance, was a calorie? Previous enlighteners had not bothered to define the unit. In the poster he developed with DATSCH, Winckel did; a calorie was eight shirtless workers lifting a 425-kilogram weight 1 meter high (fig. 3).[53] Visual nutritional epistemologies such as this reached their height in the 1928 *Ernährungsausstellung* (Nutrition Fair) in Berlin. Organized by Winckel with support from several federal ministries, the Hygiene Museum, food industry representatives, home economists, physicians, scientists, and social reformers, the exhibit looked back to wartime hunger in order to discern the path forward into a future of nutritional abundance and health. That turned out to mean making the nutrition problem visible on a grand scale. The exhibit ran for one hundred days and occupied a space of forty thousand square meters (about nine football fields), with half of that reserved for scientific displays. Organizers sought to present "the embodiment of scientific discussions, technical-industrial presentations, and economic demonstrations" in a way that would both entertain and educate.[54] A visual extravaganza, the exhibit was widely reviewed; Walter Benjamin, a feuilletonist for the *Frankfurter Zeitung*, commented on it as follows:

> Here is a vegetable oracle, a very Delphi, where you need only aim a pointer at a particular month for a colored sign to pop up prophesying the menus to come . . . There are wooden charts with little lights going on and off, illustrating crop cultivation in different seasons or the process of metabolism in the human body . . . How out of date the dry charts of old-fashioned statistics seem to us now, with their unattractive lines. The whole earth, with its bushes and trees and fields, houses and homes, men and beasts, is all just good enough to enter into the vocabulary of this wonderful new and unspoiled sign language . . . In every corner and in a thousand different forms, food performs its somersaults and does all its other tricks.[55]

In the fight to educate the German stomach, in short, entertainment and viewer experience had now gained the upper hand.

As a project that grew out of and built upon nutrition science, nutritional enlightenment thus underwent massive changes from the mid-nineteenth to the mid-twentieth century. At the beginning of the period, a discursive disconnect between male and female writers—the former focused on food science, the latter on food practice—hampered communication with German women. Disagreement about how to extract nutritional prescription from nutrition science, moreover, led to conflicting messages in the literature. Then came the war, which radically remade the project. Not only did male and

---

[52] Dirk Reinhardt and Uwe Spiekermann, "Die 'Zeitschrift für Volksernährung' 1925–1939: Geschichte und bibliographische Erschließung," in *Materialien zur Ermittlung von Ernährungsverhalten*, ed. Adolf Bodenstedt (Karlsruhe: BFE, 1997), 79–80, 88, 91–7, 117, 120, 124, 136, 143, 146, 151.

[53] Winckel, ed., *Nahrung und Ernährung* (cit. n. 51), 134.

[54] For the exhibition catalogue, see *Die Ernährung: Ausstellung für zweckmässige Ernährung Berlin 1928* (Berlin: W. Büxenstein, 1927), 19.

[55] Walter Benjamin, "Food Fair: Epilogue to the Berlin Food Exhibition," in *Walter Benjamin: Selected Writings, vol. 2 (1927–1934)*, ed. Michael W. Jennings, Howard Eiland, and Gary Smith (Cambridge, Mass.: Belknap Press of Harvard Univ. Press, 1999), 136–8.

Figure 3. *"What is a calorie? A calorie is the quantity of energy that lifts 425 kilograms 1 meter high." Max Winckel, Weimar Germany's leading nutritional enlightener, used visual narrative to communicate basic concepts such as the calorie to lay viewers. Reprinted from Winckel,* Nahrung und Ernährung *(cit. n. 51), 134.*

female authors now find themselves working together for the state in front of a national audience, they also collectively turned away from the meat-centered, high-protein dietary prescriptions of old. During the Weimar years, finally, the project of nutritional enlightenment aligned itself with democracy, even as wartime hunger continued to shape its message. Activists like Max Winckel presented good nutrition as a duty of democratic citizenship, but declined to prescribe a top-down, one-size-fits-all *Normalkost*. Instead, he organized venues in which multiple voices exchanged a variety of views in a nutritional marketplace where not just words but sounds and images competed to capture Germans' attention. Despite the cacophony, however, two linked themes bearing unmistakable traces of the war emerged. On the one hand, individuals who chose to eat poorly (too much fat, meat, alcohol, and sugar) consigned themselves to illness in peacetime and the nation to hunger in war. On the other hand, individual choices to eat well (a largely lacto-vegetarian diet with German-sourced ingredients) produced nutritional abundance and health for all.

CONCLUSION

By the late 1920s, then, not only had nutrition become central to German citizenship and survival, but German scientists—often co-opting their message from life reformers—had emerged as key players in constructing and managing nutrition as a field of both social and scientific action. In the nineteenth century, they developed laboratory and field methods that encased nutrition in a quantitative and chemical framework. When urban hunger emerged as a social problem at century's end, that framework helped them develop and dominate *Volksernährung* as a discursive field. Here, scientific actors wielding quantitative-chemical tools appointed themselves as the best hope for maximizing the nation's industrial power, albeit not without significant challenge from rogue epistemological players, such as German life reformers. When war broke out in 1914, *Volksernährung* was decisively nationalized, and nutritional experts entered government service to manage nutritional scarcity, with limited success. Scarred by the hunger years and the failure of state-run nutritional planning, nutritional experts then turned in the Weimar years to changing the way Germans ate. They developed increasingly sophisticated programs of nutritional enlightenment that, they hoped, would ensure nutritional abundance and health in perpetuity for both individuals and the nation. Nutrition science, in other words, operated from its inception as a tool of national governance. Indeed, after their rise to power in 1933, the Nazis seized on nutrition as a central area of scientific research and social policy, racializing and radicalizing the entire field. When nutritional scientists helped calculate down to the last calorie how food would be distributed and denied across Germany and German-occupied Europe, they proved the inextricability of nutrition science from modern biopolitics.

The Scientific Lives of *Chicha*:
The Production of a Fermented Beverage and the Making of Expert Knowledge in Bogotá, 1889–1939

by Stefan Pohl-Valero*

ABSTRACT

From the end of the nineteenth century into the first decades of the twentieth, Colombian *chicha* (a fermented beverage made from maize) was at one and the same time alcohol and food, a product produced and consumed on a large scale in an urban setting, and an object of intense scientific scrutiny, with multiple meanings and transformations. In this article, I argue that industrial chicha production, local scientific practices, and food policies and regulations crossed paths in Colombia, influencing each other. I explore the situated and diverse practices of knowledge production about the relationship between chicha, bodies, and society, as well as their actual implications in, and mutual effects with, matters of food governance. In laboratories and hospitals, but also in the places of production and consumption of chicha, Colombian scientists produced toxicological, physiological, nutritional, and statistical knowledge about this beverage, shaping diverse racialized perceptions of local poor populations and their capacities to achieve national progress. In turn, the material transformations in the production of chicha (related to modern urban configurations), the industrial producers of this beverage, as well as the emerging practices of governance over chicha and the bodies of its consumers, influenced the making and use of knowledge claims about this same food product. With this case study, I call attention to some insights that history of science and medicine can offer to the fields of biopolitics and food history, and their interconnections. In methodological terms, I propose a biographical approach that follows some of the multiple historical lives of chicha, both as an object of science and as a commodity.

* School of Medicine and Health Sciences, Universidad del Rosario, Bogotá, Colombia; stefan .pohl@urosario.edu.co.

The results of the research conducted for this article are part of the project "Historia sociocultural de la nutrición en Colombia durante el siglo XX," supported by the Universidad del Rosario Research Fund (Small Grants: IV-FPC003). Previous drafts of this article were presented at the faculty symposium of the Universidad del Rosario School of Human Sciences, the Center for the History of Science at the Universidad Nacional Autónoma de México, and the Department of the History of Science at Harvard University. The feedback I received in all these places was most useful. I am also thankful to Adriana Alzate, Ana María Otero, and Agustí Nieto-Galan, and to the *Osiris* editors and referees for their comments and suggestions. The final draft of this article was completed during a research stay at the University of North Carolina at Chapel Hill. I thank the Fulbright Commission in Colombia and Universidad del Rosario for this scholarship.

INTRODUCTION

Across the Americas, chicha has been a "generic term for a variety of fermented beverages based upon yellow and white maize, agave juice, yuca (manioc), peach palm, ripe plantain, banana, sugar cane juice with adjuncts such as pineapple, fruits, cinnamon, sweet potato, palm fruit, etc."[1] In the Eastern Cordillera of the Andes, where the high plateau of Bogotá is situated, the term chicha is usually applied to a maize-based, mildly alcoholic beverage produced since the pre-Columbian era by the indigenous Muiscas.[2] One particularity of this product of the Colombian Andes is the use of cane juice, which was possible only after the arrival of sugar cane to the region (as a result of the so-called Columbian Exchange). Additionally, the types of corn and procedures involved in the production of chicha underwent several changes, especially since the end of the nineteenth century and in the urban context of Bogotá.[3] At that time, this beverage was central to the diet of urban and rural poor populations around this city.[4] The industrial production of chicha was further transformed during the first decades of the twentieth century in order to increase production capacity and economic profit.[5]

Parallel to these transformations of ingredients, techniques, and artifacts in the production of urban chicha, the local practices of science were framed in analogous global exchanges of ideas, procedures, and objects. By the end of the nineteenth century, local modern medicine was an activity in which different scientific theories about the nature of diseases and about the functioning of the human body coexisted and were disputed, thereby shaping new and sometimes contested forms of expertise about matters of governance, and discourses about nation building and racial identity.[6] Besides the local well-established traditions of medical geography, anatomo-clinical medicine, and neo-Hippocratic dietetics, as well as epistemic virtues based on personal experience and in situ observations, by the end of the nineteenth century the circulation and appropriation of scientific theories such as bacteriology, evolution, and experimental physiology, as well as laboratory procedures and their technologies, were helping to shape new styles of knowledge production about the relationship between environment, local population, and national progress.[7] The scientists involved in these activities

---

[1] M. Atacador-Ramos, M. S. M. Azmey, T. Basuki, D. S. Dahiya, T. D. Ekmon, J. A. Ekundayo, A. Escobar, et al., "Indigenous Fermented Foods in Which Ethanol Is a Major Product Type and Nutritional Significance of Primitive Wines and Beers and Related Alcoholic Foods," in *Handbook of Indigenous Fermented Foods*, ed. Keith H. Steinkraus (New York, N.Y.: Marcel Dekker, 1996), 363–508, on 406.

[2] María Clara Llano Restrepo and Marcela Campuzano Cifuentes, *La chicha, una bebida fermentada a través de la historia* (Bogotá: Instituto Colombiano de Antropología e Historia, 1994).

[3] Liborio Zerda, "Estudio químico, patológico e higiénico de la chicha," *Anales de la Instrucción Pública en la República de Colombia* 14 (1889): 3–36, on 8–11.

[4] Manuel Cotes, *Régimen alimenticio de los jornaleros de la Sabana de Bogotá: Estudio presentado al Primer Congreso Médico Nacional de Colombia* (Bogotá: La Luz, 1893), 28–31.

[5] Antonio María Barriga Villalba, "Nueva discusión acerca de la chicha," *Revista de Higiene. Órgano del Departamento Nacional de Higiene* 18 (1937): 56–61.

[6] See, among others, Diana Obregón, "Building National Medicine: Leprosy and Power in Colombia, 1870–1910," *Soc. Hist. Med.* 15 (2002): 89–108; Jason McGraw, "Purificar la nación: Eugenesia, higiene y renovación moral-racial de la periferia del Caribe colombiano, 1900–1930," *Revista de Estudios Sociales* 27 (2007): 62–75; Stefan Pohl-Valero, "'La raza entra por la boca': Energy, Diet, and Eugenics in Colombia, 1890–1940," *Hispanic Amer. Hist. Rev.* 94 (2014): 455–86; and Mónica García, "Debating Diseases in Nineteenth-Century Colombia: Causes, Interests, and Pasteurian Therapeutics," *Bull. Hist. Med.* 89 (2015): 293–321.

[7] Mónica García and Stefan Pohl-Valero, "Styles of Knowledge Production in Colombia, 1850–1920," *Sci. Context* 29 (2016): 347–77.

were increasing their political power in order to define the nature of health problems and the way the state should address them.[8]

In an important sense, from the end of the nineteenth century through the first three decades of the twentieth, local scientific practices, industrial chicha production, and food policies and regulations had crossed paths, influencing each other. As I argue in this article, in laboratories and hospitals, but also in the places of production and consumption of chicha, Colombian scientists produced toxicological, physiological, nutritional, and statistical knowledge about this beverage, shaping, in turn, diverse racialized perceptions of local poor populations and their capacities to achieve national progress. In their efforts to deploy public policies destined to produce healthy and productive workers, these scientists approached the "chicha question" from different and sometimes contradictory angles. For some, the industrial elaboration of chicha produced a toxin that was causing a specific disease and "racial degeneration" of the local population; for others, this beverage was rich in nutrients, allowing the urban working class to have a cheap source of energy for performing work and thereby avoiding "physiological degeneration." My main argument is that, to some extent, these paradoxical scientific meanings about the relationship between chicha, bodies, and society were related to the different ways in which local scientists articulated laboratory work, fieldwork, and metrological networks, as well as to the different appropriations they made of the concept of human "degeneration."[9] I further argue that the material transformations in the production of chicha (related to modern urban configurations), the industrial producers of this beverage, as well as the emerging practices of governance over chicha and the bodies of its consumers, influenced, in turn, the making and use of knowledge claims about this same food commodity.[10]

In the large literature on biopolitics in Latin America, scholars have seldom explored in detail the production and application of knowledge and standards about food and nutrition, or the body conceptions related to these activities, in their efforts to understand how modern science shaped new political rationalities that took the administration of life and populations as its main subject.[11] To be sure, from Foucauldian and

---

[8] Obregón, "Building National Medicine" (cit. n. 6).

[9] For the relationships between different styles of knowing and the practices of science related to fieldwork and lab work, as well as for the historical construction of the lab-field border, see Robert E. Kohler, *Landscapes and Labscapes: Exploring the Lab-Field Border in Biology* (Chicago: Univ. of Chicago Press, 2002). For the multiple meanings around the notion of "degeneration" at the end of the nineteenth century, see, for Europe, Daniel Pick, *Faces of Degeneration: A European Disorder, c. 1848 - c. 1918* (Cambridge: Cambridge Univ. Press, 1993); Sandra Caponi, *Loucos e degenerados: Uma genealogia da psiquiatria ampliada* (Rio de Janeiro: Fiocruz, 2012); and Claude-Oliver Doron, *L'homme altéré: Races et dégénérescence (XVII-XIXe siècles)* (Ceyzérieu, Fr.: Champ Vallon, 2016). For Latin America, see, among others, Nancy Stepan, *The Hour of Eugenics: Race, Gender, and Nation in Latin America* (Ithaca, N.Y.: Cornell Univ. Press, 1991); Dain Borges, "Puffy, Ugly, Slothful and Inert: Degeneration in Brazilian Social Thought, 1880–1940," *Journal of Latin American Studies* 25 (1993): 235–56; Julia Rodríguez, *Civilizing Argentina: Science, Medicine, and the Modern State* (Chapel Hill: Univ. of North Carolina Press, 2006); and María Fernanda Vásquez-Valencia, "Degenerados, criminosos e alienados: Para uma história do conceito de degeneração na Colômbia, 1888–1950" (PhD diss., Universidade Federal de Santa Catarina, 2015).

[10] Here, I take advantage of Sheila Jasanoff's proposal to explore "how knowledge-making is incorporated into practices of state-making, or of governance more broadly, and, in reverse, how practices of governance influence the making and use of knowledge." See Sheila Jasanoff, ed., *States of Knowledge: The Co-production of Science and Social Order* (London: Routledge, 2004), 3.

[11] For a recent overview on studies of biopolitics in Latin America, see Hilderman Cardona and Zandra Pedraza, eds., *Al otro lado del cuerpo: Estudios biopolíticos en América Latina* (Bogotá: Ediciones Uniandes, 2014).

other social and cultural perspectives, attention has been given to discursive analysis of food and national identities during the nineteenth and the first half of the twentieth century, and to the implementation of food regulations, nutrition policies, and food assistance in the context of the construction and modernization of Latin American nation-states.[12] These questions have been adequately analyzed in relation to modern—and often Eurocentric—perceptions of civilization, race, class, and gender. For example, and as Enrique Ochoa has recently summarized in his bibliographical essay on food history in Latin America, at the end of the nineteenth century, "social Darwinian thought linked indigenous foods to backwardness and European foods to modernization, leading to an array of policies generally focused on women, and aimed at promoting the consumption of foods associated with Europe."[13]

However, the contextual and heterogeneous practices of science that helped shape these discourses and processes have received little attention, or have been addressed only in very general ways.[14] Nor has attention been paid to the interconnections between the history of knowledge making about food and the material history of food products.[15] Indeed, the historiography of chicha pathologization and criminalization in Colombia is a case in point. The historical narratives about these processes have been framed in two exemplary moments. One occurred in 1889, when "chichismo" was coined by physicians Liborio Zerda and Josué Gómez as a local disease—different from alcoholism—with a specific chemical etiology and an identifiable clinical picture.[16] The second was in 1948 after the murder of liberal politician Jorge Eliécer Gaitán; with the civil violence that this event produced (called *El Bogotazo*), some doctors and politicians declared that the origin of this "state of political exacerbation and criminality" lay in a pathological condition of the poor that had resulted from the consumption of chicha.[17] The next year, a new national law was passed, drawn up by Jorge Bejarano, the minister of hygiene, that banned the urban production and sale of chicha.[18] By assuming a causal connection between these two historical moments (with a distance of 60 years), local historiographers have argued that "modern scientific

---

[12] See, among others, Jeffrey Pilcher, *¡Que vivan los tamales! Food and the Making of Mexican Identity* (Albuquerque: Univ. of New Mexico Press, 1998); Paulo Drinot, "Food, Race and Working-Class Identity: Restaurantes Populares and Populism in 1930s Peru," *Americas* 62 (2005): 245–70; Sandra Aguilar-Rodríguez, "Cooking Modernity: Nutrition Policies, Class, and Gender in 1940s and 1950s Mexico City," *Americas* 64 (2007): 177–205; Gretchen Pierce and Áurea Toxqui, eds., *Alcohol in Latin America: A Social and Cultural History* (Tucson: Univ. of Arizona Press, 2014); and Sören Brinkmann, "Leite e modernidade: Ideologia e políticas de alimentação na era Vargas," *História, Ciências, Saúde–Manguinhos* 21 (2014): 263–80.

[13] Enrique Ochoa, "Food History," in *Oxford Bibliographies: Latin American Studies*, 28 October 2011, https://doi.org/10.1093/obo/9780199766581-0057.

[14] Some recent exceptions are José Buschini, "La alimentación como problema científico y objeto de políticas públicas en la Argentina: Pedro Escudero y el Instituto Nacional de la Nutrición, 1928–1946," *Apuntes* 79 (2016): 129–56; Joel Vargas Domínguez, "Metabolismo y nutrición en el México posrevolucionario: Eugenesia y clasificación de la población mexicana entre 1927 y 1943" (PhD diss., Universidad Nacional Autónoma de México, 2017); and Pohl-Valero, "'La raza entra por la boca'" (cit. n. 6).

[15] As Peter Atkins has recently argued, the "material biographies" of food products deserve "more attention from both historians of food and historians of science." See Peter Atkins, "The Material Histories of Food Quality and Composition," *Endeavour* 35 (2011): 74–9, on 74.

[16] Zerda, "Estudio químico, patológico" (cit. n. 3); Josué Gómez, "Chichismo: Estudio general, clínico y anatomo-patológico de los efectos del uso y abuso de la chicha en la clase obrera de Bogotá," *Anales de la Instrucción Pública en la República de Colombia* 14 73 (1889): 36–148.

[17] República de Colombia, "Decreto 1839 de 1948," *Diario Oficial* 84, no. 26741 (1948): 1.

[18] Jorge Bejarano, *La derrota de un vicio: Origen e historia de la chicha* (Bogotá: Iqueima, 1950).

knowledge" throughout the first half of the twentieth century was a homogeneous and monolithic body of discourse that defined chicha as a barbaric and unhealthy beverage; this "knowledge" in turn structured an array of public policies that were cumulatively and effectively deployed in order to control and eventually ban the production and consumption of this "traditional" beverage.[19]

While building on this literature, in this article I want to call attention to some insights that the history of science and medicine can offer to the subjects of biopolitics and food history. As already mentioned, these insights are related to detailed explorations of the situated, diverse, and contested practices of knowledge production about the relationship between food, bodies, and society, as well as these practices' actual implications in, and mutual effects with, matters of food governance. At the same time, I want to explore some of the interconnections between the history of knowledge making about chicha and the material history of this food product.[20] In methodological terms, I propose a biographical approach that follows some of the multiple historical lives of chicha, both as an object of science and as a commodity, from the nineteenth century to the end of the 1930s.[21] By following how these lives were historically configured and interconnected, I want to produce a narrative that overcomes disciplinary histories of science and cultural assumptions of food. Rather than using chicha as a gateway to a history of physiology, or nutrition, or toxicology in Colombia (in a historical context where many of these scientific disciplines did not exist institutionally or professionally), I want to approach chicha as an object of science that crossed these disciplinary boundaries.[22] On the other hand, I want to avoid the assumption that chicha always has been the same thing. Historians and anthropologists tend to use words such

[19] Restrepo and Cifuentes, *La chicha* (cit. n. 2); Óscar Iván Calvo Isaza and Marta Saade Granados, *La ciudad en cuarentena: Chicha, patología social y profilaxis* (Bogotá: Ministerio de Cultura, 2002); Carlos Ernesto Noguera, "Luta antialcoólica e higiene social na Colômbia, 1886–1948," in *Cuidar, controlar, curar: Ensaios históricos sobre saúde e doença na América Latina e Caribe*, ed. Gilberto Hochman and Diego Armus (Rio de Janeiro: Fiocruz, 2004), 99–123. To be sure, this literature has convincingly argued that besides the efforts by the local beer industry to demonize chicha, in order to control the local market of fermented popular drinks, some physicians and their scientific arguments played an important role in the historical construction of chicha consumption as a social pathology and in promoting beer as a modern and hygienic drink.

[20] Good examples of these historiographical entanglements between history of science and food history are the recent works of E. C. Spary on the emergence of the food industry and food sciences in the French context of the late eighteenth century; and of Peter Atkins on the multiple lives of milk in the British context of 1850–1950, from a material perspective. See Peter Atkins, *Liquid Materialities: A History of Milk, Science and the Law* (Farnham: Ashgate, 2010); and E. C. Spary, *Feeding France: New Sciences of Food, 1760–1815* (Cambridge: Cambridge Univ. Press, 2014).

[21] Scholars in both history of science and food history have argued that biographical approaches can shed light on the historicity and contextual contingency of the very reality of scientific objects and food products. See Arjun Appadurai, ed., *The Social Life of Things: Commodities in Cultural Perspective* (Cambridge: Cambridge Univ. Press, 1986); and Lorraine Daston, ed., *Biographies of Scientific Objects* (Chicago: Univ. of Chicago Press, 2000).

[22] As Hans-Jörg Rheinberger has argued, focusing on scientific objects may have several implications in the way we produce historical narratives about science. First, these narratives are not "pursuing the [historical] development of concepts, disciplines, institutions, or individual researchers," but rather, "we have to locate ourselves between boundaries: boundaries between representational techniques, experimental systems, established academic disciplines, institutionalized programs, individualized projects." Second, our studies move away from disciplinary histories and are located "on a crossdisciplinary level." And third, "talking about the trajectory of research objects gives voice to things as active participants in a conquest of transindividual dimensions in which the subjects concerned with these things are not the only players." See Hans-Jörg Rheinberger, "Cytoplasmic Particles: The Trajectory of a Scientific Object," in Daston, *Biographies of Scientific Objects* (cit. n. 21), 273.

as "ancestral," "indigenous," "traditional," and "artisanal" when referring to this beverage of the Colombian Andes, regardless of the historical period and production settings they study.[23] I have restricted my temporal window of analysis until the end of the 1930s. The complex interplay of political events, social crises, economic interests, scientific arguments, and international influences in the organization of the Colombian health system that led to the prohibition of chicha in 1949 lies beyond the scope of this essay and deserves further research.

THE TOXICOLOGICAL LIFE OF CHICHA

At the end of the nineteenth century, Colombian medicine was becoming increasingly institutionalized and professionalized. A faculty of medicine was created at the National University in 1868; the Colombian Society of Medicine was established in 1873; and in 1890, the National Academy of Medicine became a Colombian consultative body. In 1886, the government also created the Central Board of Hygiene as the first step in the centralization of public health in Colombia.[24] In these institutions, physicians began to articulate new scientific theories, ways of knowing, and procedures intended to capture the nature of diseases and their effects on the Colombian population.[25]

It was within this institutional and epistemological context that a specific disease caused by chicha consumption was fully defined and described for the first time. In 1889, the physician Liborio Zerda, working at the chemical laboratory of Bogotá's school of medicine, gave a chemical explanation of the etiology of a new local disease that he called *chichismo*.[26] At the same time, his colleague Josué Gómez, working at the city's Charity Hospital, described the specific symptoms and anatomical and clinical signs that characterized the bodies with this disease. According to Gómez, all these "morbid manifestations" were "exclusively the effect of toxic, alkaloid or acrid principles, which are produced during the terrible fermentation of maize for the preparation of the chicha."[27] Both physicians stressed that chichismo was different from other diseases related to alcohol or maize consumption (such as alcoholism and pellagra), and predicted that the consumption of chicha represented a national problem that would bring "terrible consequences" characterized by the "racial degeneration" of the Colombian population.[28]

Zerda stated that his interest in conducting a chemical study of chicha to clarify the etiology of the new disease was inspired by his former professor of medicine, José María Merizalde. At the end of the 1820s, Merizalde had studied this beverage, affirming that his "experience" had convinced him of the "nutritive" character of this "Colombian wine": "the vigor that the [Andean] Indians acquire with chicha is not inferior to the one that the Europeans acquire with wine and beer, for an Indian and his wife can transport more than 12 arrobas of weight from Honda to Bogotá on their backs, as if

---

[23] As already mentioned, the material transformations of chicha production have not been considered in detail by local historians and anthropologists.

[24] Emilio Quevedo, Germán Pérez, Néstor Miranda, Juan Carlos Eslava, and Mario Hernández, *Historia de la medicina en Colombia: Tomo III. Hacia una profesión liberal (1863–1918)* (Bogotá: Tecnoquímicas, 2010).

[25] García and Pohl-Valero, "Styles of Knowledge Production" (cit. n. 7).

[26] Zerda, "Estudio químico, patológico" (cit. n. 3).

[27] Gómez, "Chichismo: Estudio general" (cit. n. 16), 113.

[28] Zerda, "Estudio químico, patológico" (cit. n. 3), 36.

they were carrying a bunch of straw."[29] In his efforts to rationalize the possibilities of progress and civilization of the recently created and regionally diverse Republic of Colombia,[30] Merizalde's interpretation of some aspects of national reality were framed in a humoral physiology. He classified—and hierarchized—the moral and physical characteristics of population in terms of regional eating habits. The inhabitants of the lowlands of Casanare, for example, were "cruel" and "bellicose" in nature, due to their great consumption of meat. The people of Cundinamarca, with a diet based on vegetables (including chicha), were, on the contrary, "more human," "tranquil," and "healthier." He also explained how regional diets, besides the effects of climate, influenced the humoral constitutions of inhabitants and their attitudes toward work and obedience.[31]

Besides the general interest about the regional geographical characteristics that supposedly defined the racial differentiation of Colombian populations, some observers in the decade of 1850 also stressed the influence of regional diets.[32] For example, traveler Manuel Ancízar mentioned that the rural inhabitants of the Andean highlands had a diet based on "vegetables and chicha" that helped produce people of "sober constitution," who were more "obedient, industrious and honest" than the inhabitants of other regions.[33] These discourses reflected that chicha was an object of observation whose effects on the progress and civilization of the nation were ambiguous. Certainly Merizalde, like all doctors throughout the nineteenth century, condemned the abuse of alcohol and pointed out that drunkenness could lead to the "ruin of the State," and that the immoderate consumption of chicha was part of this problem. However, chicha passed between being a nutritious food—what we could call a liquid bread—and an alcoholic beverage. Furthermore, these considerations of chicha depended on the specific places of its production and consumption. In general terms, the aforementioned observations between chicha, health, and national progress were focused on rural areas, where the production process of the beverage could differ from that produced at urban sites.

The chemical study of chicha done by Zerda in 1889 was focused, on the contrary, on the urban and industrial production of chicha (as we will discuss in the next section) at a moment when industrialization and modernization were beginning to capture the imagination of local elites, and when concerns about the "working class question" slowly began to arise.[34] It was also framed in a different way of producing knowledge. As Óscar Iván Calvo Isaza and Marta Saade Granados have noticed, Zerda stressed that empirical observations of chicha consumers (as those made by Merizalde) were no longer enough to provide true knowledge about the nature of this beverage and its effects on the body. If the hospital had become the ideal place to do a systematic

[29] José Félix Merizalde, *Epitome de los elementos de higiene, o, De la influencia de las cosas físicas i morales sobre el hombre, i de los medios de conservar la salud* (Bogotá: Pedro Cubides, 1828), 360–1.

[30] At that postindependence moment, the country was called "Gran Colombia," and included the present-day territories of Colombia, Panamá, Venezuela, and Ecuador.

[31] For details of these dietetic-political discourses, see Stefan Pohl-Valero, "Food Science, Race, and the Nation in Colombia," in *Oxford Research Encyclopedia of Latin American History*, ed. William Beezley (Oxford: Oxford Univ. Press, 2016), 5–6.

[32] On geography, racial differentiation, and nation building, see Nancy P. Appelbaum, *Mapping the Country of Regions: The Chorographic Commission of Nineteenth-Century Colombia* (Chapel Hill: Univ. of North Carolina Press, 2016), among others.

[33] Manuel Ancízar, *Peregrinación de Alpha* (Bogotá: Echeverría Hermanos, 1853), 107.

[34] Santiago Castro-Gómez and Eduardo Restrepo, eds., *Genealogías de la colombianidad: Formaciones discursivas y tecnologías de gobierno en los siglos XIX y XX* (Bogotá: Editorial Pontificia Universidad Javeriana, 2008).

study of the morbid manifestations caused by chicha, it was the laboratory and its in-struments that made it possible to clarify the causes of these manifestations.[35] Never-theless, the starting point for Zerda's chemical research was the assumption that chicha was the central cause of a specific disease. At the beginning of his study, Zerda affirmed that the *enchichado* was a "true type" of sick person, characterized by a par-ticular "pigmentation of the skin," a "sad, languid and stupid look," and dried and dis-ordered hair. The enchichados "lost their energy and desire to work," their arms and legs became very rigid, and their bodies "exhaled a very bad, putrid smell."[36] All these characteristics, Zerda claimed, showed that chicha was undoubtedly the cause of moral and physical degenerations observed, according to him, in many poor people in urban settings. As an example of this situation, he described the tragic death of a "mulato" that he had known since childhood. Although this person was "intelligent, with an easy capacity to speak eloquently, . . . and had . . . moral notions . . . received from a careful education," just as Zerda had, when he "became a shoemaker, entered into popular life and started to drink chicha," he underwent a complete transformation: "His taciturn appearance, his vague, indifferent, and almost stupid gaze were the result of that vice; his eloquence became difficult, his ideas became obscure . . . The color of his skin changed, and the icteric tint of his eyes was repugnant. . . . He died at last, but without delirium or any other symptoms of alcoholism so frequent in drinkers of spir-its drinks."[37]

This "typical" case of an enchichado—incarnated in a racialized and urban body—was portrayed by Zerda as the main clue for "suspecting" that chicha contained a toxin that caused a specific disease different from alcoholism.[38] His observations were fol-lowed by the laboratory work of a chemical analysis to identify and isolate the toxin produced during the fermentation of chicha. At the time of his research, laboratory in-frastructure was very scarce in Colombia. Zerda was helped by his colleague Javier Tapia, professor of chemistry at the faculty of medicine, and the director of the faculty's laboratory, founded in the 1880s.[39] In 1887, Tapia was designated "chemical analyst" to Colombia's Central Board of Hygiene. Zerda stressed that the chemical analysis of chicha was very difficult, due to its complex composition, and that he and Tapia had followed and adapted recent procedures developed by European chemists analyzing other fermented beverages.[40] The instruments and reagents necessary for the chemical analysis of chicha were multiple and hard to obtain in the Colombian context. These items included devices such as a suction pump and porous porcelain vessels (to perform the procedure devised by the German chemist F. W. Zahn to filter organic liq-uids), dislodging devices from the "Cloez system" (to separate the chemical compo-nents), watch glasses, filters, and a microscope. Many chemical products were also men-tioned, such as neutral solvents, absolute alcohol, pumice, alkaloid reagents, crystals of potassium, permanganate, concentrated sulfuric acid, litmus paper, caustic potash, am-monia, and benzine, among others.[41]

[35] Isaza and Granados, *La ciudad en cuarentena* (cit. n. 19), 37.
[36] Zerda, "Estudio químico, patológico" (cit. n. 3), 15–16.
[37] Ibid., 15.
[38] Ibid., 23.
[39] At that time, the profession of chemistry was not institutionalized in Colombia. Local universities only offered careers in law, medicine, and civil engineering.
[40] Zerda, "Estudio químico, patológico" (cit. n. 3), 20.
[41] Ibid., 20–6.

In 1888, the government decided to create a single chemical laboratory from several small ones then placed in different Bogotá institutions. The large laboratory was installed in the building of Santa Inés, the seat of the faculty of medicine. Ninety-five boxes ("bultos") of instruments and reagents were sent by the National Institute of Agriculture, and seven boxes by the Ministry of Development ("Ministerio de Fomento"). The University of Rosario sent the remains of the "National Laboratory" created in 1847 by the French chemist Bernardo Carlos Lewy on behalf of the government. Tapia was in charge of this unified laboratory, which was described as a "very incomplete one." This site lacked a chimney, necessary for the expulsion of the gases produced in chemical reactions. Nor did the laboratory have a water supply, although a new aqueduct with iron pipes passed in front of the building.[42] It was also stressed that the chemical reagents for the laboratory ran out very quickly, and that the budget to buy more was very reduced.[43] Nevertheless, in this laboratory, and under these material conditions, the first chemical analysis of chicha was done in Colombia.

After laborious work, Zerda was able to define the chemical composition of two types of chicha: "chicha flor" and "second-class chicha." The first one was made in factories following a long and complicated procedure (see next section). According to Zerda, this "chicha flor" was not intended for "general consumption, since it is expensive and does not yield great profits for the manufacturer." Chicha flor was mixed with "mitaca" (a mixture of the sediments produced during the preparation of "chicha flor," and water and cane juice) and left for four to five days for additional fermentation.[44] This "second-class" chicha was the one produced and sold for general consumption in the "chicherías" of Bogotá. In both types of chicha, alongside varying amounts of protein, carbohydrates, minerals, alcohols, and acids, a "toxic alkaloid" was found. This substance was identified by Zerda as a ptomaine formed by the putrefaction of maize, due to bacterial action during the urban production of chicha.[45] The term ptomaine had been coined by Italian toxicologist Francesco Selmi some years before Zerda's work and provided a renewed optimism about chemical pathology in relation to bacteriology.[46]

Once he had isolated the ptomaine, Zerda injected it into different animals and observed its physiological effects. Meanwhile, Gómez conducted a clinical study of "our indios enchichados," providing a differential diagnosis of this disease.[47] Since the clinical picture of enchichados was similar to the one of animals injected with the ptomaine isolated from chicha, "chichismo" was defined by Zerda and Gómez as a specific

---

[42] República de Colombia, *Informe presentado al Congreso de la República en sus sesiones ordinarias de 1888 por el Ministro de Instrucción Pública* (Bogotá: La Luz, 1888), 29–30.

[43] República de Colombia, *Informe que el Ministro de Instrucción Pública presenta al Congreso de Colombia en sus sesiones ordinarias de 1894* (Bogotá: La Luz, 1894), lxxix.

[44] Zerda, "Estudio químico, patológico" (cit. n. 3), 11.

[45] The chemical study of ptomaines, and its relation to pathology, bacteriology, and physiology, was an emerging topic of scientific medicine at the time of Zerda's research. For contemporary book-length accounts of this, see A. M. Brown, *The Animal Alkaloids, Cadaveric and Vital; or the Ptomaines and Leucomaines Chemically, Physiologically and Pathologically Considered in Relation to Scientific Medicine* (London: Hirschfeld Bros., 1887); and Victor Vaughan and Frederick Novy, *Ptomaines and Leucomaines, or the Putrefactive and Physiological Alkaloids* (Philadelphia: Lea Brothers, 1888).

[46] Neil Morgan, "From Physiology to Biochemistry," in *Companion to the History of Modern Science*, ed. R. C. Olby and G. N. Cantor (London: Routledge, 1990), 496–7.

[47] Gómez, "Chichismo: Estudio general" (cit. n. 16), 85.

pathological state with a chemical etiology.[48] Consequently, these physicians stated—as a scientific fact based on the rigor of chemical experimentation and clinical observation—that local populations, despite "the vigor of their physical organization and the clarity of their intelligence," would be destined to suffer a "racial degeneration" if they continued consuming chicha "as it is produced today."[49] This process of degeneration was different from the one caused by chronic alcoholism.[50] Instead of dementia, "delirium tremens," and violent actions, as the effects judged to be caused by high and constant consumption of alcohol, the enchichados tended to become passive, pathetic, and lazy people; they were "harmless, but useless" members of society.[51] Zerda calculated an average of 7 percent alcohol content in chicha, but considered the physical and mental degenerations of the enchichados to be related to the toxin identified as ptomaine.

As in Europe, diverse discourses of "degeneration" were (re)emerging in Latin America at the end of the nineteenth century. Besides climatic determinism, which was, since the eighteenth century, a central argument for the alleged racial inferiority of the inhabitants of tropical Latin America, at the end of nineteenth century other scientific theories were structuring the debates about progress, civilization, and degeneration.[52] Interestingly enough, for the Colombian case—and especially for the Andean region around Bogotá—these debates, which were later articulated with the local eugenics movement, underwent a turning point around chicha and urban life.[53] Indeed, at the crossroads of personal experience, organic chemistry, experimental physiology, and clinical medicine—and the material conditions that allowed these practices—the toxicological life of urban chicha emerged; this life was undoubtedly related to notions of racial degeneration. But in the interstices of the chemical study of chicha, we can also trace the forgotten material life of an urban product that, nevertheless, persists in our memories today as an artisanal beverage produced and consumed in popular places of sociability and political action called *chicherías*, which were systematically condemned by local elites.[54] As we will see in the next section, the toxicological life of chicha, and its relationship

---

[48] The idea that chichismo was a specific pathological state was, however, questioned by other physicians, who argued that patients, like the ones observed by Gómez and Zerda, were actually suffering from a mixture of alcoholism and a local type of pellagra (a disease believed at that time to be caused by the consumption of spoiled maize).

[49] Zerda, "Estudio químico, patológico" (cit. n. 3), 36.

[50] For the process of the medicalization of alcoholism in Colombia at the turn of the twentieth century, see María Fernanda Vásquez, "Degeneración, criminalidad y heredo-alcoholismo en Colombia, primera mitad del siglo XX," *Saúde e Sociedade* 27 (2018): 338–53.

[51] Luís E. García, *Diagnóstico diferencial entre el alcoholismo crónico y el chichismo* (Bogotá: Fernando Pontón, 1889), 40. García was a medical student working with Zerda and Gómez.

[52] The circulation and appropriation of August Morel's "theory of degeneration" in the fields of legal medicine and psychiatry in Colombia (1888–1950) has recently been analyzed by Vásquez-Valencia, "Degenerados, criminosos e alienados" (cit. n. 9).

[53] In the first decades of the twentieth century, geographical, anatomical, evolutionary, hereditary, physiological, nutritional, and statistical arguments were systematically used to debate the possible "racial problems" of poor Colombian populations, resulting in proposals for a series of social engineering and eugenic measures. For recent historiographical discussions on this, see Catalina Muñoz, "Estudio introductorio: Más allá del problema racial; El determinismo geográfico y las 'dolencias sociales,'" in *Los problemas de la raza en Colombia: Más allá del problema racial; el determinismo geográfico y las "dolencias sociales,"* ed. Muñoz (Bogotá: Editorial de la Universidad del Rosario, 2011), 11–58; and Pohl-Valero, "'La raza entra por la boca'" (cit. n. 6), on 457–60.

[54] For historical analysis of "chicherias" as popular places of chicha and food consumption and political sociability, and as sites selected for police control, see Mauricio Archila, "El uso del tiempo libre de los obreros," *Anuario Colombiano de Historia Social y de la Cultura* 18–19 (1990–91): 145–84;

with notions of degeneration, was inseparable from the massive urban production of this beverage carried out in factories, which were also called chicherías.

THE URBAN AND INDUSTRIAL LIFE OF CHICHA

In the same period as the material infrastructure and epistemic virtues for the making of science were changing and affecting Zerda's research, something similar was happening to the production of chicha. Since colonial times, the indigenous production of chicha (made with native maize and water) had been modified by the introduction of sugar cane in the region. As the German traveler Alfred Hettner observed in 1888, in most parts of the Colombian Andes, cane juice had become an "essential ingredient" for chicha production.[55] Of course, this situation was possible only after the arrival of sugar cane to the region as a result of the "Columbian Exchange." The use of this new ingredient may have been designed to facilitate the corn fermentation and to increase the alcohol content of chicha. But it can also be related to local geographical characteristics. It is interesting to observe that nineteenth- and early twentieth-century descriptions of chicha production in the Andean regions of Ecuador, Peru, and Bolivia seldom mentioned the use of cane juice.[56] The eastern region of Bogotá is surrounded by hot, low-lying valleys, which made the extensive cultivation of sugar cane near this city possible. As mentioned by Colombian lawyer and traveler Phanor Eder, this region of Cundinamarca was one of the Colombian places where "the greatest area of cane" was under cultivation "for the purpose of making not sugar but syrup," which was then "fermented with corn and called *chicha*."[57]

As Rebecca Earle has recently argued about the Columbian exchange, "the process of culinary transformation that occurred in the centuries after 1492 took place in the context of hierarchical colonial and postcolonial power structures that valorized certain peoples and their foods, and disdained others."[58] Certainly, chicha was part of this process of cultural valuation. The negative perception that many elite people had about chicha had been structured around moral, economic, racial, gendered, and hygienic ideas and arguments since the colonial era.[59] To some extent, the material transformations in the production of chicha also helped shape these discourses. For example, Zerda believed that during the colonial period, the use of cane juice had made indigenous

Alberto Mayor Mora, "24 horas en la vida de dos artesanos de 1914," *Revista Colombiana de Sociología* 2 (1994): 15–58; and Max S. Hering Torres, "Chicha, bailes y tropeles: Cultura denunciante," chap. 3 in *1892: Un año insignificante; Orden policial y desorden social en Bogotá de fin de siglo* (Bogotá: Crítica, 2018).

[55] Alfred Hettner, *Reisen in den columbianishen Anden* (Leipzig: Duncker & Humblot, 1888), 87.

[56] J. B. Avequin, "On la Chicha, el Flangourin, el Pulque," *Chemist* 5 (1857–8): 288–93; Manuel A. Velásquez and Ángel Maldonado, "Contribución al estudio del maíz y de la chicha de maíz," in *Contribución al estudio de la materia médica peruana: Fascículo I*, ed. Velásquez and Maldonado (Lima: Sanmarti, 1922), 9–70; Hugh C. Cutler and Martin Cardenas, "Chicha, A Native South American Beer," *Botanical Museum Leaflets, Harvard University* 13 (1947): 33–60.

[57] Phanor James Eder, *Colombia* (London: T. Fisher Unwin, 1913), 148.

[58] Rebeca Earle, "The Columbian Exchange," in *The Oxford Handbook of Food History*, ed. Jeffrey M. Pilcher (Oxford: Oxford Univ. Press, 2012), 341–57, on 349.

[59] On the negative and positive perceptions about chicha by late colonial authorities and physicians, see Adriana Alzate Echeverri, "La chicha: Entre bálsamo y veneno; Contribución al estudio del vino amarillo en la región central del Nuevo Reino de Granada, siglo XVIII," *Revista Historia y Sociedad* 12 (2006): 161–90.

chicha become "unnerving and harmful."[60] He also believed that urban production of chicha, "as it is produced today," was an important aspect to be considered by authorities.

For this reason, Zerda gave a detailed description of, and comparison between, the production of chicha at the beginning and end of the nineteenth century, and then discussed urban and rural production methods. For the early nineteenth-century chicha, he used the detailed information given by Merizalde in 1828. According to Merizalde, chicha was produced in three general steps: soaking, grinding and cooking, and fermenting. For fifteen days, soft yellow maize grains, coarsely ground, were soaked in a mixture of water and a small amount of cane juice. In stage two, the resulting dough was ground to produce a "masato," which was wrapped in *payaca* leaves and cooked for twenty-four hours. With stage three, this cooked masato, mixed with more cane juice and water, was left in a cold place for three days and then mixed with larger volumes of water and cane juice. After six days of fermentation, the beverage was ready to consume. Merizalde wrote that 17.5 kilograms of maize and 220 kilograms of cane juice were required to produce 485 liters of chicha. The maize grinding was done on flat stones, and the soaking and fermentation processes took place in clay pots ("múcuras") with a capacity of fifty-four liters.[61]

According to Zerda, this procedure had survived in rural areas and small towns, but in big cities such as Bogotá, it had changed drastically. He visited the "chicheria or chicha factory" of "N.," since a detailed "description of the procedure for the fabrication of chicha" was not only a "curious story, but a very important aspect from a scientific point of view."[62] Zerda reported that instead of clay pots, this factory used big wooden barrels with iron bottoms, with a capacity of two hundred and fifty to three hundred liters. Also, yellow maize had been replaced by a white variety called *yucatán*. The soaking process now took longer (eighteen days or more), and it was only made with water—no cane juice was added. For this process, six wooden barrels were used. One hundred and twenty-five kilograms of maize mixed with water were added to the first of these, and after three days the content was transferred to a second barrel. This procedure was repeated six times, facilitating a continuous production of large amounts of maize dough in an operation called "cambio." The dough was then ground several times to produce masato. Blocks of one kilogram of masato were then wrapped in payaca leaves and cooked in a special oven for forty-eight hours before being ground and sieved with water and cane juice, producing a liquid mass that was left to ferment for twenty-four hours. The fermented masato was then placed into wooden barrels, mixed with water and cane juice, and left for four to five days to ferment. Sometimes, payaca leaves were added to the barrels to aid fermentation. The resulting product was the "chicha flor" mentioned earlier.[63]

As in the laboratory, the urban factory of "N." deployed diverse instruments, substances, and tacit knowledge for achieving a complex process of maize and cane juice fermentation. In a sense, Zerda translated the chemical laboratory to this specific setting, carrying out several temperature and density measurements on the masato during different stages of production. The sense of smell was central. According to Zerda,

---

[60] Zerda, "Estudio químico, patológico" (cit. n. 3), 8.

[61] Merizalde, *Epitome de los elementos* (cit. n. 29), 357–60. Merizalde used colonial and indigenous weight and volume units such as "almudes," "totuma," "medical pounds," "votijas," "carga," and so forth, interchangeably. For clarity, I have converted these units into present-day metric ones.

[62] Zerda, "Estudio químico, patológico" (cit. n. 3), 8.

[63] Ibid., 9–11.

during the process of cambio, a "spontaneous" alcoholic fermentation took place, detectable by the smell and heat coming out of the barrels. In the first barrels the smell was "alcoholic and characteristic of chicha," but later, the smell was "acetic," and finally it became "putrid." From barrel to barrel, the temperature increased from 24°C to 46°C (the ambient temperature being 16°C).[64] The producers of chicha undertook similar observations and perceptions. For example, the addition of payaca leaves to aid fermentation was, arguably, informed by close inspection of the smell and temperature of the masato. But in this case, these procedures were intended not to follow the toxic alkaloid mentioned earlier, but to produce a specific and complex substance (chicha flor) that was in high demand, and in massive quantities, by the sellers of "second-class chicha."

A central argument to explain the existence of chicha's toxin was the use of yucatán maize. Zerda claimed that it was during the soaking process, which was intended only to soften the yucatán maize grains, that ptomaine was produced.[65] The reasons why chicha factories began to use this type of maize are unknown. Later commentators mentioned that "yucatán maize" was the "final product of a succession of hybridizations," and probably of recent origin in Colombia. It was a type of maize mainly cultivated in the "upper Magdalena river valley" to the southwest of Bogotá, in a region of intermediate elevation (350 to 1350 meters above sea level).[66] Some agronomists also mentioned that this type of maize was very "desirable due to its hard cover that resists winter."[67] Although, in 1828, Merizalde had recommended that yucatán maize should not be used for chicha manufacture,[68] for four decades, starting at the end of the nineteenth century, this type of maize was the one used in Bogotá's chicha factories. Consequently, this change may be related to both agricultural and urban transformations; the characteristics and availability of this type of corn may have favored its use in the context of growing demand for chicha production in the city of Bogotá.

At the end of the nineteenth century Bogotá was transforming from a colonial city into a modern one. Besides a relatively rapid demographic increase (from about 35,000 inhabitants in the 1820s to about 90,000 in the 1880s), the city experienced many changes in its administrative and spatial organization.[69] In relation to the production of chicha, these transformations included the implementation in 1882 of a new food supply system ("sistema de abastos"), which attempted to manage increasing urban demand for food products and to organize—and in some cases municipalize—the marketplaces of Bogotá. The "Plaza of Nariño" was designated for the sale of maize and cane juice, among other products such as wood, coal, and cattle.[70] This marketplace was located in the western part of the city, in the district or "parroquia" of San Victorino, where most of the city's food supplies arrived.[71] The "western road," which connected

[64] Ibid., 12–13.

[65] Ibid., 11–13.

[66] L. M. Roberts, U. J. Grant, Ricardo Ramírez, W. H. Hatheway, and D. L. Smith, *Races of Maize in Colombia* (Washington: National Academy of Sciences, 1957), 70 and 95–6.

[67] Bernardino Marín, *Cartilla agraria colombiana* (Bogotá: Imprenta Eléctrica, 1914), 68.

[68] Merizalde, *Epitome de los elementos* (cit. n. 29), 358.

[69] Germán Mejía Pavony, *Los años del cambio: Historia urbana de Bogotá, 1820–1910* (Bogotá: CEJA, 1998), 247.

[70] William García Ramírez, *Plaza Central de Mercado de Bogotá: Las variaciones de un paradigma, 1849–1953* (Bogotá: Editorial Pontificia Universidad Javeriana, 2017), 50.

[71] Sandra Jinneth Sabogal Bernal, *Imagen y memoria de la transformación urbana de San Victorino* (Bogotá: Universidad Nacional de Colombia, 2013), 23–31.

Bogotá with the upper Magdalena river valley, where maize yucatán and cane juice were cultivated and produced, ended precisely at this market.

For a population of approximately 100,000 inhabitants in 1891, Bogotá had about 120 chicherías, which produced and sold a total of some 25,000 liters of chicha every day.[72] The historical sources that registered this number of chicherías did not distinguish between places of production and places of consumption. But, as already mentioned, the word chichería was used both for factories where "chicha flor" was produced, and for places where "second-class chicha" was made and sold. In any case, we can roughly calculate that the city required a daily supply of 920 kilograms of maize, 8,800 kilograms of cane juice, and 100,000 liters of water to produce chicha and feed a large sector of the population whose diet was based mainly on this beverage. In this context of urban transformations, the chicha factories were able to obtain the huge supplies they required. Arguably, the Plaza de Nariño was the main source of maize and cane juice, and the new "Plaza de Maderas" was probably the source of wood for cooking the masato. Given the lack of an aqueduct system, the required water may have been obtained directly from the rivers that crossed the city. As we will see, just as Bogotá's food supply system was increasingly regulated by the municipal authorities, the diet of its inhabitants also began to be an object of scientific interest and social concern.

THE NUTRITIONAL LIFE OF CHICHA

In 1890, the railroad of "La Sabana" was inaugurated, connecting Bogotá with the western town of Facatativa (situated 39.6 kilometers away).[73] This place was a storage point for many agricultural products from the western region of Cundinamarca (the state surrounding Bogotá), which were then transported to Bogotá. The train had replaced the mule as the means of transport, bringing increasing amounts of maize and cane juice needed for chicha production to the city. A year before the railroad's inauguration, the press announced with pride that the first steam engine built entirely in Colombia had been completed; this took place in the "Pradera" iron factory. This machine was destined to operate a laminator to produce rails.[74] Indeed, a few years earlier, a locally built rail (made by the same iron factory) was exhibited in the streets of Bogotá for public entertainment for months. As I have analyzed elsewhere, the images surrounding the steam engine symbolized not only a desire for national progress and civilization through

---

[72] A map of Bogotá, produced in 1891, listed and located some 120 chicherías across the city. See C. Clavijo R., "Plano topográfico de Bogotá levantado por Carlos Clavijo, 1891," Archivo Museo de Bogotá. When the map was officially published in 1894, the list and locations of chicherías were suppressed. For details of this map, see Stefania Gallini and Carolina Castro Osorio, "Modernity and the Silencing of Nature in Nineteenth-Century Maps of Bogotá," *Journal of Latin American Geography* 14 (2015): 91–125, on 106–13. I have not found registers on the amount of chicha production and consumption at the end of the nineteenth century. However, in 1911, an engineer studying the water supply of the city reported that Bogotá had some 175 chicherías that produced 40,000 liters of chicha every day. He also mentioned that for producing 1 liter of chicha, 4 liters of water were required. See Cristobal Bernal, *Ensayo sobre abasto de aguas para Bogotá* (Bogotá: Imprenta Eléctrica, 1911), 8–9. Taking into account the number of chicherías and the population of Bogotá in 1891 and 1911, and the average production of chicha at the beginning of the twentieth century, I calculated that 25,000 liters of chicha per day were produced in 1891.

[73] República de Colombia, *Informe del Ministro de Fomento al Congreso de la República de Colombia en sus sesiones ordinarias de 1890* (Bogotá: Antonio María Silvestre, 1890), 15.

[74] Rafael Nieto París, *Riel de hierro y máquina de vapor: Fabricados en La Pradera* (Bogotá: Echeverría, 1889).

technology and industrialization, but also the beginning of a new and energetic frame-work for understanding social reality. As I have explained elsewhere: "Social thermody-namics . . . sought to restore the strength of an impoverished, indigenous and mestizo population that was consistently thought to be weak and racially inferior but capable of physiological and hereditary improvement. Such a regeneration, which would allow the achievement of national progress, was framed against a conceptual horizon that under-stood the human and social body in terms of a heat engine that transformed energy."[75]

From this energetic perspective, food, including chicha, began to be understood as the "fuel" required for the proper functioning of the human machine.[76] With the inten-tion of producing knowledge about the relationship between food, human bodies, and national progress, some local scientists began to collect information about the living conditions of the "working class." Falling under the rubric of the science and culture of modern nutrition,[77] these investigations produced a detailed account of the diet of the working class. Thanks to nutritional coefficients elaborated in the laboratory, these eat-ing habits were then translated into macronutrients; when compared to international standards, the values offered knowledge concerning the conditions of nourishment, and enabled dietary recommendations to be made.[78] This nutritional hygiene contrasted with the dietetics of the first half of the nineteenth century in several respects. As already mentioned, the "frugal" and plant-based diet of Andean populations was seen as a pos-itive aspect of the constitution of their bodies, since it favored their moral disposition to work, obedience, and peace. In the new nutritional framework, a qualitative apprecia-tion of a given food was displaced by quantitative measurements of its nutrients. A diet rich in proteins and calories (including abundant meat) became an indicator of progress and civilization—and the lack of these nutrients was perceived as a source of racial de-generation.[79] We can ask, then, how contemporaries interpreted the fact that the diet of those who were beginning to be called the "working class" was based mainly on chicha?

After Zerda's chemical study of chicha, some local scientists departed the confines of hospital and laboratory for the houses, canteens, and workplaces of the "working class" on the high plateau of Bogotá. For example, in 1893, physician Manuel Cotes conducted a survey of the living conditions of two hundred workers of this region.

[75] Pohl-Valero, "'La raza entra por la boca'" (cit. n. 6), 456.

[76] For the European context of the nineteenth century, it has been widely explored how the metaphor of the body as an engine ruled by the laws of thermodynamics entailed one of the essential features of modernity, modifying the concepts of the human organism (as a laboring system), food (as fuel), and industrial society. See, among others, Anson Rabinbach, *The Human Motor: Energy, Fatigue, and the Origins of Modernity* (Berkeley and Los Angeles: Univ. of California Press, 1992); Philipp Sarasin and Jakob Tanner, eds., *Physiologie und industrielle Gesellschaft: Studien zur Verwissenschaftlichung des Körpers im 19. und 20. Jahrhundert* (Frankfurt am Main: Suhrkamp, 1998); and Stefan Pohl-Valero, *Energía y cultura: Historia de la termodinámica en la España de la segunda mitad del siglo XIX* (Bogotá: Editorial Pontificia Universidad Javeriana, 2011). For the local appropriation of this metaphor in the Colombian context, see Pohl-Valero, "The Energetic Body: Machines, Organisms and Social Thermodynamics in Colombia's Path to Modernity," in *The Routledge History of Latin American Culture*, ed. Carlos Salomon (New York, N.Y.: Routledge, 2018), 225–42.

[77] Harmke Kamminga and Andrew Cunningham, eds., *The Science and Culture of Nutrition, 1840–1940* (Atlanta, Ga.: Rodopi, 1995).

[78] On nutritional standards, see Elizabeth Neswald, David F. Smith, and Ulrike Thoms, eds., *Setting Nutritional Standards: Theory, Policies, Practices* (Rochester, N.Y.: Univ. of Rochester Press, 2017).

[79] As Colombian physician Andrés Carrasquilla assured in 1889, the "physical and moral degener-ation of countries such as Ireland is related to the lack of meat in their diets, and the most civilized countries are those that consumed more meat." See Andrés Carrasquilla, *Atrepsia* (Bogotá: La Luz, 1889), 11.

These workers were classified into three categories ("road builders and carriers," "stone masons," and "day laborers"), and their daily diet was described in detail (table 1).[80] Between 50 and 78 percent (in weight) of what these people ate and drank every day was chicha.

Other studies also reported on the diet of the children of the working class. In general terms, children between 3 and 12 years old consumed a cup (110 cc [cubic centimeter]) of chocolate and a piece of bread (in the morning); a glass (110 cc) of chicha (at noon); a glass of chicha, a plate of mazamorra, and two cooked potatoes at 3:00 p.m.; and another glass of chicha and a piece of bread later in the afternoon.[81] Writing in 1892, from an economic perspective, the lawyer Ramón Vanegas Mora also described the amount of money that the working class of Bogotá and its surroundings spent on chicha. He drew up statistical tables of the salary, annual expenses, and alimentary consumption of three categories of workers, both male and female, classified according to income and work (agriculturists, masons, and different crafts). For example, those on an average salary of sixty cents a day had a diet consisting of a plate of maize porridge, a piece of bread, and a glass of chicha in the morning, at a cost of ten cents in a tavern; a plate of porridge, one or two servings of potatoes and rice, a piece of bread, a glass of chicha, and very occasionally a minimum amount of meat, at noon (fifteen cents); and at night, a glass of chicha and a piece of bread. In a higher income category, a working family composed of 5 members spent, on average, 1.40 pesos a week on chicha, and 7.55 pesos in total on food (including candles and coal). After meat, chicha was the highest food expense.[82]

Alongside all this information from the field, some physicians resorted to the laboratory to produce nutritional knowledge. As mentioned by Andrés Carrasquilla in 1889, to translate food into its nutritional components, and "to draw possible conclusions from them," it was necessary to conduct a chemical analysis of local foods in the laboratory.[83] Due to the lack of studies on these matters, physicians drew upon the values obtained in Europe for chocolate, bread, and maize (which served in the analysis of maize porridge). For chicha, Zerda's chemical study was used. Carrasquilla estimated that from the 330 cc of chicha they consumed every day, children obtained 6.63 grams of "nutritive materials" and 2.06 grams of protein.[84] Cotes was more precise when drawing up table 2.[85]

All these researchers replicated the idea of the "degenerative" effects of chicha. Cotes mentioned, for example, that "chicha contains, according to [Zerda's] analysis, nutritious principles which would make it acceptable [as food], . . . were it not for the toxic element which it contains."[86] The lawyer Vanegas also noted that "while chicha helps maintain strength, perhaps because of the substances it contains, [this beverage] also produces, as many say, the saddest brutalization [in its consumers]."[87] Nevertheless, in these works on the living conditions of the working class, the results of Zerda's laboratory analysis were mainly understood in nutritional terms, not in toxicological

[80] Cotes, *Régimen alimenticio* (cit. n. 4), 23–32.
[81] Carrasquilla, *Atrepsia* (cit. n. 79), 5.
[82] Ramón Vanegas Mora, *Estudio sobre nuestra clase obrera* (Bogotá: Torres Amaya, 1892), 23.
[83] Carrasquilla, *Atrepsia* (cit. n. 79), 6.
[84] Ibid., 7.
[85] Cotes, *Régimen alimenticio* (cit. n. 4), 34–5.
[86] Ibid., 14.
[87] Vanegas Mora, *Estudio* (cit. n. 82), 23.

Table 1. *Percentage of types of food and drink in daily diet of inhabitants of high plateau in Bogotá, Colombia, 1893. Reprinted from Manuel Cotes,* Régimen alimenticio *(cit. n. 4).*

Working category	Porridge (grams)	Meat (grams)	Bread (grams)	Chocolate (grams)	Hard liquor (grams)	Chicha (grams)
road builders and carriers	600	50	360	15	100	3575
stone masons	800	120	360	15	30	1300
day laborers (men)	600	–	360	40	–	3575
day laborers (women)	400	–	240	40	–	2275

ones. In their analysis, a glass of chicha was represented in terms of the quantity of carbohydrates, proteins, and minerals it contained. This was so much the case that, in the dietary regimes these doctors recommended to improve a diet they considered very deficient, chicha stood out; they recommended a daily ration of 300 grams for children and 1625 grams for adults.[88] The implementation of these "rational diets" in both rural and urban areas was portrayed as a social and economic imperative that governments should promote in order to increase the "productive power of the country" and avoid the "physiological degeneration" of the working bodies of the nation.[89]

Undoubtedly, chicha was central to the diet of a sector of Bogotá's population that began to be called "working class." This category was assigned to poor people (mostly indigenous and mestizos) employed in a variety of rural and urban work activities, including artisans, who earned a daily wage, rather than to industrial, salaried employees. "Our indios enchichados," as Gómez called the people he observed in the Charity Hospital, were part of this new social category. In this sense, these studies on the living conditions of Bogotá replaced the category of "enchichados" with that of "working class." In the same way, the nutritional meaning of chicha was privileged over the toxicological one, something replicated by some scientists in the following years. As physician José María Lombana Barreneche mentioned in 1907, the chemical study of "Dr. Zerda" offered a double meaning on chicha: "as a drink it has the same pernicious effects like any other alcohol; as food it has important qualities; it is the basis of the diet of our humble workers, . . . [who] transform it into the energy with which they cultivate our fields, or in other tasks." He also acknowledged that "toxic substances" could be produced during the soaking of maize, but that due to the cooking of the masato, any of these substances were destroyed. Furthermore, Lombana Barreneche stressed that the "limited intellectual development of our Indians," and of poor people in general, was not caused mainly by chicha consumption, but was more likely caused by their lack of education and terrible living conditions.[90]

From this socio-nutritional perspective, the extraordinary amount of chicha consumption began to be explained not only as result of an "inveterate and widespread

[88] Carrasquilla, *Atrepsia* (cit. n. 79), 30; Cotes, *Régimen alimenticio* (cit. n. 4), 45.
[89] Cotes, *Régimen alimenticio* (cit. n. 4), 41–2.
[90] José María Lombana Barreneche, "Correspondencia sobre la chicha," *Revista médica de Bogotá: Órgano de la Academia Nacional de Medicina* 27 (1907): 359–61, on 360–1.

Table 2. *Amount (in grams) of nutrients, minerals, and alcohols contained in daily consumption of chicha by different types of high plateau workers in Bogotá, Colombia, 1893. Reprinted from Manuel Cotes,* Régimen alimenticio *(cit. n. 4).*

Working category	Daily consumption of chicha (grams)	Grams of proteins	Grams of carbohydrates	Grams of fats	Grams of minerals	Grams of alcohols
Road builders and carriers	3575	22.08	48.08	–	8.93	228
stone masons	1300	6.25	17.19	–	3.25	83.20
day laborers (men)	3575	22.08	48.08	–	8.93	228
day laborers (women)	2275	14.26	30.55	–	6.20	152

custom,"[91] weak morals, or a generalized "repulsion" for water among poor people, but also as a consequence of their poor dietary conditions.[92] In the absence of a nutritious diet, the working body sought a substitute in the affordable chicha. The elite's desired modern and industrial Bogotá should be inhabited by healthy, productive, and disciplined bodies, and this required an urban, rational administration based on, among other things, an adequate food supply system and public health regulations. In connection with the development of industrial food production, laboratory practices, urban transformations, and modern bodies, the scientific lives of chicha were now deployed within a new set of meanings. They represented both a cause of racial degeneration, and a source of cheap calories that could help counteract this very same biological process. In the next decades, these tensions around the meanings of modern chicha would intensify in a context in which a tax on chicha production became one of the main income sources of the city administration.

THE SCIENTIFIC LIVES OF CHICHA IN THE FIRST DECADES OF THE TWENTIETH CENTURY

In 1899, the physician José Joaquín Serrano did a study on the hygienic conditions of Bogotá. The mortality rate, especially among children, was very high. At that time, the science of bacteriology was shaping a new model of public hygiene in Colombia, based on a fight against microorganisms and the use of vaccines. Serrano's work reflected this model of public hygiene.[93] He was also very concerned with the social aspects of health such as living and dietary conditions. Drawing on the work of Zerda and Gómez, Serrano pinpointed chicha as a possible public health issue that should be managed by market hygienic regulations. Furthermore, he recommended to municipal authorities the creation of an urban board of hygiene, a laboratory, and a group of sanitary inspectors who

---

[91] Zerda, "Estudio químico, patológico" (cit. n. 3), 34.

[92] Since colonial times, several observers had mentioned that "common people" preferred to drink chicha instead of water. The reasons adduced for this preference were multiple, ranging from popular beliefs to bacteriological accounts of the dangers of water consumption from the late nineteenth century onward. See Alzate Echeverri, "La chicha: entre bálsamo y veneno" (cit. n. 59).

[93] José Joaquín Serrano, *Higienización de Bogotá* (Bogotá: Zalamea, 1899).

would oversee, among other things, the production and sale of food in the city. These inspectors would exhort the authorities to take into account and apply the new bacteriological, chemical, and nutritional knowledge produced by local experts. Serrano stressed that this knowledge remained in medical journals, and had not reached the public.[94] Effectively, the regulation of chicha, and alimentary regulations in general, seemed either too difficult for municipal authorities to execute, or of little interest to politicians. The production and consumption of chicha seemed not to have changed at all during the final years of the nineteenth century; the eating habits of the working class also seemed not to have changed. In fact, city authorities provided a daily ration of three glasses of chicha and two cups of porridge, in addition to a small quantity of potatoes, panela, and meat to the prisoners of Bogotá's Panopticon (prison).[95]

In the next three decades, many of Serrano's recommendations were gradually implemented in Bogotá. A municipal hygienic board was created, and its inspectors regularly visited food markets and slaughterhouses. Many samples of food products began to be analyzed in the municipal laboratory, and places of its production and sale had to follow hygienic standards, at least on paper. Nutritional and temperance campaigns were rolled out in public schools and poorhouses, and the dietary conditions of the working classes were further analyzed.[96] With respect to chicha, since the 1910s, the chicherías (as places of both production and sale) required a "health patent" that had to be annually renewed for continued operation. In one specific sector at the city center, chicherías were forbidden with the intent to maintain public order and avoid scandals and bad smells.[97] The municipal council of Bogotá resorted to Zerda's chemical analysis and assured in 1916 that, with the "racial degeneration" caused by alcohol abuse, chicha consumption was especially worrying: "the fabrication of chicha is very unclean, and it produces terrible toxins, coming from the ptomaines originated by the series of fermentations to which maize is submitted, with serious damage to the health of consumers and to the sanitary state of the city."[98] Consequently, the next year thirty-two chemical analyses of chicha were carried out, and one hundred and ninety-four health patents were granted to chicherías. However, the presence of ptomaine was not reported in these analyses.[99]

Indeed, the very existence of ptomaines in the production of chicha was questioned at that time by some local scientists. In a textbook used for teaching food chemistry in the normal schools of Colombia, chemist Rafael Zerda Bayón argued that the chemical

[94] Ibid., 10.

[95] Salomón Higuera, *Observaciones sobre la alimentación suministrada a los presos* (Bogotá: Torres Amaya, 1892), 6.

[96] Zandra Pedraza, *En cuerpo y alma: Visiones del progreso y de la felicidad* (Bogotá: Ediciones Uniandes, 1999); Ingrid Johanna Bolívar, "Discursos estatales y geografía del consumo de carne de res en Colombia," in *El poder de la carne: Historias de ganaderías en la primera mitad del siglo XX en Colombia*, ed. Alberto Flórez-Malagón (Bogotá: Editorial Pontificia Universidad Javeriana, 2008), 230–89; Pohl-Valero, "'La raza entra por la boca'" (cit. n. 6).

[97] For details on the hygienic campaigns against chicha in Bogotá between the 1910s and 1930s, see Isaza and Granados, *La ciudad* (cit. n. 19), 107–51.

[98] Consejo de Bogotá, "Acuerdo 14 de 1916 Por el cual se provee a la moralidad y salubridad de la capital de la República por medio de la reglamentación del expendio de bebidas embriagantes y se dictan otras medidas sobre el alcoholismo," Régimen legal de Bogotá D.C., Secretaría Jurídica Distrital de la Alcaldía Mayor de Bogotá D.C., accessed 19 February 2020, http://www.alcaldiabogota.gov.co /sisjur/normas/Norma1.jsp?i = 8983&dt = S.

[99] Cenon Solano, "Informe del Director de Higiene y Salubridad al honorable Consejo Municipal," *Registro Municipal de Higiene* 7 (1918): 1307–18, on 1310 and 1317.

analysis of chicha produced in the past could not give "conclusive results," because there was "no uniformity in its preparation." This situation, stressed Zerda Bayón, occurred in one and the same chichería, and even more if all chicherías were considered as a whole. This chemist also did not accept the physiological experiments carried out on animals injected with the chicha toxin, since these animals did "not have the habit of consuming chicha." For Zerda Bayón, chicha was a very "nutritious food," and the duty of the authorities should be "to control the manufacture of chicha," but not "to present obstacles to the development of a traditionally national industry."[100]

Meanwhile, municipal authorities began to collect a tax on chicha, first on sales, and later on production.[101] The reasons for this tax may be understood as either a hygienic measure to try to reduce chicha consumption (as proposed by some hygienists), or as an administrative procedure to generate municipal revenues on a widely consumed product (as assumed by local authorities).[102] Either way, during the 1930s, the principal monetary income for the Department of Cundinamarca came from the chicha tax. As its general administrator stated in 1939, each liter of chicha paid 2 cents of taxes, producing an overall income of 1.7 million pesos for this item during the year 1938. From the 85.5 million liters of chicha produced that year in Cundinamarca, Bogotá produced 29.7 million for a population of about 400,000 inhabitants.[103] Seven factories produced this amount of chicha, and the product was then distributed by trucks to the thousands of chicherías across Bogota.[104] A few years earlier, the municipal public health system of Bogotá had been reorganized, creating a sanitary board ("Consejo de sanidad") and five departments of "Epidemiology, Statistics, Laboratories and Propaganda," "Social Protection and Security," "Sanitary Police and Food Inspection," "Child Protection," and "Sanitary Engineering."[105] This health system's operation depended, to a large extent, on the tax incomes that the city produced.

José María Barriga Villalba, the director of the municipal laboratory at that time, carried out several chemical analyses on chicha, but he found no traces of ptomaine or other toxic substances. According to him, the increasing demand for chicha in Bogotá had transformed its production processes, in turn preventing the final product from containing ptomaine: "The modern desire to obtain fast profits, the rise in labor costs, and the increase in consumption, are the only agents that, by their nature, have caused a favorable modification in the hygienic production of this beverage; in the

[100] Rafael Zerda Bayón, *Química de los alimentos, adaptada a las necesidades económicas e higiénicas de Colombia* (Bogotá: Imprenta del Comercio, 1917), 359–65.

[101] For a compilation of the decrees, laws, and norms on chicha taxation in Cundinamarca between 1916 and 1934, see República de Colombia, Departamento de Cundinamarca, *Renta de fermentadas* (Bogotá: Asamblea de Cundinamarca, 1934).

[102] As Henry Yeomans has recently argued, "the use of excise taxation in contemporary Western societies is marked by the curious coexistence of the state's fiscal objective of raising revenue with often-articulated behavioral objectives relating to lowering or altering public consumption of certain commodities." See Henry Yeomans, "Taxation, State Formation, and Governmentality: The Historical Development of Alcohol Excise Duties in England and Wales," *Soc. Sci. Hist.* 42 (2018): 269–93, on 269.

[103] Alejandro Hernández Rodríguez, *Memoria: Administración General de Rentas de Cundinamarca* (Bogotá: Litografía Colombia, 1939), 64–5. The other departmental incomes for 1938, related to consumption products, came from distilled liquors (900,000 pesos), beer (800,000 pesos), and tobacco (1.3 million pesos).

[104] Ibid., 262.

[105] Manuel Antonio Rueda, "Informe del señor Director de Higiene municipal de Bogotá," *Registro Municipal* 55 (1935): 19–30, on 19.

present-day procedure, the formation of ptomaine is very difficult."[106] The "industrial" production of chicha in those years used "white corn flour, yucatán," "cement hoppers" for the maceration of the flour, "autoclaves" to cook the masato, "mills" to turn the cooked masato into a dough, and "tanks" for the fermentation process. In addition to cane juice, the masato was mixed with "guarapo" (fermented cane juice). The whole process took twelve days.[107]

Barriga also stressed that hygienists should not address chicha consumption purely from an "alcoholic perspective," but principally from a nutritional one. He calculated the amount of energy, measured in calories, that the "human machine" needed to perform work in Bogotá. According to him, under the climatic conditions of high altitude (Bogotá is 2600 meters above sea level) and lack of oxygen, the local working body required a diet rich in carbohydrates to achieve an energetic equilibrium and avoid physiological degeneration.[108] In that sense, chicha was a "first-class thermogenic food," and its consumption was the "natural result" of the human organism trying to adapt to the atmospheric conditions of the high plateau of Bogotá.[109] These asseverations made by Barriga in 1937 triggered a heated debate among medical experts about the scientific criteria that should be used to address the chicha question. As the hygienist and chicha opponent Jorge Bejarano declared, "the chemical arguments [adduced as favorable to chicha] cannot, on this occasion, guide the hygienic and social criteria" for addressing national problems.[110] Bejarano rejected the idea that the consumption of chicha should be understood as a "physiological need" to meet the lack of calories from workers' diets, and appealed to statistics from hospitals, mental institutions, and the Legal Medicine Office of Bogotá in order to demonstrate that chicha consumption was related to high rates of criminality, death, and even congenital birth abnormalities. The response of Barriga was that these figures were related to "multiple causes" impossible to reduce to a single factor.[111]

Although Bejarano's scientific authority in the field of public health was widely recognized at that moment,[112] some of the first official publications of the ministry of hygiene, created in 1938, reproduced and reinforced the nutritional ideas of Barriga about chicha. For example, in 1939, this ministry published and widely distributed a booklet about the "nourishment of the working class of Bogotá"; after a dietary survey of more than two hundred and twenty-five families, the average amount of calories, nutrients, and vitamins consumed was calculated. After explaining the experiments done

[106] Barriga Villalba, "Nueva discusión" (cit. n. 5), 59.

[107] "Procesos seguidos actualmente en la fabricación de la chicha," *Revista de Higiene: Órgano del Departamento Nacional de Higiene* 18 (1937): 53.

[108] For a discussion of the local debates about racial degeneration and high-altitude physiology, see Stefan Pohl-Valero, "¿Agresiones de la altura y degeneración fisiológica? La biografía del 'clima' como objeto de investigación científica en Colombia durante el siglo XIX e inicios del XX," in "Historias alternativas de la fisiología en América Latina," ed. Pohl-Valero, special issue, *Revista Ciencias de la Salud* 13 (2015): 65–83.

[109] Barriga Villalba, "Nueva discusión" (cit. n. 5), 60.

[110] Jorge Bejarano, "Química y chichismo," *Revista de Higiene* 18 (1937): 62–4, on 64.

[111] Antonio María Barriga Villalba, "Nueva defensa de la chicha," *Revista de Higiene: Órgano del Departamento Nacional de Higiene* 18 (1937): 65–9.

[112] Bejarano was the president of the Organizing Committee of the Tenth Pan American Sanitary Conference, held in Bogotá in 1938. Years later, in 1947 and between 1948 and 1949, he was the minister of hygiene. For details of the scientific and political life of Bejarano, see Rodrigo Ospina Ortiz, "Jorge Bejarano: Un intelectual orgánico del Partido Liberal, 1888–1966" (master's thesis, Universidad Nacional de Colombia, 2012).

by Zerda in 1889 on the toxicity of chicha, physician Francisco Socarrás, the author of the study, stressed that the "recent works done by Dr. Barriga show that in this beverage, as it is produced today, the toxic principles studied by Zerda are not found."[113] Furthermore, Socarrás declared that chicha should be understood as food, and not just as an alcoholic beverage, and should therefore be included in their nutritional calculations. On average, the caloric intake of the working class of Bogotá doubled, if chicha consumption was considered, from a daily amount of 2094 calories to 4161.[114] The recommendations for a "rational diet" of 3132 calories per day, at a cost of 38 cents, included, besides 70 grams of meat, 1 egg, and 200 grams of milk, the consumption of 1 liter of chicha.[115] As stated by the ministry of hygiene, this study was very important to "make understandable to the public the basis of the biological policy that the National Government is carrying out."[116]

For municipal administrators in the 1930s, besides the toxicological, hygienic, and/or nutritional meanings of chicha, studying this beverage entailed additional tasks that were central to their duties. For example, they had to ensure that the alcohol content of chicha did not exceed 5 percent by volume, yet they stressed that the fermentation process continued after chicha left the factories for its distribution. Consequently, these administrators pushed for clarification of the hygienic regulation, asking exactly when and how its alcohol content should be measured.[117] The prohibition of chicherías in certain parts of the city and the hygienic campaigns against chicha were also a matter of concern. From the municipal administrative point of view, these measures encouraged the clandestine production and distribution of this beverage, causing serious problems for tax collection.[118] Furthermore, the chicha industry had acquired considerable economic and political power, and was able to influence the decisions of the Municipal Council of Bogotá on the regulation of chicha. In fact, in 1936, a political scandal involving several municipal administrators and chicha industrialists came to public light. The main chicha factories of Bogotá had organized a consortium that bribed several municipal authorities to control the collection of taxes on chicha, and to award health patents to chicherías.[119]

CONCLUSION

With the changing industrial production of chicha and its interconnections with different ways of knowing the nature of this beverage—and its effects on the workers' bodies that consumed it—this article sheds light on the complex and contingent articulations between science, experts, and the governance of food in a specific historical context. By following, at the same time, the toxicological, industrial, and nutritional lives of chicha, I present a historical narrative that tries, on the one hand, to go beyond a schematic analysis of food biopolitics; and on the other, to overcome disciplinary

---

[113] José Francisco Socarrás, *Alimentación de la clase obrera en Bogotá* (Bogotá: Imprenta Nacional, 1939), 23.

[114] Ibid., 32.

[115] Ibid., 78.

[116] Ibid., 3.

[117] Enrique Marroquín, *Rentas de Cundinamarca: Informe 1942* (Bogotá: Imprenta Nacional, 1943), l–li.

[118] Ibid., xlvi.

[119] "La detención del ex-secretario de Hacienda," *El Tiempo* 17 (23 April 1936), 1 and 7.

histories of science and the way historians usually compartmentalize scientific activities of the past.

In the first place, the knowledge claims about chicha were paradoxical. This food product was regarded both as a "poison" that was causing racial degeneration in poor, urban populations; and also as a nutritional answer to this same biological condition. In addition to the cultural valuations about chicha that the historical actors of this case study may have made, these meanings were also shaped by the historically situated activities of doing science and producing chicha. At the end of the nineteenth century, clinical observation, laboratory experimentation, and fieldwork were articulated in different ways by different physicians in order to understand—and define—the chicha question. This question was framed in an incipient process of urban modernization and industrialization of Bogotá, which in turn was transforming the scale, ingredients, techniques, and devices to produce chicha. The establishment of taxes on chicha in the state of Cundinamarca in the 1910s, as well as the deployment of practices of governance over this product, structured additional knowledge claims about chicha and helped configure new forms of food expertise. In the 1930s, hygienists and chemists, but also administrators and even industrialists, mobilized different arguments and influences in order to define the adequate criteria to address the chicha question. Interestingly enough, the reason given for the nontoxicity of chicha at the end of the 1930s was not related to the achievements of food regulations and controls, but to the autonomous transformations of chicha's industrial production, which aimed at increasing productivity and economic profit. Similarly, if we believe the statistics of the time, chicha consumption patterns barely changed between 1889 and 1939.

In the second place, this case study may help us reconsider the way in which historians usually organize the activities of science in the so-called global south. Historian of British tropical medicine David Arnold argued some years ago about the need to take non-Western contributions to modern science seriously. In line with the developing field of postcolonial science studies, he proposed that instead of thinking of metropolitan nutrition science as imported to the British colonies at the beginning of the twentieth century, a metropolitan laboratory style and a colonial field style of nutritional research coexisted.[120] As Dana Simmons has further argued, "where metropolitan studies were concerned with managing health and productivity, colonial medicine sought to explain the causes of widespread death. Metropolitan nutrition science sought to optimize the body; colonial studies sought to identify its pathologies."[121] To be sure, at the end of the nineteenth century Colombia was not a colony, but neither is it historiographically considered a center for the making of scientific knowledge.[122] Beyond the discussion of whether local scientists produced genuine and original knowledge, I want to point out that this case study addresses the blurred historical boundaries between fieldwork and laboratory work, and between metropolitan and

[120] David Arnold, "The 'Discovery' of Malnutrition and Diet in Colonial India," *Indian Economic and Social History Review* 31 (1994): 1–26.

[121] Dana Simmons, "Starvation Science: From Colonies to Metropole," in *Food and Globalization: Consumption, Markets and Politics in the Modern World*, ed. Alexander Nützenadel and Frank Trentmann (Oxford: Oxford Univ. Press, 2008), 173–92, on 174.

[122] For a discussion of the influences and problems of the "applicability of the postcolonial studies label to Latin America," see Eden Medina, Ivan da Costa Marques, and Christina Holmes, eds., *Beyond Imported Magic: Essays on Science, Technology, and Society in Latin America* (Cambridge, Mass.: MIT Press, 2014), 3–6.

colonial science. Indeed, the scientific biography of chicha touches all these historio-graphical divisions between the medical, natural, and social sciences, without forcing us to separate them. In our case, metropolitan/colonial, laboratory/field, working body/diseased body, and natural science/social science dichotomies are much less sta-ble than usually assumed.

Historicizing "Indian Systems of Knowledge":
Ayurveda, Exotic Foods, and Contemporary Antihistorical Holisms

*by Projit Bihari Mukharji**

ABSTRACT

Some recent authors have argued that "Indian Systems of Knowledge," such as Ayurvedic medicine, cannot be historicized. They argue that Ayurvedic medicine must be understood as a "system" and with reference to its "metaphysical foundations." Food has often played an important part in these antihistoricist arguments about traditional South Asian medicines. In this article, I first describe and historicize these antihistoricisms by delineating both their colonial origins and their recent nationalist appropriations. I also argue that history of science needs to distinguish between different types of antihistoricisms emerging from different academic and political contexts. I then move on to show how food history actually can be deployed to subvert these antihistoricist claims. I pursue three interrelated inquiries to support my case. First, I demonstrate that the category of "food" is inappropriate for the textual heritage of Ayurveda, and that we need to be more sensitive to specific technical categories, such as *anupana*, *pathya*, and *dravya*, within which foodstuffs were accommodated. Second, I demonstrate that new foods, especially exotic New World foods, were absorbed into each of these technical categories recognized in Ayurveda. Finally, I show that these new foods did not simply leave the categories themselves untouched. The embodied experiences of the scholar-physicians' palates substantially transformed the allegedly disembodied, ahistorical categories they wrote about. I argue, then, that far from being an ahistorical fossil as the proponents of antihistorical arguments would have us believe, Ayurvedic medicine was a rich, heterogeneous, and historically dynamic tradition, and food history is singularly well placed to testify to that dynamism.

INTRODUCTION

History of science, it has recently been alleged, is incapable of studying the Ayurvedic medical tradition. "Non-Western systems," the argument proposes, cannot be "viewed

* History and Sociology of Science, University of Pennsylvania, 326 Claudia Cohen Hall, 249 S. 36th St., Philadelphia, PA 19104-6304, USA; mukharji@sas.upenn.edu.

An earlier version of this article was workshopped at the Sciences Beyond the West Reading Group at the University of Pennsylvania. I am grateful to the members of the group for the feedback I received on that occasion.

as historical or social artifacts."[1] They must instead be understood exclusively with reference to "their metaphysical and philosophical foundations."[2] Such antihistorical claims that deny the historicity of non-Western peoples and knowledges had once been the favored turf of European Orientalists.[3] In recent times, however, they have been rediscovered by South Asians themselves. Ayurveda has thus come to be increasingly framed as a unique and exemplary instance of a distinctive "Indian Knowledge System."[4]

There are two major motors for this rediscovery.[5] On the one hand, a strident Hindu nationalism has a visceral antipathy to acknowledging history. With the right wing nationalists capturing state power in India, such antihistorical espousals of an ancient, unbroken, and essentially high-Sanskrit foundation for all South Asian intellectual traditions have received vocal and lucrative state support.[6] On the other hand, a global market for Ayurvedic pharmaceuticals has produced an advertising discourse that harkens back to an ancient, unchanging tradition ad nauseam.[7]

In such denials of history, all kinds of difference are smothered and erased. It is not chronology alone that is denied. Geographic variations too are denied. Social exclusions are dismissively rejected simply as misplaced political correctness. Perhaps most importantly, any significant roles for people, ideas, and things entering South Asia from beyond its shores after the so-called metaphysical and philosophical foundations had been laid are entirely trivialized. Most ironically though, the very doctrinal differences that constitute the lifeblood of any thriving philosophical culture are negated by reference to a singular, ahistorical, and allegedly "foundational" doctrine.

What is most remarkable about these striking claims is neither their novelty nor their intellectual coherence. Rather, it is the fact that despite their seemingly obvious intellectual and political problems, such claims have now moved from the world of advertisers and politicians to the pages of respected academic journals. Indeed, the specific instance with which I commenced this discussion was in fact published in no less a venue than *Isis*. What is more, this new prestige for neo-Orientalist thinking arises not in the absence of serious historical scholarship on non-Western intellectual traditions—as one might say was the case in the age of formal empires of the nineteenth

---

[1] Sundar Sarukkai, review of *Doctoring Traditions: Ayurveda, Small Technologies, and Braided Sciences*, by Projit Bihari Mukharji, *Isis* 108 (2017): 935–36.

[2] Ibid.

[3] Eric R. Wolf, *Europe and the People without History*, 2nd ed. (1982; repr. with foreword by Thomas Hylland Eriksen, Berkeley and Los Angeles: Univ. of California Press, 2010); Thomas R. Trautmann, "Does India Have History? Does History Have India?," *Comp. Stud. Soc. Hist.* 54 (2012): 174–205.

[4] P. Ram Manohar, "Ayurveda as a Knowledge System," in *Indian Knowledge Systems*, ed. Kapil Kapoor and Avadhesh Kumar Singh (New Delhi: DK Agencies, 2005), 156–71. See also Makarand R. Paranjape, *Healing across Boundaries: Bio-Medicine and Alternative Therapeutics* (New Delhi: Routledge, 2015).

[5] David Hardiman, "Indian Medical Indigeneity: From Nationalist Assertion to the Global Market," *Soc. Hist.* 34 (2009): 263–83.

[6] Banu Subramaniam, "Archaic Modernities: Science, Secularism, and Religion in Modern India," *Social Text* 18 (2000): 67–86. See also Renny Thomas and Robert M. Geraci, "Religious Rites and Scientific Communities: *Ayudha Puja* as 'Culture' at the Indian Institute of Science," *Zygon* 53 (2018): 95–122.

[7] Maarten Bode, *Taking Traditional Knowledge to the Market: The Modern Image of the Ayurvedic and Unani Industry, 1980–2000* (Hyderabad: Orient Longman, 2008).

century—but rather in the teeth of a wealth of historical scholarship that has repeatedly demonstrated the inaccuracy of such neo-Orientalist frameworks.[8]

It is against this new respectability being given to intellectually dubious and politically dangerous antihistoricisms that I seek to write in this article. The history of food, I argue, is for a variety of reasons an excellent tool by which to challenge these neo-Orientalist postures. To begin with, food is one of the most concrete, ubiquitous, and conspicuous substances through which "difference" is constructed. The foods of no two historically distinctive societies are identical. Geographic and chronological differences produce endless variety and flux in what people eat. Approaching medicine, health, and therapeutics through food therefore provides one of the most redolent ways in which to map difference and change, and thus confound spurious antihistorical arguments. More importantly and ironically, however, food has come to be marshaled in a particularly powerful way by certain types of Ayurvedic antihistoricism. This is particularly the case where the antihistoricism lurks in the shadows of claims to "holism." Food history can therefore be a powerful and effective way to expose the politics implicit in such pretensions to holism.

This article will be divided into five sections. The first two of these delineate and unpack the new antihistoricisms against which I write, whereas the final three sections focus on how different aspects of food history can actually confound and expose the untenability and inaccuracy of these antihistorical narratives. In other words, I first describe the way new forms of antihistoricism have overdetermined the notion of an Ayurvedic "tradition" by mapping it onto two other notions of "systems" and "civilizations." Then, I demonstrate how claims about holism in these antihistorical narratives have mobilized food in their service. Having thus provided a map of what I seek to challenge, I proceed to food history per se. In section three I disaggregate the unified notion of "food" in Ayurveda, wherein I pay particular attention to three technical categories— namely, *anupana*, *pathya*, and *dravya*. In so doing, I also posit a critique of how we deploy the notion of tradition-as-system in the history of Ayurveda. Thereafter, in the next (fourth) section, I provide concrete cases to show how new foods coming from outside of South Asia were absorbed into each of these three particular categories of the Ayurvedic tradition. This will demonstrate that there is a historically shaped material culture of therapeutics that can be mapped without necessarily always referring to "metaphysical and philosophical foundations." In the final and fifth section, I push further and argue that these historically contingent material realities actually reshaped the very "theoretical" framework that informed the textual traditions of classification. Thus, far from history being held captive by the abstractions of metaphysics and philosophy, I show how metaphysical abstractions were deeply shaped by the concrete and historical sensory experiences of Ayurvedic scholars.

[8] Dagmar Wujastyk and Frederick M. Smith, *Modern and Global Ayurveda: Pluralism and Paradigms* (Albany: State Univ. of New York Press, 2008); Kavita Sivaramakrishnan, *Old Potions, New Bottles: Recasting Indigenous Medicine in Colonial Punjab (1850–1945)* (Hyderabad: Orient Longman, 2006); Rachel Berger, *Ayurveda Made Modern: Political Histories of Indigenous Medicine in North India, 1900–1955* (New York, N.Y.: Palgrave MacMillan, 2013); Bode, *Taking Traditional Knowledge* (cit. n. 7); Projit Bihari Mukharji, *Doctoring Traditions: Ayurveda, Small Technologies, and Braided Sciences* (Chicago: Univ. of Chicago Press, 2016); David Hardiman and Mukharji, *Medical Marginality in South Asia: Situating Subaltern Therapeutics* (Abingdon, UK: Routledge, 2013); Joseph S. Alter, *Yoga in Modern India: The Body Between Science and Philosophy* (Princeton, N.J.: Princeton Univ. Press, 2010).

ANATOMY OF AYURVEDIC ANTIHISTORICISMS

In 2003, the Indian Institute of Advanced Study at Shimla, one of the premier research institutions of the Indian state, organized a seminar on "Indian Systems of Knowledge." Its avowed intention was to reacquaint the academy with the "continuous and cumulative intellectual traditions of India" dating from the time of the earliest Vedic texts, which the organizers alleged had been entirely forgotten in modern academia. The first of the two volumes that were published from the seminar included three essays on medicine. All of them were on Ayurvedic medicine. No mention was made of any of India's other recognized nonbiomedical systems such as Unani Tibb, Siddha, or Sowa Rigpa. The reasons for this omission were not difficult to fathom. According to the volume, "Indian Knowledge Systems" were entirely identified with a classical Sanskrit textual corpus.

The first of the three essays informs us that "ancient Indian wisdom in medicine, like in many other fields of human endeavor, comes from the time honored Vedic Wisdom."[9] This is contrasted with Western medicine, regarding which we are told that "the history of medical thought in the West is a succession of errors in the ascending road of progress."[10] Even more explicitly, the author asserted that "modern medicine started five thousand years ago on the banks of the river Nile as magic, sorcery, witchcraft and mumbo-jumbo."[11] By contrast, the "science of Ayurveda" was apparently composed "10,000 to 15,000 years ago," though apparently "in one sense [it has] no beginning [i.e., it is eternal]."[12] Another author, in the same volume, is even more explicit in his antihistoricism. He alleges that "external frameworks of reference," rather than the "traditional viewpoint," have limited our ability to understand Ayurveda. By "traditional viewpoint," he signaled an exploration of the "epistemic foundations" of Ayurveda taken as a "knowledge system."[13]

The volume titled *Indian Knowledge Systems* is important because institutions directly connected to the Indian state sponsored its publication, and it engendered a programmatic statement. In the years since, however, such statements have gradually crept out of the repertoire of state propaganda and made their way onto forums enjoying greater critical credibility. One of the authors in the *Indian Knowledge Systems* volume has, for instance, repeated his claims about Ayurveda being "an example of a complete knowledge system"; he wrote this in a volume published by one of the leading global academic publishers.[14] He went on to add, "Ayurveda is a rather unique example of a knowledge system that developed in the Indian cultural milieu."[15] Going even further, he asserted that "Ayurveda is truly the most ancient model of integrative medicine" and that "the three-dimensional epistemological and ontological premises

---

[9] B. M. Hegde, "Modern Medicine and Ancient Indian Wisdom," in Kapoor and Singh, *Indian Knowledge Systems* (cit. n. 4), 141–56, on 142.

[10] Ibid., 144.

[11] Ibid., 141.

[12] Ibid., 142.

[13] Manohar, "Ayurveda as a Knowledge System" (cit. n. 4), 157.

[14] P. Ram Manohar, "Science, Spirituality and Healing: In Search of Definitions," in Paranjape, *Healing across Boundaries* (cit. n. 4), 14–27, on 20. For Manohar's article in *Indian Knowledge Systems*, see "Ayurveda as a Knowledge System" (cit. n. 4).

[15] Manohar, "Science, Spirituality and Healing" (cit. n. 14), 20.

on which Ayurveda has been erected as a knowledge system can help us sketch a panoramic view of reality."[16] The antihistoricism in the *Isis* book review with which I commenced this article has to be read in the context of this movement of fringe anti-historicisms into academically respectable forums.

Before challenging these claims, it is worth examining their anatomy in a little more depth. Two terms seem to repeatedly overdetermine the notion of an Ayurvedic tradition in these antihistorical claims. These are respectively the notion of a "system" and the notion of an "Indian civilization." In very different ways, both these terms allow the authors to selectively highlight some texts and some ideas, while ignoring other texts and other ideas.

Let us take the notion of a "system." As David Arnold points out, the idea that Ayurveda is a system has a nineteenth-century colonial genealogy. The framework derived principally from the influence of Linnaean botany upon the early European Orientalists who had taken up the study of South Asian medical traditions.[17] Later, South Asians themselves adopted the framework.[18] The usage, however, was not innocuous. It made very specific assumptions about the tradition. As Guy Attewell explains, "The use of the word 'system' . . . consolidates the impression of continuity, connoting internal coherence, discreteness, completeness, and homogeneity."[19] Rachel Berger has recently suggested that what is really necessary is for scholars to pursue the history of how the notion of Ayurveda being a system "was imposed upon Ayurveda as a way of easing its entry into formal politics."[20] The notion of Ayurveda being a system therefore not only creates a particular image of the tradition as a coherent, discrete, and internally homogenous entity, but it is also part of the colonial and postcolonial history through which Ayurveda has been inserted into a modern sphere of Hindu nationalism.

Above all, the notion of a system allows authors to draw idiosyncratic boundaries around the "system" without reference to the historical and lived contexts of authors or texts. This is where the notion of an Indian civilization acts in tandem with the notion of a system. Authors can choose what belongs to an "Indian civilization" or the relevant "metaphysical and philosophical background," and what does not.

The idea of "civilization," as Brett Bowden points out, is an "imperial idea" whose own genealogy is tightly bound up with the history of imperialism.[21] It allowed particular texts, ideas, ways of being, and pasts to be exalted above others. Attending to the imperial genealogy throws light on the ways in which the idea of a "civilization" always relies upon its opposite and hence is thoroughly relational; however, attending to the national histories of "civilization" allows us to appreciate the concept's function as a tool for disciplining, homogenizing, and marginalizing within the boundaries of the so-called civilization.

---

[16] Ibid., 21.

[17] David Arnold, "Plurality and Transition: Knowledge Systems in Nineteenth-Century India," Princeton History of Science Seminar, Princeton University, 23 October 2003, http://www.princeton .edu/~hos/Workshop%20I%20papers/Arnold%20History%20of%20Science%20paper.htm.

[18] Ibid.

[19] Guy N. A. Attewell, *Refiguring Unani Tibb: Plural Healing in Late Colonial India* (New Delhi: Orient Longman, 2007), 24.

[20] Berger, *Ayurveda Made Modern* (cit. n. 8), 36.

[21] Brett Bowden, *The Empire of Civilization: The Evolution of an Imperial Idea* (Chicago: Univ. of Chicago Press, 2009). For the national uptake of this imperial category, see David Ludden, "History Outside Civilization and Mobility in South Asia," *South Asia: Journal of South Asian Studies* 17 (1994): 1–23.

Why, for instance, does "Indian civilization" have to be founded entirely on the contents of the Vedic corpus? What about non-Vedic, non-Sanskritic pasts? How do those, such as Dalits, non-Hindus, women, and others who are structurally and historically distanced from these Vedic pasts, respond to these repeated espousals of the exclusively Vedic roots of the Indian civilization? Moreover, how and why should we assume that people living in vastly different places over several millennia read Sanskrit texts in exactly the same way? Even a cursory acquaintance with the *Charaka Samhita*, the oldest extant text of the Ayurvedic tradition, will demonstrate that the text itself is rife with multiple possible readings. Even key chapters present multiple rival views on particular technical points without attempting to reconcile everything.[22] Besides, there are serious differences, even at the level of basic understandings of the body, between the *Charaka Samhita* and the second oldest text, the *Susruta Samhita*.[23] Furthermore, the symbolic influence of these texts was hardly matched by their actual availability or accessibility. Like any other classic, they were claimed to be read considerably more often than they were actually read.[24] But the espousals of an "Indian civilization" and Ayurveda as a "knowledge system" allow such difficult questions to be easily jettisoned in favor of simple, selective, and superficially unifying criteria.

However, the question remains: Why are such intellectually and politically problematic antihistoricisms being entertained within critical academic circles? Within South Asia, of course, the powerful support of the state itself is a crucial factor. As Venera Khalikova has recently pointed out, there is a clear process of "Ayurvedicalization," where the state and its allied discursive apparatus is recasting all forms of nonbiomedical therapeutics in a narrow straitjacket of what it considers to be authentically part of the "Ayurvedic system."[25] Beyond South Asia, the ascendance of such antihistoricisms is more complex. Aside from the global play of populist identity politics, a tokenistic response from the Euro-American academic establishment in response to this populism is only part of the story. A more serious problem arises out of the difficulties in historical circles of distinguishing between different forms of antihistoricism.

The ontological turn and perspectivalism emerging mainly out of Latin America,[26] certain strands of indigenous studies attending to Native American epistemologies in North America,[27] and the Aboriginal experiences of Australia and New Zealand,[28] to name only three intellectual formations, have all produced their own versions of antihistoricism. Some of these have been thickly enmeshed in struggles for the empowerment of severely marginalized peoples, ontologies, and experiences. These antihistoricisms are quite different from those in contemporary India that resonate with strident majoritarian nationalism. Indeed, in India itself, from the anticolonial savant

---

[22] Mukharji, *Doctoring Traditions* (cit. n. 8).

[23] Dominik Wujastyk, *The Roots of Ayurveda: Selections from Sanskrit Medical Writings* (London: Penguin, 1998).

[24] Mukharji, *Doctoring Traditions* (cit. n. 8), 145–6.

[25] Venera R. Khalikova, "Institutionalized Alternative Medicine In North India: Plurality, Legitimacy, and Nationalist Discourses" (PhD diss., Univ. of Pittsburgh, 2017).

[26] Eduardo Viveiros de Castro, "Perspectival Anthropology and the Method of Controlled Equivocation," *Tipiti: Journal for the Society for the Anthropology of Lowland South America* 2 (2004): 3–22.

[27] Vine Deloria Jr., *Red Earth, White Lies: Native Americans and the Myth of Scientific Fact* (Golden, Colo.: Fulcrum, 1997).

[28] Bain Attwood, *Rights for Aborigines* (Crow's Nest, NSW, Australia: Allen & Unwin, 2003); Linda Tuhiwai Smith, *Decolonizing Methodologies: Research and Indigenous Peoples* (Dunedin: Univ. of Otago Press, 1999).

M. K. Gandhi to the postcolonial scholar Ashis Nandy,[29] there have been many antihistoricisms that did not fit smoothly into majoritarian nationalism. But the contemporary political realities of India, as well as the factual reality of the appropriation of Ayurveda by right wing political discourse, give the new Ayurvedic antihistoricism very different politics from that of an Eduardo Viveiros de Castro or a Vine Deloria Jr.

I will conclude this section by reiterating the urgent need for historicizing different antihistoricisms and resisting the urge to flatten their very different contexts, meanings, and politics by noticing simply their structural homogeneity. Majoritarian, statist, and bigoted antihistoricisms, such as that framing much of contemporary Ayurveda, have to be carefully and relentlessly distinguished from the antihistoricism that stands in solidarity with marginal voices and peoples. Antihistoricisms too, like all knowledge, must be "put in their place."[30]

FOOD FOR AYURVEDIC HOLISM

It is one of those serendipitous ironies of history that the man who coined the term "holism" is remembered in South Asian history as the racist South African politician who was responsible for having pushed the young London-trained attorney Mohandas K. Gandhi to finally lose faith in the imperial legal system and embark on a path that would one day make him "Mahatma" (literally, Great Soul) Gandhi.[31] Few today who tirelessly espouse the "holism" of Ayurveda know this historical genealogy of the term. Reminding ourselves of the genealogy serves two functions. First, it recalls how relatively recent the notion of "holism" itself is. Second, it calls attention to the range of politics, many of them far from the usual antiestablishment ring of much contemporary holism, that was historically associated with the term.

"Medical holism" in the twentieth century, Charles Rosenberg tells us, "implies an emphasis on inclusiveness and integration, on pattern and interdependence—and a symmetrical disdain for the sufficiency of elucidation and manipulating discrete mechanisms in managing the healthy or diseased body."[32] However, these broad contours, Rosenberg continues, vary according to the type of medical holism. He identifies four holisms that have been available in the twentieth century and insists that only one of them can be called a "historical holism." The other types are not necessarily historical. In fact some of them are decidedly ahistorical. I argue that most contemporary Ayurvedic holisms are ahistorical. They mostly present a timeless Ayurvedic system and a timeless human body. The only way time features in these holisms is in the form of a decay of holistic thinking.

The same essay in *Indian Knowledge Systems* that contrasted the "mumbo-jumbo" of the Nilotic past with the "eternal science of Ayurveda" also asserted that "the Indian system maintains that the change of lifestyle is the best insurance against precocious

[29] On Gandhi's antihistoricism and its relationship to nationalism, see Gyan Prakash, *Another Reason: Science and the Imagination of Modern India* (New York, N.Y.: Oxford Univ. Press, 2000); Ashis Nandy, *The Illegitimacy of Nationalism* (New York, N.Y.: Oxford Univ. Press, 1994).

[30] Suman Seth, "Putting Knowledge in Its Place: Science, Colonialism, and the Postcolonial," *Postcolon. Stud.* 12 (2009): 373–88.

[31] J. C. Smuts, *Holism and Evolution* (New York, N.Y.: Macmillan, 1926). For a balanced account of Gandhi's relationship with Smuts, see Ramachandra Guha, *Gandhi Before India* (New York, N.Y.: Knopf Doubleday, 2014).

[32] Charles E. Rosenberg, "Holism in Twentieth-Century Medicine," in *Holism: Greater Than the Parts: Holism in Biomedicine, 1920–1950*, ed. Christopher Lawrence and George Weisz (New York, N.Y.: Oxford Univ. Press, 1999), 335–56, on 336.

diseases." This assertion is followed up by an unattributed Sanskrit verse that in turn is translated as follows: "Daily eat food in moderation, but that which pleases you [*sic*], work very hard, do not tell lies, cheat others, or backbite people."[33] Another recent academic publication describes Ayurveda's "holistic treatment" by stating that such a treatment "includes diet, nondrug therapies (like yoga and pancakarma) and lifestyle changes."[34] Clearly, in both these comments, food becomes the very first site through which Ayurvedic holism is established, along with a range of moral and ethical injunctions that follow. Needless to say, history has no part in these formulations. Holism is appended to a systemic and ahistorical view of the Ayurvedic tradition.

It is this primacy and prominence of food in constituting claims about Ayurvedic holism that I take up here. While holism becomes one of the key modalities through which the systematicity of the Ayurvedic tradition is expounded and established, food is the foremost instance through which the amorphous notion of holism is evidenced in the context of Ayurveda. Consider, for instance, another example of the way Ayurveda is rendered as a system via claims of holism, and the holism itself is established with reference primarily to the role of food in pathogenesis. Describing Ayurveda as "a holistic science" in a peer-reviewed journal, the author writes the following: "In India, food is considered to be the origin of life, hence, diseases are also directly related to the food one consumes. For this reason, treatment is also to be given through the means of nourishment."[35]

This centrality of food in renditions of Ayurvedic holism is, expectedly, even more prominently exhibited in a large and global genre of books on Ayurvedic cooking. Thus, Usha Lad and Vasant Lad's best-selling book, *Ayurvedic Cooking for Self-Healing* declares the following: "The Ayurvedic science of food and diet is vast and comprehensive, and influences every aspect of one's life. Ayurveda is the eternal science of life. The tradition of Ayurveda extends more than five thousand years and has been practiced continuously to the present day. Although India has seen many different rulers and aggressors, it has never lost its integrity and essential nature as evidenced by Ayurveda's ancient tradition and continuity."[36] Likewise, the hugely popular Deepak Chopra and his colleagues write in their Ayurvedic cookbook that Ayurveda "offers a holistic approach to living that is based upon a fundamental principle: the choices you make are metabolized into your body."[37] Comments like these overdetermine an image, framed by the idiom of holism, of Ayurveda as a system grounded in the importance of food within a global market of books on "wellness."[38]

Very recently, this emphasis on food in Ayurvedic holism has evolved further by producing fields of academic research into "Ayurvedic neutraceuticals" and "Ayurnutrigenomics." A "nutraceutical," a portmanteau term composed by blending "nutrition"

[33] Hegde, "Modern Medicine and Ancient Indian Wisdom" (cit. n. 9), 153–4.

[34] Darshan Shankar and Padma Venkat, "Health Sciences in India: Traditional Health Sciences and Their Contemporary Application," in *Encyclopaedia of the History of Science, Technology, and Medicine in Non-Western Cultures*, ed. Helaine Selin, 3rd ed. (Dordrecht: Springer, 2016), 2176.

[35] Guru Nitya Chaitanya Yati, "Ayurveda–A Holistic Science," *Ancient Science of Life* 12 (1992): 286–8, on 286.

[36] Usha Lad and Vasant Lad, *Ayurvedic Cooking for Self-Healing* (Delhi: Motilal Banarsidass, 2005), 13.

[37] Deepak Chopra, David Simon, and Leanne Backer, *The Chopra Center Cookbook: Nourishing Body and Soul* (Hoboken, N.J.: John Wiley & Sons, 2002), 5.

[38] For a history of "wellness," see J. W. Miller, "Wellness: The History and Development of a Concept," *Spektrum Freizeit* 1 (2005): 84–102.

and "pharmaceutical," is described in one research article as being a "food or a part of food that provides medical or health benefits including the prevention and treatment of a disease."[39] By comparison, Ayurnutrigenomics is a blending of Ayurveda, nutrition research, and genomics.[40] It would be out of place to pursue a fuller history of these fascinating new fields of research, but suffice it to say that the global market in wellness, the regulatory difficulties in marketing Ayurvedic pharmaceuticals in Western markets,[41] and the strident support for a particular type of Ayurvedic antihistoricism in Indian governmental institutions have all contributed to the emergence of fields such as Ayurnutrigenomics.

The emergence of these new fields of research in tandem with Ayurvedic cookbooks and antihistorical accounts of Ayurveda have all thus served to consolidate an image of Ayurveda as a system that transcends or defies history. In making, establishing, and disseminating these claims to ahistorical systematicity, food plays an important and conspicuous role. A generalized idea that Ayurveda is some sort of a coherent, discrete, and internally articulated system that achieved its foundation and fulfillment in classical Sanskrit texts is explained and evidenced by repeatedly recounting the centrality of food in Ayurvedic ideas about pathogenesis.

The implicit, and occasionally explicit, assumptions on which these claims are built are an utter denial of historical specificity. Differences between Ayurvedic texts or among the various regions of South Asia are as completely absent from such claims, as is any appreciation of changes over time. It appears as though South Asians and Ayurvedic scholars continued to subscribe to exactly the same views on food and health for over two millennia.

Contrary to this obvious fiction, food historians have repeatedly documented the enormous changes over time, as well as the very significant variations in diet among various communities living in the subcontinent.[42] Not only are there very significant regional variations in diet—between say, Kashmir and Tamilnadu, or Punjab and Assam—but even in the same locality diets have varied widely according to caste, religion, gender, and so forth. Indeed, Arjun Appadurai has identified South Asia as a specifically vibrant site of "gastro-politics" where "the semiotic properties of food take on a particularly intense form," becoming both "the medium, and sometimes the message, of conflict."[43] Food history therefore acquires in South Asia, as perhaps elsewhere, a particularly acute power for identifying social and historical specificities. Hence, it has unquestionable value in pushing back against the Ayurvedic antihistoricisms that I have described above.

ANUPANAS, PATHYAS, AND DRAVYAS

Any history of food in the Ayurvedic tradition must begin by disaggregating the very category of food. The antihistorical espousals of an Ayurvedic system and its alleged

---

[39] Pramod C. Baragi, B. J. Patgiri, and P. K. Prajapati, "Neutraceuticals in Ayurveda with Special Reference to Avaleha Kalpana," *Ancient Science of Life* 28 (2008): 29.

[40] Subhadip Banerjee, Parikshit Debnath, and Pratip Kumar Debnath, "Ayurnutrigenomics: Ayurveda-Inspired Personalized Nutrition from Inception to Evidence," *Journal of Traditional and Complementary Medicine* 5 (2015): 228–33, https://doi.org/10.1016/j.jtcme.2014.12.009.

[41] Herbert Schwabl, "It Is Modern to Be Traditional: Tradition and Tibetan Medicine in the European Context," *Asian Medicine* 5 (2009): 373–84.

[42] Colleen Taylor Sen, *Feasts and Fasts: A History of Food in India* (London: Reaktion, 2014).

[43] Arjun Appadurai, "Gastro-Politics in Hindu South Asia," *American Ethnologist* 8 (1981): 494–511.

emphasis on food operate by completely flattening the distinct and different technical terms used within the tradition. These terms conceptualize food and its relationships to the body and health in very different ways. In this section, I describe and distinguish three of these technical terms for edible matter—namely, *anupana*, *pathya*, and *dravya*.

Birajacharan Gupta, one of the most eminent Ayurvedic physicians of the fin-de-siècle era, who served as the physician to the nominally independent kings of Cooch Behar, defined *anupana* simply as "that which is consumed slightly after the taking of a medicine."[44] Also, an early English to Bengali dictionary authored by Ramkamal Sen, who was himself an Ayurvedic physician, had described *anupana* as a "vehicle" for medicines.[45] Indeed, historical anthropologist Joseph Alter describes *anupanas* as "fluid vehicles used for the administration of medicines":

> The fluid itself has medicinal properties, but the property of the fluid is distinct from the medicine as such. Thus the most common "vehicles" are water, honey, ghee, butter, sugar, jaggery, buttermilk, and milk. Each vehicle is correlated both with the medicine it is used to transport and with the ailment in question. Usually the medicine is seated in—or suspended in—the vehicle, but in a number of instances the *anupana* . . . is consumed shortly after the medicine is taken.[46]

These definitions are, interestingly, at some variance from the classical, textual sources of the Ayurvedic tradition. The *Astangahridaya* or "Heart of Medicine" by Vagbhata provides a good baseline for examining early understandings of the concept. The text is thought to date from around 600 CE and remained enormously important for Ayurvedic physicians until the nineteenth century. Dominik Wujastyk asserts that throughout the medieval period, medical students literate in Sanskrit memorized the entire text, and it formed the "core of their medical education."[47] Its discussion of *anupana* is therefore crucial to understanding earlier practices. Vagbhata described *anupana* as the particular drinks to be taken after particular types of meals. Thus, he advised that one drinks cold water as *anupana* after a meal of foods made from barley or curd. After consuming foods made from crushed wheat, however, it was advisable to drink tepidly warm water as *anupana*. A meal of spinach was similarly to be followed by a drink made from yogurt, while excessively slim people were advised to consume wine after a meal. Those suffering from consumption were advised to drink meat stock as *anupana*.[48] Therefore, while avoiding a single pithy definition, Vagbhata seemed to be indicating that *anupana* was a specific type of drink to be taken after a specific meal. One recent translator has rightly chosen to translate the notion as "postprandial drinks."[49] The only further clarification Vagbhata himself offered was that the proper *anupanas* were those that were independent of the qualities of the food consumed, while not being opposed to these qualities. *Anupanas*, he further proffered,

---

[44] Birajacharan Gupta, *Banaushadhidarpan* [Mirror of forest medicines], vol. 1 (Calcutta: S. C. Auddy, 1908), 137.

[45] Ram Comul Sen, *Dictionary in English and Bengalee*, vol. 2 (Serampore: Serampore Press, 1834), 429.

[46] Alter, *Yoga in Modern India* (cit. n. 8), 264.

[47] Wujastyk, *The Roots of Ayurveda* (cit. n. 23), 236.

[48] Bagbhat, *Astanga Hridaya Sanhita Or Bagbhata*, ed. Binodlal Sen and Pulinkrishna Sen, 3rd ed. (Calcutta: Pulinkrishna Sen, 1932), 65.

[49] R. Vidyanath, *Illustrated Astanga Hrdaya: Text with English Translations and Appendices* (Varanasi: Chaukhamba Surbharati Prakashan, 2013), 163–4.

"produced enthusiasm, and satisfaction, whilst properly distributing the essences of the food consumed, giving firmness to the body, and breaking down the accumulated bundles of food and permitting its digestion."[50] Clearly, the idea here is that of a postprandial drink that aids digestion; this is a notion very different from that of a "vehicle for medicine."

I do not mean to suggest that the notion of a "medicinal vehicle" was unique to nineteenth-century Bengali physicians. There were certainly earlier authors, such as Chakrapanidatta, who voiced similar opinions. To pursue the evolution of this doctrine would be a project in its own right that I cannot do justice to here. It is also beyond the scope of this article to detail how the notion of *anupana* may have evolved differently in different parts of South Asia. But it is clear that in the Bengali-speaking regions in the east, by the turn of the twentieth century, it had evolved into a significantly different and hugely important concept.

So important was the notion that it became a stock feature in modern Bengali literature, appearing as a crucial component of Ayurvedic practice. In Rajsekhar Basu's humorous short story, "Chikitsa Sankat" (Treatment crisis), an Ayurvedic physician is described as prescribing a long list of *anupanas*.[51] In Bibhutibhusan Bandyopadhyay's critically acclaimed historical novel *Icchamati*, we find an Ayurvedic physician describing *anupanas* and emphasizing that "medicines do not cure diseases, the right *anupanas* do."[52] In Manoj Basu's *Brishti, Brishti* (Rain, rain), an Ayurvedic physician is shown to be repeatedly changing the prescribed *anupanas* while retaining the same medicine.[53]

This last aspect is in fact one of the more striking features of the Ayurvedic notion of *anupana*. For instance, according to Vangasena, a medieval Bengali author, *haritaki* (the medicinal fruit of *Terminalia chebula*) consumed with different *anupanas* in different seasons prevented different diseases. In the rainy season, one had to take it with an *anupana* of rock salt. In autumn, the *anupana* had to be sugar. In early winter, the necessary *anupana* was dry ginger, while in late winter it was the "long pepper," *pippali*. Finally, in spring and summer, the respective *anupanas* to accompany haritaki were honey and molasses.[54] *Anupanas* therefore could include a range of different foods and drinks and worked as a sort of catalyst. Their identity as an *anupana* arose relationally and over time from their consumption immediately after a medicine, rather than from purely innate characteristics.

Unlike *anupanas* that were "medicinal vehicles" that potentiated particular drugs, *pathya* was defined in the *Charaka Samhita*, the oldest Ayurvedic text, as follows: "That which is not harmful to paths (of the body) and is according to liking . . . The entities have their effects according to dose, time, preparation, place, body constitution, pathology, and properties."[55] According to the *Ayurveda Sarasamgraha* (Essential collection of Ayurveda), a nineteenth-century digest, the physician was advised

[50] Bagbhat, *Astanga Hridaya Sanhita* (cit. n. 48), 65.
[51] Rajsekhar Basu, "Chikitsa Sankat," in *Gaddalika* (Calcutta: Brajendranath Bandyopadhyay, 1924), 40–69.
[52] Bibhutibhusan Bandyopadhyay, *Icchamati* (Calcutta: Mitra O Ghosh, 1959), 155.
[53] Manoj Basu, *Brishti, Brishti* (Calcutta: Bengal Publishers Pvt., 1957).
[54] R. Vidyanath, *Illustrated Astanga Hrdaya: Text with English Translations and Appendices* (Varanasi: Chaukhambha Surbharati Prakashan, 2013), 53.
[55] Priyavrat Sharma, *Caraka Samhita (Text with English Translations)*, vol. 1 (Varanasi: Chaukhamba Orientalia, 2014), 173.

to "first examine the disease, then give medicines and thereafter prescribe *pathya*."[56] According to an early twentieth-century Ayurvedic author, the original meaning of the term *pathya* had simply been a "way," "method," or "expedient"—namely, *upaya*. Thus, strictly speaking, any food or drink taken as a way to preserve the body could be designated as a *pathya*. However, this broad meaning, our author clarified, had long since become redundant. Instead, those foods that "a physician planned with a view to the eradication of an existent disease in a patient were commonly understood to be *pathyas*."[57]

Pathyas had to be prepared in particular ways and had specific names. One popular *pathya*, for instance, was the *bilepi*. It was prepared by boiling barley or lentils. The primary ingredient was initially boiled with four times as much water. The mixture was kept boiling till three out of four portions of the water had boiled away. Thereafter, the remaining liquid was filtered out and the soft, mushy remainder fed to the patient. Bilepi was often prescribed for fever patients. *Manda* was another type of *pathya*. Its preparation was roughly similar to the way one prepared bilepi, apart from the fact that one had to start with fourteen times as much water as there was solid material.[58] There were other types of *pathya* as well. *Tarpana* was another form of *pathya*. This was prepared by crushing parched rice with water and mixing it with the juice of pomegranates and raisins. Yet another form of *pathya* was known as *vrihi*. It was prepared by placing finely minced meat in a porcelain jar, sealing the jar with dough, and then placing the sealed jar in neck-deep water held in a metal trough. The water in the trough was thereafter boiled for four full hours on a slow, steady fire. Finally, the jar was opened, and the meat strained by a fine cloth. The meat thus processed was vrihi.

Pathya was therefore a special type of food prepared for the patient's consumption. It supplemented the medicines that the patient was given. From their ingredients to their preparation, they were entirely distinct from *anupanas*.

Finally, there were the *dravyas*. Most English works tend to gloss the word *dravya* simply as substance, but the translation is grossly inadequate. The *Charaka Samhita*, the oldest Ayurvedic classic, stated that the number of *dravyas* in the universe was nine. These were respectively the five subtle elements or *panchamahabhutas*—namely, earth (*kshiti*), water (*ap*), air (*marut*), fire (*tej*), and ether (*vyoma*), alongside the soul (*atma*), the heart-mind (*mana*), duration (*kala*), and direction (*dika*). Modern authors, however, bracketed off this definition by stating that these were merely the "fundamental *dravyas*" (*muladravya*), whereas what people usually referred to as *dravyas* were effectively "active *dravyas*" (*karyadravya*). One popular classical definition for the latter that continued to be deployed by several modern authors was that a *dravya* "shelters the *gunas* [qualities] and *kriya* [actions/potencies]."[59] This expansive definition allowed foodstuff to be classified together with such clearly nonfood—and often highly abstract—entities as "darkness," "breeze from a yak-tail fan," parasols, and so forth.

Notwithstanding the enigmatic and seamless mix of food and nonfood items within the category of *dravya*, the basis of its classification is an elaborate system that maps

[56] Gopalchandra Sengupta, *Ayurveda Sarasamgraha* [Essential collection of Ayurveda], vol. 1 (Calcutta: Bhubanmohan Gangopadhyay, 1871), 10.

[57] Bipinbihari Sengupta, "Deshiya Pathya" [National pathya], *Ayurveda Bikash* 2 (1914): 12.

[58] Ibid., 13–14.

[59] Sushilkumar Sensharma, *Drabyagun-samhita*, vol. 1 (Calcutta: Kalpataru, 1937), 1.

the actions of each *dravya* upon the human physiology. The classification is calibrated along five distinct axes, namely, *rasa* (flavor), *vipaka* (transformed flavor), guna (quality), *veerya* (potency), and *sakti* or *prabhava* (specific or differential potency). All of these five axes have their own specific values, and *dravyas* are classed accordingly.[60] These axes of classification are frequently described and discussed in numerous Ayurvedic texts, especially those dealing with materia medica.

The rasa could be classified with one of six values—that is, sweet, sour, salty, acrid, bitter, and astringent.[61] Vipaka could either have one of three values (sweet, sour, or acrid) or, according to some texts, could only be either heavy or light.[62] Guna could hold one of twenty-five values such as blunt, sharp, mobile, immobile, soft, hard, and so forth.[63] Regarding the veerya values, as in the case of vipaka, there are severe disagreements. Some accept only two possible values, that is, hot and cold,[64] while others adopt eight values.[65] Finally, the prabhava axis is largely enigmatic and defies specific values.[66] Considered together, not only is the system of classification labyrinthine in its complexity, but also seemingly comprehensive in its coverage.

Yet ironically, the system when applied, at least in Bengal, was almost always applied in a somewhat superficial and inchoate manner. For the majority of the *dravyas*, the values for all the five axes were never stated. Hence, texts such as Narayan Das Kaviraj's *Rajballabhiya Dravyaguna*, written in the eighteenth century and hugely popular in Bengal, practically never mentioned more than three of the five classificatory values of each *dravya*. Of a type of small fish, for instance, it merely stated that it was heavy, caused constipation, and was helpful in a disease called *grahani roga*.[67] Likewise, regarding the flesh of the tailorbird, Kaviraj stated that it had sweet rasa, rough guna, and destroyed wind and phlegm.[68] Clearly the comprehensiveness of the system was only theoretical and not to be seen in practice.

Dravya, unlike *pathya* or *anupana*, is not meant only for the ill. It is defined with a view to the preservation and enhancement of health. Naturally, *dravya* is also the largest of the three categories we have been discussing. Yet, crucially, it does not exclusively include foods. When we seek to recast it exclusively as food, or distinguish the food from nonfood, we violate the implicit assumptions that underwrite the coherence of the category. If we refuse to inflict such violence upon the category, we must admit that *dravya*, *pathya*, and *anupana* are all distinct entities and not simply "food" by various names.

Each of these categories has its own rich and complex history. Each of these histories is tied up with the histories of specific genres of texts. *Anupana*, for instance, generated its own devoted texts such as the *Anupana Darpana* (Mirror of *anupanas*). This text was translated and expanded by later authors. *Pathya* similarly produced its own texts such as the *Pathyapathya*, which also developed and evolved along its own lines.

---

[60] Nagendranath Sengupta, *Dravyaguna-Sikshya*, 12th ed. (Calcutta: Nagendra Steam Printing, 1934).
[61] Priya Vrat Sharma, *Introduction to Dravyaguna (Indian Pharmacology)* (Varanasi: Chaukhambha Orientalia, 1976), 24.
[62] Ibid., 38–9.
[63] Sensharma, *Drabyagun-samhita* (cit. n. 59), 1:3.
[64] Sharma, *Introduction to Dravyaguna* (cit. n. 61), 49.
[65] Narayan Das Kaviraj, *Rajballabhiya Dravyagunah*, ed. Rajkumar Sen (Calcutta: Upendranath Mukhopadhyaya, 1891), 132.
[66] Sensharma, *Drabyagun-samhita* (cit. n. 59), 1:23.
[67] Kaviraj, *Rajballabhiya Dravyagunah* (cit. n. 65), 66.
[68] Ibid., 80.

But neither of these two traditions could match the voluminous tradition of *Dravyaguna Samgrahas* (Collections of the qualities of *dravyas*). This latter, that evolved out of earlier encyclopedic genres, continued to grow and diversify well into the twentieth century.

Instead of overdetermining the notion of a singular Ayurvedic tradition and its allegedly holistic engagement with food, we can think of the Ayurvedic tradition as an internally heterogeneous and plural tradition constituted by multiple specific strands. Each of these strands engenders its own specific textual tradition of writing and develops particular technical categories with reference to its own textual heritage and practical concerns. Attending to these specific strands of constitutive textual traditions helps to move the vacuous and abstract iterations of a singular Ayurvedic tradition onto a firmer, more specific, and emphatically more historic basis.

This does not mean eschewing the philosophical or metaphysical concerns of the Ayurvedic tradition. Far from it. Rather, it means limiting, if not avoiding, the anachronistic and baseless projections of contemporary agendas onto a fictionally abstracted Ayurvedic tradition, and attending to the actual, concrete, and relevant philosophical concerns of specific texts and authors. A foundational move toward such sensitivity would entail being careful about the conflation of discrete technical categories such as *dravya*, *anupana*, and *pathya*.

NEW *DRAVYAS*, NEW *PATHYAS*, NEW *ANUPANAS*

One of the most concrete and conspicuous ways in which to challenge the ahistorical image of an Ayurvedic system and its unchanging foodways is to simply map the incorporation of a range of new foods into Ayurvedic texts. In this section I map how new foods entering South Asia through European contact were rapidly adopted into Ayurvedic texts. Taking my cue from the discussion in the previous section, I focus on the entry of these new foods into the categories of *anupanas*, *pathyas*, and *dravyas*.

The arrival of Vasco da Gama on the southwestern coast of South Asia in 1498 eventually brought the region into contact with the Iberian Empires in the New World. In 1510, Portugal established its colony in Goa. On the eastern coast, especially in Bengal, there developed other Portuguese settlements and even some small, independent principalities ruled by Portuguese mercenaries.[69] One of the most momentous consequences of these Portuguese settlements was that they became bridgeheads for the extension of the Columbian Exchange to the subcontinent.

The sixteenth and seventeenth centuries witnessed the introduction of a number of New World foods into South Asia. Food historian Colleen Taylor Sen credits the Portuguese for the introduction into the subcontinent of a wide range of American foods such as potatoes, chilli peppers, okra, pineapples, papayas, guavas, custard apples, cashews, peanuts, and much else.[70] The *Dravyaguna* texts of the period rapidly incorporated these new foods into their own classificatory schemas. The *Rajballabhiya Dravyaguna*, which, as we have seen above, was used in Bengal, also included several of these new foods. Interrogating these incorporations is therefore a good way of analyzing the flexibility and strength of Ayurvedic medico-dietary paradigms.

[69] Sanjay Subrahmanyam, *The Portuguese Empire in Asia, 1500–1700: A Political and Economic History* (Hoboken, N.J.: John Wiley & Sons, 2012).
[70] Sen, *Feasts and Fasts* (cit. n. 42), 212.

Udaychand Dutt, a nineteenth-century Ayurvedic author, described the *Rajballabhiya Dravyagunah*, occasionally also referred to as the *Rajballabhnighantu*,[71] as the foremost reference book on the subject in Bengal.[72] Though its exact date of composition and its author remain unknown, G. Jan Meulenbeld has placed the text firmly in the eighteenth century. In the same century, the text seems to also have been revised by a physician named Narayan Das Kaviraj.[73] By the end of the nineteenth century, several printed editions of the text, sometimes relying on different manuscripts, had appeared in print and therefore continued to be in use by practicing Kavirajes (physicians practicing within the broadly Sanskritic tradition of medicine) at the time.[74]

One of the most conspicuous New World foods discussed in the *Rajballabhiya Dravyagunah* was the potato. The Portuguese introduced the potato into South Asia in the early seventeenth century. But cultivation was limited to the western coast. British merchants brought the tuber to Bengal, in the east, sometime later. Governor General Warren Hastings promoted potato cultivation in the second half of the eighteenth century. By 1842 potatoes were being grown close to the British capital at Calcutta.[75] Today, Bengalis are the second largest consumers of potatoes after the Irish.[76]

The *Rajballabhiya Dravyagunah* mentioned two different types of potatoes—namely, *khandakarna* and *pindalu*. The former name is possibly a corruption of *khanda-kanda* that would have meant "piece of tuber." The latter part of the name simply means "round edible root." The khandakarna is described as "phlegm-uprooting, acrid when cooked, and the vanquisher of bile" (*kaphocchedi katupaakascha pittajit*). The pindalu, by comparison, was described as "phlegm-causing, heavy, and wind-aggravating" (*kaphakaram guru vaataprakopakanam*).[77] Alongside these two types of potatoes we also find mention of the sweet potato, another New World import. This latter is called *hastikanda*. It was said to be "*raktapitta*-destroying, heavy, sweet-flavored (*swadurasa*), cool (*sheetal*)," and to increase "breast milk and semen."[78]

Besides these tubers, the *Rajballabhiya Dravyagunah* also includes what is perhaps the most successful New World import into South Asia—namely, chilli peppers. Traditionally, Bengali cuisine had used black peppers, and in many parts of Bengal, particularly in villages, the name for black peppers and chilli peppers continues to overlap up to this day. This makes detecting references in the historical record to the latter, as distinct from the former, a little more challenging. In the *Rajballabhiya Dravyagunah*, however, the task is made slightly easier. The text refers to two kinds of peppers (*maricha*). One of these references was to "dry peppers" (*sushka maricha*) and might equally have meant black peppers or dried chilli peppers, both of which are still widely

[71] G. Jan Meulenbeld, *A History of Indian Medical Literature*, vol. 2A (Groningen: Egbert Forsten, 2000), 342; see also 340.

[72] Udoy Chand Dutt, *The Materia Medica of the Hindus: Compiled from Sanskrit Medical Works* (Calcutta: Thacker, Spink, 1877), xii.

[73] Meulenbeld, *Indian Medical Literature* (cit. n. 71), 2A:340; see also 343.

[74] Narayan Das Kaviraj, *Dravyagun Darpan*, ed. Abhaycharan Gupta (Calcutta: Anglo-Indian Union Yantra, 1865); Kaviraj, *Rajballabhiya Dravyagunah* (cit. n. 65).

[75] B. D. Sharma, "The Origin and History of Potato in India," in *History of Agriculture in India (Up to c.1200 AD)*, ed. Lallanji Gopal and V. C. Srivastava (New Delhi: Concept Publishing, 2008), 149–58, on 150.

[76] Sen, *Feasts and Fasts* (cit. n. 42), 213.

[77] Kaviraj, *Rajballabhiya Dravyagunah* (cit. n. 65), 50–1.

[78] Ibid., 50. *Raktapitta* is a type of affliction recognized in Ayurveda and nowadays is often translated as "hemorrhage."

used in Bengali cooking.[79] But the second reference was to "juicy peppers" (*aardra maricha*). This latter clearly referred to green chilli peppers, since black peppers are exclusively consumed dry. Of the green chilli peppers, it is said that they were flavorful and heavy when cooked (*swadupaak, gurupaak*) and that they expelled phlegm.[80]

Finally, there was the custard apple (*nonaphal* or *ataphal*). The fruit was said to be good for the heart, possessed of a sweet fragrance, and effective in reducing phlegm and wind. It was one of the few items whose external characteristics (i.e., the fragrance) were included in the description.[81]

While *dravya*, being the most capacious category, certainly absorbed the largest number of new foods, it was not the only category to do so. New foodstuffs also entered the *pathya* and *anupana* categories.

A number of new *pathyas* were included in an appendix to Nagendranath Sengupta's *Dravyaguna Sikshya* (Teachings on the qualities of *dravyas*).[82] Sengupta was one of the best-known and the most prolific Ayurvedic physicians of the fin-de-siècle era and also owned a thriving Ayurvedic medical college. His text, first published in 1900, had gone into twelve editions by 1934. His inclusion of several new *pathyas* cannot therefore be overlooked as a marginal innovation.

He included sago among the *pathyas* he described. Sengupta explicitly stated that sago was an exotic food found mainly in "various islands of the Indian ocean and places like Borneo." He then went on to describe the trees and comment on how islanders living where the plant grew consumed it, before stating that it was "easily digestible, nutritious, and hearty."[83] Likewise, Sengupta also described barley as a good *pathya*. Once again, he was explicit about its exotic origins and stated that it grew mainly in temperate countries.[84]

Interestingly, though Sengupta stopped short of explicitly labeling some other foods as *pathyas*, he included them in a separate unnamed section right alongside *pathyas* such as sago and barley, indicating their utility for convalescents. Among these items, he included coffee, arrowroot, and cocoa. For each of these items he stated their countries of origin before outlining their history and physical features and finally describing how they affected the patient's body. For cocoa, for instance, he stated that "Mexico, in the Americas, was its birthplace," before stating how it had since been cultivated in countries such as Guatemala, Honduras, Brazil, Peru, and so forth. He then offered a physical description of the plant and its beans. Finally, he detailed its physical effects at length: "Its qualities are similar to tea and cava. The difference is that while those last-mentioned drinks stimulate, they do not build true strength [as cocoa does]."[85]

Sengupta was far from being the only Ayurvedic author of his time to use barley, sago, cocoa, and the like as *pathyas*. Dhirendranath Ray's *Rog o Pathya* (Illness and *pathya*), which was first published in 1933 and went into its third edition by 1950, also made extensive recommendations of these foods as *pathya*.[86]

[79] Ibid., 104.
[80] Ibid., 103.
[81] Ibid., 55.
[82] Sengupta, *Dravyaguna-Sikshya* (cit. n. 60).
[83] Ibid., 469.
[84] Ibid., 466.
[85] Ibid., 474.
[86] Dhirendranath Ray, *Rog O Pathya* [Illness and pathya], 3rd ed. (Calcutta: Prabartak Publishers, 1950).

Anupana, as a category, is often the most elusive since it is more rarely discussed in texts. While Ayurvedic texts often include a large number of medicinal recipes, the *anupanas* are frequently left unspecified, with the authors merely stating that relevant *anupanas* were necessary. Texts explicitly devoted to *anupanas*, though extant, were fewer in number than texts devoted to *dravyas* or *pathyas*. Yet, notwithstanding these challenges, we can still find references to exotic, postclassical foods being used as *anupanas*.

Amritalal Gupta, an eminent nineteenth-century Ayurvedic physician who had authored a text on *anupanas* as well as several influential textbooks, provides some excellent examples of *anupanas* whose ingredients were unknown in classical India. For children suffering from a type of worm (*krimirog*), for instance, a juice made from the seeds of pineapples was to be given as *anupana*.[87] Likewise, for those suffering from an affliction known as *swetapradara* (sometimes translated as leucorrhoea), the proper *anupana* was a juice made from the leaves of marigold.[88] Now, both pineapples and marigolds originated in southern or Central America and were most likely brought to South Asia by the Portuguese.

Exotic, postclassical foods brought to South Asia by Europeans, mostly from the New World, were therefore absorbed into all three of the categories we have been discussing. The examples I have provided here are neither exhaustive nor chronologically organized. They are intended as testaments bearing out the dynamism and historicism of both the Ayurvedic tradition in general and its engagement with food more particularly. It is precisely this historical dynamism that is obscured by the vague generalities that clothe the antihistoricism of Ayurvedic holisms.

HOTNESS OF CATEGORICAL THOUGHT

In this final section, I extend this argument about the historical dynamism of the Ayurvedic tradition, and especially its engagement with foods, a step further. The absorption of new, exotic, and mostly New World foods into the Ayurvedic repertoire, I argue, did not leave the theoretical edifice untouched. It was not the case that the new foods were simply fitted into preexisting classificatory schemas. Instead, the new material reacted upon the classificatory system itself, thereby radically transforming it.

Since the classificatory schema is most complex in the case of *dravyas*, it is here that the temptation to adopt an ahistorical, "systemic" image of the tradition is most attractive. Hence, in this section, I focus on *dravyas* alone, rather than on the three categories of *dravya*, *pathya*, and *anupana* that I have discussed in the foregoing sections.

Dravya, as we have seen above, was theoretically classified in terms of a complex fivefold system of rasa, vipaka, guna, veerya, and prabhava. Early modern texts such as the *Rajballabhiya Dravyagunah*, the standard reference work on the subject in eighteenth-century Bengal,[89] however, regularly underdetermined this fivefold classification. The entire set of five values were rarely stated for any of the specific *dravyas*. At best, two or three values were spelt out. Among these, the value that was in fact most consistently mentioned and formed the real backbone of the classificatory system was the rasa (flavor) value.

---

[87] Amritalal Gupta, *Ayurveda Sikshya* [Ayurvedic teachings], vol. 4, 2nd ed. (Calcutta: Kalika, 1914), 1294.
[88] Ibid., 4:1295.
[89] Kaviraj, *Rajballabhiya Dravyagunah* (cit. n. 65).

Each *dravya* could have one of six possible rasa values, namely, *madhura* (sweet), *amla* (sour), (*labana*) salty, (*katu*) acrid, (*tikta*) bitter, and (*kashaya*) astringent.[90] The determination of these values depended upon the physician/author's own sense of taste. The madhura (sweet) flavor, for instance, was described as that which "gives birth to stickiness, makes the body oily/tender (*snigdha*) and supple (*mridu*)." In addition, upon tasting, "it gives rise to happiness (*prasannata*) in the senses, and delight (*alhada*) in the interior/mind (*antara*)."[91] Katu (acrid) flavor is similarly described through the physician's perception as that which "on contact, irritates and produces piercing pain in tongue, and stimulates secretions with burning from mouth, nose, and eyes."[92]

The classification clearly relied upon a kind of radical empiricism whereby the physician/author's own sense of taste became the basis for classifying foods as belonging to one group or another. Those foods that produced similar effects in the physician's mouth were placed in the same group. But this is precisely where historical change operated—that is, on the physician's own palate.

The *Charaka Samhita*, the most ancient of the Ayurvedic texts, had classified mustard oil (*sasharpam tailam*) as having a katu flavor.[93] Katu thus seems to have indicated the flavor that most of us would today call "pungent." By the early modern period in Bengal, however, katu was coming to gradually be defined by the taste of the new green chillies introduced by the Portuguese. The *Rajballabhiya Dravyagunah*, as we have seen above, had mentioned green chillies. It did not, however, assign them a specific rasa value, stating merely that they were flavorful. Yet elsewhere in the same text while defining the katu-rasa it equated katu with the Bengali word *jhal*.[94] Today, the taste of green chillies is indeed referred to in Bengali as jhal.

Contemporary authors tend to translate the Sanskrit word katu as "acrid." Acridity, especially when read in light of the definition given in the *Charaka Samhita*, could potentially indicate multiple and somewhat different flavors on the palate. Mustard oil, for instance, is certainly acrid in the sense of being pungent. In contemporary Bengali, such pungency as is found in mustard oil (which remains an important and regularly consumed cooking medium in the Bengali kitchen) is most likely to be described by the word *jhanj*. Similarly, the tartness associated with certain fruits is also occasionally described as katu, but is neither jhanj nor jhal.[95] Finally, on occasion, a sharp mint flavor associated with *pudina* (spearmint) is also referred to as katu.[96]

The explicit equation of katu-rasa with the jhal flavor specific to chillies was an innovation that began to emerge in the eighteenth century. Later authors such as Debendranath and Upendranath Sengupta, two brothers who coauthored and translated a number of Ayurvedic texts in the late nineteenth century, for instance, explicitly identified green chillies as having katu-rasa. Sushilkumar Sensharma, writing even later, in 1937, expressly acknowledged that green chillies had been brought to the region from

---

[90] Sensharma, *Drabyagun-samhita* (cit. n. 59), 1:3–4.
[91] Ibid., 1:4. See also Sharma, *Caraka Samhita* (cit. n. 55), 1:187.
[92] Sharma, *Caraka Samhita* (cit. n. 55), 1:187.
[93] Ibid., 1:219–20.
[94] Kaviraj, *Dravyagun Darpan* (cit. n. 74), 39.
[95] See, for instance, Bibhutibhushan Bandyopadhyay, "Debjan" [The divine vehicle], in *Bibhutibhushan Rachanabali* [Collected works of Bibhutibhushan Bandyopadhyay], vol. 8 (Calcutta: Mitra O Ghosh, 1971). See also, Sen, *Dictionary in English and Bengalee* (cit. n. 45), 2:384.
[96] Gupta, *Banaushadhidarpan* (cit. n. 44), 1:108.

South America and were not mentioned in the classical texts, before affirming that they were possessed of katu-rasa.[97]

Since foods were classified with reference to their rasa or flavor, and since this in turn determined the medicinal potency of the food, the refiguration of katu-rasa as the flavor of chillies in effect radically reoriented one particular classificatory axis. Put another way, if one thinks of katu-rasa as the pungency of mustard oil and groups all similar pungent foods together, then one acquires a particular set of foods. By contrast, if a person imagines katu-rasa to be the hotness of chillies and similarly groups hot foods together, one obtains a very different set. Likewise, tartness or the quality or state of mint would produce still other groupings.

The historical dynamism of the theoretical system of food classification in the Ayurvedic tradition operates therefore through the everyday historicity of the physician's own palate. As Judith Farquhar points out in her fascinating study of the sense of taste in Chinese medicinal meals, "the logic that connects the flavors to the powers of medicines has room for the sensed responses of the lived body."[98] The human body itself connects to history, among other things, through its palate. Tastes develop, adopt, adapt, and change as the larger movements of history transform people's specific, local foodways.

It would be trite to point out that this physicality and embodied nature of Ayurvedic *dravya* classifications clearly runs counter to the antihistoricist insistence upon a disembodied and purely abstract classificatory schema. What is worth considering, however, is why, given this robustly embodied nature of Ayurvedic knowledge, do contemporary peddlers of antihistoricist claims feel the need to deny such embodiment? As Steven Shapin and Christopher Lawrence have pointed out, this denial of the embodiment of knowledge is in itself a late modern phenomenon even in the West. In the Western instance, they argue that the denial has much to do with the demise of the humoral theories of the human constitution, the desacralization of knowledge, and the professionalization and bureaucratization of knowledge making.[99] It is worth pondering to what extent these processes are mirrored beyond the West, and whether the developments in India have similar or different historical drivers. Moreover, what the shared protestations of disembodiment by both the ideologues of Western science and "Indian Systems of Knowledge" underline is how, notwithstanding their own shrill insistence on "Indianness" and "traditionalism," these two groups of ideologues in fact share in a global late modern epistemic culture where the link between the body and knowledge production is forcefully suppressed.

New World foods such as chillies did not simply enter the Ayurvedic tradition and fit neatly into existing, well-defined classificatory grids. They transformed the classificatory "system" itself. This transformation happened not by explicit textual innovation, but by the everyday gustatory experiences and memories that shaped how physicians and scholars read and interpreted texts. Thus, as they began to eat chillies as part of their diets, they gradually came to refigure the pungency of katu-rasa with the hotness of chillies.

---

[97] Sensharma, *Drabyagun-samhita* (cit. n. 59), 1:146.

[98] Judith Farquhar, *Appetites: Food and Sex in Post-Socialist China* (Durham, N.C.: Duke Univ. Press, 2002), 65.

[99] Steven Shapin and Christopher Lawrence, "Introduction: The Body of Knowledge," in *Science Incarnate: Historical Embodiments of Natural Knowledge*, ed. Shapin and Lawrence (Chicago: Univ. of Chicago Press, 1998), 1–19, on 15–16.

CONCLUSION

In this article, I have challenged the twin claims that the Ayurvedic therapeutic tradition is not amenable to historical inquiry and that it must be understood with reference to some ahistorical "metaphysical foundation." I have shown that such antihistoricist claims are rooted in older Orientalist legacies but are equally a product of the new Hindu nationalism as well as the global market for Ayurvedic and "alternative" products.

These developments have in turn consolidated an image of the Ayurvedic tradition as an ahistorical system. Chronological, geographical, and doctrinal differences and variations have been silenced. David Hardiman, following Romila Thapar, has called this a "syndicated Ayurveda" whereby "groups—or syndicates—with certain vested interests sought through combination, organization and publicity, to establish a particular, limited notion of their practice that set it apart from other forms of practice."[100] The increasing acceptance of the image of the tradition as a system propagated by this syndicated Ayurveda in parts of academia is, at least in part, a consequence of the difficulties in distinguishing various stripes of antihistoricism emerging out of various distinct political contexts. Thus, the Hindu majoritarian antihistoricism of syndicated Ayurveda often gets confused, in parts of the academy, with subaltern antihistoricisms emerging out of Latin America, Aboriginal studies, Native American studies, and so forth.

Food has played a particularly important role in syndicated Ayurveda's espousal of tradition-as-system. It has been used to ground the ahistorical sense of an Ayurvedic tradition in a Vedic past and yoke it firmly to an idea of an ancient "Indian culture." Some authors have gone further still and have argued that all extant South Asian foodways are in fact derived from this "Ayurvedic system."[101] Such frameworks operate by dwelling in abstract imprecisions and dismissing more specific, contextually and textually grounded images of "tradition" as historical and therefore untenable. Writing against this approach, I have argued that the Ayurvedic "tradition," far from being an ahistorical system, is in fact an internally heterogeneous, geographically varied, and extremely historically dynamic entity.

I have shown that far from having a monolithic view of foods based in a handful of classical Sanskrit texts, the core Ayurvedic texts had always disaggregated the very category of "food" into multiple different technical categories. Among these I have described the three broad categories of *dravya*, *pathya*, and *anupana*. None of them exactly map onto or exhaust what we would today call "food." Yet they all substantially engage with edible matter.

All three of these categories also absorbed a number of exotic foods introduced into South Asia by Europeans. Most of these foods, ranging from pineapples to chillies, marigolds to cocoa, came from the New World. These foods gradually entered each of the three categories of *dravya*, *pathya*, and *anupana* in the period between roughly the eighteenth and the twentieth centuries.

Finally, I have argued that the entry of these New World foods did not leave the classical classificatory system untouched. The system had long been underdetermined in

---

[100] Hardiman, "Indian Medical Indigeneity" (cit. n. 5).

[101] The authors of one popular cookbook claim, for instance, that "Indian cooking is based upon the therapeutic principles of the ancient Ayurvedic science of life." See Amadea Morningstar and Urmila Desai, *The Ayurvedic Cookbook: A Personalized Guide to Good Nutrition and Health* (Delhi: Motilal Banarsidass, 1994), v.

practice and application, but the advent of the New World foods also subtly but substantially transformed it. I demonstrate how katu-rasa, one of the five basic flavors used to classify foods, was reimagined as the heat of chillies rather than the pungency of mustard. Thus, foods tasting like chillies were classified together, rather than with those tasting like mustard oil.[102] The allegedly ahistorical and disembodied categories of *dravya* classification were in fact firmly located upon the medical scholar's tongue.

Food, I have thus argued, acted upon the systematizing classical textual tradition through the mediation of the physician or scholar's own palate. In other words, food became the way in which the body itself responded to history, and as a result reoriented how the scholar or physician read his textual corpus.

Food, more than anything else, then, helps to push back against the image of the Ayurvedic tradition-as-system and its alleged antihistorical moorings. The Ayurvedic tradition, like any other living intellectual and practical tradition, was a rich, varied, internally diverse, and chronologically evolving phenomenon equipped with a particularly sensitive historical palate.

---

[102] It is worth reiterating that this specific transition—that is, from the pungency of mustard to the heat of green chillies—might well have varied in different parts of the subcontinent. My example pertains specifically to the historical region of Bengal. The general principle of how new foods and their embodied tastes transformed classical classificatory schemas, however, I would argue, can be generalized for all of South Asia, and perhaps even further.

Local Food and Transnational Science:

New Boundary Issues of the Caterpillar Fungus in Republican China

by Di Lu*

ABSTRACT

This article focuses on new boundary issues that have emerged from the encounter of modern science from abroad and local foodstuffs exemplified by the caterpillar fungus in Republican China (1912–49). The caterpillar fungus was believed in premodern Chinese society to be able to reversibly transform from a blade of grass to a worm, thereby crossing boundaries between two species. It had different uses, ranging from a culinary ingredient to a medicinal substance, and in this way also crossed boundaries of identity. At the beginning of the twentieth century, scientific scholarship from Japan began to bring new perceptions of the fungus to Chinese society through translation. Modern science expanded human vision into the microscopic structure of the caterpillar fungus, and deconstructed it into two nontransformable species grouped with other similar species. The Chinese term for it also entered the Japanese language. However, the category of the term was broadened, crossing the boundary between the caterpillar fungus and other similar species, thereby indicating semantic boundaries of shared vocabulary. As local food or material culture in Republican China engaged scientific attention, the caterpillar fungus as a disenchanted wonder of nature sometimes transformed into a scientific wonder, eliciting new explorations within different scientific boundaries. The new scholarship led to tensions and negotiations between domains of knowledge about this organism but would not necessarily drive out the vernacular culinary or medical expertise. The emergent boundary issues overall depict both rupture and continuity in modern Chinese food knowledge.

What is the caterpillar fungus? The Englishman James Everard Home (1798–1853) collected specimens during the period 1841–46 when he was captain of the *North Star*; at the beginning of that time, he was also engaged in the Sino-British Opium War (1839–42).[1] He once presented the museum of the Royal College of Surgeons of England with a "series of specimens of *Sphaeria Sinensis*, Berk., tied up in a bundle

* The Zvi Yavetz School of Historical Studies, Tel Aviv University, Tel Aviv, 6997801, Israel; ludi@mail.tau.ac.il.

I am grateful to Dr. Vivienne Lo, Dr. Michael Heinrich, and Ms. Penelope Barrett (PhD candidate) of University College London for their insightful suggestions and generous support in many ways. My deep gratitude also goes to the editors and anonymous reviewers who provided learned and perceptive comments on the original manuscript. I am greatly indebted to Tel Aviv University for hosting me as a Zvi Yavetz Fellow.

[1] "Obituary: Capt. Sir Jas. Everard Home, Bart.," *The Gentleman's Magazine and Historical Review* 41 (1854): 423; Emil Bretschneider, *History of European Botanical Discoveries in China*, vol. 1

with silk, as sold in the market of Canton." Probably based on Home's information, the museum remarked that these "are used medicinally in cases where the powers of the system have been reduced by over-exertion or sickness."[2] The species "*Sphaeria Sinensis*," now commonly known as *Cordyceps sinensis* (or *Ophiocordyceps sinensis*) or the caterpillar fungus, initially appeared as an aphrodisiac in a fifteenth-century Tibetan medical text, and then started attracting the attention of Chinese authors from the early eighteenth century.[3] A 1757 Chinese text explains that in winter the caterpillar fungus stays in the ground, is able to move, and resembles a piliferous old silkworm; while in summer, its hair grows out of the ground and turns into a blade of grass together with the body. If not gathered in summer, it would turn into a worm again in the coming winter. This little organism, sweet and balanced, can protect the lung, benefit the kidney, stop bleeding, disperse phlegm, and eliminate coughing due to exhaustion.[4] However, contemporary biology explains its formation as the fungus's infection of the larvae of some moth species belonging to the family Hepialidae, with the consequent outgrowth in the form of fruiting bodies that emerge from the heads of the larvae.[5]

Although the focus of the museum's 1860 explanation is squarely on medicine, the caterpillar fungus was actually also an esteemed foodstuff, aphrodisiac, tonic, transformable natural curiosity, profitable product, and so forth. The French Jesuit missionary Dominicus Parennin had eaten duck simmered with this substance in Beijing in 1720, which thereby restored his extremely feeble body.[6] Frederick J. Simoons's 1991 monograph on Chinese food culture discloses that "recently in a Friendship Store in Canton frequented by foreigners, we found cans (380 gm.) of 'Stewed Cordyceps Sinensis with Chicken in Soup,' a product manufactured by the China National Medicines and Health Products Import and Export Corporation, Chungking Branch (Szechwan)."[7] The caterpillar fungus in these two cases can hardly be simply treated as a dietary ingredient or medicinal substance.[8] To date, historical scholarship on Chinese food has seldom given attention to changing natural knowledge about food in connection with the globalization of modern science.[9] The multiple roles of the caterpillar fungus nevertheless enable historians to examine Chinese food across

---

(London: Sampson Low, Marston, 1898), 362; compare these to William H. Flower, *Catalogue of the Specimens Illustrating the Osteology and Dentition of Vertebrated Animals, Recent and Extinct, Contained in the Museum of the Royal College of Surgeons of England, Part I* (London: Printed for the College, 1879), 205.

[2] *Catalogue of the Contents of the Museum of the Royal College of Surgeons of England, Part 1, Plants and Invertebrate Animals in the Dried State* (London: Taylor and Francis, 1860), 23.

[3] Daniel Winkler, "Caterpillar Fungus (*Ophiocordyceps sinensis*) Production and Sustainability on the Tibetan Plateau and in the Himalayas," *Asian Medicine* 5 (2009): 291–316.

[4] Wu Yiluo, *Bencao Congxin* (1757; repr., Shanghai: Shanghai Kexue Jishu Chubanshe, 1982), 36.

[5] Yongjie Zhang, Erwei Li, Chengshu Wang, Yuling Li, and Xingzhong Liu, "*Ophiocordyceps sinensis*, the Flagship Fungus of China: Terminology, Life Strategy and Ecology," *Mycology* 3 (2012): 2–10.

[6] Jean-Baptiste Du Halde, ed., *Lettres édifiantes et curieuses, écrites des missions étrangères*, vol. 10 (Lyon: J. Vernarel, 1819), 470–85.

[7] Frederick J. Simoons, *Food in China: A Cultural and Historical Inquiry* (Boca Raton, Fl.: CRC, 1991), 323–4.

[8] Eugene N. Anderson devotes a whole chapter to "traditional medical values of food"; see Anderson, *The Food of China* (New Haven, Conn.: Yale Univ. Press, 1988), 229–43.

[9] For examples, see Kwang-Chih Chang, ed., *Food in Chinese Culture: Anthropological and Historical Perspectives* (New Haven, Conn.: Yale Univ. Press, 1977); Hsing-Tsung Huang, *Science and Civilisation in China: Fermentations and Food Science* (Cambridge: Cambridge Univ. Press, 2000);

categories of knowledge. Though some have outlined the history of this organism in Eurasia by around the end of the nineteenth century, the involvement of modern science in tensions and negotiations among boundaries of natural knowledge about the organism in modern Chinese society remains little explored.[10] And it is already known that the Chinese state's pursuit of science as modernity upheld the discursive power of science in the first half of the twentieth century, which promoted the deconstruction of local natural knowledge.[11] Through a case study of the caterpillar fungus, this article probes new boundary issues that emerged from the encounter of local food with modern science in Republican China, and traces them to changing scholarship in nineteenth-century Japan and its influence on the Chinese people within transnational networks of knowledge.

PRODUCTION AND CONSUMPTION OF THE CATERPILLAR FUNGUS IN REPUBLICAN CHINA

According to Sherman Cochran, science played a strategic role in modern Chinese consumer culture.[12] However, the promotion of native edible substances or tonics in modern Chinese food culture would not necessarily invoke the power of science. Traditional accounts of the potency and transformative ability of the caterpillar fungus, for example, were able to provide impetus for the consumption of this product. However, deconstruction of its transformation in a scientific context, which has not yet been found in modern Chinese commercial advertisements, would perhaps undermine its sales. Nevertheless, the scientific enterprise in modern Chinese society wielded a growing influence on the public understanding of natural objects. The noted scholar Hu Shi, who had studied at Cornell University and Columbia University from 1910 to 1917, wrote in 1923: "In the recent three decades, a term has obtained a supreme position in China; one dares not look down upon or sneer at it whether he/she understands it or not, and whether he/she is fogyish or revolutionary. That term is 'science.'" Further, he pointed out that this "science" enjoyed nearly unanimous admiration throughout the country.[13] Under the influence of the style of this "science," as stressed by Hu Shi, the knowledge surrounding the caterpillar fungus was undergoing reconstruction.

New perceptions of the caterpillar fungus emerging in Chinese society in the first half of the twentieth century were initially elicited by Japanese scholarship. And

---

John A. G. Roberts, *China to Chinatown: Chinese Food in the West* (London: Reaktion, 2002); and Seung-Joon Lee, *Gourmets in the Land of Famine: The Culture and Politics of Rice in Modern Canton* (Stanford, Calif.: Stanford Univ. Press, 2011).

[10] Carla Nappi, *The Monkey and the Inkpot: Natural History and Its Transformations in Early Modern China* (Cambridge, Mass.: Harvard Univ. Press, 2009), 141–6; Bhushan Shrestha, Weimin Zhang, Yongjie Zhang, and Xingzhong Liu, "What is the Chinese Caterpillar Fungus *Ophiocordyceps sinensis* (Ophiocordycipitaceae)?" *Mycology* 1 (2010): 228–36; Di Lu, "Transnational Travels of the Caterpillar Fungus in the Fifteenth through Nineteenth Centuries: The Transformation of Natural Knowledge in a Global Context," *Asian Medicine* 12 (2017): 7–55.

[11] Sean Hsiang-Lin Lei, *Neither Donkey nor Horse: Medicine in the Struggle over China's Modernity* (Chicago: Univ. of Chicago Press, 2014), 91–6, 141–66. See also Ding Fubao, *Huaxue Shiyan Xinbencao (Xu)* (Shanghai: Wenming Shuju, 1909), 1–2; Chen Cunren, ed., *Zhongguo Yaoxue Dacidian* (Shanghai: Shijie Shuju, 1935), 1, 7–8; and Ding, *Chouyin Jushi Zizhuan* (Shanghai: Gulin Jingshe Chubanbu, 1948), 18–19.

[12] Sherman Cochran, *Chinese Medicine Men: Consumer Culture in China and Southeast Asia* (Cambridge, Mass.: Harvard Univ. Press, 2006), 109–15.

[13] Hu Shi, "Kexue Yu Renshengguan Xu," in *Hushi Wenji*, book 3, ed. Ouyang Zhesheng (Beijing: Beijing Daxue Chubanshe, 1998), 151–65.

Japan's acquisition of the scientific norms of European civilization benefited much from *rangaku* (Dutch learning), which spread among Japanese intellectuals in the context of the country's commercial exchange with the Netherlands, the only European country allowed to trade with Japan from the 1640s to 1854.[14] Though Dutch learning did not proceed smoothly due to ideological and political issues, European biological and medical knowledge was still periodically translated into Japanese.[15] A series of international and domestic incidents, including China's defeat in its two opium wars with European powers, and the conclusion of the Treaty of Kanagawa between the United States and Japan in 1854, provoked Japanese reflections on national destiny and modernization, and prompted the expansion of Japan's openness to the world as well as to Western science and technology.[16] In 1869, the newly established Meiji government set out to promote the German medicine it officially recognized.[17] Scientific communities and institutionalized scientific research arose in Meiji Japan.[18] And, as Morris Low indicates, Meiji science was tied to the "pursuit of national interests and profit," and "mobilized under an ideology aimed at building a nation-state."[19] Contrary to the booming of German medicine or modern science, native Kampo medicine suffered official oppression and fell into a dilemma of legitimacy several years after the Meiji restoration.[20] Meanwhile, the natural substances used in both Chinese and Kampo medicine became objects of chemical, biological, and pharmacological research supported by, for example, the Pharmaceutical Society of Japan.[21] Such new scholarship

---

[14] For the history of Dutch learning in Japan and its influence on Japanese scholarship, see Sugita Genpaku, *Rangaku Kotohajime* (1815; repr., Tokyo: Tenshinrō, 1869); Grant R. Goodman, *Japan: The Dutch Experience* (London: Athlone, 1986); and Federico Marcon, *The Knowledge of Nature and the Nature of Knowledge in Early Modern Japan* (Chicago: Univ. of Chicago Press, 2015), 127–39.

[15] Goodman, *Japan: The Dutch Experience* (cit. n. 14), 190–222. For examples, see Noro Genjō, *Oranda Honzō Wage* (Tokyo: National Diet Library, 1742–50); Hirokawa Kai, *Ranryō Yakukai* (Heian: Hayashi Gonbee, 1806); Kō Ryōsai, *Ranhō Naiyō Yakunōshiki* (Osaka: Shōendō, 1836); and Tsuboi Shinryō, *Shinyaku Hyakuhinkō* (Tokyo: Shimamura Risuke, 1866).

[16] Robert Hans van Gulik, "Kakkaron: A Japanese Echo of the Opium War," *Monumenta Serica* 4 (1940): 478–545; John K. Fairbank and Kwang-Ching Liu, eds., *The Cambridge History of China*, vol. 11, *Late Ch'ing, 1800–1911*, pt. 2 (Cambridge: Cambridge Univ. Press, 1980), 340–3; Masayoshi Sugimoto and David L. Swain, *Science and Culture in Traditional Japan* (Tokyo: Charles E. Tuttle, 1989), 291–346; Bob Tadashi Wakabayashi, "Opium, Expulsion, Sovereignty: China's Lessons for Bakumatsu Japan," *Monumenta Nipponica* 47 (1992): 1–25; Marius B. Jansen, *The Making of Modern Japan* (Cambridge, Mass.: Harvard Univ. Press, 2002), 257–93.

[17] John Z. Bowers, *When the Twain Meet: The Rise of Western Medicine in Japan* (Baltimore: Johns Hopkins Univ. Press, 1980), 105–7; Yoshio Izumi and Kazuo Isozumi, "Modern Japanese Medical History and the European Influence," *The Keio Journal of Medicine* 50 (2001): 91–9. For the German influence in the making of Meiji Japan, see Hoi-eun Kim, "Made in Meiji Japan: German Expatriates, German-Educated Japanese Elites and the Construction of Germanness," *Geschichte und Gesellschaft* 41 (2015): 288–320.

[18] Yuasa Mitsutomo, "The Growth of Scientific Communities in Japan," *Japanese Studies in the History of Science* 9 (1970): 137–58; James R. Bartholomew, *The Formation of Science in Japan: Building a Research Tradition* (New Haven, Conn.: Yale Univ. Press, 1989), 49–67.

[19] Morris Low, *Science and the Building of a New Japan* (New York, N.Y.: Palgrave Macmillan, 2005), 7–8.

[20] Shigeo Sugiyama, "Traditional Kampo Medicine: Unauthenticated in the Meiji Era," *Historia Scientiarum* 13 (2004): 209–23.

[21] This society launched the *Yakugaku Zasshi* (Journal of the Pharmaceutical Society of Japan) in 1881, which was devoted to modern pharmaceutical research. See also Yakazu Dōmei, "Meiji Jidai Ni Okeru Kanyaku no Yakurigakuteki Kenkyū Gyōseki To Sono Shiteki Kōsatsu: Shutoshite Inoko Yoshitoshi No Kanyaku Kenkyū O Megutte," *Nihon Tōyō Igaku Zasshi* 13 (1962): 111–19; and Yasuo Otsuka, "Chinese Traditional Medicine in Japan," in *Asian Medical Systems: A Comparative Study*, ed. Charles M. Leslie (Berkeley and Los Angeles: Univ. of California Press, 1976), 322–40.

on natural (medicinal) substances even affected some Chinese students educated in Japanese schools in the 1900s, though the caterpillar fungus and other insect-fungi had as yet received little attention from the field of pharmacology.[22]

The caterpillar fungus retained its popularity in Chinese society. As the twentieth century commenced, caterpillar fungus found growing in Tibet began to be recorded as a medicinal product in local chronicles; this suggested increasing local attention to its medicinal or commercial value.[23] Geographically considered, when this product from Tibet was transported eastward, it had to first pass through Sichuan, Yunnan, and/or Qinghai, where it also grew. According to an investigation in 1919, for example, the caterpillar fungus was then already a special export product of Yushu, Qinghai. Many people were collecting it there and trading with merchants; the collecting activities even bred discontent among local headmen, who thought that the "pulse" of the land was thus being broken by the impact of digging for this product, and many flocks and herds were dying in consequence.[24] In 1937, the caterpillar fungus was still listed among the medicinal substances that constituted a significant percentage of the special local products of Yushu.[25] Doubtless the trade in this product contributed to the local economy in Yushu, despite tensions between local headmen and merchants. Compared with Tibet and Qinghai, however, Sichuan and Yunnan were more widely known as the areas where it grew naturally.

An extensive survey of products in postal delivery areas of China, conducted by the General Post Office of the Ministry of Communications during the period from the spring of 1934 to February 1936, enables an overview of the production and nationwide dissemination of the caterpillar fungus in the mid-1930s. With this investigation came a book published in 1937, whose prefaces indicated its aim of giving a general idea of available Chinese products so as to facilitate their procurement. Underlying the investigation was the belief that the flourishing of local commodities on the national market would help rescue the war-beleaguered national economy from shortages and crises, while simultaneously improving the postal services.[26] The investigation, though excluding a few regions such as today's western Tibet, reveals the importance of the Yangtze River in transporting the caterpillar fungus from its production areas to southeastern China. The bulk of the caterpillar fungus was produced in today's Sichuan and then Yunnan provinces.[27] In particular, the caterpillar fungus from Maogong, Kangding, and Lijiang was treated as a representative product of Sichuan and Yunnan, and thus was photographed for the investigation.[28] The significance of geographical information on its areas of growth is rooted in, as Emily T.

[22] For example, see Wang Huanwen, "Bukuryō No Seibun Ni Ju Te," *Yakugaku Zasshi* 327 (1909): 461–72.

[23] For examples, see Duan Pengrui, "Yanjing Xiangtuzhi" (first published 1909); Liu Zanting, "Dingqingxian Tuzhi" (ca. 1917); and Liu Zanting, "Wucheng Xianzhi" (ca. 1921), all in *Zhongguo Difangzhi Jicheng* [Collection of local chronicles of Tibet] (repr. together, Chengdu: Bashu Shushe, 1995); 391–424, on 405; 555–578, on 573; and 129–156, on 150, respectively.

[24] Zhou Xiwu, "Yushu Diaochaji," in *Zhongguo Fangzhi Congshu* [Northwest region, book 37] (1919; repr., Taipei: Chengwen Chubanshe, 1968), 149–50, 180.

[25] Ma Hetian, *Ganqingzang Bianqu Kaochaji* (Shanghai: Shangwu Yinshuguan, 1947), 374–5, 386–8.

[26] *Zhongguo Tongyou Difang Wuchanzhi* (Shanghai: Shangwu Yinshuguan, 1937), 5–6.

[27] Ibid., 529–43, 607, 655–65, 1089.

[28] Ibid., 527, 609, 651; cf. Shina Shōbetsu Zenshi Kankōkai, *Shinshū Shina Shōbetsu Zenshi*, vol. 2 (Tokyo: Tōa Dōbunkai, 1941), 383.

Yeh and Kunga T. Lama state, "the fact that the larvae-fungus complex cannot be cultivated," which means that "nonhuman nature determines where it can and cannot be found."[29]

By the mid-1930s, price lists for medicinal products from Kangding and Lijiang demonstrate that the caterpillar fungus was significantly more expensive than most or all of the other medicinal plants and fungi, but was cheaper than medicinal animal products such as bear bile.[30] In about 1935, a drugstore of repute in Shanghai sold the caterpillar fungus for 44.24 *yuan/jin*, about 4.9–7.4 times the prices in Kangding, where it fetched 6.00–9.00 yuan/jin.[31] By comparison, common strains of rice sold in Shanghai in 1935 were a fraction of the price, around 0.067–0.080 yuan/jin.[32] A variety of other sources from the 1930s–40s both confirm the production of the caterpillar fungus in west China, and show how it sold well in Shanghai, Hong Kong, and some other cities in southeastern China.[33] In 1947, the export volume and prices of medicinal products from Chongqing (then a part of Sichuan) were reported to have recently increased significantly; some merchants trading between Guangdong and Chongqing were even willing to spend about twice as much money for the caterpillar fungus as they had before.[34] Clearly, over the decades leading up to the Communist victory, economic interests facilitated a west-east trend in the transportation of the caterpillar fungus to economically more developed areas. In November 1948, the governor of Changdu (a city in today's eastern Tibet) issued a proclamation whose first demand was the opening up of the mountains, previously forbidden by local lamaseries; this change would allow people to collect natural products (including the caterpillar fungus) there.[35] The announcement doubtless boosted collection of the caterpillar fungus and thus promoted the growth of the local economy. But today, scientists find that overexploitation and climate change are threatening the sustainability of the ecology and economy of this product.[36]

The growth and domestic circulation of products was then accompanied by lively consumer demand for them in Chinese society. In many cases, there was no distinct boundary between medicinal products and food, and tonics for improving health, rather than treating illness, were embedded throughout the region, and sold in drugstores, dispensaries, food companies, and restaurants.[37] When traditional physicians

[29] Emily T. Yeh and Kunga T. Lama, "Following the Caterpillar Fungus: Nature, Commodity Chains, and the Place of Tibet in China's Uneven Geographies," *Social & Cultural Geography* 14 (2013): 318–40.

[30] *Zhongguo Tongyou Difang Wuchanzhi* (cit. n. 26), 607, 662.

[31] Hu Anbang, *Shiyong Yaoxing Zidian* (Shanghai: Zhongyang Shudian, 1935), 40. One jin was then equal to 500 grams.

[32] *Shanghai Jiefang Qianhou Wujia Ziliao Huibian (1921–1957)* (Shanghai: Shanghai Renmin Chubanshe, 1958), 217.

[33] For examples, see "Jinsannian Xikang Shexiang Chongcao Chukou Tongji," *Guoji Maoyi Qingbao* 1 (1936): 74; Zhuang Xueben, *Qiangrong Kaochaji* (Shanghai: Liangyou Tushu Yinshua Gongsi, 1937), 127–8; "Chuanxi Diaochaji," in *Zhongguo Bianjiang Shehui Diaocha Baogao Jicheng*, collection 1, book 5 (1941; repr., Guilin: Guangxi Shifan Daxue Chubanshe, 2010), 494–5; and Long Yun, *Xinzuan Yunnan Tongzhi*, book 4 (1944; repr., Kunming: Yunnan Renmin Chubanshe, 2007), 126–9.

[34] "Benshi Guoyao Zhangchao Fanlan," *Zhengxin Xinwen* 646 (1947): 7.

[35] "Changdu Gelunlalu Gaoshi Xiaochu Zangkang Liangzu Jiexian," *Shenbao*, 23 November 1948, section 2.

[36] Kelly A. Hopping, Stephen M. Chignell, and Eric F. Lambin, "The Demise of Caterpillar Fungus in the Himalayan Region due to Climate Change and Overharvesting," *PNAS* 115 (2018): 11489–94.

[37] Fan Yajun, "Zibu Yu Jiankang: Shenbao Buyao Guanggao De Shehui Wenhuashi Yanjiu, 1873–1945" (master's thesis, Nanjing University, 2012), 14–36.

used the caterpillar fungus in medical treatments in late Qing China, merchants also explored its commercial value by developing new products.[38] In the Republican period, the caterpillar fungus continued to be sold in drugstores in Suzhou and other domestic and overseas cities.[39] By the 1930s, it had become a staple product of today's eastern Tibet and western Sichuan, finding much favor with wealthy people in Guangdong, Fujian, Shanghai, and Nanjing, despite its high price.[40] The value of the caterpillar fungus was greatly exploited in Shanghai, the commercial capital of China. The food company Guanshengyuan, for example, once advertised its new tonic food called the caterpillar fungus-duck on New Year's Day, 1925.[41] Some shrewd restaurants also served tonic dishes involving the use of the caterpillar fungus, such as *chongcao ruge* (caterpillar fungus-young pigeon).[42] These dishes count as variations on the caterpillar fungus-duck combination described by Parennin about two centuries earlier. Some gourmets actively introduced the recipes for these dishes to the public. One of them promoted such a recipe with additional reference to the magical transformation of the caterpillar fungus (from a blade of grass to a worm), and a premodern medical record of its potency, in the journal *Changshou* (Longevity), published in Shanghai.[43] Even in 1946, the latest guidebook to Shanghai listed the duck stewed with the caterpillar fungus as a famed tonic dish produced in local Sichuan-style restaurants and allegedly only sold to frequent customers.[44]

So profitable and popular was the caterpillar fungus that some speculative merchants began to sell it even though it was beyond the scope of their original business. For example, a snow fungus (*Tremella fuciformis*) company in Shanghai advertised on 4 January 1928 that it sold not only the snow fungus, but medicinal substances and tonics such as the caterpillar fungus.[45] In another advertisement on 23 September of the same year, a Sichuan store in Shanghai claimed it also sold the caterpillar fungus and some other specialist products of Sichuan.[46] It is worth adding here that the *guohuo* (national products) movements prospering in Republican China and delivered in highly nationalistic and anti-imperialist tones were infused with propaganda about how the products could benefit national economic interests.[47] Against this background, the

---

[38] For examples, see Wang Shixiong, "Suixiju Chongding Huoluanlun," in *Wangmengying Yixue Quanshu*, ed. Sheng Zengxiu (1862; repr., Beijing: Zhongguo Zhongyiyao Chubanshe, 1999), 173; Zhang Naixiu, *Zhangyuqing Yi'an* (ca. 1905; repr., Shanghai: Shanghai Kexue Jishu Chubanshe, 1963), 128–9; "Jishou Chongcaogao," *Shenbao*, 19 December 1881, section 6; and "Shenqi Chongcaogao," *Shenbao*, 7 November 1884, section 6.

[39] Curtis G. Lloyd, "Cordyceps sinensis, from N. Gist Gee, China," *Mycological Notes* 54 (1918): 766–80; "Xuzhongdao Guoyao Zongfenhao Shijia Lianhe Jintian Dajianjia," *Shenbao*, 30 September 1932, section 17.

[40] "Xikang 'Chongcao' Chukou Jushu," *Fangzhou* 12 (1934): 12; Ran, "Chongcao," *Shusheng Zhoubao* 51 (1937): 16.

[41] "Xinfaming Dongchongcaoya Shangshi," *Shenbao*, 1 January 1925, section 19.

[42] See, for examples, "Nanyuan Jiujia," *Shenbao*, 16 November 1928, section 21; "Weiya Jiulou Xinfengji Shangshi," *Shenbao*, 21 September 1929, section 16; and "Yanhualou Jiujia Zhi Zibu Dunpin," *Shenbao*, 1 November 1929, section 25.

[43] Shen Xi, "Dongchong Xiacao Weiya," *Changshou* 144 (1935): 350.

[44] Leng Xingwu, *Zuixin Shanghai Zhinan* (Shanghai: Shanghai Wenhua Yanjiushe, 1946), 107.

[45] "Shutongsen Yinerzhuang Jianjia Zhanqi," *Shenbao*, 4 January 1928, section 21.

[46] "Sichuan Shangdian (Yizhou Jinian) Yiner Dajianjia," *Shenbao*, 23 September 1928, section 13.

[47] Karl Gerth, "Consumption as Resistance: The National Products Movement and Anti-Japanese Boycotts in Modern China," in *The Japanese Empire in East Asia and its Postwar Legacy*, ed. Harald Fuess (Munich: Iudicium, 1998), 119–42; Gerth, *China Made: Consumer Culture and the Creation of the Nation* (Cambridge, Mass.: Harvard Univ. Press, 2003), 125–202.

largest national products exhibition held in China at that time took place in Shanghai between 1 November 1928 and 3 January 1929. Two official representatives of Sichuan and Yunnan promoted the caterpillar fungus among visitors as one of their most prized local medicinal products.[48] As the consumer market for the caterpillar fungus was not confined to China, a company that was probably hoping to expand its overseas market advertised its caterpillar fungus stocks and futures in the weekly for Shanghai's Consulting Institute for International Trade on 20 June 1946.[49]

The persistent circulation of the esteemed caterpillar fungus is a perfect example of the dynamic production and consumption of local medicinal and edible products in Republican China. Despite the Nationalist government imposing certain legal restrictions on the sale and use of Chinese medicinal substances, there was a substantial gap between expectation and realization before the outbreak of the full-scale Sino-Japanese War in 1937, due to the resistance of medicine merchants, local governments' dereliction of duty, and such.[50] During the war, the Nationalist government had temporarily softened the restrictions, mainly because of medication shortages.[51] Meanwhile, the Communist regime encouraged the employment of both native and imported medicinal substances in its wars with the Japanese army and the Nationalist government.[52] These circumstances, together with the power of tradition, ensured the ongoing use of Chinese medicinal substances. Traditional physicians continued to apply the caterpillar fungus in their medical practices.[53] Indigenous knowledge about the caterpillar fungus and many other medicinal substances also circulated through a variety of medical and popular publications and schools, and persisted throughout the Republican period.[54] Moreover, many medicinal substances, like the caterpillar fungus, could

---

[48] Dong Shaoshu, "Zhonghua Guohuo Zhanlanhui: Dong Shaoshu Zhi Yanci," *Shenbao*, 26 December 1928, section 14; Li Kuian, "Zhonghua Guohuo Zhanlanhui: Sichuan Daibiao Li Kuian Zhi Baogao," *Shenbao*, 29 December 1928, section 13. More than ten thousand invited guests and fifty thousand tourists attended the exhibition; see Hong Zhenqiang, "1928 Nian Zhonghua Guohuo Zhanlanhui Lunshu," *Huazhong Shifan Daxue Xuebao (Renwen Shehui Kexueban)* 45 (2006): 83–8.

[49] Guoji Maoyi Zixunsuo, "Chukou Xiaoxi," *Jinchukou Maoyi Xiaoxi* 122 (1946): 1.

[50] Di Lu, "Minguo Shiqi Yaoshang He Putong Yaopin Guanli Fagui De Zhiding Yu Tuixing," *Jindai Zhongguo* 27 (2017): 77–102.

[51] Wen Xiang, *Yizhi Yu Chaoyue: Minguo Zhongyi Yizheng* (Beijing: Zhongguo Zhongyiyao Chubanshe, 2007), 102–8.

[52] Yang Lisan, "Di Shiba Jituanjun Yezhan Houqinbu Yang Lisan Buzhang Zai Yaoping Cailiaochang Gongzuo Huiyi Shang De Zongjie," in *Liudeng Dajun Weisheng Shiliao Xuanbian*, ed. He Zhengqing (1941; repr., Chengdu: Chengdu Keji Daxue Chubanshe, 1991): 27–30; Jin Jin, ed., *Zhongguo Renmin Jiefangjun Yaocai Gongzuo Shi* (Beijing: Zong Houqin Bu Weisheng Bu, 1997), 29–32, 66–7, 119, 151; John R. Watt, *Saving Lives in Wartime China: How Medical Reformers Built Modern Healthcare Systems amid War and Epidemics, 1928–1945* (Leiden: Brill, 2013), 77–95.

[53] Lu Jinsui, *Jingjing Yihua* (1916; repr., Taiyuan: Shanxi Kexue Jishu Chubanshe, 1999), 1382; Ding Ganren, *Ding Ganren Yi'an* (1927; repr., Shanghai: Shanghai Kexue Jishu Chubanshe, 1960), 106; Qin Bowei, *Qianzhai Gaofangan* (1938; repr., Fuzhou: Fujian Kexue Jishu Chubanshe, 2007), 43; Shi Jinmo, *Zhuxuan Shi Jinmo Yi'an* (1940; repr., Beijing: Huaxue Gongye Chubanshe, 2010), 46–7.

[54] For indigenous knowledge about the caterpillar fungus in Republican medical texts, see, for examples, Wenming Shuju, *Yaoxing Yizhi* (Shanghai: Wenming Shuju, 1919), 9; Xie Guan, *Zhongguo Yixue Dacidian* (Shanghai: Shangwu Yinshuguan, 1921), 668–9; Hu Fangxi, "Zengbu Bencaoshi: Dongchong Xiacao," *Zhongyi Zazhi* 16 (1925): 7; Lu Peng, *Yaowuxue Jiangyi* (ca. 1929; repr., Beijing: Zhongguo Zhongyiyao Chubanshe, 2016), 23; Weisheng Baoguan, *Zhongyao Dacidian* (Shanghai: Weisheng Baoguan, 1930), 86; Zhang Shanlei, *Bencao Zhengyi* (1932; repr., Fuzhou: Fujian Kexue Jishu Chubanshe, 2015), 102–3; Hu, *Shiyong Yaoxing Zidian* (cit. n. 31), 40; "Daodi Yaocai," *Liangyou* 158 (1940): 12; Cai Luxian, *Zhongguo Yiyao Huihai*, book 1 (Shanghai: Zhonghua Shuju, 1941), 537; and Zhou Zhilin, *Bencao Yongfa Yanjiu* (1941; repr., Shanghai: Zhonghua Shuju, 1948), 756–7. The Shanghai Specialist School of Chinese Medicine (1916–48) is a representative school that

be used as food or as culinary ingredients, and thereby could circumvent the state regulation mentioned above, just as happens in Europe and China today.

Following the Communist victory, under particular political, social, and economic conditions, Chinese medicine began to enjoy a much higher social standing than it had in the Republican period.[55] Traditional understanding of the caterpillar fungus continued, and appeared in, for example, the 1963 Chinese national pharmacopeia.[56] In 1953, Chairman Mao had also presented some caterpillar fungus as a gift to a teacher he had once had.[57] In some cities of early Communist China, like Chongqing, the dish known as steamed duck with caterpillar fungus featured in the celebrated local cuisine.[58]

CHANGING PERCEPTIONS OF THE CATERPILLAR FUNGUS IN NINETEENTH-CENTURY JAPAN

The flourishing of European natural history in Japan's Edo and Meiji periods boosted new passions for observing, describing, and collecting native or exotic natural objects.[59] The Linnaean classification system, beginning to take root in Japan in the early nineteenth century, also prompted the equivalence between some East Asian and European scientific (Latin) names for indigenous organisms.[60] As the caterpillar fungus continued to be exported to Japan and sold in Japanese drugstores in the early nineteenth century, some local naturalists and physicians were no longer satisfied with learning about it from previous Chinese and Japanese records.[61] They sought to discover this natural curiosity in their own country, though it did not inhabit Japan and hence had never been found there. But this trend led to new natural history discoveries and reflections in the encounter between East Asian and European academic traditions. In his 1801 collection of drawings, the physician Yuzuki Tokiwa grouped the caterpillar fungus from China together with some similar insect-fungi growing in Japan under the name of kasō tōchū; he illustrated their different morphological characteristics and particularly recorded the former as imported.[62] Federico Marcon points out that by the late Edo period, "accurate and detailed illustrations of plants and animals developed as a new cognitive apparatus to identify species and solve the old

---

devoted itself to education in Chinese medicine in the Republican era. For the history of this school, see *Mingyi Yaolan: Shanghai Zhongyi Xueyuan (Shanghai Zhongyi Zhuanmen Xuexiao) Xiaoshi* (Shanghai: Shanghai Zhongyiyao Chubanshe, 1998).

[55] Kim Taylor, *Chinese Medicine in Early Communist China, 1945–63* (London: RoutledgeCurzon, 2005), 30–62, 151–3.

[56] Weishengbu, *Zhonghua Renmin Gongheguo Yaodian* (Beijing: Renmin Weisheng Chubanshe, 1964), 77.

[57] Li Shuntong, *Daifang Shuwu Wenji* (Xiangtan: Xiangtan Daxue Chubanshe, 2013), 207.

[58] Chongqingshi Yinshi Fuwu Gongsi, *Chongqing Mingcaipu* (Chongqing: Chongqing Renmin Chubanshe, 1960), 38–9.

[59] Nishimura Saburo, *Bunmei No Naka No Hakubutsugaku: Seiō to Nihon*, vol. 1 (Tokyo: Kinokuniya Shoten, 1999), 129–35; Itō Mamiko, "19 Seikinihon No Chi No Chōryū: Edo Kōki ~ Meiji Shoki No Hakkajiten, Hakubutsugaku, Hakurankai," *19 Seikigaku Kenkyū* 6 (2012): 59–78; Jung Lee, "Provincialising Global Botany," in *Worlds of Natural History*, ed. Helen A. Curry, Nicholas Jardine, James A. Secord, and Emma C. Spary (Cambridge: Cambridge Univ. Press, 2018), 433–46.

[60] Itō Keisuke, "Tōyō Shokubutsugaku No Ichi Daikaikaku Wonasazaruka Karazu," *Shokubutsugaku Zasshi* 19 (1888): 177–81; Kitamura Siro, "The Japanese Studies on the Chinese Plants," *Acta Phytotaxonomica et Geobotanica* 1 (1989): 119–22.

[61] Fujii Kansai, *Zōho Shuhan Hatsumō* (Tokyo: Yamashiroya Sahei, 1829), 347–50.

[62] Yuzuki Tokiwa, *Hakurai Kasōtōchū Zu* (Tokyo: National Diet Library, 1801), 2–13.

problem of matching Chinese names with actual plants and animals."[63] Yuzuki's drawing and his use of the Japanese term kasō tōchū, which contains the same two pairs of characters in the Chinese term xiacao dongchong (summer grass winter worm), also contributed to the solving of the "old problem" of matching Chinese names with actual organisms. But for the Japanese, kasō tōchū broadly denoted a group of insect-fungi rather than merely the caterpillar fungus.

Yuzuki's record received attention from the herbalist Ohara Momohora (1746–1825), who learned about the geographical origin and medicinal properties of the caterpillar fungus from some Japanese and Chinese accounts. He agreed with Yuzuki that similar organisms grew in Japan, as some Japanese publications had reported discoveries of such organisms around ditches and courtyards in 1805, 1808, and 1824. In his posthumous manuscript, which contains illustrations of eleven specimens of such native organisms, Ohara suspected that some of the insect-fungi found in Japan were *semihana* (*chanhua* in Chinese, which literally means flowers on cicada); this was a medicinal substance that had long been used in China.[64] Another herbalist, named Mizutani Toyofumi (1779–1833), once also depicted specimens of such insect-fungi, or "flowers on cicada," in his drawings of insects and animals.[65] The specimens mainly differ from each other in the morphological characteristics of their fruiting bodies. From the early nineteenth century to the early twentieth century, discoveries of insect-fungi were occasionally being made in Japan.[66] Meanwhile, the term kasō tōchū, or its inverted form tōchū kasō of Chinese origin, was also often used in a broadened sense to denote insect-fungi in relevant Japanese publications.[67] To differentiate it from its Chinese homonym, some Japanese authors accentuated the geographical origins of the organisms in question when they used the terms. For example, Fujii Kansai stated that he had seen both the Chinese caterpillar fungus sold in Japanese drugstores and similar organisms native to Japan. The entry for tōchū kasō in his 1829 text on materia medica explicitly identifies two kinds of such organisms: one is *hakurai* (imported), while the other is *kazusan* (produced in Japan).[68]

However, hakurai is an ambiguous expression, because it does not specify where the caterpillar fungus was imported from. In the late nineteenth and early twentieth centuries, some Japanese scientists began to use the more specific term *kansan* or *shinasan* (produced in China) to refer to the caterpillar fungus from China, or *Cordyceps sinensis* or *Sphaeria sinensis*.[69] Clearly, the attempts to seek a "Japanese" caterpillar fungus coincided with reflections on new relationships between names and entities. The discoveries of similar organisms in Japan also presented new findings on the

[63] Marcon, *The Knowledge of Nature* (cit. n. 14), 228.

[64] Ohara Momohora, *Momohora Ihitsu*, vol. 3 (Wakayama: Sakamotoya Kiichirō, 1833), 29–36.

[65] Mizutani Toyofumi, *Mushimujina Shashin* (Tokyo: National Diet Library, ca. 1833), 88–90. However, Mizutani's drawings of the insect-fungi lack captions.

[66] Ezaki Teizō, "Fukuoka Agata Yamegun San Kasō Tōchū Nijute," *Kyushu Teikuni Daigaku Nōgakubu Gakugei Zatsushi* 3 (1929): 221–31.

[67] For example, see Oda Seisuke, "Tōchū kasō," *Konchū Sekai* 2 (1898): 465.

[68] Fujii, *Zōho Shuhan Hatsumō* (cit. n. 61), 347–8.

[69] Kurita Manjirō, "Zoku Shina Hakubutsu Ikō (Shōzen)," *Tōkyō Chigaku Kyōkai Hōkoku* 11 (1889): 29–32; Shirai Mitsutarō, *Shokubutsu Yōikō* (1914; repr., Tokyo: Oka Shoin, 1925), 364–6. At the beginning of the twenty-first century, the Japanese scholar Okuzawa Yasumasa uses the word *kōgi* (broad sense) as an addition to the term tōchū kasō, serving the purpose of disambiguation; see Okuzawa Yasumasa, "Tōchū Kasō (Kōgi) Torai No Rekishi To Yakubutsu To Shite No Juyō," *Nihon Ishigaku Zasshi* 53 (2007): 178–9.

geographical distribution of insect-fungi. The transformation of the Chinese term xiacao dongchong, or dongchong xiacao, to the Japanese term kasō tōchū, or tōchū kasō, together with the broadened meaning of its identifications in the Japanese context, point to the semantic boundaries of the same word, and indicate a Japanization of the category for the Chinese caterpillar fungus. This accords with Benjamin A. Elman's analysis of the adaptation of Chinese medicine and appropriation of Chinese thoughts and learning before the late nineteenth century in Japan.[70] The altered category crossed the boundary between the caterpillar fungus and other insect-fungi. It also counts as a response to European natural history, because it could accommodate European natural knowledge about the fungi parasitic on insects.

The introduction of scientific information on insect-fungi testifies to European influence and to a pluralistic understanding of such organisms in nineteenth-century Japan. Even in the late nineteenth century, some Japanese still supported the transformation theory of the caterpillar fungus, and/or applied the theory to native insect-fungi.[71] Nevertheless, since the early nineteenth century, some Chinese knowledge about the caterpillar fungus had become a target for criticism. The naturalist Masushima Ranen mentioned this organism in his 1811 book on fungi. He related the "grass" to *kin* (fungi) and emphasized that the formation of the "winter worm summer grass" was absolutely not caused by the extremely absurd transformation, but by fungal infections of dead insects underground. Still, he valued Chinese medical knowledge about the caterpillar fungus, and suggested not abandoning it with the fallacious transformation theory. This concern for medical utility explains why he particularly quoted a related Qing Chinese medical record.[72] Some late nineteenth-century Japanese botanical and entomological articles also sometimes set out to inform readers about the true nature of the caterpillar fungus and similar organisms. For example, stimulated by an inquiry about the caterpillar fungus and its transformation, Miyoshi Manabu, then studying botany at the Imperial University of Tokyo, published a review article in *Shokubutsugaku Zasshi* (The Botanical Magazine) in 1888.[73] He aimed to help readers abandon their belief in fallacious ideas. After quoting Ohara's account, he turned to a few Chinese and English publications, including Mordecai C. Cooke's mycological monograph.[74] Miyoshi treated tōchū kasō as a taxonomic group of organisms, and enumerated nine species of fungi belonging to the genus *Torrubia*. According to the article, twenty-five species of the insects on which these fungi grew, such as *Hepialus virescens*, had been discovered; and both the insects and fungi were distributed around the world. With this assertion, the caterpillar fungus not only lost its ability to transform, but also lost its value as being a rare fungus.

---

[70] Benjamin A. Elman, "Sinophiles and Sinophobes in Tokugawa Japan: Politics, Classicism, and Medicine during the Eighteenth Century," *East Asian STS* 2 (2008): 93–121.

[71] For example, see Umeno Takizō and Mitani Yūshin, *Chikugo Chishi Ryaku* (Kurume: Kinbundō, 1879), 45.

[72] Shirai, *Shokubutsu Yōikō* (cit. n. 69), 361–2.

[73] Miyoshi Manabu, "Tōchū Kasō No Ben," *Shokubutsugaku Zasshi* 2 (1888): 36–40. For a brief chronicle of Miyoshi's life, see Andou Yutaka, "Shokubutsu Gakusha Miyoshi Manabu Kenkyū Shiryō: IV," *Kiyoizumi Jogakuin Tankidaigaku Kenkyū Kiyō* 13 (1995): 67–90.

[74] This monograph is *Fungi: Their Nature, Influence, and Uses*, which, however, had been published in different editions before 1888. For the original account in its first edition, see Mordecai Cubitt Cooke, ed. Miles Joseph Berkeley, *Fungi: Their Nature, Influence, and Uses* (London: Henry S. King, 1875), 246–7.

A few years later, in 1894, Yasuda Atsushi, a botany student at the Imperial University of Tokyo, reported his identifications of two species of parasitic fungi.[75] One was "*Isaria arachnophila*, Ditm.," found growing on a trapdoor spider; the other was "*Torrubia militaris*, Fr.," found growing on some species belonging to the order Lepidoptera. Yasuda discovered them in Japan, and generally called them tōchū kasō. Like Miyoshi, he criticized the transformation theory as a fallacy, though the emphasis of his articles was on macroscopic and microscopic descriptions of the specimens, which were given to support his identifications. He employed mycological terms to describe their morphological structures, such as *shijitsutai* (stroma), *hōshi* (spore), *kinshi* (mycelium), and *hachiretsushi* (ascospore).[76] Besides, he also used the characters such as *ka* (family), *zoku* (genus), and *tane* (species) to describe their taxonomic ranks. The concept of species and taxonomic hierarchy, and the application of microscopic observation in identifying species, doubtless originated in modern European biology. In his articles, the two specimens had formed as follows: fungal spores infected underground insects, developed into mycelium inside the insects, and eventually killed them; after having occupied the interior of the dead insect bodies, the mycelium then grew out of the bodies and formed visible fruiting bodies. Yasuda's identifications and theoretical explanations of the formation of the fungi embody the tensions between East Asian and European perceptions of nature. In particular, the microscope, which spoke for the epistemic virtue of what Lorraine Daston and Peter Galison call "mechanical objectivity," enabled Japanese biologists to "see" inaccessible and invisible regions of nature.[77] The power of new scientific instruments (e.g., the microscope and telescope), perceived by modern Europeans as "evidence of the superiority of their age over antiquity," and aiding "fresh and truthful observations," was also adopted into the powerful rhetoric of modern science in Japan.[78]

Like Miyoshi and Yasuda, Oda Seisuke, who had been trained at an agricultural school, also criticized the old transformation theory in his short, exoteric article about the diversity of native insect-fungi. The article, directly entitled "Tōchū Kasō," and published in the magazine *Konchū Sekai* (Insect world) in 1898, introduces the biological nature, taxonomic positions, and habitat of insect-fungi, and gives a scientific explanation of their formation.[79] In contrast, an 1889 article by the naturalist Kurita Manjirō primarily focuses on the caterpillar fungus. But Kurita also associated it with similar Japanese insect-fungi, and still called the latter kasō tōchū.[80] Kurita first wrote of its fungal nature, its identity as a famous Chinese medicinal substance, and its

---

[75] Yasuda Atsushi, "Chitsutō Ni Kisei Suru Tōchūkasō Ni Ju Te," *Shokubutsugaku Zasshi* 8 (1894): 337–40; Yasuda, "'Kisa Nagi Take' (Tōchū Kasō No Isshu) Torrubia militaris, Fr.," *Shokubutsugaku Zasshi* 8 (1894): 410–11. Yasuda graduated from the university in 1895; see *Imperial University of Tōkyō: The Calendar* (Tokyo: Imperial University, 1898): 333.

[76] The English terms "stroma" and "ascospore" are not my own translations but are directly cited from Yasuda's articles.

[77] Lorraine Daston and Peter Galison, *Objectivity* (New York, N.Y.: Zone Books, 2007), 115–90.

[78] Albert Van Helden, "The Birth of the Modern Scientific Instrument, 1550–1770," in *The Uses of Science in the Age of Newton*, ed. John G. Burke (Berkeley and Los Angeles: Univ. of California Press, 1983), 65; Jennifer Tucker, *Nature Exposed: Photography as Eyewitness in Victorian Science* (Baltimore: Johns Hopkins Univ. Press, 2005), 187.

[79] Oda Seisuke, "Tōchū Kasō," *Konchū Sekai* 2 (1898): 465. For Oda's educational background, see "Daiichikai Zenkuni Gaichū Kujo Shūgyōsei Seimei," *Konchū Sekai* 3 (1899): 397–8.

[80] Kurita Manjirō, "Zoku Shina Hakubutsu Ikō (Shōzen)," *Tōkyō Chigaku Kyōkai Hōkoku* 11 (1889): 29–32.

scientific name by referring to John Lindley's *The Vegetable Kingdom* (1853). Then he quoted related records from three materia medica texts in the English, Chinese, and Japanese languages respectively.[81] The popular transformation theory did not receive his direct criticism. And Kurita seemed to avoid acting as a judge of true or fallacious knowledge, rather endeavoring to tolerate and bring together knowledge from different cultures. However, the terms "Sphaeria Sinensis, Berk." and "*kinzoku*" (fungi) introduced at the beginning of the article already indicated the priority of European scholarship on the natural properties of the caterpillar fungus in his mind.

THE SHAPING OF A SCIENTIFIC CATERPILLAR FUNGUS IN REPUBLICAN CHINA

After receiving the caterpillar fungus as a gift from one of his Sichuan friends, along with advice on its culinary and medical purposes, the Confucian scholar Yu Yue (1821–1907), a native of Zhejiang, praised it as a *lingyao* (panacea), with the ability to transform between a winter worm and a summer grass, and to exist beyond life and death.[82] Chinese literati lamented and extolled the virtues of the caterpillar fungus in this way and associated it with their personal experiences and reflections on the potency and immortality of their culture.[83] From the beginning of the twentieth century, however, challenges to previous narratives of the caterpillar fungus began to appear in China.

Humiliated by defeat in the Sino-Japanese War (1894–95), Chinese central and provincial governments "sought Japanese expertise on topics relating to modernization," such as finance, science, education, and engineering; and "for most Chinese, Japanese imperialism was not yet seen as a problem."[84] Education in China underwent profound transformation in the 1900s. By 1905, as Benjamin A. Elman indicates, "the new Qing Ministry of Education was staunchly in favor of science education and textbooks based on the Japanese scientific system"; and the *Nongxue Bao* (Journal of Agriculture, Shanghai), published from 1897 to 1906, was among the many periodicals and books that mediated "Japanese-style science and technology" for the Chinese.[85] In August 1900, the Japanese sinologist Fujita Toyohachi published a Chinese translation of Oda Seisuke's 1898 article on tōchū kasō in *Nongxue Bao*.[86] It propagated new and disenchanted natural knowledge about insect-fungi among the

---

[81] The three texts are Frederick Porter Smith's *Contributions towards the Materia Medica & Natural History of China* (London: Trübner, 1871), Zhao Xuemin's *Bencao Gangmu Shiyi* (Qiantang: first printed by Zhang Yingchang, finalized ca. 1803), and Fujii Kansai's *Zōho Shuhan Hatsumō* (Edo: Yamashiroya Sahei, 1829). Kurita recorded the title and the author of the English book as *Shina Yakuhin Bikō* and Sumisu (a Japanese transliteration of the English word "Smith"). Besides, Kurita added that he referred to the seventy-third page of Smith's text. These clues, together with the content of the quotation, lead us to Frederick Porter Smith's 1871 book on materia medica and natural history.

[82] Yu Yue, "Chunzaitang Shibian," in *Xuxiu Siku Quanshu*, book 1551, ed. Gu Tinglong (1899; repr., Shanghai: Shanghai Guji Chubanshe, 2002), 559.

[83] For examples, see Zhang Weiping, "Guochao Shiren Zhenglue," in *Xuxiu Siku Quanshu*, book 1713, ed. Gu Tinglong (1819; repr., Shanghai: Shanghai Guji Chubanshe, 2002), 1–401, on 3; Zhang Shu, "Suyangtang Shiji," in *Xuxiu Siku Quanshu*, book 1506, ed. Gu Tinglong (1842; repr., Shanghai: Shanghai Guji Chubanshe, 2002), 119–415, on 270; and Fan Xinghuan, "Cao Fugu Dongchong Xiacao Shi," *Shaoxing Yiyao Xuebao* 32 (1910): 9.

[84] June T. Dreyer, *Middle Kingdom and Empire of the Rising Sun: Sino-Japanese Relations, Past and Present* (Oxford: Oxford Univ. Press, 2016): 53.

[85] Benjamin A. Elman, "Toward a History of Modern Science in Republican China," in *Science and Technology in Modern China, 1880s–1940s*, eds. Jing Tsu and Benjamin A. Elman (Leiden: Brill, 2014): 15–38, on 27–9.

[86] Oda Seisuke, "Dongchong Xiacao," trans. Fujita Toyohachi, *Nongxue Bao* 114 (1900): 484–5.

Chinese. But the title, *Dongchong Xiacao*, would lead readers to think it was about the caterpillar fungus consumed in Chinese society. Fujita was employed in Shanghai until 1919 by a founder of *Nongxue Bao* to translate Japanese sources for the journal, which, established against the background of a social movement directed at modernizing Chinese agriculture, placed emphasis on both classical Chinese agricultural knowledge and newer European, American, and Japanese agriculture and applied sciences.[87] In view of the emphasis of the journal, Fujita's translation actually digressed from the journal's object, though in two later translations he focused on parasitic insects and toads, both highly relevant to crop protection.[88] It is reasonable to speculate about Fujita's possible intention of overturning the long prevailing stories of the caterpillar fungus's magical transformation and treating this organism instead as an ordinary example of an insect-fungi.

Three years later, in 1903, *Nongxue Bao* published a relatively long translation entitled *Dongchong Xiacao Shuo* (On winter worm summer grass), which was originally written by the Japanese botanist Itō Tokutarō (1866–1941).[89] The translation starts with a discussion of more than ten specimens of the caterpillar fungus brought from Tibet to Japan by the Buddhist monk Kawaguchi Ekai (1866–1945), who then presented them to Itō for identification. Itō described the appearance of these specimens, explained the life cycle as an irreversible process of fungal infection, and then succinctly reviewed the history of European studies of this species. But he further stressed that sixty-two such fungal species had been discovered around the world. Like Oda or Fujita, he treated dongchong xiacao or tōchū kasō as a group of insect-fungi rather than a single species, and claimed the existence of similar species native to Japan. Besides the caterpillar fungus, Kawaguchi also presented Itō with some plants collected in Sikkim Himalaya.[90] Itō seemed to have a greater interest in the caterpillar fungus, since he wrote an article exclusively on it. His intention, as indicated in the article, was to expose the errors of the popular old theory of its formation, so that, now the scientific theory of its life cycle had become clear, people should not continue to believe the erroneous theory.

Following the publication of such articles and popularizations of scientific knowledge, Chinese intellectuals also gradually engaged in the dissemination of new facts about the caterpillar fungus. In 1905, a Chinese author published his translation of the last chapter of Miyoshi Manabu's book *Shokubutsugaku Jikken Shoho* (Introduction to botanical experiments, 1899) in *Nüzi Shijie* (The female world), a journal dedicated to female education and women's rights.[91] The chapter gave an outline of nineteenth-century European classification of flowering, flowerless, and seedless plants,

---

[87] Zhang Kai, "Wunonghui, Nongxue Bao, Nongxue Congshu Ji Luo Zhenyu Qiren," *Zhongguo Nongshi* 1 (1985): 82–8; Douglas R. Reynolds, *China, 1898–1912: The Xinzheng Revolution and Japan* (Cambridge, Mass.: Harvard Univ. Press, 1993): 116; Li Yongfang, "Tengtian Fengba: Qingmo Xifang Nongxue Yinjin De Xianxingzhe," *Shehui Kexue* 8 (2012): 142–9.

[88] "Jishengchong Baohuqi," trans. Fujita Toyohachi, *Nongxue Bao* 114 (1900): 485; "Ji Chanchu," trans. Fujita Toyohachi, *Nongxue Bao* 114 (1900): 485–6.

[89] Itō Tokutarō, "Dongchong Xiacao Shuo," trans. unknown, *Nongxue Bao* 231 (1903): 440–4.

[90] Itō Tokutarō, "Notes on Some Himalayan Plants Collected by the Rev. Keikai Kawaguchi in 1902," *Botanical Magazine* 17 (1903): 157–9.

[91] Miyoshi Manabu, "Zhiwuyuan Goushefa," trans. Zhiqun, *Nüzi Shijie* 3 (1905): 21–6; Manabu, "Zhiwuyuan Goushefa (Continued)," trans. Zhiqun, *Nüzi Shijie* 6 (1905): 31–46; compare these to Manabu, *Shokubutsugaku Jikken Shoho* (Tokyo: Keigyōsha, 1899), 134–41. For the objective of the journal, see Jin Songcen, "Nüzi Shijie Fakanci," *Nüzi Shijie* 1 (1904): 1–3.

in which the caterpillar fungus was listed as a representative species of the ascomycetes fungi. In 1913, a set of short articles by Ya Bo, who was studying at the College of Agriculture, Imperial University of Tokyo, appeared in the *Kexue Conghua* (Collected narratives of science) column of *Datong Zhoubao* (Great harmony weekly, Shanghai). The first of these articles deals with the *zhenxiang* (truth) about the caterpillar fungus.[92] Ya Bo wrote that scientific investigation revealed the mycelial infection of underground butterfly larvae was the true reason for the formation of the caterpillar fungus, as well as *jinchanhua* (golden flowers on cicada, an insect-fungus). To reinforce the authenticity of this new scientific explanation, he mentioned the microscope and encouraged readers interested in natural history to carry out microscopic observations. Such a statement obviously indicated the discursive power of microscopy.

The Republican period witnessed the increasing impact of scientific discourse on the Chinese intellectual community. The entry for dongchong xiacao in the first edition of *Ciyuan* (Origins of [Chinese] terms, 1915), also the first modern Chinese comprehensive encyclopaedia, deals only with fungi, insects, and parasitism, totally ignoring premodern Chinese accounts.[93] This is consistent with one of the main principles for its compilation, which was scientism in the conceptualization of natural objects and phenomena.[94] The illustration in the entry shows the fungus growing out of a mature insect rather than a larva. This indicates that dongchong xiacao in the entry does not refer to the Chinese caterpillar fungus (*Cordyceps sinensis*) but some other insect-fungi, which also reflects the influence of Japanese scholarship.[95] Moreover, a 1928 illustrated popular science article on the caterpillar fungus, aimed at children, not only mentions other insect-fungi but also denies the reality of transformation between different species.[96] Pan Jing, who had studied in France at the end of the 1900s, introduced the caterpillar fungus as a famous Sichuan foodstuff, and additionally invoked related biological research in his 1931 book.[97] Lu Wenyu's 1932 article in the *Guangzhi Xingqibao* (Weekly for spreading wisdom) refutes traditional ideas of the oddities of the caterpillar fungus by referring to Matsumura Jinzō's *Shokubutsu Meii* (Collection of botanical terms), Nishimura Suimu's *Semi No Kenkyū* (A study of cicada), and Adolf Engler's system of plant classification. It also mentions a few other fungal species of the genus *Cordyceps*, the medical and culinary uses of the caterpillar fungus, and his brother's experience of eating steamed caterpillar fungus in Sichuan.[98] Quite a number of articles exemplifying scientific authority involved in perceiving the caterpillar fungus were published.[99] But modern science did not enjoy superiority over, for example, the culinary preparation, edibility, and medicinal properties of the fungus.

---

[92] Ya Bo, "Dongchong Xiacao Zhi Zhenxiang," *Datong Zhoubao* 2 (1913): 1.

[93] Lu Erkui, ed., *Ciyuan* (Shanghai: Shangwu Yinshuguan, 1915), 303.

[94] Wang Jiarong, "Ciyuan, Cihai De Kaichuangxing," *Cishu Yanjiu* 4 (2010): 94, 130–40.

[95] The entry also says that the infected insects include the *lougu* (mole cricket), which *Cordyceps sinensis* actually does not infect.

[96] Ren Shou, "Dongchong Xiacao," *Ertong Shijie* 22 (1928): 33–6.

[97] Pan Jing, *Qiaoshan Zazhu* (n.p.: privately printed, 1931), 131. For Pan's educational experience in France, see Wang Huanchen, ed., *Liuxue Jiaoyu: Zhongguo Liuxue Jiaoyu Shiliao*, book 2 (Taipei: Guoli Bianyiguan, 1980), 631, 688–9.

[98] Lu Wenyu, "Xinnong Jianwen Suibi," *Guangzhi Xingqibao* 154 (1932): 4–6.

[99] For examples, see "Dongchong Xiacao Jiujing Shi Shenme Dongxi," *Xiao Pengyou* 569 (1933): 39; "Dongchong Xiacao," *Zhishi Huabao* 4 (1937): 26–7; Zhu Peiran, "Dongchong Xiacao Yu Maoyan," *Guoxun* 156 (1937): 100; Tao Bingzhen, *Kunchong Manhua* (Shanghai, 1937), 63–4; and Zhiren, "Dongchong Xiacao: Dongwu Hu, Zhiwu Hu," *Juequn Zhoubao* 3 (1946): 9.

The blend of scientific and indigenous knowledge about the caterpillar fungus also indicates the influence of local food or material culture on science communication.

The dissemination of exotic, scientific "truth" in modern China did not proceed smoothly. In the 1910s, several articles on the caterpillar fungus, published in, for example, *Tongsu Jiaoyu Bao* (Journal of popular education, Shanghai) and *Xinmin Bao* (Journal of new citizens, Shanghai), lack any scientific knowledge.[100] In 1924, the painter Zhu Fengzhu's article on *buke siyi* (the incredible) in a Shanghai magazine even actively promoted the transformation theory of the caterpillar fungus on the basis of his observation of this wonder from Sichuan. Zhu strongly asserted that the caterpillar fungus transcended the categories of animals and plants; and when compared with bats, another organism that crossed the boundaries of birds and beasts, the caterpillar fungus no longer seemed so implausible. To induce readers to accept his opinion, he confidently suggested readers buy samples from drugstores and examine them with their own eyes.[101] Zhu's view could claim to be verifiable through observation, because what would be seen largely depended on, according to Lorraine Daston and Peter Galison, "what it [the subjective self] hoped to see."[102] In particular, as an author lamented in *Shenbao* (Shanghai news) in the same year, the caterpillar fungus was still popularly considered among the Chinese to be a magical, transformable organism.[103] Furthermore, the scientific theory of its life cycle had not yet been directly confirmed by a continuous one year or longer field observation in cold alpine environments or laboratories, which also contributed to the survival of the transformation theory in Chinese society.

Scientific research on the caterpillar fungus in China appeared in the Republican period. This period featured the growth of scientific professionalization and institutionalization, as well as the rise of scientific nationalism against the background of creating a vigorous, united, and modern Chinese nation through science.[104] Deng Shuqun pioneered Chinese mycological research on this fungus. In 1932, Deng, then working at the Science Society of China in Nanjing, reported his identifications of some fungi in southeastern areas of China, among which were specimens of "Cordyceps sinensis (Berk.) Sacc." obtained from a drugstore in Sichuan in 1928.[105] Two years later, his identifications and descriptions of the caterpillar fungus and the other forty-one fungal species were published.[106] This time he gave more detailed descriptions of its geographical range and natural habitat, produced an illustration of its fruiting bodies, and added an account of the structural characteristics of its stromata,

---

[100] "Dongchong Xiacao," *Tongsu Jiaoyu Bao* 1 (1913): 1; Chai Zifang, "Dongchong Xiacao," *Xinmin Bao* 2 (1915): 33–4.

[101] Zhu Fengzhu, "Buke Siyi Zhi Chonglei," *Hong Zazhi* 2 (1924): 1–6.

[102] Daston and Galison, *Objectivity* (cit. n. 77), 34.

[103] Li, "Ji buchongcao," *Shenbao*, 4 April 1924, section 8.

[104] Zuoyue Wang, "Saving China through Science: The Science Society of China, Scientific Nationalism, and Civil Society in Republican China," *Osiris* 17 (2002): 291–322.

[105] Shu Chun Teng, "Additional Fungi from Southwestern China," *Contributions from the Biological Laboratory of the Science Society of China: Botanical Series* 8 (1932): 1–4; Di Lu, "Recording Fungal Diversity in Republican China: Deng Shuqun's Research in the 1930s," *Archives of Natural History* 46 (2019): 139–52.

[106] Shu Chun Teng, "Notes on Hypocreales from China," *Sinensia* 4 (1934): 269–98. Deng's description of the caterpillar fungus in this article was later assimilated into his 1939 mycological monograph; see Shu, *A Contribution to Our Knowledge of the Higher Fungi of China* (n.p.: National Institute of Zoology & Botany, Academia Sinica, 1939), 41.

perithecia, and spores. Deng's publications laid a partial foundation for Pei Jian's 1947 article introducing up-to-date scientific knowledge of the caterpillar fungus and snow fungus, both of which the Chinese then alleged to be tonics suitable for everyone. Pei's article reveals scientific attention to local food or material culture. A research fellow of the Institute of Botany, Academia Sinica, Pei provided an illustration of new specimens of the caterpillar fungus, which, in contrast with Deng's, additionally showed anatomical and microscopic structures such as the transverse section and asci.[107] Both Deng and Pei's illustrations included scale bars, adding to the unprecedented accuracy in morphological representation of the caterpillar fungus in China.

The caterpillar fungus also incited related chemical or pharmaceutical research. In the 1940s, Tang Tenghan and his collaborators at the West China Union University (Chengdu) reported their analysis of chemical constituents in the caterpillar fungus, which might assist further exploration of bioactive constituents.[108] Also in Chengdu, Yang Shoushen, principal of the Military Academy of Veterinary Medicine, published his preliminary study of "Cordycepin," a fat-soluble crystal extracted by chemical methods from specimens of the caterpillar fungus growing in Lijiang. Beyond toxicity testing of Cordycepin in animals, he also carried out in vitro experiments to determine its antibacterial properties, because he hypothesized that *Cordyceps sinensis* must generate some substance that inhibited the growth of rival microorganisms in the larvae. He expected this study to contribute to research on issues relating to bacterial infections in animals.[109] These institutionalized studies did not interact with native culinary or medical knowledge about the caterpillar fungus, but were devised and performed in scientific contexts. They also demonstrated no disapproval of the medical, tonic, or culinary value of the caterpillar fungus. Some later historians even felt discontented with the neglect of native medical knowledge, valuable for scientific inquiry, in such chemical or pharmacological research on local medicinal substances in the Republican period.[110] Compared with microscopic observations, such research delved deeper into the interior of the caterpillar fungus, and thereby created new boundaries of knowledge beyond the reach of traditional empirical knowledge.

The localization of modern science in China was characterized in part by the combination of scientific practice and local natural products. In his 1909 research article on a Chinese fungus, Wang Huanwen, then studying pharmacology in Japan, wrote that the Japanese pharmacologist Nagai Nagayoshi once told him it was reasonable for native people to perform (scientific) research on domestic natural products.[111] Nearly forty years later, the principal of the National Specialist School of Materia Medica (Nanjing) formulated four missions for the school, the first of which was aimed at special Chinese medicinal substances and their effective constituents.[112] As

[107] Pei Jian, "Yiner He Xiacao Dongchong," *Kexue Shijie* 16 (1947): 102–4.
[108] Tang Tenghan, Wang Zhaowu, and Chen Xuhuang, "Dongchong Xiacao (Chongcao) Zhi Chubu Yanjiu," *Zhongguo Yaoxuehui Huizhi* 3 (1945): 1–4.
[109] Yang Shoushen, "Dongchong Xiacao Junsu (Cordycepin) Zhi Chubu Yanjiu Baogao," *Guofang Kexue Jianbao* 2 (1948): 711–17.
[110] Chen Xinqian and Zhang Tianlu, *Zhongguo Jindai Yaoxueshi* (Beijing: Renmin Weisheng Chubanshe, 1992), 216–17.
[111] Wang Huanwen, "Bukuryō No Seibun Ni Ju Te" (cit. n. 22), 472.
[112] Meng Xinru, "Yaowu Kexue Zhi Guoqu Ji Guoli Yaoxue Zhuanke Xuexiao Zhi Shiming," *Yaoxun Qikan* 5 (1947): 1–4.

indigenous material culture often sparked scientific attention toward local food, remedies, tonics, and economic plants, state-aided modern science promoted the reconstruction of knowledge about the caterpillar fungus and many other natural objects. In the Chinese physician Chen Cunren's 1935 dictionary of Chinese materia medica, which was intended to stimulate scientific research on native medicinal substances, the entry for the caterpillar fungus presented itself as an integration of some modern biological knowledge (from Chinese and Japanese sources) and vernacular knowledge about the plant's ancient names, production areas, appearance, medicinal properties, medical applications, culinary preparations, and so forth.[113] The popularity of Chen's dictionary helped spread a new eclectic intellectual face of the caterpillar fungus in society.[114] Similar to the entry, some other Republican records dedicated to a scientific caterpillar fungus also demonstrated that modern science would not necessarily drive out vernacular culinary or medical knowledge.[115] Even in Britain, the news reporting James W. Spreckley's "three bundles of the Chinese fungus, *Cordyceps sinensis*" presented to the Department of Botany of the British Museum, and published in the scientific journal *Nature* in 1930, still informs that it was a "celebrated drug" and was "said to bestow energy and to be partaken of with stewed duck."[116]

CONCLUSION

Anna L. Tsing's anthropological study of the delectable matsutake mushrooms as examples of "interspecies entanglements" indicates the intersections between "science and vernacular knowledge" and "international and local expertise."[117] The caterpillar fungus counts as an interspecies complex that, however, embodies parasitism rather than the mutualism represented by matsutake and pine trees. It is a complex that crosses boundaries of species through time, but explanations of its formation varied in different periods and cultures. It also crosses boundaries of identity. In Republican China, the caterpillar fungus appeared in the tonic food produced and sold by some restaurants and food companies. It was a natural plant in its geographic range, and also an expensive commodity in the hands of merchants. Some physicians used it as a traditional medicinal substance; some scientists treated it as an object of scientific investigation; and some officials related it to national economic interests in a political and nationalistic context. Moreover, some conservative intellectuals, consumers, or practitioners of Chinese medicine believed in its ability to transform from a blade of grass to a worm; at the same time, some proponents of natural history and fungal

[113] Chen, *Zhongguo Yaoxue Dacidian* (cit. n. 11), 303–6.

[114] The entry of the caterpillar fungus in the dictionary had also been extracted and published separately in periodicals; see, for example, Chengren, "Dongchong Xiacao," *Xiandai Yiyao Zazhi* 1 (1945): 15–18. For the popularity of Chen's dictionary, see Shu Shan, "'Chongcao' Ji Qita: Lüetan Xichui De Miaoyao," *Tanfeng* 11 (1937): 511–13; and Chen, *Yinyuan Shidai Shenghuo Shi* (1973; repr., Shanghai: Shanghai Renmin Chubanshe, 2000), 262–5.

[115] See, for examples, Zhang Lu, "Dongchong Xiacao Zhi Yanjiu," *Kunming Jiaoyu Yuekan* 3 (1919): 1–2; Xu Ke, *Kangju Biji Huihan* (1933; repr., Taiyuan: Shanxi Guji Chubanshe, 1997), 324; Sun Zulie, "Tanpian: Dongchong Xiacao," *Minsheng Yiyao* 62 (1941): 26; Zhujun, "Yanxia Hua Chongcao," *Nong Zhi You* 10 (1947): 16–17; and Zong Zhen, "Dongchong Xiacao," *Kexue Shidai* 3 (1948): 37.

[116] "News and Views," *Nature* 126 (1930): 856; cf. "News: The Department of Botany of the British Museum," *North-China Daily News*, 22 December 1930, section 7.

[117] Anna L. Tsing, *The Mushroom at the End of the World: On the Possibility of Life in Capitalist Ruins* (Princeton, N.J.: Princeton Univ. Press, 2015), vii, 287.

microscopy criticized traditional accounts of its natural properties and formation. In some cases, the two categories of audiences possessed different interests and aims, which did not simply set boundaries of knowledge about food, but also shaped somewhat incommensurable intellectual worlds. An essentialist view of this incommensurable categorization is nonetheless not always advisable, as many Republican actors, like Chen Cunren, actively engaged in the integration of scientific and local knowledge.[118]

According to Hiromi Mizuno, Imperial Japan (1868–1945) aspired "to be recognized by the West as a modern, civilized nation, as the Western powers were, and to celebrate the nation's particularity to build a national identity"; and modern science was linked with imperial mythology, "the absolute core of its national identity."[119] Tong Lam also states that since the beginning of the twentieth century, Chinese cultural and political elites have shared a myth that "modernity is purely rational and that the triumph of science and reason is a self-evident, natural, and unproblematic process."[120] The powerful rhetoric of modern science embodied in the shaping of the truthfulness of the caterpillar fungus in China was initially imported from Japan through translation at the beginning of the twentieth century. With the dissemination of scientific knowledge and the caterpillar fungus in Japan, significant changes in Japanese perceptions of the fungus emerged and persisted throughout the nineteenth century. This natural curiosity was incorporated into the category of insect-fungi, which crossed the boundary between the caterpillar fungus and other similar insect-fungi; it was also deconstructed into two different and untransformable species grouped with other similar species in the European natural order; microscopic structures were invoked in support of the scientific theory of its formation; and interestingly, the Chinese term for it also entered the Japanese language, with its meaning being broadened to encompass other similar insect-fungi, indicating the semantic boundaries of shared vocabulary. As this new scholarship prevailed in Republican China, the caterpillar fungus would sometimes transform into a scientific wonder that prompted new facts within different scientific boundaries. However, modern science did not dispel indigenous culinary, medical, or other forms of empirical knowledge about the caterpillar fungus. This case study of the fungus and related boundary issues reveals both rupture and continuity in knowledge about food in Republican China's pursuit of science as modernity.

---

[118] Volker Scheid, *Currents of Tradition in Chinese Medicine, 1626–2006* (Seattle: Eastland, 2007), 202–8; Bridie Andrews, *The Making of Modern Chinese Medicine, 1850–1960* (Vancouver: Univ. of British Columbia Press, 2014), 112–205; Erik J. Hammerstrom, *The Science of Chinese Buddhism: Early Twentieth-Century Engagements* (New York, N.Y.: Columbia Univ. Press, 2015); Jia-Chen Fu, *The Other Milk: Reinventing Soy in Republican China* (Seattle: Univ. of Washington Press, 2018), 109–28.

[119] Hiromi Mizuno, *Science for the Empire: Scientific Nationalism in Modern Japan* (Stanford, Calif.: Stanford Univ. Press, 2009), 2.

[120] Tong Lam, *A Passion for Facts: Social Surveys and the Construction of the Chinese Nation-State, 1900–1949* (Berkeley and Los Angeles: Univ. of California Press, 2011), 8.

Hungry, Thinking with Animals:
Psychology and Violence at the Turn of the Twentieth Century

*by Dana Simmons**

ABSTRACT

Edward L. Thorndike (1874–1949), at the turn of the twentieth century, set up animal hunger as a model system for understanding human motivation and learning. Hungry animals participated in over a hundred years' worth of experiments designed to characterize human emotions and behavior. Hunger, along with electric shocks, became standard tools for producing psychological effects, such as motivation, excitement, fear, learning. Scientists deprived kittens, monkeys, chicks, turtles, children, and soldiers of food for four, eight, twenty-four, or forty-eight hours to observe the variable effects. I want to think through the meaning and context of this choice. What is the nature of hunger as an epistemic tool and as a model system? Why did hunger appeal to Thorndike and his colleagues at the turn of the twentieth century as a reasonable and productive relation with their animal subjects? What preexisting relations made hunger an obvious choice? What relations, in the end, did hunger experiments produce? I am interested in how hunger, as a model system, helped to establish a field of behavioral-physiological-neuroscientific knowledge. I am even more interested in what the traces of these model systems, and the animals within them, can tell us about the history of hunger. In the global nineteenth century, hunger was a tool for social violence.

The first ones to be hungry, on purpose and experimentally, were the cats. The cat was in a wooden box, and around her hung strings and wooden buttons and levers. She could smell fish outside. She had not eaten since the previous daylight. Edward Thorndike, who made the box and who held the fish, called the cat's state "utter hunger." Driven by hunger and the smell of fish, the cat swiped and clawed and pushed against the box and all the things around her. She moved impulsively, randomly, reaching at

* Department of History, University of California, Riverside, 900 University Avenue, Riverside, CA 92521 USA; dana.simmons@ucr.edu.

I am deeply grateful to Emma Spary and Anya Zilberstein, W. Patrick McCray, and Suman Seth for offering this forum, collegial support, and close, critical readings. I have benefitted enormously from the insight and feedback of many colleagues. Participants in the 2018 Society for the History of Recent Social Science (HISRESS) meeting, particularly Philippe Fontaine, Jamie Cohen-Cole, Jeff Pooley, Nancy Campbell, Katerina Liskova, and Susanne Schmidt, gave essential early feedback, as did my wonderful University of California, Riverside, colleagues Jade Sasser, Chikako Takeshita, and Juliet McMullin. Fellow members of the Nutrire Collab also offered important comments. Finally, many thanks to the two anonymous reviewers and especially to Sheila Dean for her care and encouragement.

whatever she could grasp. All at once, the box opened, and the cat leaped out toward the fish smell. The man offered her a tiny morsel; this was her "reward." The man picked the cat up and returned her to the box. Still driven by utter hunger, the cat began again to move about until the box door gave way, again and again, as she gained very small pieces of fish, until the time when she no longer returned to the box. At the end of the day, at last, the cats could eat "abundant food to maintain health, growth and spirits, but commonly some what less than they would of their own accord have taken."[1]

Dogs came after the cats, though the dogs howled loudly at night when the man left them hungry, and their cries awoke William James and his family sleeping upstairs. The dogs, like the cats, lived experimentally in James's basement, which James had lent to his postgraduate student as no suitable space could be found at Harvard University. Because of their howls, the dogs could not live in "utter hunger." They exercised in the wooden boxes in the morning, when they had not yet eaten, and they "made great effort for a bit of meat," if somewhat irregularly.[2]

Edward L. Thorndike's experiments at the turn of the twentieth century set up animal hunger as a model system for understanding human motivation and learning. Hungry animals participated in over a hundred years' worth of experiments designed to characterize human emotions and behavior. Nonhuman hunger was central to the formation of both comparative psychology and neuroscience. Hunger, along with electric shocks, became standard tools for producing psychological effects, such as motivation, excitement, fear, learning. Hunger (mostly mouse hunger) remains an important model system today in the fields of behavioral genetics and neurochemistry.

Hunger became an epistemic tool, designed to produce behaviors and emotions. Scientists deprived kittens, monkeys, chicks, turtles, children, and soldiers of food for four, eight, twenty-four, or forty-eight hours to observe the variable effects. Hunger became a standard tool, in part because its intensity could be controlled on an objectively measured scale, as could shocks of electrical voltage or hours of deprivation. I want to think through the meaning and context of this choice. What is the nature of hunger as an epistemic tool and as a model system? Why did hunger appeal to Thorndike and his colleagues at the turn of the twentieth century as a reasonable and productive relation with their animal subjects? What preexisting relations made hunger an obvious choice? What relations, in the end, did hunger experiments produce?

Why hunger? What brought Thorndike to introduce hunger to these formative experiments in comparative psychology and why does hunger remain so central to this field? I am interested in how hunger, as a model system, helped to establish a field of behavioral-physiological-neuroscientific knowledge. I am even more interested in what the traces of these model systems, and the animals within them, can tell us about the global history of hunger.

In this article, I think through hunger as a *situation*. When Thorndike built wooden "problem boxes" in William James's basement and put food-deprived cats in them, he created a new kind of situation (fig. 1). Thorndike set out to build an objective, experimental model for animal learning. In his view, animals (and people) had no innate personality or intelligence. Mental life was formed entirely by responding to

---

[1] Edward L. Thorndike, *Animal Intelligence: Experimental Studies* (New York: Macmillan, 1911), 27n1.
[2] Ibid., 59.

Figure 1. Edward L. Thorndike's puzzle-box B1. Robert Mearns Yerkes papers, 1822–1985 (inclusive), Manuscripts & Archives, Yale University Library.

novel conditions, including new environments, internal sensations, challenges, encounters. Repeated encounters with the same situation led to new mental functions.

Thorndike appears to have invented the *situation* as a psychological unit of analysis: "In general the term situation is used for any total set of circumstances in the outside world and in one's body by which the mind is influenced."[3] That is, a situation, in his usage, is a total combination of circumstances, both inside and outside the body, which together cause a reaction in the mind. "The situation may . . . be the whole state of mind, the circumstances or thing in its context, the entire 'attitude' or 'set' of mental life, as well as the particular fact in its focus."[4] Situations are an aggregate of coincidental elements, each stimulating particular connections of thought, feeling, or movement. Thorndike designed his experimental situations as instruments to direct and control the subject's responses. For him, situations were technologies for shaping the mind.

Thorndike's cat boxes were also *situations* in the sense suggested by Lauren Berlant's affect theory: "A situation is a state of things in which something that will perhaps matter is unfolding amid the usual activity of life. It is a state of animated and animating suspension that forces itself on consciousness, that produces a sense of the emergence of something in the present that may become an event."[5] A situation forces its participants to "adapt to an unfolding change."[6] In using "situation" to describe a

---

[3] Edward L. Thorndike, *The Elements of Psychology* (New York, N.Y.: A. G. Selier, 1907), 17.
[4] Ibid., 206.
[5] Lauren Berlant, *Cruel Optimism* (Durham, N.C.: Duke Univ. Press, 2011), 5.
[6] Ibid., 10.

psychological state of openness and emergence, Berlant may be drawing on an etymology that points back to Thorndike. He intended his situations to be contained in boxes and mazes. Berlant suggests that a situation is not a container, but a state of affective, psychological, and historical suspension: "This kind of attention to the becoming-event of something involves questions about ideology, normativity, affective adjustment, improvisation, and the conversion of singular to general or exemplary experience."[7] This is what models do; they normalize, mediate, improvise, and convert singular experiences to general. We will see that despite Thorndike's efforts to contain his model, to box his hungry cats in, there were constant slippages that call into question what exactly the situation was.

A situation is also a place where *situated knowledges* are made. Feminist science and technology scholars study situations as social ecologies of knowledge production. Alison Wylie writes of "our location in hierarchical systems of power relations that structure our mental conditions of life." This situation, she writes, "shapes our identities and epistemic capacities."[8] Adele Clarke and Joan Fujimura describe a situation as a set of processes and relationships: "We seek not only an ecology of knowledge, including an ecology of the contents of scientific knowledge, but also an ecology of the conditions of its production—an ecology of scientific activity/practice/ work."[9] Thinking about a situation as an ecology allows us to attend to interdependencies and flows across domains. Situations include epistemic things, living beings, and technologies. Different participants configure their situation differently. This view of the situation locates knowledge in relations of gender, race, and class, and ways in which these relations produce ways of working and being. "Concepts," writes Charles W. Mills, "orient us toward the world."[10]

Thorndike's problem boxes were part of the history of the situation as a scientific object. It turns out that this history, the history of situated knowledge, is bound up with the history of hunger and learning in the early twentieth century. This article explores the hungry cats' situations, expanding in ever-widening circles from Thorndike's problem boxes, to their world-historical surroundings. The cats' situations revolve around the two epistemic objects of hunger and learning. Living things in the situation included cats, dogs, rats, students, and those tagged with "simple minds." The situations included laboratory tools, namely problem boxes and learning technologies, designed to produce certain effects and relationships. Intervening in this situation were psychologists, utilitarians, eugenicists, and settlers of occupied Native American lands. All these animals and people were situated in a landscape of violently changing food webs, ecologies of prosperity and hunger. In the global-historical situation, hunger was a tool for social violence.

This article describes a situation—an ecology of knowledge production—in the United States at the turn of the twentieth century. Edward Thorndike's hunger experiments

---

[7] Ibid., 6.

[8] Alison Wylie, "Feminist Philosophy of Science: Standpoint Matters," *Proceedings and Addresses of the American Philosophical Association* 86 (2012): 47–76, on 62.

[9] Adele E. Clarke and Joan H. Fujimura, "What Tools? Which Jobs? Why Right?," in *The Right Tools for the Job: At Work in the Twentieth-Century Life Sciences*, ed. Clarke and Fujimura (Princeton: Princeton University Press, 1992), 3–45, on 4. See also Donna Haraway, "Situated Knowledges: The Science Question in Feminism and the Privilege of Partial Perspective," *Feminist Stud.* 14 (1988): 575–99.

[10] Charles W. Mills, *Black Rights/White Wrongs: The Critique of Racial Liberalism* (New York, N.Y.: Oxford Univ. Press, 2017), 63.

are the touchstone of this inquiry. They send us outward in search of their context and meaning. I examine hunger as an epistemic tool, in the same way that we might examine the history of the galvanometer or the calorimeter. Epistemic tools and concepts are always situational, material, conceptual, and political. In this sense, hunger is no different from heat transfer or electromagnetism. Just as steam engines, piece work, and industrial improvement societies shaped the history of nineteenth-century physical sciences, human sciences were shaped by their material and ideological context. Here I sketch a situational history of twentieth-century human sciences, food deprivation, and nonhuman hunger. I engage with methods and practices that might facilitate that goal: thinking with the animal, speculative extrapolation, gestural knowledge, and material-semiotic-affective infrastructures.

I am following Nikolas Rose and Joelle Abi-Rached's call for "thinking with the animal" in scientific models of behavior.[11] Rose and Abi-Rached encourage us not to dismiss animal models as merely "artificial" constructs of scientific imagination. Animal model systems are also "setups to make visible, to elicit, features of the lives and potentialities of animals that previously were difficult to discern."[12] To dismiss them is to miss what we might learn from them: perhaps not what the model-builders intended. Instead of following experimenters' avowed anthropomorphism, we might practice historical "zoomorphism."[13] Thinking with the animal means treating animals as participants, actants, as well as objects or instruments, in scientific knowledge making. I suggest that animal models show us some of the features and potentialities of hunger in the twentieth century.

I began this article with a piece of what Steven Shaviro calls "speculative extrapolation." Shaviro suggests that scientists and humanists both practice a form of controlled free imagination, constructing hypotheses and testing them to see whether they work.[14] I do not know whether Thorndike's cat was female. I do not know if she would recognize him as a "man," or if her experience matches his description of it. The anecdote above is an attempt to extrapolate the hungry animal's subject position, and to test this extrapolation against the known evidence. It matters to me to begin this article's narrative with a cat and her situation. I prefer to risk an imperfect speculation rather than to reproduce only the voice of a scientist-narrator (whose account of the same events appears immediately below.) I seek to pay attention to those who were hungry.

Thinking with animals implies attending to animals' gestural knowledge, what James Griesemer calls "tactile, muscular, kinesthetic experience."[15] Model animals each have particular chronologies, rhythms, intensities, and orientations. Laboratory observations leave us clues about them. Scientific reports and notebooks retain traces of animals' and observers' gestures in different model situations. How animals react, navigate, and live in a situation points to gestural knowledge beyond that which the experimenter sought to capture. In turn, thinking with animals may open a heuristic to a broader

---

[11] Nikolas Rose and Joelle M. Abi-Rached, *Neuro: The New Brain Sciences and the Management of the Mind* (Princeton, N.J.: Princeton Univ. Press, 2013), 104.

[12] Ibid., 86.

[13] Ibid., 104.

[14] Steven Shaviro, *Discognition* (London: Repeater, 2016), 11–12.

[15] James Griesemer, "Three-Dimensional Models in Philosophical Perspective," in *Models: The Third Dimension of Science*, ed. Soraya de Chadarevian and Nick Hopwood (Stanford, Calif.: Stanford Univ. Press, 2004), 433–42, on 440.

world-historical view of hunger. Thorndike's cat experiments reflect and condense a deep social and political history. These experiments help us to think historically about food deprivation, its material and political effects. A history of animal models helps us to sense hunger as what Michelle Murphy calls a "material-semiotic-affective infrastructural presence."[16]

A SIMPLE CASE OF LEARNING

In 1897, Edward Thorndike put a young cat in an uncomfortable situation:

> If we take a box twenty by fifteen by twelve inches, replace its cover and the front side by bars an inch apart, and make in this front side a door arranged so as to fall open when a wooden button inside is turned from a vertical to a horizontal position, we shall have means to observe [a] *simple case of learning*. A kitten, three to six months old, if put in this box when hungry, a bit of fish being left outside, reacts as follows: it tries to squeeze through the bars, and bites at its confining walls. Some one of all these promiscuous clawings, squeezings, and bitings turns round the wooden button, and the kitten gains freedom and food. By repeating the experience again and again, the animal gradually comes to omit all the useless clawing, etc., and to manifest only the particular impulse (e.g., to claw hard at the top of the button with the paw, or to push against one side of it with the nose) which has resulted successfully. It turns the button round without delay whenever put in the box. It has formed an association between the situation, "confinement in a box of a certain appearance," and the impulse to the act of clawing at a certain point of that box in a certain definite way.[17]

Many questions arise from this experimental description. What elements define this setup as a model of learning? How is it a "simple case"? Why kittens, and why were they hungry? What does it mean to think of this setup, described as "confinement in a box of a certain appearance," as a "situation"?

One set of clues can be found in the history of Thorndike's menagerie and his academic trajectory. Thorndike's model system for learning, which was hungry kittens in a box, became a paradigm for American educational practice. Learning, Thorndike told his readers, was governed by situations, not by culture or personality. Certain situations could become technologies for producing effects in the mind. Thorndike's experiments also shaped the new discipline of comparative psychology. Hunger became a tool for producing psychological knowledge, and a model for how to stimulate learning.

Having carried out animal experiments for a year in William James's basement, Thorndike moved to complete his doctorate at Columbia University, which was more accommodating with a graduate stipend and on-campus laboratory space. He wrote to his future wife, Bess, that he was impatient to install his "menagerie" at Columbia; he was, he wrote, "hungry for work."[18] He was ambitious and eager to challenge prevailing assumptions about animal intuition and intellect. As the child of a New England traveling minister, he was raised in a culture of self-control and diligence; his prodigious publication record testifies to his professional discipline. He thought of himself

---

[16] Michelle Murphy, *The Economization of Life* (Durham, N.C.: Duke Univ. Press, 2017), 7.
[17] Thorndike, *Elements* (cit. n. 3), 202 (emphasis mine).
[18] Geraldine M. Jonçich, *The Sane Positivist: A Biography of Edward L. Thorndike* (Middletown, Conn.: Wesleyan Univ. Press, 1968), 118.

as a disrupter, bringing scientific rigor, laboratory experiment, objective measurement, and statistical analysis to a field dominated by anecdote and speculation. His contemporaries praised his thesis on "Animal Intelligence" as a foundational work in the rising field of experimental comparative psychology.[19] Senior scholars in his field, however, did not appreciate his brash dismissal of work preceding his own.[20]

Thorndike identified hunger as a solution to psychologists' perceived lack of scientific objectivity. Hunger offered a controllable and quantifiable experimental variable. This variable could be measured using everyday equipment, scale balances (to weigh food), and clocks (to record duration of fasting and speed of activity). Subjected to a standard rate of food deprivation, animals presumably would respond with consistent behaviors. Animals in a state of "utter hunger" could be run repeatedly through a puzzle-box and produce coherent results. Such results required no interpretation or subjective judgment; all an experimenter needed, said Thorndike, was a clock. "Facts . . . may be obtained by any observer who can tell time."[21] Hunger made objective psychology possible.

These experiments turned animal psychology into laboratory work. Thorndike had to create his own experimental setup for animals. Just as he repurposed common technologies (clocks, boxes) as experimental instruments, he repurposed domestic animals as experimental subjects. He first installed his menagerie in his Cambridge, Massachusetts, boarding house, whose landlady voiced strong objections. From there the animals moved to William James's basement, then on to an attic in the new Columbia psychology building in New York City. At various times, his attic lab housed chicks, kittens, dogs, a monkey, and even a tank of minnows. All were domesticated animals. Chicks, cats, and dogs already depended on human food and care, and were accustomed to human infrastructure. Thorndike's experimental schedule must not have differed too much from that of a household pet, locked indoors and fed once each day according to human rhythms of industry, work, and consumption.

Thorndike tested his model for learning and intelligence on simple minds, including those of kittens, chicks, children. Simple-minded subjects allowed him, he thought, to observe the operation of learning at its most basic. Before building cat puzzle-boxes at Harvard, Thorndike traveled to a mental institution to study unconscious cues in young children. He gave pieces of candy to three-year-olds if they guessed correctly a number or letter he was thinking. When the authorities denied him further access to girls and boys, he turned to animals. Instead of candy, the animal subjects who succeeded in their task received a morsel of food to relieve their hunger.

What applied to animals, at the most basic level, applied equally well to humans. "These simple, semi-mechanical phenomena, multiple response, the cooperation of the animal's set or attitude, with the external situation, the predominant activity of parts or elements of a situation . . . which animal learning discloses, are the fundamentals of human learning also."[22] To learn was not to think or intuit, but to respond

[19] Margaret Floy Washburn, *The Animal Mind: A Text-Book of Comparative Psychology* (New York, N.Y.: MacMillan, 1908), 11.

[20] Wesley Mills, "The Nature of Animal Intelligence and the Methods of Investigating It," *Psychological Review* 6 (1899): 262–74.

[21] Thorndike, *Animal Intelligence* (cit. n. 1), 28.

[22] Edward L. Thorndike, *Educational Psychology: Briefer Course* (New York, N.Y.: Teachers College, Columbia Univ., 1919), 136.

to *situations*, which connect specific feelings, sensations, and bodily movements. Thorndike's model operated at a basic level, directing simple feelings to aggregate along a particular path. Learning did not involve thought or wisdom; it meant building up mental connections, "associations," bit by bit. The key to learning, in his mind, was the capacity to form mental connections between ideas, actions, and things.

Thorndike made much of the fact that his kittens only gradually became better at opening the puzzle-box's trap door. The smooth curve of their improvement suggested, he thought, that they stumbled across the right solution purely by blind and fumbling chance. He found "no sign of abstraction, or inference, or judgment" in kittens' repeated attempts to open the door.[23] "The cat does not look over the situation, much less think it over, and then decide what to do."[24] He contemptuously dismissed observers who sought proof of animal intellect, memory, or rationality; such attempts, he scoffed, were as ridiculous as a zoologist looking for claws on a fish.[25]

Higher-level learning differed from animal learning by quantity, not quality. Complex human intellect was "an extended variation from the general animal sort"; intelligent people simply were able to form many more connections than animals or simple-minded folk. "[The] intellectual evolution of the race consists in an increase in the number and speed of formation of such associations."[26] He thought this capacity to form connections heritable, and hoped to subject it to scientific breeding.

Thorndike was a committed eugenicist and insisted on hereditary difference: "In the same way and for the same reason that tall parents have tall children or dark-haired parents dark-haired children, so also stupid parents have stupid children, hot-tempered parents have hot-tempered children, and musical parents, musical children."[27] Inherited qualities determined children's capacities to make associations and learn. Thorndike did not hesitate to draw racist conclusions from this premise. He reportedly told a popular audience that psychology's first task following the First World War would be to investigate "the problem of the mental and moral qualities of the different elements of the population of the United States." Thorndike asked: "What does this country get in the million or more Mexican immigrants from the last four years. What has it got from Italy, from Russia, from Scotland and Ireland?"[28] Late in his career, he sat on the board of the American Eugenics Association and chaired the Subcommittee on Psychometry of the Eugenics Research Association.[29] The kittens' simple minds came to model "feeble-minded" victims of eugenic segregation and violence.

Within a decade after publication of Thorndike's thesis, animal behavior labs spread to Harvard, Clark University, Cornell, Johns Hopkins, and the University of Texas. Psychologists in all of these labs adopted hunger as an epistemic tool. Hunger became a standard instrument, alongside new introductions such as Willard Small's animal maze (1900) and Robert Yerkes's electric shock apparatus (1907). James B. Watson's

---

[23] Thorndike, *Animal Intelligence* (cit. n. 1), 75.
[24] Ibid., 74.
[25] Ibid., 75.
[26] Ibid., 294.
[27] Thorndike, *Elements* (cit. n. 3), 195.
[28] Quoted in Jonçich, *Sane Positivist* (cit. n. 18), 375.
[29] "Letter from Field Secretary, American Eugenics Association to Fair Associations Asking Education Exhibit Space" (ca. 1930); "Sub-Committee on Psychometry," *Eugenics Research Association 16th Annual Meeting* (1928), 4, Image Archive on the American Eugenics Movement, Dolan DNA Learning Center, Cold Spring Harbor Laboratory, records 704 and 255.

early experiments were derivative of Thorndike's. Psychologists tested hungry turtles, mice, rats, rhesus monkeys, and crows. This setup became standard to the extent that a young psychologist in 1911 could state simply and without elaboration, "hunger was used as a motive."[30] Through the twentieth century, hunger remained a foundational tool for comparative psychology.

Thorndike's kittens left an even longer-lasting imprint on the American educational system. In 1899, he was hired to bring his scientific, experimental rigor to the newly affiliated Columbia Teachers College. At Teachers College, he trained generations of American educational leaders. His textbooks, dictionaries, and teaching and testing materials extended his influence far wider. His *Thorndike-Barnhardt Junior* and *Intermediate Dictionaries*, containing selections of frequently used words, still remain in publication in the early twenty-first century. Education scholar Ellen Condliffe Lagemann, exaggerating somewhat on purpose, writes, "one cannot understand the history of education in the United States during the twentieth century unless one realizes that Edward L. Thorndike won and John Dewey lost."[31]

Teaching tools based on this educational model, many designed by Thorndike, spread across the United States. Thorndike "devised rating scales to standardize and measure children's proficiency in hand-writing, spelling, drawing, history and English comprehension, and sold millions of arithmetic textbooks that stressed drill, repetition and the 'overlearning' of basic skills."[32] He was deeply involved in designing the US Army alpha and beta tests for incoming recruits during the First World War.[33] He applied the same analytic zeal to children's education, disaggregating each skill into its smallest component tasks and exercising them one by one. The cat experiment showed, he claimed, that learning was cumulative, not holistic. Each task must be broken down into components. Teaching must focus exclusively on tasks with future use value.

Thorndike excoriated classical humanistic education, which presumed to offer rewards beyond the limits of individual texts or languages. He had no time for nebulous claims on behalf of general culture.[34] He single-handedly scoured word frequencies in a library of core English books, beginning with the Bible, so that his dictionaries would present only words that children were most likely to encounter every day. And encounter they did, through the laborious exercises that many American students today still undergo, as they copy vocabulary words five or ten times in a row. "Use value" guided both content and method: exercise, repetition, and reward. This was a division of intellectual labor. Thorndike's teaching technologies were to education what Frank Galbraith's motion studies and the sciences of work were to industrial labor. "More than any other person," writes education scholar Stephen Tomlinson, "it was Thorndike who, from this institutional power base, shaped the curriculum, pedagogy, and

---

[30] Vinnie C. Hicks, "The Relative Values of the Different Curves of Learning," *Journal of Animal Behavior* 1 (1911): 138–56, on 142.

[31] Ellen Condliffe Lagemann, "The Plural Worlds of Educational Research," *Hist. Ed. Quart.* 29 (1989): 185–214, on 185.

[32] Stephen Tomlinson, "Edward Lee Thorndike and John Dewey on the Science of Education," *Oxford Review of Education* 23 (1997): 365–83, on 373.

[33] John Carson, *The Measure of Merit: Talents, Intelligence, and Inequality in the French and American Republics, 1750–1940* (Princeton, N.J.: Princeton Univ. Press, 2007), 206–11.

[34] Tomlinson, "Edward Lee Thorndike" (cit. n. 32), 373.

organizational structure of the American school as well as the basic aims and methods of university-based inquiry."[35]

When schoolchildren rewrite their multiplication tables twenty times for a teacher's treat, when policy makers disparage humanist claims for the richness of general education, when education is sold as a set of transferrable skills, we have entered the cat box. We become part of Thorndike's model system.

PLEASURE AND PAIN

Why were there hungry kittens in a box? Another set of clues can be found in the history of utilitarian psychology.

The cat box was an instrument for learning, Thorndike explained, because it connected specific actions to feelings of satisfaction and discomfort. When one of the cat's random clawings and squeezings flipped the latch, and the same action led, over repeated trials, to similar success, the cat came to associate that action with the satisfaction of a small piece of fish. The action connected to pleasurable feelings (clawing the latch and eating fish) was strengthened; other actions connected to discomfort or annoyance (like the continuing sensation of hunger) were weakened. As cats or humans encounter the same situation over and over, these connections are "stamped in." Thorndike called this relationship between satisfaction, discomfort, and learning through repeated experiences, the Laws of Effect and Exercise. His advice to teachers sums it up: "Exercise and reward desirable connections; prevent or punish undesirable connections."[36]

Thorndike almost certainly chose to experimentalize hunger out of unacknowledged debt to Herbert Spencer and David Ferrier. Spencer defined hunger in his *Principles of Psychology* (1855) as one of the simplest "units of feeling." Most states of mind, he claimed, were "compounds" of a few basic mental units. A few "really simple" feelings were unique in that they were "not decomposable by introspection" into smaller units.[37] Hunger was one of them. Hunger appeared as an original, natural, basic feeling, a low entry in the hierarchy of mental functions.

Spencer held an evolutionary theory of mental life, from primitive, simple feelings to complex mental compounds. Hunger appeared at the primitive end of Spencer's evolutionary account; at the first stage of mental development, "Mind is present probably under the form of a few sensations, which, like those yielded by our own viscera, are simple, vague and incoherent."[38] Mental evolution, thought Spencer, was a process of compounding simple feelings into ever more complex aggregates.[39] Visceral feelings, simple and vague, "play but subordinate parts in the actions we chiefly class as mental."[40] Hunger, thirst, nausea, and other visceral feelings "cohere little with one another and thus integrate but feebly into groups." They do not connect easily with other sensations and feelings to form more complex ideas and emotions. Hunger was thus relegated to the lowest rung on Spencer's ladder of mental evolution.

---

[35] Ibid., 376.
[36] Thorndike, *Educational Psychology* (cit. n. 22), 142.
[37] Herbert Spencer, *The Principles of Psychology* (New York, N.Y.: D. Appleton, 1890), 163.
[38] Ibid., 189.
[39] Ibid., 488.
[40] Ibid., 187.

Hunger, wrote Thorndike, belonged to the class of the "vaguer sensations" (note his unacknowledged cribbing from Spencer), to which at first no connections are attached.[41] Hunger was an "elementary sensation," one of those "feelings so simple or minute as to be unanalyzable into simpler ones." At first, in infancy, elementary sensations have no connections to other sensations or acts at all; they are "uninfluenced by previous experiences." An infant's sensations are "pure."[42] Hence the kittens were of very young age; and hence hunger was central to Thorndike's model system.

Thorndike also drew inspiration from British psychologist David Ferrier, director of neurology at the West Riding Lunatic Asylum. Ferrier was involved in attempts in the 1870s to localize sensations and muscle movements in specific brain areas. He ablated or applied electrical stimulation to different areas of monkey and dog brains to observe the results.[43] Ferrier acknowledged an intellectual debt to Herbert Spencer's pleasure-pain theory of learning.[44] Actions that produce pleasure, he wrote, "tend to continuance and repetition," whereas actions associated with hurt and pain "are checked and avoided."[45]

Ferrier sought to redirect animals' voluntary movements, to produce "learning," by the application of pleasure and pain. The experimental tools he described in his book *The Functions of the Brain* (1880) involved burning animals' skin and submitting them to hunger. Ferrier proposed that even a hungry animal can be taught by painful experience to avoid food. "A hungry dog," he wrote, "is impelled by the sight of food to seize and eat." But this gratification can be "neutralized and counteracted" by an even stronger sensation of pain. Ferrier recounted the story of a hungry dog that was struck with a strong whip each time it came near to food. Soon enough it ceased to eat. Thereby "the dog is said to have learnt to curb its appetite."[46]

For Ferrier, relief of hunger stood on the "pleasurable" end of the sensory spectrum. In his anecdote about the dog learning to curb its appetite, its pleasure in food was curbed by painful interventions. By contrast, he burned and whipped his animals to produce sensations of pain. Thorndike did not employ tools for inflicting pain on his kittens, dogs, and chicks. However, he adopted Ferrier's equivalence of pain-punishment-avoidance and pleasure-reward-repetition. He deprived his experimental subjects and then gave them food; this, for Thorndike, stood in for pleasure.

Thorndike wholly adopted Spencer and Ferrier's psychological utilitarianism. Cathy Gere shows brilliantly how an eighteenth-century political philosophy was transmogrified into a founding principle of modern American psychology. Edward L. Thorndike was the man who did it. "Thorndike succeeded in founding utilitarian psychology anew . . . reinvented as pure, value-free laboratory science, with no connections to politics and history."[47] I am grateful to Gere for so perceptively drawing a path straight from Bentham to behaviorism, with Thorndike as its conveyor. He turned a philosophy based on a balance of pain and pleasure into a naturalized model system; in effect, he

---

[41] Thorndike, *Animal Intelligence* (cit. n. 1), 27.

[42] Ibid., 21.

[43] Michael Hagner, "The Electrical Excitability of the Brain: Toward the Emergence of an Experiment," *J. Hist. Neurosci.* 21 (2012): 237–49, on 245.

[44] Cathy Gere, *Pain, Pleasure, and the Greater Good: From the Panopticon to the Skinner Box and Beyond* (Chicago: Univ. of Chicago Press, 2017), 153.

[45] David Ferrier, *The Functions of the Brain* (New York, N.Y.: G. P. Putnam's Sons, 1880), 290.

[46] Ibid., 309. See Gere, *Pain, Pleasure* (cit. n. 44), 153.

[47] Gere, *Pain, Pleasure* (cit. n. 44), 169.

turned utilitarianism into an experimental science. This, too, explains why hunger was part of his model for mental life. Ideologies of hunger, learning, and capitalism formed an implicit part of Thorndike's experimental situation.

A century of utilitarian philosophers held up hunger as a tool for learning; specifically, this was a tool for impressing laborious and thrifty behaviors upon spendthrift and shiftless people. Hunger was the philosopher's whip. The Reverend Joseph Townsend, a fellow traveler of Jeremy Bentham's, thought that charity aid injured the poor by preventing them from learning the lessons of hunger: "Hunger will tame the fiercest animals, it will teach decency and civility, obedience and subjection, to the most brutish, the most obstinate and the most perverse."[48] Robert Thomas Malthus famously argued that hunger and famine should teach the poor to exercise moral restraint, to postpone marriage until they had the means to feed a family.[49]

Beside moral decency and diligence, hunger stimulated investment in private property. A "love of property," Herbert Spencer proposed, "evolved" from hunger. If a man were struck with hunger while wandering about, and if he were to remember where he had last encountered some uneaten foodstuffs, Spencer imagined, he would return to that very place to satisfy himself. His feelings would attach themselves to memories of particular things (foods) and places, which eventually he would want to own for himself. This same pattern recurs over and over, and possessiveness is born. "A subsequently-recurring hunger . . . will establish an organized connection between the remembrance of such remaining food and the various states of consciousness produced by a return to it. Thus will be constituted an anticipation of a return to it—a tendency to perform all such actions accompanying a return to it as are not negatived by satiety—a tendency, therefore, to take possession of it."[50] Once the desire to possess goods evolves, its reach spreads to shelter, clothing, tools, and, finally, to all varieties of goods. Satisfaction of hunger grows into all the many satisfactions of a civilized capitalist lifestyle. "It results that the excitement of possession will grow into one of a new kind, uniting into a large but vague aggregate the various excitements to which it ministers."[51] Spencer offered a natural formula for private property, perhaps even more direct than John Locke's famous untilled field.

Echoing Spencer, Thorndike referred to acquisition and possession as "original tendencies concerned with food getting." He derived possessiveness from young children's tendency to grasp objects and put them in their mouths. Hunting and food gathering explain the "original tendency" to "pounce," "grab," and "seize" at things.[52] The desire for property makes a subtle appearance as a logical, evolutionary outgrowth of hunger.

Built into the cats' situation, then, was the assumption that hunger drove animals and people to work. Human industry, the desire to work and to achieve, came from the promise of satisfaction. Hunger's discomfort was a stimulus to civilization. But Thorndike's experimental setup belies this claim.

---

[48] Cited in ibid., 101.
[49] Thomas Robert Malthus, *An Essay on the Principle of Population* (1798; repr., Cambridge: Cambridge Univ. Press, 1992).
[50] Spencer, *Principles* (cit. n. 37), 488.
[51] Ibid.
[52] Thorndike, *Educational Psychology* (cit. n. 22), 17.

PUNISHMENT AND REWARD

Thorndike's thesis stimulated a consequential and long-lasting debate over punishment and reward. Which made animals (or children) learn best? Which was more efficacious, or more humane? Was it better to prevent and punish, or reinforce and reward? In animal psychology, punishment came to stand in for electric shocks; and reward meant to give food to the deprived and hungry.

Harvard psychologist Robert Yerkes wrote a scathing critique of Thorndike's method, which was by then (1907) the dominant paradigm in animal behavior studies: "Usually in experiments with mammals hunger has been the motive depended upon. The animals have been required to follow a certain devious path, to escape from a box by working a button, a bolt, a lever, or to gain entrance to a box by the use of teeth, claws, hands, or body weight and thus obtain food as a reward." Yerkes objected to this method as both inconsistent and cruel. Experimenters could not be certain that their animals felt as hungry at the beginning of a test run as at its end; nor could they know if one animal's hunger was equivalent to another's. For hunger to function as a consistent motive, it must be so strong as to damage the animal and its abilities. For these reasons, "the use of the desire for food as a motive in animal behavior experiments . . . [is] almost worthless in the case of many mammals."[53]

An animal in a state of "utter hunger" (like Thorndike's) would be constitutionally unable to perform complex acts. The experiment itself produced an incapable subject. More than this, Yerkes strenuously objected to hunger on moral grounds. It was, he argued, "inhumane."

> However prevalent the experience of starvation may be in the life of an animal, it is not pleasant to think of subjecting it to extreme hunger in the laboratory for the sake of finding out what it can do to obtain food. Satisfactory results can be obtained in an experiment whose success depends chiefly upon hunger only when the animal is so hungry that it constantly does its best to obtain food, and when the desire for food is equally strong and equally effective as a spur to action in the repetitions of the experiment day after day.[54]

Yerkes submitted that electric shocks were superior to food deprivation on grounds of consistency and humanity. He exercised his animal subjects, dancing mice, in a "discrimination box" designed to test visual ability. In order to escape a narrow, confining corridor, the mouse had to pass through one of two white, gray, or black boxes. When it entered the "wrong" box it received an electric shock of a voltage "disagreeable but not injurious" to the animal. Over repeated tests, the mouse gradually came to choose the correct box more often than not. Yerkes vaunted the reliability of electricity as compared to food rewards: "The experimenter cannot force his subject to desire food; he can, however, force it to discriminate between conditions . . . by giving it a disagreeable stimulus every time it makes a mistake."[55] He took pains to justify the "humaneness" of this practice: he regulated the current carefully, so as to prevent injury; the shocks were brief and went off at intervals; and his mice remained in

---

[53] Robert M. Yerkes, *The Dancing Mouse: A Study in Animal Behavior* (New York, N.Y.: Macmillan, 1907), 98.
[54] Ibid., 99.
[55] Ibid.

perfect health for months.[56] As strange as it may appear to a twenty-first century reader, Yerkes weighed food deprivation against painful electric shocks, and judged the latter best.

Yerkes translated these experimental methods into the language of utilitarian psychology. Electric shocks were "punishment"; food was a "reward" for hungry animals. These terms became common shorthand in the field. Yerkes himself considered "the method of punishment . . . more satisfactory than the method of reward, because it can be controlled to a greater extent."[57] Decades of papers in the *Journal of Animal Behavior* (which Yerkes edited) tested the relative merits of punishment and reward. Mildred Hoge and Ruth Stocking of Johns Hopkins University ran rats through a visual discrimination box, some in a state of hunger: "Punishment was a light electric shock; the reward, milk-soaked bread. The rapidity of learning in the two cases was taken as an indication of the value of the method."[58] Rats punished by shocks made correct choices somewhat more quickly than hungry rats incentivized by food. Hoge and Stocking recommended both punishment and reward for rapid learning.

Yerkes's former student John D. Dodson compared the two tools, hunger and electric current, at various levels of intensity to determine the optimal setup for learning. He ran rats through a discrimination box under varied conditions of duress. The animals got faster and more accurate as they became hungrier and hungrier. Rats deprived of food for forty-one hours outperformed less hungry rats. Past forty-one hours without food, though, their performance declined. They appeared disturbed and "assumed the hump of a starving animal."[59] Likewise, animals improved their performance when subjected to increasingly strong shocks, up to a point of severity beyond which their performance fell off. (This curve of optimal drive strength became known as the Yerkes-Dodson Law.)[60] Dodson compared "a curve of relative values of different degrees of hunger and a curve of the relative values of different strengths of electrical shock." He found that electric shocks produced the fastest learning times. Dodson wondered whether this had to do with the difference between pleasure and pain, or whether rats simply were primed to flee a dangerous situation more quickly than to seek out food. In any case, punishment trounced reward.[61]

A few years later, Fred A. Moss determined that seventy-two hours of food deprivation proved an even stronger incentive to get rats running than a twenty-eight-volt electric shock. Moss declared that his results would decide the "battle" raging "from time immemorial . . . in the field of education, concerning the comparative worth of punishment and reward as incentives in learning."[62]

As a field, psychologists decided in favor of both. No sooner had Yerkes published his work on dancing mice, *A Study of Animal Behavior*, than nearly every animal

[56] Ibid., 100.

[57] Ibid., 99.

[58] Mildred A. Hoge and Ruth J. Stocking, "A Note on the Relative Value of Punishment and Reward as Motives," *Journal of Animal Behavior* 2 (1912): 43–50, on 43.

[59] John D. Dodson, "Relative Values of Reward and Punishment in Habit Formation," *Psychobiology* 1 (1917): 231–76, on 265.

[60] Robert M. Yerkes and John D. Dodson, "The Relation of Strength of Stimulus to Rapidity of Habit-Formation," *Journal of Comparative Neurology* 18 (1908): 459–82.

[61] Dodson, "Relative Values" (cit. n. 59), 237, 276.

[62] Fred A. Moss, "Study of Animal Drives," *Journal of Experimental Psychology* 7 (1924): 165–85, on 178.

apparatus began to employ both methods.[63] Hunger and electricity, alongside the puzzle maze, became standard equipment for behavioral psychologists from John B. Watson to B. F. Skinner, from the 1910s to the 1950s and beyond.

Punishment and reward. I am stuck considering how the relief of hunger came to represent a reward. Animals deprived of food for one, two, or three days, running and digging and swiping at levers to relieve their discomfort; these were models for psychological pleasure. Even Yerkes, who considered starvation unpleasant and inhumane, called food incentives for deprived animals "the method of reward." Both hunger and electric shocks imposed pain and discomfort. In one case, relief was quick (the shock ceased); in the other, relief (in the form of food) appeared only after animals solved a maze or puzzle. What kind of reward was this? Work hard, driven by hunger, and you will earn a taste of pleasure—only to be pulled away to work again. Reward, in this utilitarian psychology, meant a little less suffering, contingent on successful work, and lasting only for a short while.

NEVER SATISFIED

Thorndike's psychology hung on utilitarian claims about the power of satisfaction, "exercise and reward." The cat's pleasurable feelings, when it opened the door and reached its fishy prize, imprinted upon patterns of movement. The cat, like children in the classroom, learned by exercise and reward. But thinking with the animals leads me to believe that Thorndike's narrative was misleading. *His cats, in fact, were never satisfied.*

When Thorndike reprinted his doctoral dissertation a dozen years after the fact, he added a telling footnote. The dissertation made much of the usefulness of "utter hunger" as an experimental tool. By 1911, he felt a need to defend his setup: "I have been accused of experimenting with starving or half-starved animals."[64] To demonstrate his probity, Thorndike revealed, in that footnote, what the cats ate and when. This is how we know that the same cats repeated multiple iterations of the problem-box experiment. The cats experienced hunger and discomfort, then relief, over and over.

This suggests a contradiction: if the cat's hunger gets satisfied when it opens the box, how would it still be hungry on the next round? Experimental consistency required that "the animal should be as hungry at the tenth or twentieth trial as at the first." Thorndike explained his solution: "to attain this [consistency] the animal was given after each 'success' only a very small bit of food as a reward (say, for a young cat, one quarter of a cubic centimeter of fish or meat)."[65] That quarter-centimeter cube was the reward, the relief, the satisfaction that was meant to produce learning. But in fact, it was also a prerequisite for further experimental work. The "reward" was not one. Thorndike designed the cat's reward so as to maintain its hunger.

[63] "The reward for successful choice was food; as punishment for failure the electric shock was used"; see D. B. Casteel, "The Discriminative Ability of the Painted Turtle," *Journal of Animal Behavior* 1 (1911): 1–28, on 1. "I used the well-known discrimination method, combining the motives of punishment and reward"; see H. M. Johnson, "Visual Pattern-Discrimination in the Vertebrates-II. Comparative Visual Acuity in the Dog, the Monkey and the Chick," *Journal of Animal Behavior* 4 (1914): 340–61, on 342.

[64] Thorndike, *Animal Intelligence* (cit. n. 1), 27.

[65] Ibid., 27.

What kind of satisfaction was this? Willard Small picked up this question while elaborating on Thorndike's experiments in 1898 and 1899 as a graduate student at Clark University. Small built puzzle-boxes for hungry rats to break into (later, he introduced his animals to the Hampton maze). Hunger was their "motive"; food inside the box their reward. An early series of his tests failed catastrophically. One of his rats died and the other refused to move. This, Small believed, was above all a "pedagogical failure"; he had not brought the animals to full satisfaction before starting up the test again. "The quick succession of experiments, followed in each case by deprivation of the fruits of their labor, was bad method. Nothing could be worse pedagogically, at least from a human standpoint." The failed rats were not able to form strong mental connections between hunger, puzzle solving, and pleasure. "To establish an association train of which the motive and first term is hunger, and the end and last term is satisfaction of hunger, the train ought to be fully realized each time."[66] In other words, for food to truly be a reward, the hungry rats had to feed until they were fully satisfied. They had to feel real pleasure at the end of their work.

John B. Watson disagreed. "The rat does not reason, 'I was not allowed fully to satisfy my hunger when I went to the food just now; therefore I really do not care to make the effort a second time.'"[67] Watson repeated Small's experiments as a graduate student at the University of Chicago. He set test boxes containing bread before hungry rats and observed how quickly they managed to enter. Like Thorndike, Watson allowed successful rats to taste only a small amount of food "for an instant" before immediately starting another test run. He saw no reason to allow the animals to sate their hunger. "Small is possibly applying here somewhat too much of his own conscious processes to the associative powers of the rat. If the rat is successful in overcoming the difficulties keeping it from the food, and is allowed to eat of the food for a short time, both terms of the 'association train' are completed and the rat is instantly ready to repeat the same procedure until his hunger is fully satisfied. Such was certainly the case with my rat."[68] In other words, one could trick the rat into going back to work by giving it the slightest hint of satisfaction.

Watson complained that some psychologists (namely, Yerkes) had maligned the hunger method. "It is not fair to talk of the cruelty and inhumanity of keeping the animal hungry, as has been done by several writers. . . . There is not the slightest difficulty in keeping the animal in perfect condition and at the same time hungry enough to work properly."[69] Satisfaction and reward did not have to be fulfilled in order to motivate animals to work. Even a promise, a taste, was enough. It may be worth mentioning that Watson spent most of his subsequent career in advertising, a field dedicated to stimulating desire for delayed gratification.[70]

[66] Willard S. Small, "An Experimental Study of the Mental Processes of the Rat," *Amer. J. Psychol.* 11 (1900): 133–65, 139n3.

[67] John B. Watson, *Animal Education: An Experimental Study on the Psychical Development of the White Rat, Correlated With the Growth of Its Nervous System* (Chicago: Univ. of Chicago Press, 1903), 9.

[68] Ibid.

[69] John B. Watson, *Behavior: An Introduction to Comparative Psychology* (New York, N.Y.: Henry Holt, 1914), 58.

[70] Rebecca Lemov, *World as Laboratory: Experiments With Mice, Mazes, and Men* (New York, N.Y.: Hill and Wang, 2005), 30–1.

B. F. Skinner eventually would carry delayed gratification to an almost absurd level. Skinner trained hungry pigeons in the 1950s to press a lever hundreds of times to receive a single small pellet of food.[71] The longer the delay, he claimed, the more his animals grew "increasingly compulsive" in their activity.[72] In this way, Skinner made explicit Thorndike's deferral of animals' satisfaction. Skinner conditioned his pigeons to work indefinitely toward a deferred reward.

Thorndike's cats, like Watson's rats and Skinner's pigeons, were never truly satisfied. By his own admission, his cats were kept always hungry by design. They were meant to feel neither relief nor reward, but rather a promise of future satisfaction: the odor of fish, tiny pieces, incomplete meals. Even at the close of their workday, the cats were not given what they hungered for: "After the experiments for the day were done, the cats received abundant food to maintain health, growth and spirits, but commonly some what less than they would of their own accord have taken."[73] Thorndike designed the cat's feeding schedule to replicate the industrial time clock of working hours and meals. (How did Thorndike know what the cats would have eaten of their own accord? Were they his pets before becoming his subjects? Watson later weighed his animals to establish a baseline "maintenance ration.") In any case, Thorndike then left them, still hungry, without food for fourteen hours in preparation for the following day's work.

What does this tell us about the utilitarian promise of hunger, and about the parameters of the cats' situation? Thorndike presented the cat's experience as a closed circle of discomfort, movement, and relief. Learning happened in an eternal present: the imprinting of a movement-satisfaction, the solving of a single problem. As mentioned above, he denied that solving one kind of problem could inspire solutions in a different, unrelated area. Yet his narrative was undermined by his own cats' experience. They lived in a constant state of low-level hunger and dissatisfaction, primed to perform experimental work for the future promise of a taste or a smell.

As Cathy Gere observes, Thorndike worked to purge his utilitarian model system of its political implications, to render it natural and scientifically rigorous. The frame opened by his revealing footnote puts this effort to the lie. Utilitarian tales about desire, work, and reward were disproven by these experimental cats who were lured to work by false promises of a future satisfaction that never arrived.

The cats' experience suggests that to understand this experiment, we need to open that frame. We know that the cat's hunger was not contained within the problem box. Their hunger was perpetuated by the very "reward" that was meant to represent the satisfaction due to a job well done. The promise of a reward was always deferred.

If we think with the animals and their hunger, Thorndike's experimental and explanatory frame breaks down. We see the situation differently. The cats found themselves in what Berlant calls a "becoming-event"; this was "a state of things in which something that will perhaps matter is unfolding amid the usual activity of life."[74] Elements of this becoming-event include the cats, the attic laboratory, settler colonial

---

[71] B. F. Skinner, *Science and Human Behavior* (New York, N.Y.: Simon and Schuster, 1953), 31. See Paul E. Meehl, "Needs (Murray, 1938) and State-Variables (Skinner, 1938)," *Psychological Reports* 70 (1992): 407–50.

[72] Gere, *Pain, Pleasure* (cit. n. 44), 174.

[73] Thorndike, *Animal Intelligence* (cit. n. 1), 27.

[74] Berlant, *Cruel Optimism* (cit. n. 5), 5.

occupation, utilitarian politics, eugenics, and violence. The cat's situation was political and historical.

THE SITUATION

Thorndike's cats were part of the history of hunger around the turn of the twentieth
century. In the last thirty years of the nineteenth century, between thirty and sixty million people perished in a global wave of famines, exacerbated by climate anomalies,
colonial violence, and economic liberalization.[75] New print technologies, especially
photographic reproductions, brought terrible images of famished children to a mass
media audience. Hunger, in the late nineteenth century, became a "figure of human-
itarian concern" and a flashpoint of political activism.[76] Foreign correspondents and
nationalist politicians publicized colonial abuses inflicted on peoples suffering from
mass starvation.[77]

At the very moment when Thorndike carried out his animal experiments, the United
States government implemented a program of Native American containment and reeducation through hunger. In 1887, ten years before Thorndike's first experiment, the
US Congress initiated a vast program of human engineering that linked hunger, private property, labor, and learning. The Dawes Act mandated the enclosure of unincor-
porated Native American territories into private allotments, annexed to the United
States. The programmatic destruction of bison hunting economies, mass death, and
forced displacement from fertile lands, had rendered many Native North American
peoples dependent on government rations. Conditions on the Plains began to decline
precipitously in the 1870s.[78] With the Dawes Act, Congress sought to complete the
task of assimilating Native Americans into an individualist, agrarian mode of existence, while appropriating "extra" lands for white settler occupation.[79] Allotments enclosed Native lands and peoples.

Four years later, in 1891, Congress mandated schoolhouse education of Native
American children. If a family refused to send children to Indian schools, which were
often boarding schools miles from home, Congress authorized the Indian Bureau to
withhold that family's rations.[80] Native Americans were starved of food as a means of
forcing them to learn. An Indian agent at the New Mexico reservation of the Mesca-
lero Apaches wrote in 1897 that "the deprivation of supplies . . . worked a change" in
families unwilling to let their children go. "Willing or unwilling every child five years
of age was forced into school."[81]

---

[75] Mike Davis, *Late Victorian Holocausts: El Niño Famines and the Making of the Third World*
(London: Verso, 2002), 7.
[76] James Vernon, *Hunger: A Modern History* (Cambridge, Mass.: Harvard Univ. Press, 2007), 17.
[77] Ibid., 31.
[78] James Daschuk, *Clearing the Plains: Disease, Politics of Starvation and the Loss of Indigenous
Life* (Regina, Canada: Univ. of Regina Press, 2014), 99–126.
[79] David Wallace Adams, "Fundamental Considerations: The Deep Meaning of Native American
Schooling, 1880–1900," *Harvard Educational Review* 58 (1988): 1–19, on 5.
[80] Ibid., 3.
[81] Quoted in Margaret D. Jacobs, "Indian Boarding Schools in Comparative Perspective: The Re-
moval of Indigenous Children in the United States and Australia, 1880–1940," in *Boarding School
Blues: Revisiting American Indian Educational Experiences*, ed. Clifford E. Trafzer, Jean A Keller,
and Lorene Sisquoc (Lincoln: Univ. of Nebraska Press, 2006), 202–31, on 215, citing *The Indian's
Friend* 1897, 10:10.

David Adams emphasizes the pivotal role that education played in the enclosure of Native peoples. Adams suggests that the education and Americanization of Native peoples was designed to foster a sense of possessive individualism among them. Senator Henry Dawes laid out his program for Indian reeducation thus: "Teach him to stand alone first, then to walk, then to dig, then to plant, then to hoe, then to gather, and then to *keep*."[82] Dawes could have just as well quoted from Herbert Spencer. The Dawes Act required the US president to certify that a tribe was ready to adopt a sedentary agrarian life before the Indian Bureau could convert their territory to allotments and sell the remaining land to white settlers (proceeds of which were used to fund Indian schools). Schoolhouse education was a tool to effect a change of economic and cultural disposition, and thus to open those lands quickly.[83]

Historians tend to tell the story of social engineering in this way: early twentieth-century laboratory scientists developed techniques to record, quantify, and manipulate their subjects' behavior. Their midcentury successors took these tools out into the field and invaded our social environment. I would like to reframe that story, drawing inspiration from recent work on medical and social engineering of enslaved and colonial peoples.[84] Laboratory techniques for psychological manipulation were intimately linked to projects of human engineering already underway in the colonial world of the nineteenth century.[85] This included the settler-colonial environment of the American West. They were all part of this global-historical situation.

VERY-HUNGRY AND NOT-HUNGRY

A blocky sans serif font on a slightly shaky black background announces the genre of a mid-twentieth-century educational film: "MOTIVATION and REWARD in LEARNING" (fig. 2). "Two pale albino rats," intones a deep bass voice. "What do you think is the reason for the difference in their behavior?—The one on the left is hungry."[86]

We see two rats in cylindrical wire baskets, one otherwise empty and the other lined with pet food. One rat climbs up the wire, pokes his nose through the open mesh, pushes the latch with his head, and climbs out as soon as an opening presents itself; the other raises his head in acknowledgment of the human opening his basket, and continues eating. (Here I follow the usage of the film, which genders both rats male.) Two human arms enter the frame: a male hand with a light-colored sleeve seizes the escaping rat, and a female hand with a black sleeve picks up the one that is eating. They place the rats in a wood and plexiglass box divided into chambers labeled "VERY HUNGRY" and "NOT HUNGRY."

---

[82] Quoted in Adams, "Fundamental Considerations" (cit. n. 79), 6, citing *Journal of the Thirteenth Annual Conference with Representatives of Missionary Boards*, in the *Annual Report of the Board of Indian Commissioners*, 1883, House Exec. Doc. No. 1, 48th Cong., 2nd sess., 1883–84, Serial 2191, pp. 731–732; Adams capitalized "*keep*."

[83] Adams, "Fundamental Considerations" (cit. n. 79), 19.

[84] Helen Tilley, *Africa as a Living Laboratory: Empire, Development, and the Problem of Scientific Knowledge, 1870–1950* (Chicago: Univ. of Chicago Press, 2011).

[85] Daniel Headrick, *The Tentacles of Progress: Technology Transfer in the Age of Imperialism, 1850–1940* (Oxford: Oxford Univ. Press, 1988); see especially chap. 9, "Technical Education."

[86] Neal E. Miller and Gardner L. Hart, *Motivation and Reward in Learning*, film produced by Yale Institute of Human Relations, 1948, 13:15, Prelinger Archives Collection, Internet Archive, https://archive.org/details/Motivati1948.

Figure 2. *Still image (at 0 min. and 18 sec) from mid-twentieth-century (1948) educational film (cit. n. 86). Institute of Human Relations, Yale University.*

Human hands disappear from the frame. On the far wall of each compartment are affixed a metal stirrup and a tin dish. The box's floor appears at first to be striped; closer examination reveals that the stripes are regularly spaced metal bars that open to a hidden compartment underneath. The box has more depth than we can see; the rats are balancing on a metal grid above empty air. The slats are wide enough to allow pellets to fall through. (Later scientists will use a platform suspended in the air like this to produce the rodent equivalent of human stress.)

Noses flaring, the rats case the joint. They sniff each corner and wall, stretching upward toward the open box top and bright bulb. After some minutes, having investigated all possible escape routes, Not-Hungry turns away, presses his back against the plexiglass barrier and hides his head in the shadow of the suspended tin dish. He remains there, immobile, perhaps availing himself of some privacy. Very-Hungry continues to sniff and climb. His nose flutters rapidly; his eyes, intense and black, reflect the strong light above. What can he see and smell, to his sides and below, that the viewer cannot? Can he smell his neighbor, as he did in the wire cage?

Very-Hungry stands high on two legs, sniffs the air, descends. One such descent activates the metal stirrup, and a food pellet falls into the tin dish. When Very-Hungry discovers it, he crouches, folded over his belly as if he would hide the pellet in a pouch, and eats from his front paws. He leans one paw on the dish and licks the other. The bass voice tells us: "Food would not be a reward without the drive of hunger." A montage shows Very-Hungry sniffing ever more insistently around the dish and stirrup, moving ever more directly to activate the lever. The voice: "After several more trials, which are not shown, . . . the animal has eliminated irrelevant responses. He has *learned* to press the bar efficiently." The camera pans to the right. Not-Hungry is hiding, immobile, beneath the tin dish. The voice asks: "Now, what do you think the satiated animal has learned to do?"[87]

A text slide appears, reading, "Will the satiated animal learn if we give him a drive?" At which point, the viewer encounters Robert Yerkes's contribution to the hunger-electricity-puzzle apparatus. A female hand enters the frame, turning the knob of a potentiometer until, the voice tells us, "the shock is adjusted to be annoying, but not painful." A tone sounds. Not-Hungry bristles like a scared cat, and repeatedly leaps off the electrified metal floor, high enough to leave the camera frame. On one such landing, he hits the stirrup bar and the shock cuts off. He buries his head under the tin dish, and the voice tells us that he has been "rewarded," though his fur remains stiff and bristled. And so the experiment repeats. We are told that Not-Hungry "learns even more rapidly than the hungry one, . . . because the drive produced by the electric shock is stronger than hunger."[88] To belabor their point, the producers show us rats biting through rubber tubing, turning exercise wheels, and even fighting other rats in response to repeated electric shocks. "We have demonstrated," intones the voice, "that the satiated animal is neither stupid nor lazy. All he needs is a little motivation."[89]

Motivation and Reward in Learning was produced in 1948 by psychologist Neal E. Miller and the Institute of Human Relations, Yale University. "Rat learning," writes Rebecca Lemov, "lay at the heart of the [Yale] institute's hopes for a grand theory that would explain the full range of human behavior."[90] The hungry rat is active and "learning"; the satiated rat is immobile and static. Traces of Thorndike's setup appear throughout the film: a problem box, domesticated animals, hunger, learning, reflex-reward, pain and pleasure, Situation-Response, rates of activity over time, all of which constituted a model apparatus for human psychology.

Young cats and rats who are hungry encounter a *situation*: "confinement in a box of a certain appearance." This was the experimental frame. Exercised repeatedly in the same situation, cats and rats "learned" to solve it by forming connections between specific actions (flipping a latch, pressing a lever) and satisfactions (a piece of fish or a food pellet). Successive generations of behaviorists narrowed the scope of Thorndike's model: his S-R (Situation-Response) became behaviorists' S-R (Stimulus-Response,) where the S stood for a single input instead of a set of conditions.

Ironically, at around the same time he produced a film that publicized the drive-reduction theory, Neal Miller was involved in experiments that would destabilize it. By the late 1940s, the psychological bonds between hunger and learning were loosened and recast.

Several psychologists challenged the usefulness of hunger as a model for motivation and learning. Abraham Maslow, who had worked with Thorndike as a postdoctoral researcher in 1934, declared that "the choice of hunger as a paradigm for all other motivational states is both theoretically and practically unwise and unsound."[91] John Dashiell noted that rats placed in a new nesting box almost always take time to

---

[87] Ibid., at 7 min. and 21 sec.
[88] Ibid., at 8 min. and 52 sec.
[89] Ibid., at 9 min. and 13 sec.
[90] Lemov, *World as Laboratory* (cit. n. 70), 92.
[91] Abraham H. Maslow, "Preface to Motivation Theory," *Psychosomatic Medicine* 5 (1943): 85–92, on 85. On Maslow and motivation, see Kira Lussier, "Of Maslow, Motives, and Managers: The Hierarchy of Needs in American Business, 1960–1985," *Journal of the History of the Behavioral Sciences* 55 (2019): 319–41.

explore their surroundings, even if they are hungry and food is nearby.[92] Henry Nissen exposed well-fed rats to a mild electric shock and found them willing to overcome the shock just for the sake of exploring a new territory. Nissen proposed that exploration itself should be considered "a dynamic form of behavior akin to the hunger, thirst, sex, and maternal drives."[93] When he became director of the Yerkes Primate Laboratory in the mid-1950s, Nissen deplored behaviorists' use of food deprivation as a motivator to get chimps to do experimental work.[94]

Primatologists Harry Harlow and Donald Meyer questioned whether food deprivation had any impact on learning, and whether hunger should be called a "drive" at all. Monkeys in the Wisconsin Primate Laboratory, deprived of food for zero, two, seven, or twenty-three hours, showed no difference in rates of error or activity on a discrimination learning test.[95] In a blistering commentary, Harlow reprised some of Yerkes and others' early critiques of the hunger apparatus: "The kinds of learning problems which can be efficiently measured in these apparatus represent a challenge only to the decorticate animal. It is a constant source of bewilderment to me that the neobehaviorists . . . should choose apparatus which, in effect, experimentally decorticate their subjects."[96] Though his educational film portrayed a direct correlation between hunger and learning, Neal Miller advised readers of his own scientific publications to use "caution in drawing inferences about 'drive' from consummatory behavior in both psychological and psychiatric studies."[97]

Young people in classrooms around 1948 must have encountered a doubling or entangling effect in this educational film. The film technology replicates its subject, the rat learning apparatus, on a material, sonic, and visual level. The authoritative bass voice tells students what they are learning (the content of their lesson) and how they are learning, like the rats on the screen. "Now, what do you think the satiated rat has learned to do?" With which rat were students meant to identify? The Very-Hungry Rat, diligent and active in pursuit of its end? The Not-Hungry Rat, who at first appears stupid or lazy, and then is shocked into rapid learning? Did students experience this film as punishment or reward?

CONCLUSION

Cats, rats, students, and displaced Native Americans were entangled in the histories of twentieth-century hunger. Close attention to experimental animals helps us imagine a history of hunger. Hunger experiments reproduced dynamics of eugenics, settler colonialism, and racial exclusion. The animals' surroundings and gestures leave us hints as to the nature of those dynamics. This was the situation: hard work for meager returns,

---

[92] John F. Dashiell, "A Quantitative Demonstration of Animal Drive," *Journal of Comparative Psychology* 5 (1925): 205–8, on 208.

[93] Henry W. Nissen, "A Study of Exploratory Behavior in the White Rat by Means of the Obstruction Method," *Pedagogical Seminary and Journal of Genetic Psychology* 37 (1930): 361–76, on 362.

[94] Donald A. Dewsbury, "Conflicting Approaches: Operant Psychology Arrives at a Primate Laboratory," *Behavior Analyst* 26 (2003): 253–65, on 257.

[95] Donald R. Meyer, "Food Deprivation and Discrimination Reversal Learning by Monkeys," *Journal of Experimental Psychology* 41 (1951): 10–16, on 14.

[96] Harry F. Harlow, "Mice, Monkeys, Men, and Motives," *Psychological Review* 60 (1953): 23–32, on 28.

[97] Neal E. Miller, Clark J. Bailey, and James A. F. Stevenson, "Decreased 'Hunger' But Increased Food Intake Resulting From Hypothalamic Lesions," *Science* 112 (1950): 256–9, on 259.

the promise of a reward deferred. Social violence underlay the infrastructure of hunger, learning, drive, and motivation.

Hunger today still serves as a model for a wide range of animal and human behavior, especially behavior related to work, drive, and motivation. Hunger undergirds a utilitarian ideology: hunger drives people—and other animals—to work for food. There is no more powerful natural justification for human labor. Hunger models the nature of life under capitalism.

The sciences of hunger in this story are sciences of forgetting. Forgotten was the use of hunger and education as a means of subjugation of Native peoples, colonial subjects, and black Americans. Charles W. Mills has theorized this particular form of "stripping away" as "a white epistemology of ignorance." Mills describes the willful production of "white amnesia," which erases knowledge inconvenient to dominant group histories. In the "editing of white memory, . . . the mystification of the past underwrites a mystification of the present."[98] Faced with a single animal in a puzzle-box, one forgets about the multiplicity of relations that make up an animal's life and world, and that shape the experience of hunger. One forgets the global-political relations that structure the access of millions to adequate food. One forgets the eugenic surround for Thorndike's theories of animal learning and intelligence, and the collapsing of racial hierarchies into phylogenetic hierarchies. One forgets that hunger is an effect of deprivation, not of individual efficiency and motivation. Thinking with animals, with their gestural heuristics, allows us to remember some of those things, which were forgotten.

[98] Mills, *Black Rights/White Wrongs* (cit. n. 10), 71, 65, 66.

World War II and the Quest for Time-Insensitive Foods

by Deborah Fitzgerald*

ABSTRACT

When one walks around the local supermarket, one is often struck by the proportion of highly processed foods, or what some refer to as "the middle of the store." Yet to the military in the 1940s and 1950s, these same foods represented the apogee of scientific progress—the creation of time-insensitive foods suitable for the rigors of military combat. This article explores the social and technical development of a system for producing such things for the military in World War II, including the collaborations between the military, American food firms, and university scientists. While at the beginning of the war, military and civilian food systems were quite different, by the early 1950s the two systems had effectively merged. Thanks to the supercharged food research agenda of this period, focused on achieving maximum time insensitivity in military foods, Americans can now quite easily avoid eating time-sensitive foods entirely. Here I will explore the professional networks that made this possible in the mid-1940s, as well as the challenges of standardizing something as alive as fresh food.

> The serviceman—whether he is a soldier, sailor, or airman—
> is a unique type of consumer.[1]
>
> Food research is defense activity.[2]

On the morning of October 10, 1956, representatives of the leading American shrimp processors sat down with representatives of the Quartermaster Corps of the United States Army (QMC) to discuss alterations to the military ration. Several weeks later, business leaders of the QMC Committee on Vitamins and the Committee on Fat Spreads traveled to Chicago for their own meetings with Col. John D. Peterman, commandant of the Quartermaster Corps Food and Container Institute (QC F&CI); they were eager to be of service to their country and to cultivate long-term relationships with the federal government. Throughout World War II and the Korean War, the QMC organized commercial firms to advise its efforts in making military rations. Over the years, it worked very closely with several hundred firms whose business needed to be attuned not only to the latest consumption trends, or innovations in food science

---

* Program in Science, Technology and Society, Massachusetts Institute of Technology, 77 Massachusetts Avenue, Cambridge, MA 02139, USA; dkfitz@mit.edu.
[1] J. D. Peterman, "Motivations for Department of Defense Research on Rations," *Food Technology* 10 (1956): 512.
[2] Editorial, *Food Technology* 1 (1947): 113.

and technology, but to the peculiar needs of the military.[3] There was nothing unusual about such meetings. On the contrary, scientists and commercial firms had collaborated with military leaders for many years, so intertwined were their interests. In the early nineteenth century, Nicholas Appert developed a method of food preservation by thermal sterilization expressly for the French Navy. Similarly, Gail Borden developed condensed canned milk during the American Civil War, in part because of the dangerously common phenomena of problems related to milk contamination. Yet World War II differed from these examples in scale and scope, and ultimately led to an important and unprecedented convergence of military and civilian markets in the United States.[4]

Within the history of science, World War II has been seen primarily as an occasion for vastly increased funding for scientific research as well as the growing prestige and importance of scientists in setting an agenda—not only for their academic fields, but in many ways for the state itself. The urgent demands of the military—for up-to-date weaponry, ships, and sensing technologies, and for greatly expanded educational opportunities for military personnel—have immediately increased the pace of scientific and technological change throughout history. Yet, with few exceptions, historians have rarely explored how wars have provoked the food provisioning systems to produce more, and faster, to feed both the military and civilians during wartime.[5] While the food provisioning system is more mundane, and more offscreen, than most of the other scientific innovations that war has generated, its importance is hard to exaggerate. Why? For two reasons. First, wartimes have given the state unlimited power to commandeer the food supply, both the raw material farmers produce and the processed foods developed by industry. Deciding what gets grown, raised, treated, and made palatable for the military and civilians alike, has been a central concern of the state. Second, as with so many wartime innovations and "make-dos," those established in wartime provisioning have had a tendency to persist after the war. Arrangements made to accommodate the military were often simply applied to the civilian world when the war ended. In this, the American provisioning practices during World War II greatly accelerated the transition to a national, highly standardized food system that blurred the line between intense and emergency feeding under military conditions, and ordinary dining practices among civilians.[6]

[3] For shrimp processors, see folder "Minutes of the meeting of the Industry Advisory Committee on frozen breaded shrimp," 10 October 1956, 5–6, A. Stuart Hunter to Quartermaster General, 23 October 1956, and Hunter to Quartermaster General, 1 November 1956. On fat spreads, see A. W. Harvey to Quartermaster General, 29 November 1956. All preceding in Record Group (RG) 544 (Natick Laboratories Research and Development, Administrative), box 4, National Archives, College Park, Maryland (**hereafter** cited as NACPM). The best study of this subject is Kellen Backer, "World War II and the Triumph of Industrialized Food" (PhD diss., Univ. of Wisconsin, Madison, 2012).

[4] Stuart Thorne, *The History of Food Preservation* (Totowa, N.J.: Barnes and Noble, 1986), 28–42; Sue Shephard, *Pickled, Potted, and Canned: How the Art and Science of Food Processing Changed the World* (London: Headline, 2000).

[5] Important exceptions include Lizzie Collingham, *A Taste of War: World War II and the Battle for Food* (London: Allen Lane, 2011); Ina Zweiniger-Bargielowska, Rachel Duffet, and Alain Drouard, eds., *Food and War in 20th Century Europe* (London: Ashgate, 2011); Helen Zoe Veit, *Modern Food, Moral Food: Self-Control, Science, and the Rise of Modern American Eating in the Early Twentieth Century* (Chapel Hill: Univ. of North Carolina Press, 2013); Amy Bentley, *Eating for Victory: Food Rationing and the Politics of Domesticity* (Urbana: Univ. of Illinois Press, 1998); and Harvey Levenstein, *The Paradox of Plenty: A Social History of Eating in Modern America* (New York, N.Y.: Oxford Univ. Press, 1993).

[6] On the issues raised by the industrialization of food, see Gabriella Petrick, "'Like Ribbons of Green and Gold': Industrializing Lettuce and the Quest for Quality in the Salinas Valley, 1920-1965," *Agr. Hist.* 80 (2006): 269–95; Karin Zachmann and Per Ostby, "Food, Technology, and Trust: An Introduction," *Hist. & Tech.* 27 (2011): 1–10; and Roger Horowitz, *Putting Meat on the American Table: Taste, Technology, Transformation* (Baltimore: Johns Hopkins Univ. Press, 2006).

The very idea of standardizing food is, of course, slightly odd. Standardizing objects, like utensils, or shoes, or guns, was exceedingly difficult to accomplish in the nineteenth century, even with the help of scientists and mechanicians. A spoon that had been standardized for scaled-up production might not have seemed vastly different than a spoon made by the local metalsmith. Food that was standardized was typically an entirely different substance. What has given food its unique place in our day-to-day lives are its sensual characteristics—the sweet juiciness of a ripe plum, the smelly creaminess of aged cheese, the snappiness of fresh green beans, the earthy dampness of mushrooms, or the deep umami of beef stew. The smell, taste, texture, and even the sound of plants and animals, seem the very essence of our experience in eating and drinking. And yet, to a nation at war, the sensual pleasures of food fall way down the list of priorities in procuring food for millions of soldiers all over the world, for years on end. Indeed, this very aliveness of fresh and ripe food was the enemy of wartime provisioning, and the food industry worked hard with the military to arrest the development of foods, to make them inert and insensitive to the passage of time.[7] It is impossible to overstate the importance of time insensitivity to military concerns. Time-insensitive foods could sit in a warehouse until needed by the military; could be produced whenever the cost of the ingredients was low; and could be planned and requisitioned far in advance, avoiding last-minute deployment issues. Time-insensitive foods enabled the military to prepare for future wars far in advance, creating an ever-ready economy in at least this single respect.[8]

Here I propose to explore how military needs for provisions with this unique quality of time insensitivity were met by intense collaboration between the military and scientists in the food industry and academy during and after World War II. The process was not smooth, and the goals were not always entirely clear in the fast-moving wartime environment. But by war's end, new techniques and products were being introduced at a rapid clip. Further, by effectively merging the military and civilian food markets, time insensitivity became the key feature in standardizing food, and describes a huge percentage of what Americans eat today.[9]

THE QUARTERMASTER CORPS AND FOOD RESEARCH

Since the creation of the Continental army in 1775, the office of the QMC of the United States Army had been in charge of acquiring all the materials needed by military

---

[7] Susanne Friedberg, *Fresh: A Perishable History* (Cambridge, Mass: Harvard Univ. Press, 2009); Steven Stoll, *The Fruits of Natural Advantage: Making the Industrial Countryside in California* (Berkeley and Los Angeles: Univ. of California Press, 1998). There are several excellent studies of the history of food and the senses; see Christy Spackman, "Perfumer, Chemist, Machine: Gas Chromatography and the Industrial Search to 'Improve' Flavor," *Senses and Society* 13 (2018): 41–59; Nadia Berenstein, "Designing Flavors for Mass Consumption," *Senses and Society* 13 (2018): 19–40; Sarah E. Tracy, "Delicious Molecules: Big Food Science, the Chemosenses, and Umami," *Senses and Society* 13 (2018): 89–107; and Gabriela Petrick, "Tasting History," in *Nutrition and Sensation*, ed. Alan R. Hersch (Boca Raton, Fla.: CRC/Taylor and Francis Group, 2015): 1–22. Steven Shapin's study of the tension between objective and subjective measures of a wine's "goodness" is very helpful here; see Shapin, "A Taste of Science: Making the Subjective Objective in the California Wine World," *Soc. Stud. Sci.* 46 (2016): 436–60.

[8] John C. Fisher and Carol Fisher, *Food in the American Military–A History* (Jefferson, N.C.: McFarland, 2014).

[9] The pace of innovation, particularly in chemicals, new materials, and new mechanical processes, increased markedly during and after the war. See, as examples, trade press publications that listed sections titled "New Packages and Products" and "Patents" in *Food Industries* 13 (1941): 79–81, 112–17; and "Food Patents" in *Food Technology* 18 (1964): 56–9.

personnel. During World War II, it had to figure out how to purchase and transport in a timely manner all the uniforms, mess kitchens, boots, vehicles, spare parts, bedding, medicines, tools and machines, and food that the military needed, on bases as well as in the field, in twenty-three theaters of battle, for the duration of the war. This scale of operations, and the length and number of supply lines, had never been tackled before by the United States, and the food part of this effort alone was extremely complex. The QMC Subsistence Research Laboratory in Chicago was the hub of all food-related activity. Created in 1936 with three people, the lab was "authorized to test foods and design modern packaging of foods; to prepare drafts of proposed specifications and modify those which became obsolete; to conduct studies and make analyses of various reserve and emergency rations or components thereof; to prepare informative bulletins and maintain liaison with other government agencies."[10] As the war progressed, the lab ballooned in size and importance. Supported by a large staff of nutritionists, bacteriologists, chemists, and other scientific advisors, the lab developed the recipes for military rations, worked with industry on designing appropriate packaging, consulted with the surgeon general regarding what to include or avoid in the rations, figured out what to put in the various ration packages, solicited bids from firms that could produce the chile or coffee or caramels in the rations, contracted with each of these firms, took delivery of all the components, assembled the ration packages, and delivered them to military bases all over the world. In 1946, the lab was combined with the packaging part of the operation, and the whole thing was called the QMC Food and Container Institute (QMC F&CI).[11]

During World War II, the QMC committed itself to the unprecedented effort of feeding all the American troops no matter where on earth their service took them. This was not a small thing; over the course of the war, sixteen million people served in the American military. All soldiers in the field and those on base consumed primarily foods made and packaged in the United States. While the bases tried to provide fresh foods, combat soldiers lived on food that was entirely processed to withstand rough conditions. The ration names said it all—"Ration, frigid, high altitude" or "Ration, tropical," thereby describing not only the places American soldiers found themselves, but also hinting at the qualities that the rations required to maintain their edibility. Rations had to be acceptable in flavor, with the right texture and balance. They had to be free from rancidity, mold, and insect damage, and in packaging that had not been crushed due to a helicopter drop. They had to have not been rusted open due to landing on a beach; or broken due to poor metal seams, loose cardboard, or any number of other assaults; they had to be nutritious; and soldiers had to like them enough to eat them again and again. Thus, the food provisioners and the QMC had to consider not only the edibility and safety of the food, but also the security of the packaging, as they worked to keep the soldiers in fighting trim.[12]

By 1942, the QMC had assembled an extraordinary group of scientific advisors, corporate leaders in the food and packaging industries, engineers, chemists, and dietitians, in addition to farmers, packers, growers, and transportation experts, all of

---

[10] Fisher and Fisher, *Food in the American Military* (cit. n. 8), 181; C. O. Ball, "Scientists Will Run the Farms," *Food Technology* 1 (1947): 503.

[11] Fisher and Fisher, *Food in the American Military* (cit. n. 8); Roland F. Hartman, "Quartermaster Subsistence Training Keeps Pace with Food Industry Advances," *Food Technology* 10 (1956): 509–10.

[12] Fisher and Fisher, *Food in the American Military* (cit. n. 8), 148–80.

whom worked on provisioning problems small and large. As Merritt Roe Smith famously argued in describing the development of interchangeable parts manufacturing during the American Civil War, governments will pay any price to secure whatever is needed to prevail in war, and during World War II, food was one of those things. Indeed, the wartime food provisioning system was itself an urgent, very large-scale, and broad-based effort notable for corralling and managing impressive scientific and engineering expertise, and it profoundly shaped postwar developments in the university and commercial sectors. Like the better-known Manhattan Project, the food provisioning work was likewise invisible to the citizenry, shuttered away in a diffuse but massive collection of university labs, packing plants, canneries, packaging companies, fruit farms, and hatcheries.[13]

The newly named QMC F&CI represented very important institutional arrangements for both the war effort and the future of food. During World War I, the military had started collaborating with the food industry to develop rations, and began contracting with university science and engineering departments as well, "because . . . the demands for foods in enormous quantities for the millions of soldiers, greatly increased the interest and the actual demand for manufactured or treated foods."[14] Indeed, the war encouraged strong collaborations between the government and outside laboratories, not only to create new foods that could be mass produced, but also to develop new techniques that would ensure the "keeping quality" of highly processed new foods.

At the beginning of the war, the QMC built on the experiences of World War I, striving to ship field rations that were familiar, filling, and likely to be eaten by soldiers. Several different functional rations were developed. The field ration A was a field-prepared meal using fresh meats and vegetables. Ration B consisted entirely of canned and dehydrated foods; these were standard, commercially available foods that were adapted to the military. The C ration (C for combat) was the most ubiquitous ration, consisting of three meat (M) units and three bread (B) units, all packed in round twelve-ounce cans. The menu was bland overall. Chile con carne, ham and beans, beef and vegetable stew—the ration consisted of some kind of stew, crackers, juice powder, coffee powder, candy, canned fruit, crackers, and jam. In 1943, accessory packets containing cigarettes, gum, halazone water purification tablets, toilet paper, and a can opener were included.[15]

There were also a number of special rations designed to be used under particular circumstances. Meat bars, developed in 1942 and styled after pemmican, were made of dehydrated and compressed meat, something like a cold hamburger bar. They were intended for use in Arctic regions, developed by the United States Department of Agriculture and commercial firms, and tested in Alaska. While a success in terms of flavor, compactness, and ease of preparation, they failed in keeping quality. Emergency ration D, which was a fortified chocolate bar, typically made people sick with nausea and headaches, partly because it looked like a regular candy bar but had far more calories and bulk. Nonetheless, Pan Am Airlines expressed great interest in purchasing D bars as

[13] Merritt Roe Smith, *Harpers Ferry Armory and the New Technology: The Challenge of Change*, rev. ed. (1977; Ithaca, N.Y.: Cornell Univ. Press, 1980). On the Manhattan Project, see especially Daniel Kevles, *The Physicists* (New York: Knopf, 1977).
[14] Samuel C. Prescott, "Beginnings of the History of the Institute of Food Technologists," *Food Technology* 4 (1950): 305–7.
[15] Fisher and Fisher, *Food in the American Military* (cit. n. 8), 148–9.

an emergency food. The QMC also thought this was a great idea and suggested that they contact Hershey, Nestle, and other manufacturers for further information. The K ration was a flat package designed for soldiers in action, such as paratroopers, and was used from 1942 to 1946; the Mountain ration and Jungle ration, which were designed for small groups of soldiers in unusual conditions, were only used in 1942 and 1943; and the 5-in-1 ration, designed for groups of soldiers driving through the desert, was used only for a few years. More specialized rations were also developed, including the Assault lunch and the Aircrew lunch (mostly candy since "during the initial 6 to 8 hour period in combat the soldier is emotionally upset, and does not desire nor have the time to eat."); Army Air Force Combat lunch; Parachute Emergency ration; and a hospital supplement ration. A Life-raft ration was developed in 1944, which consisted of mostly sweets, and was tested on various groups of POWs and conscientious objectors who were located either on a small island or on an actual life raft floating around the Gulf of Mexico. Despite these many special ration categories, the contents of all rations shared one central characteristic. They included only foods that were highly processed, both to withstand the passage of time and to deliver nutritional requirements.[16]

In many ways, submarines offered the biggest challenge to the QMC, given the close quarters and infrequent resupplies. Submarine kitchens were 6 by 9 feet, and were meant to keep provisions to feed 60 to 80 men for up to 75 days. Here is how one author described the challenge: "The food technologist makes his contribution to the submariner by removing the waste, peel, rind, skin, shell, water, grounds, bone, cores, pits, stems, pods, and unusable fat. He compresses the food into less space; molds it into a square storage space-saving shape, packages it in the best packaging material, reduces storage by chemical dips, irradiation, ultra-violet treatment, ozone, anti-ripening agents, anti-sprouting agents, and establishment of optimum storage temperatures and humidity." Submarines were an ideal test for new products and processes that could not be tried in the civilian market without considerable caution.[17]

The QMC also studied carefully the rations carried by soldiers from other countries. The French ration was considered very appealing, with chocolate, pate, soft cheese, strong coffee, and Gaulois cigarettes. In designing parachute rations, the QMC consulted with the former military attaché to Germany and Great Britain for guidance on making a lightweight but substantial package. The Norwegian rations, which included barley broth, blueberry soup, and porridge, were considered inappropriate

---

[16] On the Meat bar, see "Project Abstract, Deterioration in Dehydrated Pork," 20 July 1953, RG 92, entry 1003 (A-303-University of Missouri, External Research and Development Technical Reports), box 111, NACPM. The best historical account of pemmican is George Colpitt, *Pemmican Empire: Food, Trade, and the Last Bison Hunts in the North American Plains, 1780-1882* (New York, N.Y.: Cambridge Univ. Press, 2015). On the D bar, see "Results of Tests of Ration D on 100 Medical Students," 6 February 1942, RG 92, entry 1003 (Combat rations, WW II and Korea), box 70, NACPM; as one observer commented, "an improperly anchored tooth may be broken off once in a while when the D Bar is eaten." See Captain Louis A. Wright to J. H. Daugherty, 27 August 1943, RG 92, entry 1003 (Combat rations, WW II and Korea), box 71, NACPM. On Pan Am, see Purchasing Mgr., Pan Am to QM Subsistence R&D Labs, 1 February 1946; and Lt. Robert P. McLevitt to Purchasing Mgr, 8 February 1946; both in RG 92, entry 1003 (Combat Rations WW II and Korea), box 71, NACPM. On the Assault lunch, see "Proposal of Experimental Food Packet for Combat Personnel," February 1951, RG 92, entry 1003 (Combat rations, WW II and Korea), box 66, NACPM. On the Life-raft ration, see, as examples, Maurice E. Shils to William M. Fosdick, 21 October 1944; and Rohland Isker to Col. Louis F Kosch, 29 November 1944; both in RG 92, entry 1003 (Quartermaster Food and Container Institute Bulletins, Conference Files), box 2, NACPM.

[17] Arthur C. Avery, "More Miles Per Pound of Food," *Food Technology* 9 (1955): 533–5.

for American soldiers; the Turkish ration was considered far too spicy. But everyone admired the embossed cookies in a beautiful container, "like Lorna Doones," that were included in the Russian ration.[18]

Although the rations must have seemed cruelly unchanging to soldiers at the time, they were in fact frequently redesigned to accommodate soldiers' complaints and the unpredictable availability of both food and packaging materials. The QM constantly solicited feedback from the troops, and might have been chagrined by the repeated pushback from soldiers. One report from the Southwest Pacific in late 1944 offered scathing ratings for the meat and vegetable stew, corned beef hash, and dehydrated eggs, which soldiers called "slum rations." Canned vegetables were well liked—"they taste like home." Biscuits were very unpopular, and mess sergeants pleaded, "give us the components for baking our own bread!" Soldiers wanted more spaghetti and macaroni, a big surprise to the mess sergeant, as well as more peanut butter. Complaints were made of too much bully beef, and not enough cheese and chicken. Spam tended to be popular, unless it had spoiled in the can. "Lemon drink" was called battery acid by soldiers. In some areas of combat, these rations were served by the cooks on the bases, and they were repackaged in various ways. "Meat and Vegetable Stew can be improved by making a pie of it, but few cooks go to this trouble," one staffer reported. "Beef and gravy and pork and gravy are relatively new items hence popular. Few cooks take the trouble of taking off the large quantity of grease which will probably soon make these unpopular as well." Another said, "spaghetti is an old favorite and standby." One way to understand what soldiers did not like was to track what was left in inventory; according to this metric, corned beef and corned beef hash were loathed, and in some cases, as in Luzon (Philippines), accumulated to nearly one million pounds.[19]

What is especially intriguing about these arguments regarding rations is what they tell us about the military's assumptions about "standard" American eating habits. By the 1940s, distinct ethnic and regional food traditions were deeply ingrained, and proliferated all across the country. Especially before the interstate highway system and television linked people across regions, specific culinary traditions were locally supported with specialty grocery stores, restaurants, and community gatherings. The notion that there was a universal or standard American diet was, as food historians have demonstrated, a convenient myth, especially for those in charge of feeding masses of people at low cost. To some degree, American this included food norms became what the sixteen million military personnel ate during the war; food that, like Spam, was appreciated years later not only as a food item, but as one that marked a time of extraordinary challenge and emotional cost.[20]

[18] "Foreign rations, 1952" photograph, RG 92, entry 1003 (Combat rations, WW II and Korea), box 74, NACPM; Lt. Col. Paul P. I. Logan to Lt. Col. Rohland Isker, 2 June 1941, RG 92, entry 1003 (Combat rations, WW II and Korea), box. 61, NACPM; "Captured Foreign Food Items," QMC to Intelligence Branch, 26 August 1952, RG 92, entry 1003 (Combat rations, WW II and Korea), box 74, NACPM.

[19] On slum rations, see Lt. Col. D. B. Dill, "Report on Observations in Southwest Pacific and Pacific Ocean Areas, Oct.–Dec. 1944," RG 92, entry 1003 (Combat rations, WW II and Korea), box 58, p. 3, NACPM; on bread baking, see p. 6 of preceding report; on lemon juice see p. 21 of preceding report. On meat and vegetable stew, see Capt. Leo G. Voss to Capt. William R. Junk, 26 September 1945, RG 92, entry 1003 (Combat rations, WW II and Korea), box 59, p. 2, NACPM; on spaghetti, see Voss to Junk (preceding letter), p. 6.

[20] See especially Sidney Mintz, *Tasting Food, Tasting Freedom: Excursions into Eating, Power, and the Past* (Boston: Beacon, 1997); and Donna Gabaccia, *We Are What We Eat: Ethnic Food and the Making of Americans* (Cambridge, Mass.: Harvard Univ. Press, 2000).

One might also note that the QMC's yardstick for ration quality was likely quite different from that of the soldiers' themselves, and in two ways. First, as William Cronon has shown, Chicago was in the absolute center of the food industry. It was home to the largest concentration of grain, meat, and dairy concerns, as well as the stockyards, boards of trade, farming organizations, and longtime food processors and manufacturers. Here the QMC could easily take account of the raw materials needed to feed the military, and was able to meet the basic nutritional requirements of soldiers especially. Food that was fancy, unusual, or not easily mass produced was not very appealing to this operation. And second, we might consider how great food should be, how deliciously distracting, given the grisly business of war. As scholars have pointed out, the moral imperatives of "goodness" in nutritional advice, and the ambivalence people have long had about how "good food" should be either pleasurable or nutritious, but probably not both simultaneously, may have led the QMC to embrace a bland food regimen for the military.[21]

FOOD SCIENCE AND UNIVERSITIES

College programs in food science and technology were not new in World War II. In 1930, there were five such programs, but the numbers grew rapidly over the next two decades, and by 1950 there were about thirty departments. Most of them were located in the land-grant universities that had large departments of agriculture, science, and engineering, as well as in departments of home economics and nutrition. This concentration of specialties created an especially rich environment for training students in both traditional fields, as well as in more experimental, science-based techniques of food production and preservation. These were also perfect sites of recruitment for the expanding food industry, which had a nearly insatiable need for young, technically proficient graduates. And indeed, the relationship between the food industries and academic food science and technology departments was intimate from the beginning. At the Illinois Institute of Technology, for example, the Food Engineering program was in the Chemical Engineering department, and it had an advisory council comprised of representatives from leading food industries, including General Mills, American Can Company, Armour, Kroger, Beatrice Foods, Kraft Foods, General Foods, and so forth. Their goal was explicitly to "best serve the food industries."[22]

This activity closely mirrored the experience of other scientific and engineering disciplines that partnered with industry. Professors and administrators found such an arrangement appealing because it offered hands-on educational opportunities, greatly assisted departments in placing their graduates in industrial positions, and frequently created a new income stream for the university. At the Massachusetts Institute of Technology, Professor Samuel Prescott organized an academic program in food technology

---

[21] William Cronon, *Nature's Metropolis: Chicago and the Great* West (New York, N.Y.: W. W. Norton, 1992). On the morality of food see John Coveney, *Food, Morals, and Meaning: The Pleasure and Anxiety of Eating* (London: Routledge, 2006); Veit, *Modern Food, Moral Food* (cit. n. 5); and Gabriella Petrick, "'Purity as Life': H. J. Heinz, Religious Sentiment, and the Beginning of the Industrial Diet," *Hist. & Tech.* 27 (2011): 37–64.

[22] H. W. Schultz, "Educating Our Food Scientists and Technologists," *Food Technology* 18 (1964): 49–52; for a list of academic programs, see Editorial, *Food Technology* 1 (1947): 300; on Illinois Tech, see J. H. Rushton and M. E. Parker, "A Course of Study for Engineers for the Food Industry," *Food Technology* 4 (1950): 223–5.

and, in 1937, held a conference at MIT on the subject. The organizers hoped to attract about sixty people, and ended up with five hundred, an indication of the growing importance of and the opportunity presented by the new food sciences and technologies. In June 1939, the Institute of Food Technologists was established; in 1947, the trade journal *Food Technology* made its debut. Clearly, the war served to more closely align the interests of industry, the military, and university-based science and engineering.[23]

In both university and corporate food labs, scientists zeroed in on a set of processes and techniques that built time insensitivity into the foods themselves. One major effort underway at MIT concerned the irradiation of meats and vegetables. Getting fresh meats and vegetables to the war zones was expensive, as well as being an inherently inefficient use of shipping space, and was doomed to result in a great deal of loss from spoilage. One of the QMC's big priorities was finding a way to circumvent these problems with new preservation methods. Drying and dehydration were obvious and proven methods, but soldiers did not like most of the dried foods they were fed, and if they would not eat them, there was little point in sending them. Irradiation seemed like a very promising technology because when irradiated, food stopped aging for a long period of time, and, even without refrigeration, stayed "fresh." This was a very appealing solution for provisioning submarine kitchens, which were extremely small and lacking in much refrigerator space. Ultimately, irradiation was deemed a bust because it left an aftertaste in the food that was unpleasant and impossible to mask, even in bacon.[24]

Scientists were keenly aware, of course, that natural ingredients—butter, milk, whole wheat flour, cheddar cheese—were not time insensitive. They became rancid, lost their texture, separated, and underwent other unappetizing processes. As a result, food scientists increasingly looked toward both taking things out of natural food (e.g., removing fats from dairy and wheat germ from wheat to prevent rancidity), and putting things into natural food (e.g., stabilizers to keep things from separating). During the 1940s, a significant industrial capacity to meet these challenges grew in the form of chemical companies that produced food additives, either specifically for the food industry, or as an offshoot of other chemical work. Allied Chemical and Dye, for example, was a major producer of aniline, important in making dyes, photographic chemicals, and chemicals related to rubber, but a small part of the company focused

---

[23] It is very striking how much these food collaborations resembled those in physics and engineering. See David Kaiser, ed., *Becoming MIT: Moments of Decision* (Cambridge, Mass: MIT Press, 2012); and David F. Noble, *America by Design: Science, Technology, and the Rise of Corporate Capitalism* (New York, N.Y.: Knopf, 1977). There has not been a great deal of research on how dietitians, nutritionists, and other home economists might have been hired by these large food and packaging companies, but see Carolyn M. Goldstein, *Creating Consumers: Home Economics in Twentieth-Century America* (Chapel Hill: Univ. of North Carolina Press, 2012); Sarah Stage and Virginia B. Vincenti, eds., *Rethinking Home Economics: Women and the History of a Profession* (Ithaca, N.Y.: Cornell Univ. Press, 1997); Uwe Spiekermann, "Redefining Food: The Standardization of Products and Production in Europe and The United States, 1880-1914," *Hist. & Tech.* 27 (2011): 11–36; Amy Sue Bix, "Equipped for Life: Gendered Technical Training and Consumerism in Home Economics, 1920-1980," *Tech. & Cult.* 43 (2002): 728–54; and Sally Horrocks, "A Promising Pioneer Profession: Women in Industrial Chemistry in Interwar Britain," *Brit. J. Hist. Sci.* 33 (2000): 351–67.
[24] Nicholas Buchanan, "The Atomic Meal: The Cold War and Irradiated Foods, 1945-1963," *Hist. & Tech.* 21 (2005): 221–49; Karin Zachmann, "Atoms for Peace and Radiation for Safety: How to Build Trust in Irradiated Foods in Cold War Europe and Beyond," *Hist. & Tech.* 27 (2011): 65–90; Stuart Thorne, ed., *Food Irradiation* (Barking, UK: Elsevier Applied Science, 1991); Walter M. Urbain, *Food Irradiation* (Orlando, Fla.: Academic, 1986).

on making food dyes. Florasynth Laboratories made flavorings, aromatic chemicals, essential oils, and perfume compounds for the cosmetics, soap, and pharmaceutical companies, as well as for the food industry. The overlapping of processes and products for the food industry and the chemical industry only increased during and after the war; highly processed foods needed a lot of work to make them taste either "natural" or even "scientifically delicious"—that is, utterly fake but oddly addictive, like Cheetos.[25]

One of the most important additives used during the war was vitamins. The QMC tried hard to ensure that the rations were sufficient in calories and nutritional value and, given soldiers' constant need for strength and stamina, was inclined to use food as a vehicle for vitamins and other additives deemed important. Natural foods alone were seen as insufficient for general health, largely because those who worried about such things, like the QMC, or the surgeon general, could not control which natural foods people actually ate. With the rise of highly processed foods in the 1950s and 1960s, which often replaced natural foods, this problem became even more urgent. Soldiers in the field were known to throw canned food away, so the QMC used every means available to put vitamins and other nutritional elements in everything soldiers ate. So-called enriched foods could in this way deliver the nutrition that would otherwise be lost. One scientist pointed to the elimination of pellagra in the American South in the late 1930s by the compulsory addition of niacin to grits and cornmeal, which seemed like an important lesson in public health.[26] But there was no guarantee that this technique would work. The QMC developed a synthetic lemon powder meant to taste like lemonade in the rations, and its purpose was to give the soldier vitamin C. Many reported, however, that it was usually thrown away. The difficulty of getting soldiers to take nutrition seriously was a constant challenge, and Dr. Russell M. Wilder, chair of the Food and Nutrition Board of the National Research Council, which advised the QMC, commented that "what is needed . . . is insurance of the vitamin adequacy of the staple foods so that diets, no matter how carelessly chosen, will be fool-proof." Giving soldiers tablets of, say, iodized salt, was not likely to work because "a soldier of average intelligence" would think that "there is no use taking medicine because I am not sick." This was evidently what happened when trying to give soldiers quinine while serving in the tropics. If the additive was a stand-alone item not mixed into the food, and looked like medicine, it would be thrown away. And indeed, this is how manufacturers looked at it also. In trying to sell the QMC on a fortified biscuit for a ration, Quaker Oats noted that the cost of these biscuits might be a little higher than usual, but "such a product falls almost into the category of pharmaceuticals."[27]

---

[25] Allied Chemical and Dye Corporation, 35th Annual Report (1954), Wilmington Trust Co. Papers, Acc. 2118, box 1, p. 21, Hagley Museum and Library, Wilmington, Delaware (**hereafter** cited as WTCP). On Florasynth, see *Food Technology* 9 (1955): 44.

[26] Daphne A. Roe, *A Plague of Corn: The Social History of Pellagra* (Ithaca, N.Y.: Cornell Univ. Press, 1973).

[27] H. W. Bruins, "Protein-rich Foods for Overcoming Nutritional Deficiencies: The Industrial Viewpoint," *Food Technology* 18 (1964): 51–3. On lemon powder, see Brig. Gen. Georges F. Doriot to Surgeon General, 7 April 1945, RG 92, entry 1003 (Combat rations, WW II and Korea), box 62, NACPM. On vitamins, see Russell M. Wilder to National Research Council (NRC) Subcommittee on Nutrition, 17 September 1940, RG 92, entry 1003 (Combat rations, WWII and Korea), box 60, NACPM. On using salt tablets, see Lt. Col. Rohland A. Isker to Dr. Russell M. Wilder, 7 November 1940; on Quaker Oats see W. G. Mason to Quartermaster, 24 September 1940; both in RG 92, entry 1003 (Combat rations, WWII and Korea), box 60, NACPM. See also Rima Appel, *Vitamania: Vitamins in American Culture* (New Brunswick, N.J.: Rutgers Univ. Press, 1996).

Dehydration was another important way to render foods time insensitive. It was discovered by accident around 1900 in a German plant that was evaporating potatoes; however, simply drying foods (as in fruits) was a very old technique of preservation. In the United States, the earliest commercial dehydrating occurred in 1910, and World War I depended heavily upon dehydrated foods. By World War II, over four hundred companies worked on vegetable dehydration alone, although coming up with a single, universal technique proved elusive, as each fruit and vegetable required slightly different treatment. But by the mid-1950s, scientists had figured out how to dehydrate meat, and this discovery opened a vast new world of frozen-dehydrated meals: "The advice furnished by the Industry Advisory Committee on Dehydrated Meat has been very helpful in the development of new products for Armed Forces use. . . . Precooked freeze-dehydrated meat and seafood products are being developed for use in instant type meals. These include such products as a chili with beans, chicken and rice, chicken stew, turkey a la king, clam chowder, beef stew, Swiss steak, frankfurters and beans, shrimp cocktail, and others." By 1959, the four primary methods of dehydrating were fairly well established: spray drying (for liquids); hot air drying (for vegetables); vacuum drying (for juices); and freeze drying, which was very new.[28]

COLLABORATING WITH FIRMS

The QMC's process of working with the food industry was elaborate. Scientists at the Subsistence Research Lab developed the list of things each ration unit should include, and sent recipes and specifications to companies for bid. Companies bid on what it would cost to produce, say, ten thousand cans of beef stew or one million packets of soluble coffee in six weeks time. Some items already existed and could simply be purchased from manufacturers—loose cigarettes, plastic spoons, chewing gum—but other things needed to be created specifically for the military. These included virtually all the food items. In assembling the K ration in 1943, for example, the QMC developed a stable of companies that were willing and able to produce some portion of the eighteen million rations needed over a six-month period. No single firm could handle the entire order, but small and large firms could commit to deliver several times over this time period at the rate of anywhere from tens of thousands to millions of units. Records from 1950, when the QMC was supporting the Korean War, indicate this scale clearly. In producing 1,500,000 units of the C-6 ration, Carr-Consolidated Biscuit Company agreed to ship 82,500 pounds of sandwich cookies in September, October, and November. The Sol Café Manufacturing Company agreed to ship 3,175,000 packages of soluble coffee in September and October, and 1,675,000 in November. Stokely-Van Camp sent 91,664 cans of ham and lima beans, and 183,328 cans of spaghetti and ground meat, in each of

---

[28] For a history of this process, see W. D. Farnum, "What is the Matter with Dehydration?," *American Food Journal* 18 (1923): 171–3; and S. C. Prescott, "Food Technology and Defense," *Food Technology* 3 (1949): 3. On the difficulty of standardizing, see *Food Technology* 18 (1964): 117–20. For a description of dehydration techniques, see L. E. Clifcorn, "An Appraisal of New Processing Methods for Military Foods," *Food Technology* 13 (1959): 176; for quotation on dehydrated meat, see "QM Food and Container Institute Activities FY 1956, Animal Products, Research and Development, Minutes of the meeting of the Industry Advisory Committee (IAC) on QM F&CI Items, 12 July 1956," RG 92, entry 1004 (Records of IAC, Minutes of Meetings 1952–1962), NACPM. For more information on the emerging infrastructure for these brave new foods, see Shane Hamilton, *Trucking Country: The Road to America's Wal-Mart Economy* (Princeton, N.J.: Princeton Univ. Press, 2008); and Hamilton, *Supermarket USA: Food and Power in the Cold War Farms Race* (New Haven, Conn.: Yale Univ. Press, 2018).

these months. In all, sixty-two firms participated in producing these eighteen million K rations in the summer and fall of 1942. Thus, each item in the ration was the result of a very complex provisioning system developed expressly for wartime needs.[29]

Given the intense production schedule, it is not surprising that the QMC tried to specify what kinds of firms might get contracts. In addition to having "a reputation for fulfillment of contracts and agreements, both verbal and written," and being "financially responsible," potential partners had to have an assembly line and packaging equipment, as well as room to store a week's worth of rations. Further, "the firm needed to be geographically located in accordance with any Army plan for distribution of the Ration"; this meant the company's locations had to be convenient to military shipment points around the country. Clearly these requirements disqualified any small firms, and instead favored large, national firms that had production capacity, experience with large orders, a proven organizational structure, and often a distributed manufacturing operation. Offering contracts to geographically distributed firms was also beneficial in spreading the work around to many congressional districts, but also because shortages of materials like sugar or tin plate affected firms differently, and a firm that could meet production one week might not be able to do so the next. It is unclear if such requirements pushed small producers out of the market entirely, but it is difficult to imagine how they would hold on during the war without state support.[30]

Not surprisingly, there were problems. The QMC was a very fussy customer and often rejected ration items that seemed poorly made. The Baker Importing Company was chastised for their poor packaging of soluble coffee packets; the Campbell Soup Company was told their stews were too thin and the "meat particles" too small. Comments like, "If you are still following the practice of grinding frozen meat as you were about six months ago" suggest that the QMC exercised a high degree of visitation and oversight at each of their contractors' plants. In deciding which chewing gum to put into the rations, QMC staff asked for the following: "Specification CQD No. 324, gum, chewing, requires that all types of gum shall contain a sufficient amount of flavor to impart a pleasing taste sensation. There shall be a primary flood of flavor and a residual characteristic flavor after one-half hour of constant chewing." While the QMC went to great lengths to ensure that the rations were as high quality as possible, the depredations of distance, weather, climate, and the rough-and-tumble of shipping made the end product less than ideal much of the time. In World War I, scientists had developed a "hedonic scale" that measured "a continuum of preferences" and tried to zero in on what soldiers liked and disliked. And by the early 1960s, Cornell University scientists were developing a variety of mechanisms that tested the appeal of various processed foods, including a machine that "wears metal molars to test the 'chewability' of vegetables" and an "electronic nose that gives an objective measure of vegetable flavors."[31]

[29] For a description of this process, see, for example, Captain MacDonnell to Colonel Logan, 29 June 1942, RG 92, entry 1003 (Combat rations, WW II and Korea), box 77, NACPM.

[30] Quotations in attachment to ibid., p. 1.

[31] On the Baker Importing Company, see Capt. Charles G. Herman to R. K. Baker, 21 January 1941; on Campbell's Soup Company, see Woodrow W. Bailey to C. D. Trombold, Campbell Soup Co., 18 August 1943; both in RG 92, entry 1003 (Combat rations, WW II and Korea), box 61, NACPM. On chewing gum, see M. Bollman to W. C. Winokur, 11 October 1948, RG 92, entry 1003 (Combat rations, WW II and Korea), box 63, NACPM. For the hedonic scale, see David R. Peryam and Francis J. Pilgrim, "Hedonic Scale Method of Measuring Food Preferences," *Food Technology* 11 (1957): 9–14, on 9. On mechanisms for measuring the appeal of foods, see *Food Technology* 16 (1962): 97; and Herbert L. Meiselman and Howard G. Schutz, "History of Food Acceptance Research in the U. S. Army," *Appetite* 40 (2003): 199–216.

The cooperating firms had their own headaches. Hershey Chocolate, a major supplier of D ration bars, was unhappy when the military recalled a large quantity of D bars and asked the company if they could reuse them somehow. Hershey suggested that the military work them into cake mixes, ice cream, or chocolate beverages in the military kitchens; because it was so difficult to find workers, the company did not want to waste precious time unwrapping all those bars itself. Hershey was also aggrieved when the military sharply reduced its purchase of cocoa beverage powder right after the firm invested a great deal of money in new machinery for its manufacture for the wartime contract. Likewise, when pharmaceutical company Smith Kline & French did not receive their labeling machine in time, they could not complete their order for one hundred thousand bottles of water purification tablets, resulting in rations that had half as many tablets as usual. When packers who had the equipment to pack hamburger and sausage patties could not be found, the military had to quickly replace that order with more easily packable stews, a maneuver that had repercussions all down the line.[32]

Standardization was not only a trope of military life; it was a recurring theme of wartime provisioning as well. Obviously, the very idea that an organization could produce eighteen million cans of beef stew or packages of cookies presupposes a high degree of standardization and, by 1942, mechanization of production. But because there was never a single producer of all of these units, true standardization was often lacking. As one packer put it, "in spite of strict compliance with the requirements of the specifications governing the components, no one here has ever seen samples of what you may term the perfect product or an acceptable product." And indeed, when five companies each followed the recipes and instructions exactly, their final products could still look and taste different. The rations themselves were constantly changing as well in response to shortages of certain ingredients or packing materials; slowdowns in the factories; problems coordinating the production, assembly, or shipping processes; or strong negative responses from soldiers. Rations frequently lacked some item or had lots of something else—no crackers, all cookies, for example, or cigarettes but no matches, no spoons, and so forth.[33]

It was not just foods that were transformed by military rationing; packaging was similarly challenged and changed. It is likely that the demand for strong and secure packaging had never been as intense as it was during World War II. Further, the shortage of tinplate forced packers to explore alternative packaging materials, leading to such innovations as pouch-packing, waxed paperboard, and individual-unit containers, to name but a few. All packaging, from the tiny cellophane packets holding lemonade powder, to the round cans holding beef stew, to the pallets dropped from helicopters onto beaches and into jungles, had to be "weatherproof, waterproof, resistant to air pressure changes and capable of withstanding the weight of a person sitting on the package." Two examples can illustrate. The first were round twelve-ounce cans that held most components of

[32] On Hershey Chocolate, see S. F. Hinkle to Capt. Louis A. Wright, 20 April 1945; Col. Rohland Isker to Quartermaster, 2 May 1945; both in RG 92, entry 1003 (Combat rations, WW II and Korea), box 71, NACPM. On Smith-Kline, see Lt. Col. J. S. Kusanski to Commander, 21 February 1951; and Lt. Col. William A Warner to Chief, QM Purchasing Division, 24 January 1951, RG 92, entry 1003 (Combat rations, WW II and Korea), box 65, NACPM.

[33] For quotation, see Theodore Phillips, Phillips Packing Co., to Rohland Isker, 29 January 1945, RG 92, entry 1003 (Combat rations, WW II and Korea), box 62, NACPM. On the lack of standardization, see, for example, "Amendments for Ration, Individual, Combat, c-4," 12 July 1950, RG 92, entry 1003 (Combat rations, WW II and Korea), box 65, NACPM.

the C ration. Soldiers were supposed to consume two cans at each of three daily meals when away from base. They had to carry them in their pack or in their pockets, along with the other things they needed, including water, ammunition, clothing, maps, and so forth. And the cans were heavy; the meat stews were full of liquid, meat, and vegetables, and after a few days of these meals, and often exactly the same meals, it was little wonder that soldiers tended to throw them away to lighten their load. To the QMC, this was a dangerous thing to do, because a soldier who was not adequately fed was a cranky and weakened fighting unit. Soldiers and their superiors recognized a much better packing idea when they inspected Norwegian and French rations; these were comprised of flat cans much like sardine cans. Soldiers pleaded with the QMC to switch to flat cans that could fit into a pocket, but in the midst of war it was difficult not only to retool the machinery that produced cans, but also to replace the canned stews with something equally nutritious and familiar. As early as 1941, the military grappled with what turned out to be a big challenge. The only two companies that produced the round cans—American Can and Continental Can—were strongly opposed to the change because all their production momentum was invested in round cans. Round cans were cheaper to make, factory equipment was designed to produce round cans, and there were very few factories that had the requisite machinery for closing the square cans; those that did were working round the clock making cans for Spam! As one military man put it, although "it has been clearly understood for a long time that the sardine type can is a better container than the cylindrical can as far as soldiers are concerned," industry was not interested in shifting over because it was a big expense and headache. The industry probably also did not believe that the postwar civilian market would be interested.[34]

The second challenge was the collapsible metallic tube, which again was found in foreign rations and contained cheese, pate, and jam. United States soldiers and commanders alike were very keen to adopt these tubes for the ration kits, and they were strongly suggested to the QMC as early as 1943. Yet not much happened until the mid-1950s, when the Collapsible Tube Manufacturers Council organized and began a campaign to put condiments, cheese, meat pastes, peanut butter, and caviar into tubes. In 1959, the American Can Company experimented with putting meat pastes in tubes for fighter pilots. Yet, despite its seeming convenience and practicality, the metal tube failed to catch on in either the American domestic or the military market.[35]

For many industrial firms and food companies, World War II offered both an opportunity and a challenge. Most companies were very happy to have the government's business, even if it could be difficult at times. After all, the rationing of foods for civilians not only cut into their expected profits, but also threatened to disrupt the

---

[34] A. H. Anderson, "Past, Present and Future of Packaging Processed Foods," *Food Technology* 18 (1964): 153–7; Walter Maclinn to Col. D. B. Dill, 5 December 1942, RG 92, entry 1003 (Records of the Office of the QM General, QM Food and Container Institute, Conference files), box 1, NACPM. On flat versus round cans, see, for example, Charles G. Herman to Col. Harry Keeley, 5 August 1941, RG 92, entry 1003 (Combat rations, WW II and Korea), box 61, NACPM. Quotation in Col. Charles S. Lawrence to QM, 12 December 1946, folder "Rations, 5-in-One, 1946 and 1947," RG 92, entry 1003 (Combat rations, WW II and Korea), box 73, NACPM. By 1942, six companies, including Underwood, were producing square cans for the Jungle ration; see Lt. Woodrow Bailey to Col. Rohland A. Isker, 29 August 1942, RG 92, entry 1003 (Combat rations, WW II and Korea), box 75, NACPM.

[35] *Food Technology* 9 (1955): 40; L. E. Clifcorn, "An Appraisal of New Processing Methods for Military Foods," *Food Technology* 13 (1959): 176. Tellingly, the military packaged the K ration in flat cardboard boxes that were more easily carried, having shifted from metal cans to cardboard to avoid American Can's issues.

hard-won relationships between companies and their other best customers—the housewives of America.[36] During the war and after, these companies focused on fulfilling their military contracts while at the same time preparing for postwar markets. American Can Company is a case in point. Incorporated in 1901 by consolidating over sixty can-making companies, American Can by 1944 was the largest one in America. During World War II, the company faced a severe tin shortage due to the lack of domestic sources, but was able to keep its machines running by filling orders for the military. It produced not only the cans used in soldiers' rations, but also torpedoes, TNT containers, parachute flare cases, incendiary grenades, smoke-screen pots, and more.[37] Acknowledging that "American Can had flexed its muscles for the first time during World War I," the company assiduously followed both military and domestic markets in a constant search for new processes and uses for its expertise and equipment. Like other industrial firms, particularly those involved in chemical processes, American Can developed a research staff of around three hundred people by 1947 who worked in "bacteriology, biochemistry, food chemistry, food inspection, general engineering, industrial hygiene, metallurgy, non-metallic containers, organic coatings, packaging, sealing compounds, and thermal engineering." With this deep reserve of talent, and the demands placed on producers by the tin shortage, the company developed fiber milk containers, fiber frozen food boxes, and cans for motor oil, vacuum-packed coffee, and acid fruits and juices; these were all products whose properties required special coatings in the cans themselves. They also developed the individual beer can, an important social innovation in that it replaced the draft beer only available in taverns. With this, people could drink in the privacy of their own homes, an appealing dimension of the new suburban lifestyle; however, some may have given up the pleasures of neighborhood camaraderie (especially men) as part of the bargain. In 1953, canned beer accounted for half of the company's growth over the previous seven years.[38] American Can also reached deeply into the emerging habits of young people through its Home Economics Department, sending brochures and "instructional materials that explain the nutritive values and exciting menu possibilities of canned foods," as well as a forty-eight-page "High School Manual on Commercially Canned Foods." Employing the time-honored practice of having children persuade their parents of a product's value, American Can left no doubt: "As the twig is bent, so the tree is inclined."[39]

Likewise, the Container Corporation of America, which was incorporated in Delaware in 1926, dominated the nonmetal container market, specializing in folding cartons, boxboard, and corrugated shipping containers, but also cellophane, polyethylene, glassine, foil, and spiral-wound paperboard cans, to name but a few of its products.

[36] See, for example, Susan Strasser, *Satisfaction Guaranteed: The Making of the American Mass Market* (New York, N.Y.: Pantheon, 1989).

[37] While Standard and Poor said that American Can consolidated "over 60 companies," the company itself claimed the number was 123; see Standard and Poor, "American Can Company," (Investment Fact Report), 6 March 1944, acc. 2118, box 2, WTCP; American Can Company Annual Report, 1948, acc. 2118, box 2, WTCP. On sources for tin, see H. C. Wainwright and Co., "Report, American Can Company," 28 November 1951, acc. 2118, box 2, WTCP; and W. W. Geddes, "Report to Securities Committee—Container Industry (Metal)," 31 May 1949 (folder, Container Industry, 1949–1953), acc. 2118, box 15, WTCP.

[38] Lawrence M. Marks and Co., "American Can Company," 6 March 1947, acc. 2118, box 2, WTCP; Argus Research Corporation, "Can Manufacturers American Can Co. and Continental Can Co.," 30 October 1953, acc. 2118, box 2, p. 4, WTCP.

[39] American Can Company, "Annual Report" 1954, acc. 2118, box 2, pp. 29–31, WTCP.

Through vertical integration, CanCo, as it was known, was able to buy hundreds of thousands of acres of forest land from which to make cardboard, and it was energetic in buying smaller companies that possessed packaging materials and techniques that could be used to expand their market. But during the war, "a considerable portion of production consisted of specialized packaging for military equipment and supplies, and lend-lease shipments."[40]

Another company that grew up with the military was Borden. The company got its start during the American Civil War, when Gail Borden figured out how to reduce the water content of milk under vacuum by about 60 percent. The resulting milk was called either evaporated milk or condensed (if sugared), and was sold as a completely reliable and safe form of fluid milk at a time when "fresh" milk was not pasteurized and thus often carried bacteria and worse. Like the packaging companies, Borden invested early in research staff, and by 1937 was selling a dizzying number of milk-related products in a wide variety of markets. Fluid milk, in fact, declined in sales strength, but that was made up for with sales of things like dry milk, used by bakers; Dryco, an infant formula with vitamin D; beta lactose (milk sugar, which had a number of industrial uses); Casco glue and liquid cement; ice cream; caramels; a cheese flavoring that was sprayed on popcorn (Gobbles); and animal rations, to name but a few. Like a sausage maker that used "everything but the squeal" of a pig, Borden was tireless and ingenious in developing and promoting its many specialized products. For example, Borden built a major exhibit at the 1939 World's Fair in New York City, featuring a Walker-Gordon Rotolactator, "the milking Merry-go-round." This also marked the debut of Elsie the Cow, called a "trade character," who was featured across the country in dramatic promotions for Borden's products.[41]

But as one analyst astutely observed, when companies boasted about their research efforts, they were really talking about market creation. Given a company's areas of expertise and reputation, its existing physical plant, including machinery, its capitalization, and its geographic location and advantage, the question was how this company could continue to make new things that people would want to buy. Whereas innovation in many early twentieth-century industries has been described as "problem solving," innovation in the midcentury food industry could better be described as creating needs and desires in people that had not been there before, and then fulfilling them. A lot of this was packaging—fiberboard rather than glass milk containers, individualized packets of sugar, coffee, and other condiments, and cellophane wrapping of foods that shoppers wanted to inspect, such as carrots and radishes. But often it was things that no one needed until a company created and marketed them as solving

---

[40] Standard and Poor, "Investment Fact Report, Container Corporation of America," 19 March 1945 (folder, Container Corporation of America, 1944–54), acc. 2118, box 1, WTCP; Container Corporation of America, "Annual Report" 1951 (folder, Container Corporation of America, 1944-1954), acc. 2118, box 15, WTCP.

[41] The Borden Company, "82nd Annual Report," 1939; "80th Annual Report," 1937; "81st Annual Report," 1938; "83rd Annual Report," 1940, p. 9; all preceding in acc. 2118 (Borden), box 7, WTCP. Elsie also had a philosophical side, as when she confided to shareholders, "Of course you know that when I say 'I' went to these places, I really mean the cows that play my part . . . I can't go by myself because the truth is that I am not a real cow at all, but an idea—or a character something like Alice in Wonderland or Snow White"; see The Borden Company, "84th Annual Report," 1941, acc. 2118 (Borden), box 7, WTCP. See also Anna Thompson Hajdik, "A 'Bovine Glamour Girl': Borden Milk, Elsie the Cow, and the Convergence of Technology, Animals, and Gender at the 1939 New York World's Fair," *Agr. Hist.* 88 (2014): 470–90.

the problem of convenience and time pressures. Crystallized orange juice, frozen dinners, and pancake mix were just some examples; there was seemingly no end to the surprising forms of new food that industry could put on supermarket shelves.[42]

Borden had several products made especially for the military market after the War between the States. The first was called "KLIM," or milk spelled backwards. It was the spray-dried milk developed for World War I, and used heavily in World War II because it could withstand the high temperatures of the South Pacific as well as the low temperatures of the alpine regions. This product found a ready market with civilian bakeries in the years after the Second World War. The company also adapted its dehydrating equipment so that it could be used to dry not only milk, but also orange and lemon juice, eggs, coffee, and soup. As early as 1942, Borden was making plans for the possibilities of the postwar market.[43]

Second, Borden made a variety of cheeses, or cheese products as we would now call many of them. From 1941 to 1945, much of the cheese went to the military and the Lend-Lease program, through which the United States supplied the Allies with much-needed food and other necessities. However, there were also specialty cheeses, such as "Military Brand Camembert and Brie, Liederkranz, and Borden Cocktail Spreads." This was a real investment in the future; as company officials admitted, half of the company's cheese production was bound for the military and Lend-Lease markets. Americans were not big cheese eaters in the early 1940s, but Borden hoped to create that market once the war was over. Similarly, ice cream was not a household treat before the war; it was consumed in drug store soda fountains exclusively. But during the war, Borden made a big push to supply ice cream to servicemen, building a belief that ice cream was a comfort food and "morale builder." After the war, a confluence of events, including the rise of self-service grocery stores, the invention of home freezers, the difficulty that soda fountains had in hiring people, and the huge popularity of ice cream among returning servicemen, all led to the normalization of ice cream as a year-round food that, like beer, could be purchased in small, even individualized units, and enjoyed in the privacy of one's home.[44]

CONCLUSION

The years following World War II and the Korean War were extraordinarily busy for the food industry. Huge, sudden surpluses of cheap grain led to imaginative new foods; these were what we have come to call convenience foods, many of which featured novel spray-on flavors, nutritional additives, and lots of salt and fat. Foods formerly reserved for the military entered the civilian market in decidedly more stylish packaging. The cocoa powder in rations became Nestle's Quick in 1948; freeze-dried

---

[42] W. W. Geddes, "Report to the Securities Committee—Container Industry (Metal)," 31 May 1949, "Container Industry, 1949-1953," acc. 2118, box 15, WTCP. On the expansion of the mass food market, see, for example, Richard Tedlow, *New and Improved: The Story of Mass Marketing in America* (New York, N.Y.: Basic, 1990); Shane Hamilton, "The Economics and Conveniences of Modern-Day Living: Frozen Foods and Mass Marketing, 1945-1965," *Bus. Hist. Rev.* 77 (2003): 33–60; and Harvey Levenstein, *Paradox of Plenty* (cit. n. 5).

[43] The Borden Company, "85th Annual Report," 1942, acc. 2118 (Borden), box 7, pp. 9, 24, WTCP.

[44] The Borden Company, "84th Annual Report," 1941; "86th Annual Report," 1943, p. 25; "81st Annual Report," 1938, p. 10; "85th Annual Report," 1942, p. 7; "Annual Report," 1953, pp. 12–13; all preceding in acc. 2118 (Borden), box 7, WTCP.

coffee and orange juice followed soon after thanks to new vacuum technology developed for medical uses during the war; and corn and wheat became new kinds of snacks, cereals, and filler in bakery mixes. Although these trends existed before the Second World War, the years following the war offered the food industry—already sufficiently prepped following its intense war-related work—an unmatched opportunity to capitalize on the knowledge, new materials, and new techniques developed during the conflict. And by 1950, a number of companies that had supplied the military with rations could count notable growth: National Biscuit (which became Nabisco) grew by 26.4 percent, Borden by 27.7 percent, General Foods by 21.6 percent, and Corn Products by 23.6 percent. Surely, much of this growth stemmed from war business, but one might surmise that there was a little more to it. In 1949, Samuel Prescott at MIT had been studying the military-commercial nexus for many decades when he proposed "a more integrated relation between foods for defense and food fully acceptable for commercial distribution and broad civilian consumption." He added: "After all, the consumers all have the same nutritive requirements, and approximately similar tastes. Moreover, the buying public is already familiar with many types of concentrated or condensed foods." Indeed, the editor of the leading trade journal summed up the changes that had made all this possible: first, "the greater portion of our food supply no longer goes directly from agricultural producing centers to the consumer," but detours through the processing industry; second, "discoveries in nutrition science have shifted attention from preservation to conservation"; and third, food processors had started to recognize that "they can all use the same basic unit operations irrespective of the raw materials with which they start." This last was a good insurance policy in uncertain times for the company seeking maximum flexibility. In other words, the extraordinary expansion of the processed food and flavor industry was not an accidental and surprising outcome of the war—it was baked into the planning all along.[45]

The linkages between the food industry and the military continued to tighten in the 1950s and 1960s. Prescott's vision seemed to have come together. The government found that by having the food industry intimately connected to evolving military needs, the industry itself could develop "items suitable for both Navy and commercial use . . . at moderate cost to the government." In 1966, this arrangement was formalized in the creation of Research and Development Associates, a group of 250 representatives of industry, universities, government agencies, and the military that helped "the civilian food and container industries cooperate in meeting military feeding needs." From the point of view of the military, this was a brilliant and possibly unavoidable strategy. The government had no capacity to produce food itself, and depended upon the food industry to generate the rations. For the food industry, what could be better than a customer dedicated to time-insensitive foods, foods that had no "sell-by" date and could be sold to citizens as well as soldiers? It was the perfect market match. Yet, it is striking how easily commentators conflated the entire American population with eighteen-year-old

---

[45] On the effects of military demands on the civilian market, see, for example, Anastacia Marx de Salcedo, *Combat-Ready Kitchen: How the U.S. Military Shapes the Way You Eat* (New York, N.Y.: Comment, an imprint of Random House, 2015). On growth, see Laurence M. Marks and Company, "Food and Dairy Profits and Security Prices," 13 January 1950, acc. 2118, box 20, WTCP. Prescott quotation from Samuel C. Prescott, "Food Technology and Defense," *Food Technology* 3 (1949): 3. On the three big changes, see E. H. Harvey, "Food is Fundamental," *Food Technology* 1 (1947): 303.

soldiers, as if the infamous "training table" for high-school and college athletes could be smoothly transferred to families of women and men, children and grandparents, without consideration of differing nutritional needs and preferences. The tastes and demands of young men were converted into norms for everyone. The tacit assumption that soldiers and athletes are somehow "model organisms" in designing food availability and policy might have made sense for the military, but not so much for everyone else.[46]

The biggest innovation in military food during World War II did not really involve the food itself, although that would come later. It was instead generating the system of creating, producing, and distributing the food, which included overseeing the industrial production and packaging that were specifically designed to merge military and civilian needs, thereby making the food provisioning system more seamless and less reliant on fresh produce and meats. This generated a nearly unstoppable barrage of research into such products and resulted in a mammoth engine of scientific and technological prowess in the years following the war. This effort also created tight links between the military, universities, home economists and dieticians, farmers, the food industry, and entities responsible for feeding masses of people—hospitals, schools, prisons, and NGOs sponsoring food relief around the world. It should probably come as no surprise that the mass feeding of the military during the war precipitated the mass feeding of other groups as well, most of whom were even further from the eighteen-year-old soldier norm than the rest of us.

[46] First quotation from Milton E. Ryberg, "The Navy's Food and Equipment Needs Outlined by the U.S. Naval Supply Research and Development Facility," *Food Technology* 15 (1961): 15. On civilian industries, see *Food Technology* 20 (1966): 59–61.

Meat Mimesis:

Laboratory-Grown Meat as a Study in Copying

*by Benjamin Aldes Wurgaft**

ABSTRACT

This article examines an emerging form of contemporary food biotechnology, laboratory-grown or "cultured" meat, that often seeks to copy conventional "in vivo" animal flesh by using in vitro techniques. The ultimate goal of cultured meat research is to devise an alternative to the environmentally damaging and ethically undesirable infrastructure that makes "cheap" industrial-scale meat possible. Formal research into cultured meat has been underway since the early 2000s. However, after almost two decades of experiments, it is still unclear if this avenue of research will produce a viable meat product at scale, or if it is even possible to perfectly copy the physical characteristics of in vivo meat. There are technical limitations on scientists' ability to reproduce the precise textures, tastes, and overall "mouthfeel" of familiar types of meat gleaned through butchery. Cultured meat proceeds from a premise we might call "biological equivalency," the view that animal cells grown in a bioreactor will have the same characteristics as their in vivo counterparts, and it breaks from a standing approach in food science that we might call "sensorial equivalency," which seeks to reproduce not meat itself but rather the sensory experience of eating meat, usually starting with a substrate of plant cells. This article, which draws from five years of ethnographic fieldwork in the cultured meat movement, seeks to illuminate not only the historical but also the philosophical questions raised by efforts to copy meat. Drawing on the work of the intellectual historian Hans Blumenberg, this article concludes with an exploration of *mimesis* itself, understood as the imitation of nature.

What if we could get our meat by growing cells in vitro and harvesting the result, rather than by raising, killing, and butchering animals by the billions? In corporate and academic laboratories from Maastricht to Silicon Valley to Tokyo, scientists grow animal cells under carefully controlled conditions in hopes of creating meat. Their methods are scientific, in that they conduct experiments to test hypotheses about cell growth and development, or the efficacy of types of growth media, or the production of fat cells. But their ultimate goal is not to understand the growth and function of bovine or porcine muscle cells (or those of that ubiquitous jungle fowl, the chicken), or the properties of meat after slaughter and butchery. The researchers, whose backgrounds range from stem cell science to muscle physiology to meat science, hope to change the way we eat and live. Toward this end, they want to copy in vivo meat, conventionally taken from

* ben.wurgaft@gmail.com.

This article expands upon "Copy," chap. 11 in Benjamin Aldes Wurgaft, *Meat Planet: Artificial Flesh and the Future of Food* (Oakland: Univ. of California Press, 2019).

butchered cows, pigs, chickens, and other common food animals, including several species of fish. The resulting substance would be meat without the birth and growth and suffering of millions of animals each year—meat without killing.

Cultured meat workers hope that their new products, variously called "cultured meat," "clean meat," "in vitro meat," or "lab meat," among other names, will replace conventional meat partly or entirely. Debates over nomenclature have flared up, with some groups promoting "clean meat," others "cultured meat," while detractors use "lab meat" or "Franken meat" or "schmeat" as terms of opprobrium. In 2018, a cluster of start-up companies rebelled against the label "clean meat," which had been gaining traction, and decided that "cell-based" was a monicker less insulting to traditional producers of meat, whose products must be "dirty" if others can be called "clean." This battle over names is instructive. It is not simply a struggle to find the most descriptively accurate or pleasing name for a new food product. It is a marketing battle fought before there is any actual product to market. Entrepreneurs and cultured meat promoters make competing claims over which terms will best win over the hearts and stomachs of eaters, or which terms communicate the nature of lab-grown meat most transparently, thus helping to build public trust or smooth diplomatic relations with the very industry that stands to be "disrupted." This industry is conventional meat production, whose spokespeople reject the notion that their offerings are "dirty."[1] The naming issue is anticipatory, a form of futurism.

Although cultured meat research only reached mainstream media audiences around 2013, it has in fact passed through two discernible phases over the past twenty years. As Neil Stephens, Alexandra Sexton, and Clemens Driessen have shown, the first phase consisted largely of academic research, as well as public discussions in which the precise nature of laboratory-grown meat was unclear.[2] A research group at Touro College, in New York City, with a grant from NASA, grew goldfish cells and intended to explore cultured meat as a protein source for long-term space travel;[3] a team of bio-artists grew amphibian cells in order to serve "frog leg steaks" to an audience at a public event in Nantes, France;[4] and a team of Dutch researchers, with encouragement from a very determined businessman named Willem Van Eelen, secured a grant from their government and pursued laboratory work.[5] In 2008, the People for the Ethical Treatment of Animals, based in the United States, surprised the small cultured meat community by announcing a cash prize for the first team who could bring lab-grown chicken to market; such hopes of fast success exceeded the expectations of many researchers themselves. While the 2013 unveiling, by the Dutch scientist Mark Post, of

---

[1] This article uses the term "cultured meat" because I believe that it is the most scientifically descriptive: this is meat grown via tissue culture techniques. This reflects my own goal, which is not marketing but scholarly communication. For one journalistic overview of the nomenclature issue, see Chase Purdy, "Would You Eat 'Clean Meat'?," *Quartz*, 18 November 2017, https://qz.com/1086825/theres-a-debate-among-the-makers-of-cell-cultured-meat-what-do-you-call-it/. For a scholarly view, see Neil Stephens, Alexandra Sexton, and Clemens Driessen, "Making Sense of Making Meat: Analyzing Key Developments in the First Twenty Years of Tissue Engineering Muscle to Make Food," *Frontiers in Sustainable Food Systems* 3 (2019), https://www.frontiersin.org/article/10.3389/fsufs.2019.00045.

[2] Stephens, Sexton, and Driessen, "Making Sense" (cit. n. 1).

[3] Stephen Pincock, "Meat, In Vitro?" *Scientist*, 1 September 2007, https://www.the-scientist.com/notebook-old/meat-in-vitro-46164.

[4] Oron Catts and Ionat Zurr, "Disembodied Cuisine," 2003, The Tissue Culture & Art Project, accessed 14 March 2020, https://tcaproject.net/portfolio/disembodied-cuisine/.

[5] For one account of the early Dutch cultured meat research scene, and of Van Eelen's role therein, see Michael Specter, "Test-Tube Burgers," *New Yorker*, 16 May 2011.

the world's first lab-grown hamburger was undoubtedly the most important media event for cultured meat, it was not until around 2015 that a distinct phase change took place. Enterprise rather than academic research dominates this second phase, still ongoing as of this writing, although academic researchers—some of them supported by grants from nonprofit organizations that promote lab-grown meat research—are just as active as before. The second phase has coincided with the emergence of a distinctive sector of investment, research, and product development known as "food tech," in which entrepreneurs attempt to develop new forms of food products. (One famous or infamous "food tech" product is the meal-replacement beverage Soylent; others include plant protein-based meat alternatives such as the Impossible Burger and Beyond Meat.) A wave of food tech start-ups has sprung up, promising cultured meat products on faster timetables than many academic researchers previously proposed, with a number of Silicon Valley companies especially visible. While many actors in the first phase of cultured meat research had definite goals, such as ameliorating harm to livestock animals or reducing the environmental damage caused by conventional meat production, these priorities have become nearly a priori presumptions in phase two, assumed to be shared by all.

Notably, in its accounting of the problems of meat production, the cultured meat movement—for it now often styles itself not only as a research field and nascent industry, but as a movement with significant "mission-based" goals in addition to its obvious commercial ones, goals that, to quote one start-up accelerator program, pertain to "humanity's global challenges"[6]—elides any distinctions between the ways different human communities around the world breed, raise, and kill and eat animals. It targets the infrastructure of feedlots and slaughterhouses necessary to produce enough meat to support the globalizing "Western Diet"; this is industrial, or "cheap meat."

Given that it aims at nothing less than transforming the global meat industry, laboratory-grown meat is an incredibly ambitious project, capable of inducing a sense of vertigo in anyone trying to follow its story, as I did as an ethnographic researcher between 2013 and 2019.[7] While it is very new, laboratory-grown meat is also extremely mediagenic, and scores of articles about cultured meat and its architects periodically fill the pages of science magazines both on- and offline. The relative newness of cultured meat research belies the fact that the idea of enjoying meat without raising and killing animals—of eating meat that is "de-animalized" but meat nonetheless—has a long cultural and intellectual history that runs decades ahead of laboratory practice: medieval European and Chinese texts spread legends of lambs growing from plants;[8] meat falls from the sky or appears out of thin air in the Talmud; or in medieval Europe, starving peasants would dream of a magical land called Cockaigne, where food is plentiful and cooked birds fly into the mouths of eager eaters. Modern advertisements, in which grinning and anthropomorphically bipedal pigs gesture at their own tasty flesh, recall earlier fantasies of meat that wishes to be eaten, as animals generally do not. Food technologists are well aware of science fiction's much more recent fantasies

[6] See the website of Singularity University, Singularity Education Group, 2020, accessed 20 August 2019, https://su.org/about/global-grand-challenges/.

[7] I am grateful for support from the National Science Foundation via Grant #1331003, "Tissue Engineering and Sustainable Protein Development."

[8] See Benjamin Aldes Wurgaft, "Animal, Vegetable, or Both? Making Sense of the Scythian Lamb," *Lapham's Quarterly*, 5 August 2019.

about protein, as evidenced by the name Soylent, taken from the film *Soylent Green* (1973), which was adapted from the 1966 Malthusian novel *Make Room! Make Room!*, by Harry Harrison.

At the center of cultured meat research is the assumption that scientists and engineers will one day be able to reproduce in vivo meat more or less perfectly, or at least perfectly enough to satisfy eaters. But copying the flavors, textures, and other physical qualities of meat through tissue culture is technically challenging. It also raises philosophical and social questions about the definition of meat, as well as about who gets to do the defining. In this article, I explore the challenge and possible meanings of meat mimesis from the perspective of the intellectual history of copying natural things, as well as from the vantage of the history of food science. I argue that the effort to copy meat using in vitro techniques is not simply a technical affair. It involves shifting the site of meat's definition from conventional agricultural infrastructures (feedlots, slaughterhouses, supermarkets, and eater's tables) to the physical locations of "cellular agriculture," currently the laboratory and, perhaps someday, the factory. To copy meat, we must define the terms on which we copy it, which means deciding what "meat" really means.[9]

First, however, we need a more detailed understanding of the state of cultured meat research. In 2010, observing the earlier, academic phase of research, Neil Stephens called in vitro meat an "as-yet undefined ontological object."[10] It was not yet obvious what animal muscle cells grown in vitro really were, or what the public expected them to become. Were they going to be the equivalent of soy burgers or Quorn, those vegetarian replacements for the animal protein that often sit at the center of dinner plates? Could they be the equivalent of the conventional meat with which billions of eaters are familiar? Since 2010, and especially following Mark Post's 2013 hamburger demonstration, public expectations for a new form of meat seem to have taken a definite shape: animal muscle cells grown in vitro are assumed to be an approximate equivalent for conventional meat, with the important question of just how close an equivalent cultured meat will be. While there are still no cultured meat products available to consumers as of this writing in late 2019, cultured meat practitioners now imagine cultured meat not only as a replacement for conventional meat, but as something with a meat-like "ontology," in Stephens's terms. This is also the image of cultured meat that entrepreneurs present to audiences through media announcements, product pitches, and promotional talks.[11]

---

[9] The effort to copy conventional meat recalls a different sense of the word mimesis, one famously developed by the literary scholar and philologist Erich Auerbach in his 1943 *Mimesis*: "the representation of reality" or, as he puts it in his book's subtitle, "the representation of reality in Western thought." There are many versions of copying, but they all begin with a sense of the thing to be copied, that is, a representation. See Auerbach, *Mimesis: the Representation of Reality in Western Thought*" (Princeton, N.J.: Princeton Univ. Press, 2013).

[10] Neil Stephens, "In Vitro Meat: Zombies on the Menu?," *SCRIPTed: A Journal of Law, Technology & Society* 7 (2010): 394–401.

[11] For two very different examples in two genres—the promotional film and the promotional talk—see the short film made for Mark Post's 2013 hamburger demonstration by the documentary film company, The Department of Expansion, "Cultured Beef," 29 July 2013, YouTube video, 6:05, https://www.youtube.com/watch?v = gdMQND4TPqM; and Bruce Friedrich, "Market Forces and Food Technology Will Save the World," streamed live 30 January 2018, YouTube video, TEDx Talk, 17:39, https://www.youtube.com/watch?v = liZtyP2tKhA.

While some artists and designers have speculated about the multiple physical forms cultured meat may eventually take, breaking with popular conventions such as hamburger, sausage, steak, or chicken nuggets, the common wisdom in the cultured meat movement is that in order to accomplish its objectives, mimesis (in the sense of copying) is necessary.[12] That is, cultured meat products must duplicate the physical traits of conventional ones in order to either partly or fully replace the most widely consumed forms of industrially produced meat. Consumers should not be wooed to embrace strikingly new forms of animal protein. Instead of giving them pyramids or perfect spheres of pig flesh, cultured meat merchants should offer them pork chops. Instead of bovine "meat chips" (an early project by a start-up previously interested in cultured meat), consumers want hamburger, chicken strips, meatballs, fish fillets, and so forth. According to the mimetic approach to cultured meat, the modern meat dishes consumers enjoy should remain the same, but the process by which they get their meat should change radically. In place of Concentrated Animal Feeding Operations (CAFOs) and slaughterhouses, the architects of the cultured meat movement imagine either large-scale industrial facilities with giant bioreactors (a vessel or device that supports the growth of cells or biochemical reactions), or smaller production centers in urban areas, resembling beer microbreweries. They would do away with the sheer waste of raising millions or billions of animals to eat only parts of their bodies,[13] and also eliminate the suffering of those animals. They would simply never be born or hatched. Some in the cultured meat movement speculate that relatively small breeding populations of animals would be kept, perhaps tens of thousands per species, worldwide, to preserve genetic diversity.

All of this mimetic effort is meant to remedy two central problems already noted: the widespread suffering of animals in the global meat production system; and the environmental damage caused by that system, which some experts believe produces about 14 percent of annual greenhouse gas emissions. Theoretical life cycle assessments of cultured meat production, intended to gauge its environmental impact and natural resource use as compared to those of conventional meat production, have projected that the former might demand 99 percent less farmland, almost as great a reduction in water use, and require significantly (up to 45 percent) less energy; though other assessments run contrary to some of these findings, indicating the possibility of higher energy costs than conventional meat production.[14] Biological safety concerns are less often mentioned than animal ethics or the environment, but they are another reason to move away from industrial-scale meat production. CAFOs may breed antibiotic resistant disease-causing

---

[12] For examples of nonnormative meat shapes, see Koert van Mensvoort and Hendrik-Jan Grievink, eds., *The In Vitro Meat Cookbook* (Amsterdam: Next Nature Network, 2014). See also Tissue Culture and Art Project, by Oron Catts and Ionat Zurr, who work collectively as SymbioticA, based at the University of Western Australia, http://www.tca.uwa.edu.au/.

[13] From a 540-pound steer butchered in the United States, we consume on average 185 pounds of meat, or 34 percent of the animal's live weight. See Vaclav Smil, "Eating Meat: Evolution, Patterns, and Consequences," *Population and Development Review* 28 (2002): 599–639, on 601.

[14] H. Tuomisto and M. de Mattos, "Environmental Impacts of Cultured Meat Production," *Environmental Science and Technology* 45 (2011): 6117–23; and C. S. Mattick, A. E. Landis, B. R. Allenby, and N. J. Genovese, "Anticipatory Life Cycle Analysis of In Vitro Biomass Cultivation for Cultured Meat Production in the United States," *Environmental Science & Technology* 49 (2015): 11941–9. By contrast, in S. Smetana, A. Mathys, A. Knoch, and V. Heinz, "Meat Alternatives: Life Cycle Assessment of Most Known Meat Substitutes," *International Journal of Life Cycle Assessment* 20 (2015): 1254–67, the authors found that for many animal types, cultured meat would actually be more environmentally problematic than its conventional counterpart because of the high energy use that would go into making it.

microorganisms that not only affect those animals, but also jump across species barriers to local human populations. Cultured meat could potentially be produced without antibiotics (particularly if entirely sterile facilities are used), and some of its promoters imagine that it would thus help to reduce pandemic risk.[15]

Advocates for cultured meat generally agree that a mass transition to vegetarianism or vegan diets would accomplish the goals of cultured meat more easily than the tough and expensive work of creating and marketing a new technology and product. However, they also believe that such a transition is unrealistic, not only given the historically high levels of meat consumption around the globe, but also given that meat consumption seems to be on the rise in those parts of the world where the ranks of the middle classes are growing, as in China and India.[16] Despite some reductions in meat consumption among certain demographic groups in the developed world (older, health-conscious, and more affluent adults; younger vegetarians), meat consumption is generally on the rise. Economists often characterize meat as an "income-elastic" commodity, and our demand for it is rising with available funds; in an era in which Western tastes for a meat-centric meal have globalized, rising numbers of people in the middle classes mean rising numbers of meat eaters. Some advocates for cultured meat have framed this as a production shortfall problem waiting to happen, and one made worse by global warming and our gradual loss of available farmland—some 70 percent of which, by some estimates, now goes toward animal agriculture either directly, in pasturage, or indirectly, in the farming of crops for fodder.[17] This is a remarkable fact, because prior to the industrial production of meat and the great increase in meat consumption in the developed world, most human societies have consumed a diet in which animal foods played a supplementary role, with the bulk of calories being derived from plant foods.[18] When plotted on a time line of human history, current levels of per capita meat consumption in the developed world look like a shocking deviation, rather than an expression of our species' normal and timeless relationship with meat.[19]

Panic over the future of meat eating is not a novel development of the early twenty-first century. While the conversation about cultured meat emphasizes the reform of omnivory over the question of whether or not there will be enough meat to go around, the latter worry has been a persistent feature of the history of debate on the future of food supply and demography going back over two centuries. While a glimmer of our modern preoccupation with meat can be found in Plato's *Republic*, where Socrates depicts livestock as a problematic luxury food whose land requirements commit a

---

[15] On CAFOs and pandemic risk, see Mary J. Gilchrist et al., "The Potential Role of Concentrated Animal Feeding Operations in Infectious Disease Epidemics and Antibiotic Resistance," *Environmental Health Perspectives* 115 (2007): 313–16.

[16] Henning Steinfeld, Pierre Gerber, Tom Wassenaar, Vincent Castel, Mauricio Rosales, and Cees de Haan, *Livestock's Long Shadow: Environmental Issues and Options* (New York, N.Y.: Food and Agriculture Organization of the United Nations, 2006), accessed 30 August 2019, http://www.fao.org/3/a-a0701e.pdf (see esp. 9–11).

[17] For example, see the Worldwatch Institute's 2014 report *Peak Meat Production Strains Land and Water Resources*.

[18] See Smil, "Eating Meat" (cit. n. 13), 602.

[19] On the history of human omnivory, its relations to human evolution, and debates about its relevance to the future of eating animals, see Wurgaft, "The Debate: Will the World Run Out of Food?," chap. 2 in *Meat Planet: Artificial Flesh and the Future of Food* (Oakland: Univ. of California Press, 2019); and Josh Berson, *The Meat Question: Animals, Humans, and the Deep History of Food* (Cambridge, Mass.: MIT Press, 2019).

state to constantly expand its territory through war, a more proximate beginning can be found in the writings of Thomas Robert Malthus, whose 1798 *Essay on the Principle of Population* became a touchstone of modern thought on the future of food.[20] Malthusianism, which famously holds that increases in population tend to outstrip increases in food supply, was not innocent of ideas about proper diet; like many Englishmen, Malthus himself took the beefsteak as a measure of sufficiency, and like other writers whose work was informed by a concern for the average livelihood in the British colonies, he worried about an imagined rice-based Asian diet becoming standard in a future overpopulated West.[21] A similar emphasis on meat as a food definitional for proper sustenance has characterized debates on food security after Malthus. In the later twentieth century, this emphasis has been reflected not only in discussions between representatives of think tanks and policy institutes, but in ersatz forms of protein (often, algae or soy-based foods roughly in the form of a hamburger or steak) designed to replace in vivo meat, part of a "diet for a small planet," to use Frances Moore Lappé's phrase. In her 1971 book of that title, Lappé took careful note of the dismal feed conversion ratio of cows and advocated a plant-centric diet closer to the sun, in trophic terms, to achieve a more environmentally sustainable set of food practices.[22] What is most remarkable about the prominence of meat in debates over the future of food—and protein as a centerpiece in debates about population and dietary sufficiency—is that from the perspective of the long history of human carnivory, this prominence looks like sheer exaggeration. It was only in the mid-nineteenth century (decades after Malthus's *Essay*) with the beginnings of industrial-scale breeding, transport, slaughter, and processing of livestock, that meat consumption began to climb toward the levels many enjoy today.[23] While humans are natural omnivores, there is thus nothing "natural," in the sense of being driven by our biological needs, about our current levels of meat consumption—the very levels that cultured meat might help to sustain.

Proposed methods of cultured meat production usually begin with a tiny biopsy of skeletal muscle tissue taken from a healthy and otherwise unharmed donor animal. The satellite stem cells in the sample are separated and encouraged to differentiate into muscle cells (myoblasts) and to proliferate; this encouragement involves the use of both biological and chemical cues in the cell culture media, as well as mechanical stimulation. The myoblasts then join together to form myotubes, and these then join to form the larger structures of muscle fibers. Cultured meat actually begins to mimic conventional animal muscle (and thus meat) during the in vitro production process itself, because the process of stimulating stem cells attempts to duplicate the process by which muscles regenerate after injury or trauma; a healing process is repurposed as a production process. The cells form strands that tissue engineers can then shape into muscle,

---

[20] For a discussion of Malthus and his relationship with the history of debate on the future of food, see Warren Belasco, *Meals to Come: A History of the Futures of Food* (Berkeley and Los Angeles: Univ. of California Press, 2006). On the way the context of British colonialism informed Malthus's thought, see Alison Bashford and Joyce E. Chaplin, *The New Worlds of Thomas Robert Malthus: Rereading the Principle of Population* (Princeton, N.J.: Princeton Univ. Press, 2016).

[21] See Belasco, *Meals to Come* (cit. n. 20), 13.

[22] Frances Moore Lappé, *Diet for a Small Planet* (New York, N.Y.: Ballantine, 1971).

[23] On the industrialization of meat in North America and its implications for the global spread of industrial meat, see William Cronon, *Nature's Metropolis: Chicago and the Great West* (New York, N.Y.: W. W. Norton, 1991). And on the political dimensions of beef in the United States specifically, see Joshua Specht, *Red Meat Republic: A Hoof-to-Table History of How Beef Changed America* (Princeton, N.J.: Princeton Univ. Press, 2019).

and one frontier of cultured meat research is the use of scaffolds that allow the growth of more complex forms of muscle tissue. Designers of the bioreactors in which cells proliferate make an effort to duplicate the in vivo context in which cells normally grow. While cultured meat conferences often feature diagrams of industrial-scale bioreactors that resemble the tanks in beer breweries, most of the bioreactors I observed during my fieldwork were much smaller affairs of glass, steel, and plastic, often roughly the size of a domestic French press coffeemaker or smaller. Images of industrial bioreactors are one of the ways cultured meat workers communicate about their desired scale of operation, but their current scale of work more closely resembles artisanal handcraft, albeit done at the laboratory bench.

Cultured meat is premised on an idea about the nature and flavor of meat that we could call "biological equivalency." Meat's flavor and texture characteristics are presumed to come from qualities inherent in animal muscle and fat cells, and the structures they form. The implication of biological equivalency is that meat's flavor and texture characteristics are best reproduced not through the synthesis of artificial flavors and textures in concert with the use of plant cells (as has historically been the practice in meat surrogate creation), but by finding a new way to grow animal cells. Yet, despite the seeming obviousness of meat's ontology, something conveyed by colloquial English expressions such as "the meat of the matter" in which "meat" stands in for "substance" or "the essential thing," meat's qualities have presented serious challenges for anyone who seeks to copy them. This has been true across decades of efforts to reproduce meat's qualities using plant proteins, and it is likewise true for contemporary scientists working with in vitro methods and counting on biological equivalency.

As of late 2019, to the best of public knowledge, two technical hurdles hold cultured meat back. If not surmounted, either could doom the endeavor. The first is growth media. At present, the media used to feed animal cells in vitro typically include a quantity of fetal bovine serum (FBS), taken from a decidedly nonvegetarian source that is obviously counter to the goals of cultured meat. While there are vegan solutions on the market that include no FBS, they are considered too expensive for use in producing muscle tissue at scale. Thus, the search is on for a suitable growth medium that is cheap enough to replace FBS, but just as good at encouraging cell growth. Rumors circulate that one start-up company or another has devised an FBS-free serum that can be produced cheaply at scale, or that another has a technique for encouraging cell growth by environmental cues rather than through the nutrients provided by FBS, but as of this writing such claims cannot be substantiated. The second hurdle is the thickness or dimensionality of the pieces of meat to be produced. At present, it is feasible to grow muscle fibers in sheets that are relatively two-dimensional. These fibers can then be assembled into loose collections of tissue resembling hamburger or sausage meat, with the corresponding texture. However, in order to mimic the "bite" of more complex meat structures, something more is necessary. Muscle cells and the myotubes and muscle fibers they become must be grown in specific alignments with one another—in sheets much like those of in vivo skeletal muscle. Not only that, fat cells must be grown either simultaneously and in proper placement relative to muscle cells, or they must be grown separately and added later. Thickness is difficult not only because it requires the creation of scaffolding, preferably scaffolding made of material that is organic, digestible, and vegan in origin; it also requires vasculature. This is because the cells of most of our primary food animals can't live and grow more than one to two hundred microns (about the width of the widest human hair) from nutrients carried by blood. The need to create proper scaffolds and vasculatures for the growth of cultured meat is very much like the need for scaffolds and vasculatures for the production of human tissues intended for medical transplant. Indeed, cultured meat work has roots in biomedical research, and tries to translate the techniques of what is often called "regenerative medicine" to food production. Mark Post, and many of the other scientists at work on cultured meat, got their start as medical researchers (Post is a cardiologist), and see the use of in vitro methods to grow animal cells as an intuitive extrapolation from growing human cells for medical purposes.

While growth media and producing three-dimensional tissue via vasculature and scaffolding are both critical, they are necessary but insufficient conditions for the next important piece of the puzzle, namely the "scaling up" of cultured meat production from the laboratory to the factory. As of 2019, cultured meat is still a very small-scale endeavor, and only tiny samples of meat have been produced and revealed to the public thus far. The well-known high price of producing the first tissue-cultured hamburger in 2013 (over $300,000 US dollars) is often quoted by detractors of cultured meat efforts, but supporters can quote more recent and much lower estimated figures for producing similar amounts of meat to demonstrate the potential to produce meat at greater scales and at prices that compete with conventional meat. Such quotations of lower estimates are meant to serve as evidence of past progress in improving production techniques, and evidence of the potential for future progress. We are, the advocates of cultured meat imply, heading toward price parity with a McDonalds burger, though hard evidence of such movement toward parity has yet to be shown—indeed, given the impossibility of perfectly predicting technological trends, it is hard to say what such satisfying evidence would really look like.

The operating premise of biological equivalency means that early conversations about the physical properties and ultimate potential of cultured meat have often sidestepped the work of past generations of food scientists who pursued not biological, but sensorial equivalency. They worked to quantify and measure the sensations of consuming meat so that those sensations might be reproduced using an alternative protein base, usually derived from plants. Cultured meat scientists do not normally present their work as a paradigm shift in meat copying, instead simply ignoring the old paradigm and presenting their work as if it proceeded de novo. But some meat scientists have objected to the simple equation of lab-grown muscle with meat. Producing beef in a series of culture flasks is simply different than producing beef in a cow, they say; cows are not mammalian bioreactors, and bovine cells produced in bioreactors are not biologically equivalent to cells in a living animal. Muscle biologist Jean-François Hocquette suggests that it would be more accurate to call the result of tissue culturing bovine satellite muscle "artificial muscle proteins," rather than meat.[24] Meat, Hocquette writes, is not simply muscle, but the product of the process of muscular development and then of transitions that follow slaughter; meat "comes from muscle which needs to mature, a process during which important biochemical transformations gradually take place as the pH of muscle falls as a result of the absence of oxygen following the slaughter of the animal." To this point we might add food writer Harold McGee's observation that meat carries the flavor not of cells of different types (primarily muscle

---

[24] See Jean-Francois Hocquette, "Is In Vitro Meat the Solution for the Future?," *Meat Science* 120 (2016): 167–76, on 3.

and fat) alone, but also of the life history of an animal.[25] Those who believe that the flavor of an animal's meat reflects the *terroir* of the place where the animal lived and grazed lean heavily on this notion of life history. Biological equivalency tends to ignore such issues, as well as the past work of food scientists who, drawing on physics and psychology, sought to reproduce the sensation of eating particular foods.

When animal cells are encouraged to divide, multiply, and form into tissues outside the donor animal body, it raises a series of questions both scientific and, in the broadest sense of the word, philosophical. As Hannah Landecker points out in her *Culturing Life*, the advent of tissue culture techniques in the twentieth century led to new understandings of the potential plasticity and temporality of cells, as those cells can be induced to grow in vitro in ways that they do not in vivo, and to endure for longer periods of time than the individual laboratory animals from which those cells' progenitors came.[26] Similarly, I suggest that the use of tissue culture to produce animal muscle in vitro, interpreted as "meat," places pressure on our sense of animality itself, especially because many conventional understandings of meat are premised on the assumption that meat production begins with the life processes of an entire animal. The claim of cultured meat production seems to be that tissue cultures can exploit plasticity and temporality without sacrificing animal identity, implying that the latter quality is powerfully transferrable, perhaps because it is reducible to the identity of individual cells and not to the life processes of massive numbers of cells operating collectively in vivo. For most of the scientists I interviewed during my fieldwork, the question of whether or not in vitro cells deviate from in vivo cells, in terms of their animal identity, simply did not register as significant. At an exhibit on the future of food displayed at the Boerhaave Museum of the History of Science and Medicine, Leiden, in 2015, I saw a plastinated hamburger (the twin to that produced by Post's team in 2013) displayed next to a device for extracting semen from bulls. The pairing of the two artifacts was deeply suggestive, implying a kinship between two wildly different techniques for reproducing beef, one through the production of new bovines, the other through the production of analogues to their body parts. The implication was that animality is the same in both cases, that animal meat is simultaneously what animals produce and carry, what muscle cells become in vitro, and what we eat.

But biological equivalency has not eclipsed its earlier counterpart—sensorial equivalency. Many advocates for cultured meat also promote contemporary plant-based meat alternatives carefully designed to imitate the flavor (gustatory and olfactory) and textural properties of meat. As of 2019, the two most visible examples of sensory equivalency are plant-based hamburgers that are already available to consumers in restaurants or grocery stores; these include the Impossible Burger, created by Impossible Foods, and the Beyond Meat hamburger, made by Beyond Meat. In an autoethnographic essay, Alexandra Sexton describes cooking and eating a Beyond Meat Burger, whose creator defines it in terms of very basic components, ones that can be taken from plants and are essentially equivalent to the components of animal-based meat: "amino acids, lipids, carbohydrates, minerals, and water."[27] The Impossible Burger's designers often describe

---

[25] Ibid. See Harold McGee, *On Food and Cooking* (New York, N.Y.: Scribner, 1984), 121–37.
[26] See Hannah Landecker, *Culturing Life: How Cells Became Technologies* (Cambridge, Mass.: Harvard Univ. Press, 2007).
[27] Alexandra Sexton, "Alternative Proteins and the (Non) Stuff of 'Meat'" *Gastronomica: A Journal of Critical Food Studies* 16 (Fall 2016): 66–78.

it in similarly molecularizing language. The notion that meat is easily reduced to its components and that those components can be had elsewhere is intimately related to the idea that straightforward replacement of meat by a copy of meat will allow us to "build a world that's zero downside and all delicious upside," in the words of Beyond Meat's promotional copy. As Sexton shows, Beyond Meat presents their product not as a meat surrogate but as meat by other, more benign means that closely replicate the sensory qualities of a beef burger. The Impossible Burger uses a plant-derived heme, an analogue of hemoglobin (the oxygen-carrying protein in mammalian blood), to achieve a "meaty" taste, though one critic notes that both burgers seem to fall within an "uncanny valley," resembling real beef just enough for us to be very conscious of their distance from it.[28]

These plant-based burgers thus inherit the specific tradition of meat mimicry from which cultured meat deviates. They are made by extrusion, the same way that the common meat substitute, textured vegetable protein (TVP), has been produced for decades, but they differ from familiar "veggie burgers" primarily in being much more successfully meat-mimetic, according to food critics. These plant-based products also inherit a tradition of thinking about meat copying that, as historian of science Joel Dickau has demonstrated, took shape in American laboratories in the 1960s, when researchers used both human subjects and mechanical devices to measure the properties of foods.[29] Meat eating (like the consumption of many other foods) was increasingly studied through the techniques of applied physics and psychology, the goal being a quantification of the experience of eating and the reproduction of that experience by other means—mimesis via sensory or visceral equivalence rather than biological equivalence. Just so, Pat Brown, founder of Impossible Foods, has argued that we do not hunger for meat itself but rather for the experience of eating meat. (This severing of experience from its object is philosophically provocative in ways Pat Brown does not expand upon; it is like saying that Odysseus did not long for his home, Ithaka, but rather for a certain experience of Ithaka).[30] As Dickau also points out, the decades following the formation of a mimetic school of American food science saw the rise and widespread use of artificial flavors, including zero-calorie sweeteners and flavoring agents to enhance the resemblance between a surrogate food and the original it copied.[31] Needless to say, this allowed producers to offer consumers taste experiences without certain feared attributes a food might possess, such as calories, cholesterol, and saturated fat. Because red meat had developed associations with heart disease, meat was one of the products for which researchers began to seek surrogates, and they were further motivated by a perceived need to meet growing demand for protein in the developing world. Scientists at major food companies drew on extrusion techniques that had been used to make an imitation ground beef out of vegetable protein in the 1940s, and in the 1960s, "spinning" machines borrowed from the devices of the textile industry; these came into use to turn plant proteins into meat analogues. Not surprisingly,

---

[28] See J. Kenji López-Alt, "Let it Bleed (Humanely): We Taste the Vegan 'Impossible' and 'Beyond' Burgers," 31 October 2016, Serious Eats, https://www.seriouseats.com/2016/10/beyond-burger-impossible-burger-vegan-taste-test.html.

[29] Joel Dickau, "Inventing Texture: Edible Science and the Management of Familiarity, 1963–1975," *Global Food History* 3 (2017): 171–93.

[30] See the Impossible Foods short promotional film "For the Love of Meat," 3 October 2016, YouTube video, 3:02, https://www.youtube.com/watch?v = FjW2vNVZIhE.

[31] See Dickau, "Inventing Texture" (cit. n. 29), 184.

many hoped that meat surrogates, usually imitations of processed meat rather than of higher-value butchered cuts, would find eager consumers in the protein-hungry developing world. Decades later, the express goal of the new plant-based burger entrepreneurs is not to appeal to vegetarian eaters but to begin to displace the fast food hamburger itself, beginning in the developed world where their products first become available.

When compared to its plant-based precursors and contemporaries, laboratory-grown meat seems like an attempt to perform a double mimesis, because it tries to copy both the ultimate product of an industry and the natural process of muscle growth or repair. Yet these two efforts at copying take place at an artificial remove from intact animal bodies. This is because, rather than following the template of parts of the animal body, they copy industrial meat forms whose goal is convenient, hand-held edibility; hamburgers and sausages do not, as the phrase goes, "carve nature at the joints," even if the latter began with a repurposing of the intestines. As the intellectual historian Hans Blumenberg argued in a 1957 essay entitled "The Imitation of Nature," the philosophical problem of imitating nature is related to the question of where humans stand relative to the natural world—the question of the relationship between culture and nature.[32] It is also related to the more pragmatic and psychological issue of the anxiety caused by distance from a world we did not create ourselves—that is to say, the natural one that preceded us.

The problem of imitating nature is not as simple as the problem of fakeness or forgery, of realizing that our creations—however perfectly mimetic—are not nature's originals. Blumenberg argued that the modern attitude toward novelty in the realm of "the made"—the realm of human tools and technologies—represented a break from a prior attitude that had characterized Western antiquity. The legitimation crisis of modern people, and their difficulty existing comfortably in an artificial, rather than a natural, world likewise stems not only from the environmental conditions of civilized, urban, technological existence, but from this shift in attitude toward fabrication. Blumenberg begins his account in a kind of Aristotelian garden in which all technology is understood to either imitate or extend natural processes. In this garden, imitation not only reproduces natural forms or functions—the paddles on a mill's wheel recalling a dolphin's flippers—it also assures us that our artificial creations are, in some sense, part of a natural order. But Blumenberg's essay progresses toward a modern age whose inhabitants are uncomfortable creators, ill at ease with newer creations that do not emulate naturally occurring forms or processes. That which is truly de novo unsettles us.

The pivotal figure in Blumenberg's story is a humble spoon maker, one of the main interlocutors of Nicolas of Cusa's *Three Dialogues* (1450). Cusa's spoon maker proves himself inventive, for he takes as his model for making spoons not some natural form, but instead only an idea for a tool, originating in the human mind but realized, in its rough materiality, through the act of creation. Thus, the spoon maker emulates the spontaneous creative power of the divine and becomes the foundational figure for a distinctly modern way of thinking about making. For Blumenberg, modernity is characterized by the human rebellion against imitating nature, and by the desire to set ourselves up as creators whose own creations are valid, but this setting up is in fact an effort to cover up a growing sense of the groundlessness of our existence. For mimesis was always first

---

[32] Hans Blumenberg, "Imitation of Nature: Toward a Prehistory of the Idea of the Creative Being," trans. Ania Wertz, in "The End of Nature," special issue, *Qui Parle* 12 (Spring/Summer 2000): 17–54.

and foremost a matter of relation or connection to a world in which nature and being were equivalent, a world of interconnectedness. We are not at ease in our rebellion against mimesis. One of the philosophical culprits, prior to Cusa, had been the infiltration of Platonic ideas about the greater value of *methexis*, or the relationship between a particular object and the ultimate Form to which that object relates, over mimesis; methexis shifted into a regime of craft and creation that had been broadly Aristotelian and either comfortable or unconcerned with the "sufficient worth" of mimesis.[33] In other words, Platonism effectively inserts a hierarchy of making that had not been there before, and this hierarchy could readily be translated into the theological notion that imitation meant rebellion not against nature alone, but against God. Blumenberg argues, with some relevance to the impulses of technologists, that we moderns now experience *techne* as a metaphysical event, and novelty as a metaphysical need. Yet a question mark hovers over our needs, and over their propriety. *Homoiosis theoi*, or the desire to be like God, is both attractive and hard to live with.

Cultured meat occupies an interesting middle ground in Blumenberg's view of the world. It appears to conform to the Aristotelian understanding of technology in certain ways, for via techne it "extends" the natural processes of the cell and of muscle development beyond what nature affords. At the same time, it provides such an obvious opportunity for flesh to take on new forms that only an intense attachment to familiar forms could keep cultured meat within the paradigm of imitation. The scientists who conduct experiments in cultured meat production see the muscle cells they grow as identical to the muscle cells that develop in animal bodies. But in the lab, they have to work hard to maintain the in vitro conditions duplicating in vivo ones, and the sheer effort involved gives them a keen awareness of our current distance from mimesis. Despite the claims of many of its promoters, to make cultured meat is not "growing meat in the lab instead of in the animal," as if "growing" and "meat" meant the same thing in each instance. Despite the effort to create a double mimesis, imitating both process and form, cultured meat introduces a crucial, biotechnological distinction between the Aristotelian categories of *natura naturans*, or nature as a productive process, and *natura naturata*, or nature as a set of specific shapes.[34]

The challenge of copying is that this entire dynamic must be pressed into that symbolically American industrial meat form, the hamburger, when it so manifestly could overflow such a limiting mold. The imitation of nature, for Aristotle, was a principle of relation—human hands making things that depended on their prehuman antecedents. In early twenty-first-century cell culture and tissue engineering, that dependence has

---

[33] As Blumenberg points out, the Aristotelian position on mimesis was already a response to Plato, and in particular to a question within Platonism having to do with the origin of human works. Were there Forms for artificial objects, as Plato, book 10 of the *Republic* seems to imply? Plato's school, his Academy, appears to have dropped such a notion by Aristotle's time, replacing it with the idea that the cosmos built upon the Forms reflects the best of that which is to come, and that there are no "leftover" Forms that human artisans might tap (see Plato's *Timaeus*). In summary, the Aristotelian response is to deny the existence of invention that is not imitation of nature. See Blumenberg, "Imitation of Nature" (cit. n. 32), 29.

[34] For certain followers of Aristotle, the distinction between *natura naturans* and *natura naturata* took on a sexed dimension, as the former was understood to be a male productive principle, while the finished but inert product was taken to be female. This was an adaptation of Aristotle's own view that the mother's body simply provides the raw material for the process of reproduction. See Mary Garrard, "Leonardo da Vinci: Female Portraits, Female Nature," in *The Expanding Discourse: Feminism and Art History*, ed. Garrard and Norma Broude (New York, N.Y.: HarperCollins, 1992), 58–86.

become tenuous, and the presence of human will is keenly felt. Blumenberg's argument in 1957 was not that mimesis would be better for us troubled moderns than de novo invention, or that Aristotle is the key thinker out of whose work we should spin a history of techne in general or of technology in particular. His stakes were nothing less than our ability to view reality, whether organic or artifactual, grown or made, as legitimate, recognizing the human freedom called creation without getting caught up in a compulsive search for the next tool or toy. Blumenberg seemed to wish for a less psychologically vexed, and wiser, version of technological modernity, one in which we are less preoccupied with "discovering ourselves"—that is to say, discovering our own image in every new creation. This lessening of preoccupation would follow because we would feel less keenly that the word "nature" and the word "being" are synonymous (the given world is, the line of thought runs, already a perfected one). Thus, we also would no longer feel that every innovation beyond nature's replete set of creatures, plants, and land- and seascapes, constitutes an existential mistake, and cover for our mistakes by making much of the originality of our creations—once again, Homoiosis theoi. The philosophical problem cultured meat offers us is truly one of definition, for if cultured meat succeeded, we would not only have to grow accustomed to an entirely new form of meat; but, we could have to grow comfortable with our newly asserted right to redefine meat, and to shift the site of its definition from familiar agricultural sites (including industrial animal agricultural sites, like CAFOs) to the laboratory. Blumenberg's work helps us to situate that problem historically and philosophically, and reminds us that our capacity to redefine meat is not absolute, but is always limited by material circumstances, and contoured by the conditions of human culture, even as it is enabled by the techniques of tissue culture.

This is perhaps a good spirit to bring back with us from the realm of intellectual history in order to view the history and future of meat copying in generous terms. The technical difficulty of copying meat's texture and flavor perfectly does not only present us with the potential failure of mimesis, but with an opening toward something beyond natura naturans and natura naturata, toward a world in which we appreciate the limitations on our capacity to copy either sensations or parts of living things. In the case of cultured meat, perfectly mimetic processes may prove less successful, at a technical level, than ones that exploit natural cellular processes of repair and then form the resulting muscle strands into new and unfamiliar shapes. Rather than copies of natural forms, these would be new results of existing biological processes, made possible by in vitro growth and tissue engineering techniques. The resulting new products might defy the mimetic logic of simply replacing conventional meat, all the while reminding us of the limits to our powers to copy nature.

Breakfast at Buck's:
Informality, Intimacy, and Innovation in Silicon Valley

*by Steven Shapin**

ABSTRACT

This is a study of some connections between eating-together and knowing-together. Silicon Valley technoscientific innovation typically involves a coming-together of entrepreneurs (having an idea) and venture capitalists (having private capital to turn the idea into commercial reality). Attention is directed here to a well-publicized type of face-to-face meeting that may occur early in relationships between VCs and entrepreneurs. The specific case treated here is a large number of breakfast meetings occurring over the past twenty-five years or so at a modest restaurant called Buck's in Woodside, California. Why is it *this* restaurant? What is it about Buck's that draws these people? What happens at these meals? And why is it breakfast (as opposed to other sorts of meals)? This article goes on to discuss historical changes in the patterns of daily meals and accompanying changes in the modes of interaction that happen at mealtimes. Breakfast at Buck's may be a small thing, but its consideration is a way of understanding some quotidian processes of late modern innovation, and it offers a possible model for further inquiries into eating and knowing.

> I hate people who are not serious about meals.
> It is so shallow of them.
> — Oscar Wilde

EATING AND KNOWING

We take on so much as we take on food—culture as well as calories, knowledge as well as nutrients. Histories of the food sciences, or histories of the relationships between food and scientific knowledge, have typically been concerned with the constituents and powers of foods, with their fate in the body and their benefits and risks to the body. There are studies of the historical role of the sciences in managing the quality, quantity, and safety of food; in understanding the functions of food in the body; and in devising new foods. Yet there are other possible topics for historians (and social scientists) concerned with food and knowledge, and these include *eating*, and, more specifically, eating-together (*commensality*, in terms of sociological art) as a venue for

* Department of the History of Science, Harvard University, Cambridge, MA 02138, USA; shapin@fas.harvard.edu.

making, maintaining, and modifying *knowledge*. And, while there are some fine, though largely programmatic, social scientific meditations on eating practices and on the structure and the meaning of meals, studies of eating-together remain largely disengaged from studies of knowing-together.[1]

Commensality takes place in face-to-face modes of interaction which are, as is now said, high bandwidth. Taking food and drink with others makes social bonds, establishes (or aims to establish) social identities, generates (or may generate) obligations, and marks out and celebrates special moments in time and in social life. (Eating alone used to be deprecated as a sign of gluttony or misanthropy; now, while solitary feeding is moving fast toward the norm, eating alone may still occasionally be seen as a mark of social insufficiency or, alternatively, celebrated as a display of busyness.[2]) Different *sorts* of eatings-together mark out different sorts of social relationships, from the informal to the formal, from the intimate to the institutional, from the private to the public. Some eatings-together are instrumental—as occasions meant to have specific outcomes; others are just ways to "keep in touch" or maintain social ties. At most occasions of eating-together, no notice at all may be taken of aliment consumed, or it may be considered inappropriate to pay much attention to the identity of the food and drink; at others, less commonly, the nature, quality, and staging of food is the point.[3]

---

[1] The editorial introduction to this volume concisely reviews and summarizes some of the most worked-over historical frameworks for considering the relationships between food and science: E. C. Spary and Anya Zilberstein, "On the Virtues of Historical Entomophagy," in this volume. A notable sociologist of food and eating observes: "The amount of research on eating within the socio-cultural sciences . . . is slight." Eating, he writes, "has rarely been targeted as an object of research." Social scientists, he adds, have been far more concerned with production than consumption: Alan Warde, *The Practice of Eating* (Malden, Mass.: Polity, 2015), on 15–16, also 24–8; see also Warde and Lydia Martens, *Eating Out: Social Differentiation, Consumption, and Pleasure* (Cambridge: Cambridge Univ. Press, 2000); Roy C. Wood, *The Sociology of the Meal* (Edinburgh: Edinburgh Univ. Press, 1995); Anne Murcott, ed., *The Sociology of Food and Eating: Essays on the Sociological Significance of Food* (Aldershot, UK: Gower, 1983); and the classic note by Georg Simmel, "The Sociology of the Meal," trans. Michael Symons, in Symons, "Simmel's Gastronomic Sociology: An Overlooked Essay," *Food and Foodways* 5 (1994): 333–51, on 345–50 (first published 1910). More empirically focused sociological studies of modern commercial and domestic meal taking include Joanne Finkelstein, *Dining Out: A Sociology of Modern Manners* (New York: New York Univ. Press, 1989); Finkelstein, *Fashioning Appetite: Restaurants and the Making of Modern Identity* (New York, N.Y.: Columbia Univ. Press, 2014); and Jean-Pierre Poulain, *The Sociology of Food: Eating and the Place of Food in Society*, trans. Augusta Dörr (2002; repr., London: Bloomsbury, 2017). There is also a considerable anthropological literature on eating and meals in non-Western settings; for entry points; see Marshall Sahlins, *Culture and Practical Reason* (Chicago: Univ. of Chicago Press, 1976), esp. 170–9; Jack Goody, *Cooking, Cuisine and Class* (Cambridge: Cambridge Univ. Press, 1982); *Food Consumption in Global Perspective: Essays in the Anthropology of Food in Honour of Jack Goody*, ed. Jakob A. Klein and Anne Murcott (New York, N.Y.: Palgrave Macmillan, 2014); and Sidney Mintz and Christine M. Du Bois, "The Anthropology of Food and Eating," *Annu. Rev. Anthropol.* 31 (2002): 99–119. However, eating-together has attracted scarcely any attention as an occasion of knowledge making, apart from the sorts of knowledge concerned with companions' social identity.

[2] Claude Fischler, "Commensality, Society and Culture," *Soc. Sci. Inform.* 50 (2011): 528–48, on 539–40.

[3] For example, Mary Douglas, "Deciphering a Meal," in Douglas, ed., *Implicit Meanings: Essays in Anthropology* (London: Routledge, 1975), 249–75; Douglas, "Food as a System of Communication," chap. 4 in *In the Active Voice* (London: Routledge & Kegan Paul, 1982), 82–124; Douglas, "Standard Social Uses of Food: Introduction," in Douglas, ed., *Food and the Social Order: Studies of Food and Festivities in Three American Communities* (New York, N.Y.: Russell Sage Foundation, 1973), 1–39; Paul Freedman, "Medieval and Modern Banquets: Commensality and Social Categorization," in *Commensality: From Everyday Food to Feast*, ed. Susanne Kerner, Cynthia Chou, and Morten Warmind (London: Bloomsbury, 2015), 99–108. Food consumption is not a major topic in Pierre Bourdieu's hugely influential study of the fine structures of French social class: *Distinction: A Social*

Different kinds of social interaction are signaled by the presence of different foods, by different modes of presentation and consumption, by meals taken at different times of day, and by the presence of different kinds of people. Lunch "on the go" is as different from an Oxbridge college feast or a regimental dinner as is the meaning of "let's meet for coffee" from "would you like to come up for a cup of coffee?" From the silent meal in a monastery to the mess group in the military, the bonds formed by eating-together have historically been acknowledged and actively managed. Eating-together is at the center of religious life and belief—with the *seder* (the Jewish Passover meal) historically transformed into the Last Supper and then into the mysteries of Christian sacraments—the transubstantiations of communion wine and wafer. So too are the ritual occasions of fasting, of intentional not eating, which may be found at the center of political action—the fasting of suffragettes, of Irish Republican prisoners, and of Indian resistors to British colonial rule. The manner of eating is an element in making social identity and social distinction, spectacularly so in Victorian and Edwardian Britain with specialized cutlery as weapons of class destruction, embarrassment attending imperfect knowledge of the proper use of the fork and the proper placement of wine and water glasses.[4]

Many of these aspects of eating have been appreciated by social and cultural historians at least since the 1930s writings of Norbert Elias, whose treatment of manners in the making of modernity inspired a number of later historians and social scientists.[5] The history of food has recently been drawing much academic attention, while the history of *eating* and its significance remains marginal, treated in the main by social historians and historians of manners and etiquette. Such academic attention as has been given to these things has centered on the Renaissance and the early modern period, with arrangements of contemporary and the more recent past tending to escape notice or, more commonly, becoming the property of nutritionists and policy experts worried about bad eating and consequent poor health outcomes. Nor have scholars concerned with *knowledge* seen much point in considering occasions of eating, which have generally been taken as *time-out* from knowledge making. It's what thinkers do to "fuel up" before going to thought-work. Comic writers make sport of the very idea of a Friedrich Nietzsche diet book, absurdly juxtaposing the intellectual and the corporeal, the High and the Low.[6]

Critique of the Judgement of Taste, trans. Richard Nice (1979; repr., Cambridge, Mass.: Harvard Univ. Press, 1984), e.g., 79, 179–200, 278–80; this is a relative reticence that has surprised other sociologists—for example, Warde, *Practice of Eating* (cit. n. 1), 55.

[4] See the delightful, and unobtrusively learned, Bee Wilson, *Consider the Fork: A History of How We Cook and Eat* (New York, N.Y.: Basic, 2012), esp. chaps. 2 (41–71) and 6 (181–210).

[5] Norbert Elias, *The Civilizing Process: Sociogenetic and Psychogenetic Investigations*, trans. Edmund Jephcott (1939; repr., Oxford: Blackwell, 1978), esp. 45–108; see also Claude Lévi-Strauss, *The Origins of Table Manners*, trans. John and Doreen Weightman (1968; repr., New York, N.Y.: Harper & Row, 1978); Jean-Paul Aron, *The Art of Eating in France: Manners and Menus in the Nineteenth Century*, trans. Nina Rootes (London: Owen, 1975); Stephen Mennell, *All Manners of Food: Eating and Taste in England and France from the Middle Ages to the Present* (Oxford: Basil Blackwell, 1985); Margaret Visser, *The Rituals of Dinner: The Origins, Evolution, Eccentricities, and Meaning of Table Manners* (New York, N.Y.: Grove Weidenfeld, 1991).

[6] Woody Allen, "Thus Ate Zarathustra," in *Mere Anarchy* (New York, N.Y.: Random House, 2007), 141–6; see also Steven Shapin, "Lowering the Tone in the History of Science: A Noble Calling," chap. 1 in *Never Pure: Historical Studies of Science as if It Were Made by People with Bodies, Situated in Space, Time, and Society, and Struggling for Credibility and Authority* (Baltimore: Johns Hopkins Univ. Press, 2010), 1–14, on 1–2. At the risk of spoiling the joke, Nietzsche was, in fact, much concerned about diet.

It would be wrong to dismiss the sensibilities that have served to marginalize the links between eating and knowing; there are a number of famous stories about thinkers so "lost in thought" as to forget eating, and the ancient cultural *topos* that opposed the mind and the belly still retains some of its pertinence in late modernity. Both the jokiness and the cultural deflation accompanying the very idea of linking feeding to thinking belong to the topos of asceticism—as Nietzsche indeed pointed out.[7] This article describes a specific occasion, and a specific site, of commensality, one in which eating-together is reflectively linked to knowing-together, and even to practical actions based on that knowledge. The unlikeliness of this focus can also be acknowledged; for all sorts of reasons, critics might say that the eating concerned is "not relevant," that no knowledge (properly so called) is produced at these mealtimes, or that the knowledge generated is "not science." Making knowledge is what happens when feeding the stomach is not happening or, at most, the knowledge that does attend eating-together is "social knowledge"—knowledge of one's fellow diners' characters and whatever mundane knowledge is conveyed by general conversation.

The unlikeliness of telling a story about eating and knowing attaches even more strongly to the particular materials addressed here. First, this article deals with very recent history, even with passages that persist in the present and that very likely will continue to do so in the foreseeable future. However, I suggest toward the end a genealogy connecting these sorts of eatings-together to deep history while I also point to patterns of historical change that make them an index of late modernity. Second, although some of what happens on the eating occasions described here certainly involves accounts and evaluations of science and technology, most of what is pertinent about them concerns the *conditions of possibility* for making science and technology and, specifically, for transforming technoscience into goods and services that figure in the marketplace. Knowledge is indeed being produced on such occasions, even if it's not the kind of knowledge that is routinely embodied in science textbooks or that gets worked over by epistemologists.[8] The knowledge that is made, maintained, and modified at these breakfasts is of different sorts: there is knowledge of commercializable science and technology, its potential and circumstances; knowledge of the persons speaking for the science and technology, their virtues, vices, and, indeed, their apparent expertise and skills; knowledge of these persons' likely commitment and dedication to the pertinent projects; and, indeed, knowledge of their capacity for relevant sorts of social interaction, since the VC-entrepreneur interaction occurring here and now is thought to bear a family resemblance to the modes of interaction that will figure

---

[7] Steven Shapin, "The Philosopher and the Chicken: On the Dietetics of Disembodied Knowledge," in *Science Incarnate: Historical Embodiments of Natural Knowledge*, ed. Christopher Lawrence and Shapin (Chicago: Univ. of Chicago Press, 1998), 21–50.

[8] Historians of science are now familiar with encouragement to widen the scope of what is routinely taken as "science." First, many historians of science are no longer defensive about attending to "technology," to so-called "applied science" and "commercial science," or they are less confident than they once were in justifying such distinctions; second, historians are now less inclined to equate "science" solely with "what happens in the laboratory," with what is inscribed in the textbooks, or with "disembodied ideas." Bruno Latour and others have urged scholars to document the processes through which scientists secure support for their work, obtain credibility for its outcomes, and embed those outcomes in "black-boxed" artifacts and processes: Latour, *Science in Action: How to Follow Scientists and Engineers through Society* (Cambridge, Mass.: Harvard Univ. Press, 1987); see also Steven Shapin, "Invisible Science," *Hedgehog Review* 18 (Fall 2016): 34–46 (on arguments about "embedded science" and for problems attending distinctions between "hard" and "soft" sciences), www.iasc-culture.org/THR/THR_article_2016_Fall_Shapin.php.

should the business come into being and grow. I make no claims here that there is any-thing special about occasions of eating-together that involve science and technology.[9] Arguably, all forms of knowledge making have their moments of the face-to-face, and eatings-together are significant among these moments; science, technology, and the conditions for making more science and technology are not excluded.

PLACES OF EATING AND PLACES OF KNOWING

In recent decades, historians and sociologists of science have written about the *places* and *spaces* of science, but among these, only the early modern and Enlightenment cof-feehouse has really represented a place of eating and drinking as a place of face-to-face knowledge making. (The Enlightenment *salon* has also engaged historians' attention, though its alimentary elements scarcely at all.)[10] Why not consider the pertinence of a range of places, each with their characteristic modes of interaction and significances: the domestic or restaurant dinner table; the pub and tavern; the high-tech corporate cafeteria, salad and sushi bars; the Asian noodle houses lined up on Castro Street in Mountain View, California; the tea room of the University Library in Cambridge; the yet-to-be formally named rooms in modern "coworking spaces" and "incubators" where innovators share a pizza and a beer; and, of course, the late twentieth-century and con-temporary coffee shop?[11] Architects and designers of modern research facilities (and even of furniture and décor) aim to build places that facilitate the exchange of ideas, often paying great attention to constructing spaces for taking food and drink that will serve as venues for "spontaneous" and "serendipitous" interaction—sometimes mak-ing explicit gestures to early modern coffeehouse culture—but it also happens that such spaces emerge without the assistance of professional design and become iconic magnets for further intellectual interaction.[12] In the late modern culture of innovation, reflective

[9] That said, the arrangements and the alimentary contents of many late modern technoscientific feeding sites (the well-supplied cafeterias of Google and of high-energy physics installations, the pro-visioning of entrepreneurial networking events, and the catered lunches of academic departmental meetings) merit attention, as do the ascetic dietetics of heroic code-writing sessions and attitudes to-ward solitary "brown bagging": see references in nn. 11 and 12, below.

[10] A fine exception is E. C. Spary, *Eating the Enlightenment: Food and the Sciences in Paris, 1670-1760* (Chicago: Univ. of Chicago Press, 2012), though here too the focus is on the sciences of food innovation and the scientific understanding of the consequences of consuming certain foods and drinks.

[11] See, for example, Kerry Miller, "Where the Coffee Shop Meets the Cubicle," *Bloomberg Business-week*, 27 February 2007, www.nbcnews.com/id/17367175/ns/business-us_business/t/where-coffee -shop-meets-cubicle/#.WwyGgYpunIU. For focused attention on the foods judged to be proper for entre-preneurial consumption in contemporary coworking spaces and for their prescribed "relentless sociabil-ity," see Gideon Lewis-Kraus, "The Rise of the WeWorking Classes," *New York Times Magazine*, 27 Feb-ruary 2019, https://www.nytimes.com/interactive/2019/02/21/magazine/wework-coworking-office-space .html.

[12] See, for example, Kursty Groves Knight and Oliver Marlow, *Spaces for Innovation: The Design and Science of Inspiring Environments* (Amsterdam: Frame, 2016); Anahad O'Connor, "How the Hum of a Coffee Shop Can Boost Creativity," *New York Times*, 21 June 2013, https://well.blogs .nytimes.com/2013/06/21/how-the-hum-of-a-coffee-shop-can-boost-creativity/; Chloë Brown, Christos Efstratiou, Ilias Leontiadis, Daniele Quercia, Ceclia Mascolo, James Scott, and Peter Key, "The Archi-tecture of Innovation: Tracking Face-to-Face Interaction with Ubicomp Technologies," www.arxiv.org /pdf/1406.6829.pdf; Julie Wagner and Dan Watch, *Innovation Spaces: The New Design of Work* (Wash-ington, DC: Brookings Institution, 2017), https://www.brookings.edu/wp-content/uploads/2017/04/cs _20170404_innovation_spaces_pdf.pdf. Designers are reflectively aware of conditions of visibility and audibility in different spaces of innovation: for examples, see Ali Morris, "Workplaces Must Include

attention to designed or spontaneous spaces of interaction coexists with the asocial asceticism now notably associated with the geek-designed, all-purpose drink Soylent, offered as a solution to time supposedly *wasted* in eating. Researchers in business schools and in organizational sociology now recognize relationships—long appreciated by nonacademics—between commensality and the success of negotiations, and seemingly agree that deals over meals (or drinks) build trust and go better, though the academics are not agreed whether this effect is independent of the identities of the deal, the meal, the place, and the stage of discussions.[13] Both foodstuffs consumed and the spaces in which they are taken have, and always have had, a bearing on conceptions of the consuming body and the modes of physical and intellectual interaction between bodies and minds. There are historians who now recognize *food for thought* as a historical topic.

The eating-together occasion treated here is breakfast, and, specifically, breakfast eaten out rather than taken at home; the period concerned is anytime over the last three decades or so; the particular place is a restaurant in California, south of San Francisco and several miles from the campus of Stanford University; the people involved are actors in and around the entrepreneurial high-tech culture of Silicon Valley; and the breakfasts are largely instrumental occasions, whose outcomes may be decisions—or elements in a decision-making process—about which scientific and technological ideas are to have a commercial future and which are destined for history's dustbin. So, this is a sketch of what happened at a specific time, place, and type of social, cultural, and economic life, while the general manner in which these specificities are interpreted might suggest ways of treating all sorts of other science-relevant commensalities at different times, different places, and at different moments in the making, maintaining, or modification of knowledge. This is, inescapably, a case study, partly meant to encourage future studies.

BUCK'S FIZZ: SILICON BREAKFAST

The restaurant concerned is an iconic site, celebrated by members of late twentieth-century and early twenty-first century entrepreneurial culture. It is well known to many of those who write about technoscientific innovation and who seek to decipher the secrets of Silicon Valley as a global center of technoscientific and commercial innovation. The breakfasts described are a setting for the earliest stages of deals done between venture capitalists (VCs) and aspiring high-tech entrepreneurs, and are much talked

Spaces for Work and Rest to Foster Innovation," *Dezeen*, 7 December 2017, www.dezeen.com/2017/12/07/haworth-white-paper-research-workplaces-offices-foster-innovation-work-rest-spaces/; Ben Waber, Jennifer Magnolfi, and Greg Lindsay, "Workspaces That Move People," *Harvard Business Review*, October 2014, www.hbr.org/2014/10/workspaces-that-move-people. See John Bessant, "Creating the Physical Space for Innovation," *Hype Innovation Blog*, 7 September 2016, https://blog.hypeinnovation.com/creating-spaces-for-innovation, on awareness of seventeenth-century coffeehouses.

[13] For Soylent, see, among many accounts, Arwa Mahdawi, "I Tried Soylent, Silicon Valley's Favourite Foodstuff," *Guardian*, 11 September 2018, https://www.theguardian.com/commentisfree/2018/sep/11/i-tried-soylent-silicon-valleys-favourite-foodstuff-its-everything-thats-wrong-with-modern-life. For meals and deals, see Lakshmi Balachandra, "Should You Eat While You Negotiate?" *Harvard Business Review*, 29 January 2013, https://hbr.org/2013/01/should-you-eat-while-you-negot; Elizabeth MacBride, "Researchers: Does Breaking Bread Help Make a Negotiation a Success?" *Insights by Stanford Business*, 21 April 2014, https://www.gsb.stanford.edu/insights/researchers-does-breaking-bread-help-make-negotiation-success; and Kaitlin Woolley and Ayelet Fishbach, "Shared Plates, Shared Minds: Consuming from a Shared Plate Promotes Cooperation," *Psychological Science* 30 (2019): 541–52.

about as "where things begin"—where the process of turning technical and commercial ideas into material and corporate realities commences, where technoscientific concepts meet potential sources of cash.[14] The deals whose origins are traced to this particular place are said to have amounted to billions of dollars over the past several decades. Stories are told about this eating place, and tourists as well as aspiring entrepreneurs continue to go there in hopes of seeing some famous geek or VC, or, at least, of seeing the place where famous deals got done. Sites like these are, so to speak, *charismatic places*—sites endowed with power and surrounded by mystique because of the collective memory of things that happened there and that might, because of the energy still presumed to reside in them, work their effects on future events.

The restaurant is Buck's in the small town of Woodside, California. In many respects, Buck's is nothing special; one VC-customer wrote that the restaurant "looks as if it could be the anchor of any upscale retail strip in any small town west of the Rockies." They added, "all that's missing are some hitching posts and a watering trough or two."[15] (Founded in 1991, Buck's is said to have been named after the town drunk, Leo Buckstaber, and not, as it's sometimes assumed, after dollars.)[16] Though Buck's has got a number of favorable reviews by Bay Area foodies, its fame in certain circles has to do with the technoscientific and commercial events that happened in it—not with whatever it is that usually makes certain restaurants famous, like Michelin-starred cuisine or remarkable things to eat or spectacular views or spectacularly rude waiters. Buck's is open seven days a week, for breakfast, lunch, and dinner. But what makes Buck's famous is breakfast, or, less commonly, lunch—*never* dinner, and never any meal on the weekends, when it's taken over by gangs of lycra-clad mountain bikers.

The breakfast menu is quite ordinary for local eateries. Breakfast offerings include the usual range of American fare; you can have pancakes, eggs—"any style," or omelets (Californized by the avocado on top and made just of egg whites in case of cholesterol phobia)—or waffles, or just some fruit and toast or granola. Everyone has what they want, thereby avoiding the awkwardness of someone acting as host in their own home when someone has an allergy, or someone else doesn't eat what's offered for religious or moral reasons. The prices are also in the usual range for this sort of place—not as cheap as McDonald's but nothing to put off even the most liquidity-constrained Valley geek. A similar sort of restaurant, well known for reasons similar to Buck's, is one of the Hobee's chain; there are establishments in Cupertino, Mountain View, and Palo Alto used by VCs and entrepreneurs. There's also the Konditorei coffee and pastry shop in Portola Valley. There's Chef Chu's Chinese restaurant in Los Altos.[17] And posher, but also well known for deal making, is Il Fornaio in Palo Alto, favored by Stanford

[14] There are many sources—most journalistic, some scholarly, some disengaged, the majority celebratory—about the relations between VCs and entrepreneurs in Silicon Valley and other California clusters of high-tech and biotech. Several of the more star-struck popular sources are cited in notes below; see also Steven Shapin, "The Scientific Entrepreneur," and "Visions of the Future," chaps. 7 and 8 in *The Scientific Life: A Moral History of a Late Modern Vocation* (Chicago: Univ. of Chicago Press, 2008).
[15] William H. Draper III, *The Startup Game: Inside the Partnership between Venture Capitalists and Entrepreneurs* (New York, N.Y.: Palgrave Macmillan, 2011), 1.
[16] Deborah Perry Piscione, *Secrets of Silicon Valley: What Everyone Can Learn from the Innovation Capital of the World* (New York, N.Y.: St. Martin's, 2013), 154.
[17] John Boudreau, "Chef Chu's Restaurant Has Served Tech Elite, from Steve Jobs to Jerry Yang," *Mercury News* (San Jose), 14 August 2012, www.mercurynews.com/ci_21312105/chef-chu-restaurant-has-served-tech-elite-steve-jobs-los-altos.

University administrators as well as capitalists.[18] Some recent commentators say that the still-thriving Buck's has now become too famous for its own good, while others show no awareness of this new alleged "unhipness."[19]

What makes Buck's famous, and especially breakfast at Buck's, are the meetings that have happened there since its opening—meetings between technoscientific entrepreneurs and VCs, meetings that became legendary because of the high-tech companies that are said to have been launched there.[20] Jamis MacNiven, the founder, patron, and owner, is happy to point out the tables where wonderful things happened, the events that make Buck's charismatic. There are photos of him proudly pointing to table 48 where John Doerr of the celebrated VC firm Kleiner Perkins met with Marc Andreesen, and the result was Netscape. Then there is the table where PayPal was brought into being; another is the Hotmail table; and yet another is the Tesla table. There is, however, no pointing with pride to the place where Jerry Wang of Yahoo! breakfasted with VCs Bill and Tim Draper (of Sutter Hill Ventures and Fisher, Draper, Jurvetson), after which Wang decided to go with another firm.[21]

Eventually, it became impossible for anyone in the business to have breakfast at Buck's without doing a quick scan of *who else* was then having breakfast at Buck's, and speculating what their meetings might mean. The VCs who come to breakfast at Buck's may not want to be overheard by other VCs, but encounters are inevitable and not necessarily unwelcome. There is a lot of table-hopping and handshaking. "It's like Facebook with real faces," the owner said.[22] In Britain, *The Daily Telegraph* breathlessly announced that at Buck's "more deals are struck between those with money (cash-rich VCs) and those wanting it (cash-hungry entrepreneurs) than in all [of] America's other diners put together."[23] If you're an out-of-town, or out-of-country, journalist, Buck's tends to be where you go to "take the pulse" of the Valley innovation economy; if business at Buck's is good, the Valley is good.[24] Tourists come to take photos of Buck's and the famous tables, and aspiring entrepreneurs lurk in the parking lot armed "with

---

[18] Eric S. Hintz, "Historic Silicon Valley Bar and Restaurant Review," Lemelson Center, Smithsonian Institution, 9 September 2013, www.invention.si.edu/historic-silicon-valley-bar-and-restaurant-review. This piece is one of a series on "Places of Invention" produced by the Smithsonian's Lemelson Center for the Study of Invention and Innovation. The series is a web-based exhibition illustrating the importance of such informal sites in the processes of innovation, and it draws inspiration from the work of the urban sociologist Ray Oldenburg on what he called "the Third Place" (not home, not workplace): *The Great Good Place: Cafés, Coffee Shops, Bookstores, Bars, Hair Salons, and Other Hangouts at the Heart of a Community* (New York, N.Y.: Da Capo Press, 1999); see also Shachar M. Pinsker, *A Rich Brew: How Cafés Created Modern Jewish Culture* (New York: New York Univ. Press, 2018).

[19] Alexander Haislip, *Essentials of Venture Capital* (Hoboken, N.J.: John Wiley, 2011), 8.

[20] A cable TV show about "Millionaire Hangouts" confidently claimed about Buck's that "more than 90% of Internet venture deals were kicked around here": Alana Ross and Barry Ross, "S. F. Bay Area [. . .] Restaurants, Buck's," *Love to Eat and Travel Blog*, accessed 24 May 2018, http://www.lovetoeatandtravel.com/site/sfbay/peninsula/Food/bucks.htm.

[21] For the photographs, see Boonsri Dickinson, "Billions of Dollars in Wealth Were Created at this Silicon Valley Cafe," *Business Insider*, 8 February 2012, www.businessinsider.com/inside-the-silicon-valley-cafe-where-paypal-tesla-and-netscape-did-deals-2012-2?op = 1; John McChesney, "Checking a Tech Bellwether: Buck's Restaurant," *National Public Radio*, 2 August 2010, www.npr.org/templates/story/story.php?storyId = 128874569.

[22] Conner Forest, "How Buck's of Woodside Became the 'Cheers' of Silicon Valley," *TechRepublic*, 4 July 2014, www.techrepublic.com/article/how-bucks-of-woodside-became-the-cheers-of-silicon-valley/.

[23] Amanda Hall, "Silicon Giants: The Valley of Dreams," *Telegraph* (London), 24 September 2000.

[24] McChesney, "Checking a Tech Bellwether" (cit. n. 21); Connie Guglielmo, "Silicon Valley Buzz, Served with Pancakes: Signs of Recovery at Diners and Cafes," *International Herald Tribune*, 6 April 2006, 20.

photographs of venture capitalists and leap on them as they" leave, hoping one day to join them for breakfast and pitch a start-up.[25] Buck's is a pilgrims' destination for politicians, celebrities, and hugely successful techies in no current need of VC funding—Al Gore, Shimon Peres, Mike Tyson, John Cleese, Gordon Moore, and Andy Grove are all proudly photographed with the equally proud owner (fig. 1). An online list of ten of the "World's Best Millionaire Hangouts" includes Positano, St. Moritz, the Cannes Film Festival—and Buck's Diner.[26] In 1999, there was an eBay auction for breakfast at Buck's with VC Steve Jurvetson, and the winning bid, from a New York entrepreneur, was $9,400. The underbidder arranged a separate breakfast with the VC, and the price was higher.[27] The cash value of "informal" breakfast face time was quantified, and the circumstance that it was breakfast at *Buck's* must have contributed to its value. Buck's has been called "the breakfast spot of champions": it has "buzz."[28] Venture capitalist Bill Draper said that Buck's "is the distillation of much of what is so odd, special, and compelling about the Valley—a place where great ideas meet smart money."[29]

BUCK'S BREAKFASTS: THE MUNDANE AND THE MYTHIC

The mythic status of these meetings is beyond doubt. On the one hand, participants talk about Buck's as "a place where myths were born"[30]—in the sense that the high-tech companies resulting from deals made there are parts of Valley legend. On the other hand, to identify Buck's breakfasts as these companies' genuine and unique origins is problematic. In one form or another, the meetings did happen, though it's impossible reliably to ascribe the existence of Netscape or PayPal, or, indeed, any company, to what happened at a single place and time. These should be understood as origin myths, though it isn't an easy matter to make an absolute distinction between myth and inter-actional reality. Stories about seminal meetings, however burnished by legend makers, colonize the minds, and the expectations, of participants at subsequent documentable meetings. Like a lot of myths, these stories point to sacralized beginnings, some defin-able moment when everything changed for the better, when the stable order of things irrevocably changed, when the future was foreseen, and the first steps toward a new reality were taken.

Why was it *Buck's*? What was special about it? In the beginning, in the early 1990s, there was probably only one pertinent reason why it was this place, and that was the location. The restaurant was conveniently off I-280, within a few miles of VCs' offices on Sand Hill Road near the Stanford campus, and, for many VCs, it was on their way from homes in wealthy South Bay communities—Woodside itself, a bijou little community

[25] Jamis MacNiven, *Breakfast at Buck's: Tales from the Pancake Guy* (Woodside, Calif.: Liverwurst, 2004), 9–10; Dickinson, "Billions of Dollars" (cit. n. 21).

[26] "World's Best Millionaire Hangouts," *Road and Travel Magazine*, accessed 29 January 2020, https://www.roadandtravel.com/luxurytravel/2005/millionairehangouts.htm.

[27] The $9,400 figure is from MacNiven, *Breakfast at Buck's* (cit. n. 25), 6–7. A lower figure for the winning bid is cited in "Ebay Auctions Off Breakfast Meeting with Venture Capitalist," 3 February 1999, www.sfgate.com/business/article/EBay-Auctions-Off-Breakfast-Meeting-With-Venture -2949061.php; see also Cheryl Himmelstein, "Surfing on the Slippery Skin of a Bubble," *New York Times Magazine*, 25 June 1999, www.nytimes.com/library/magazine/home/19990620mag-tech-jurvetson .html.

[28] Janelle Brown, "Buck's: The Breakfast Spot of Champions," *Salon*, 25 June 1999, www.salon.com /1999/06/25/bucks/.

[29] Draper, *Startup Game* (cit. n. 15), 3.

[30] Wayne McVicker, *Starting Something: An Entrepreneur's Tale of Control, Confrontation & Corporate Culture* (Los Altos, Calif.: Ravel Media, 2005), 193.

Figure 1. Owner Jamis MacNiven (standing) with Israeli President Shimon Peres (center) at Bucks's. Other figures unidentified. (Photo courtesy of Jamis MacNiven.)

whose residents included Larry Ellison of Oracle, John Doerr of Kleiner Perkins, Gordon Moore of Intel, and Jim Breyer of Accel Partners[31]—as well as Atherton, Los Altos Hills, and Portola Valley. If you are commuting from San Francisco on I-280, it's just a slight detour from the turning to Sand Hill Road. But, after a while, location was joined as a reason by the fact that it *was* Buck's, the place where all those deals happened. The entrepreneur Bob Metcalfe of 3COM and Ethernet probably had something to do with it when in 1992— shortly after Buck's founding—he "wrote in his weekly 'Info World' column that Buck's was the new power breakfast spot for Silicon Valley."[32] Then that's what it became. You had breakfast at Buck's because that was where VCs met entrepreneurs, and, if entrepreneurs got an invitation to breakfast at Buck's, they knew that it was, or might become, serious. (MacNiven says this of the VC-entrepreneur breakfasts: "Often, we're the first meeting on a deal. We're the first-date place.")[33]

[31] Piscione, *Secrets of Silicon Valley* (cit. n. 16), 153.

[32] Seth Similof, "Lunch is Not for Wimps at Buck's of Woodside Restaurant," *Haute Living*, 11 August 2012, www.hauteliving.com/2012/08/lunch-is-not-for-wimps-at-bucks-of-woodside-bucks-of-woodside/305999/.

[33] Quoted in Carolyn Jung, "Where the Tech Deals are Made," *San Francisco Chronicle*, 18 May 2014, www.sfgate.com/restaurants/article/Where-the-tech-deals-are-made-5483804.php; see also "Buck's of Woodside—Big Business Breakfast," *jesswords.com* (blog), 20 February 2014, www.jesswords.com/bucks-of-woodside-big-business-breakfast/.

By the end of the 1990s, VCs and entrepreneurs breakfasted at Buck's because *that was what you did* when you were at the early stages of deal making. It became globally famous for these sorts of meetings. Journalists and camera crews came to record the place where the Modern Technoscientific World was brought into being. The breakfast that launched Hotmail had to be reenacted for CNN, and, when a Japanese television crew came to do the same story, MacNiven let them shoot the reenactment (fig. 2).[34] It's said, though I can't confirm it, that there is a restaurant in Hong Kong named after MacNiven that tried to replicate whatever was taken as Buck's model, hoping to attract local VCs; and that Chicago breakfast spots popular with deal makers have been explicitly likened to Buck's—"but you don't see cops at Buck's."[35] (It turns out, however, that charismatic originals are not so easy to replicate.)

MacNiven is what's called "good copy," one of the Valley's most interviewed personalities. He boasts, "Since 1995, we have had over 600 TV, radio, glossy-print, and fish wrapper press come to Buck's to speak with me and my customers."[36] It is as if he was part of the Valley secret and *in on* the secret, as if he could fix up a meeting for you with a major VC—neither of which was true, even if MacNiven says he was asked to sit in on a number of pitches and parlayed his proprietorship into TED-talk celebrity.[37] (Buck's menu comes folded into a two-page newsletter in which MacNiven offers his oddball opinions on technology, business, politics, and culture.) "I'm only the second-biggest press-whore in America," MacNiven said in 2004, ceding first place to Donald Trump (who was not then what he now is).[38] MacNiven has been variously called the mayor of Silicon Valley, its prime minister, and "the unofficial diplomat of the independent nation of Silicon Valley," in much the same way that you might say that Madame Geoffrin was in charge of the French Enlightenment, with Buck's as a late modern Silicon salon.[39]

"We reinvented the modern world"; "we are where the modern world is emerging from"; we are "really the new Athens," MacNiven boasts,[40] and some entrepreneurs

---

[34] MacNiven, *Breakfast at Buck's* (cit. n. 25), 10.

[35] Piscione, *Secrets of Silicon Valley* (cit. n. 16), 154 (for Hong Kong). For Chicago, see John Pletz, "The Breakfast Club: Captains of Tech Shun Steakhouses for Dealmaking Over Easy," *Crain's: Crain's Chicago Business*, 14 May 2012, www.chicagobusiness.com/article/20120512/ISSUE03 /305129989/captains-of-tech-shun-steakhouses-for-dealmaking-over-easy.

[36] Quoted in Dietrich Walther, *Green Business—the Billion Dollar Deal* (Bozeman, Mont.: International Center for Education and Technology, 2012), 101; see also MacNiven, *Breakfast at Buck's* (cit. n. 25), 16.

[37] MacNiven, *Breakfast at Buck's* (cit. n. 25), 7 (on sitting in on pitches); for his first TED talk, see Jamis MacNiven, "What Up Silicon Valley," filmed 7 June 2011, Munich, TED video, 17:38, www .youtube.com/watch?v=Nxx73qeZlQI.

[38] Andrea Gemmet, "All About Jamis," *Almanac*, 31 March 2004, www.almanacnews.com /morgue/2004/2004_03_31.jamis.shtml.

[39] Bill Goss, *There's a Flying Squirrel in My Coffee: Overcoming Cancer with the Help of My Pet* (New York, N.Y.: Atria, 2002), 204–5; Mark Zetter, "Silicon Valley Unofficial Mayor Talks Pancakes, Steve Jobs, Startups and Pouring Millions Into Ideas," *Venture Outsource*, n.d., ca. 2004, accessed 25 May 2018, www.ventureoutsource.com/contract-manufacturing/mayor-silicon-valley-jamis -macniven-pancakes-steve-jobs-startups-pouring-millions-ideas-videos.

[40] Variously quoted in Guy Adams, "Steve Hilton's Social Network: The Policy Wonk Renowned for His Outlandish Ideas Has Left No. 10 for the Mecca of Tech, Palo Alto," *Independent* (London), 19 May 2012, www.independent.co.uk/news/world/americas/steve-hiltons-social-network-7767061.html; and Patrick May, "Facebook Movie Strikes a Chord with Silicon Valley Tech Crowd," *Mercury News* (San Jose), 1 October 2010, www.mercurynews.com/2010/10/01/facebook-movie-strikes-a-chord -with-silicon-valley-tech-crowd/.

Figure 2. *VC Steve Jurvetson (left) and entrepreneur Sabeer Bhatia (right) reenact for CNN the founding of Hotmail at a Buck's breakfast. (Photo courtesy of Jamis MacNiven.)*

have happily agreed: at Buck's "rustic tables the information era was forged and here the next revolution [will happen], the transition to Green Valley or the birth of the clean tech revolution."[41] Venture capitalist Bill Draper, said to be MacNiven's favorite customer, wrote a book about Silicon Valley start-ups, the first part of which was titled "Breakfast at Buck's." A German entrepreneur was enchanted by it all, driven to contrast Buck's fluidity and informality with the rigidity and traditionalism of German restaurants: "The open atmosphere of Silicon Valley makes the difference." No need for fanciness and formality, he says. If you go to Buck's website, you can see references to clean-tech. What German restaurant would have that? If you want to explain the Valley and its innovatory culture to a German, you couldn't do much better than to take her to Buck's.[42] And MacNiven himself put together a self-published book celebrating himself, his restaurant, his quirky lifestyle, and Silicon Valley utopianism in general. He's what's called a character, in a Valley world that's full of characters.[43]

A STAGE-SET FOR FUTURE MAKING

Venture capitalists and other commentators say that Buck's is "an unlikely venue for serious business."[44] But when you go inside, you see right away that it's even more

[41] Walther, *Green Business* (cit. n. 36), 101.

[42] Ibid., 100–1; for a similar lament that Britain also has no iconic Buck's-like eating places for VCs and entrepreneurs, see Mike Butcher, "Where is Your Bucks of Woodside?," *TechCrunch*, 26 February 2008, www.techcrunch.com/2008/02/26/where-is-your-bucks-of-woodside/.

[43] MacNiven, *Breakfast at Buck's* (cit. n. 25).

[44] McChesney, "Checking a Tech Bellwether" (cit. n. 21).

bizarre than the VCs say, or—and this is arguably the right story—that's it's a *very* likely place for what counts as serious business in the Valley, which is technoscientific future making, visionary, often utopian stuff, business-without-a-template but surrounded by an aura of mythic self-consciousness. Buck's is designed (partly intentionally, partly accidentally) to be just the right setting for capital-meets-technoscientific-entrepreneurship. It's an artfully crafted stage-set for the business of future making.

Buck's décor, or what counts as its décor, is a mash-up of whatever has taken MacNiven's fancy over about thirty years of haphazard collecting. Some call it "Western kitsch"; MacNiven describes it as "American whacky" and a collection of "life's ironic magic"; others call it a "phantasmagoria" of toys for boys; and VC Bill Draper said that it "resembles what the Smithsonian warehouse might look like in the wake of a tornado as straightened up by the Mad Hatter"—which was his way of saying that he really liked it.[45] The only remarkable thing outside in the car park is a twenty-foot long wooden fish called Woody, but much more is inside. There's an enormous ketchup bottle, a Statue of Liberty with an ice-cream sundae for a torch, a Google personalized car number plate, the world's biggest Swiss Army knife, huge model airplanes and blimps, and a long-unknown photograph of Steve Jobs in an atypically goofy mood—wearing a Groucho Marx mustache and glasses (figs. 3, 4, 5).[46] There's a Russian cosmonaut space suit, together with MacNiven's (implausible) account of how it got there, one which involves a visiting Russian general who was breakfasting with the US Secretary of Defense, and left behind a business card saying that MacNiven should look him up if he was ever in Russia. Which he did—but the general was then on more pressing business in Chechnya. Undeterred, MacNiven went to the place where the space suits were made, armed with the general's card, and, amazingly, secured the Yuri Gagarin model (fig. 3).[47]

That was nothing compared to the chutzpah involved in an attempt to obtain still another Russian tchotchke in 1993. It was an endeavor that failed, but the failure itself has become a Buck's design feature. MacNiven had the idea that maybe the post-Soviet government would be looking to turn a profit on Lenin's now surplus-to-requirements embalmed body. So, he wrote to a Russian official, offering a sum "in the high six figures" for Lenin. MacNiven got a reply, explaining that the Russian government was not looking to sell Lenin "as of today," but wanting more precise details about the sum offered.[48] Everything about Buck's décor screams eccentricity, the vision of a unique individual, the rejection of aesthetic convention, the going-together of things that don't go

---

[45] Zetter, "Silicon Valley Unofficial Mayor" (cit. n. 39); Gemmet, "All about Jamis" (cit. n. 38); Draper, *Startup Game* (cit. n. 15), 2.

[46] McChesney, "Checking a Tech Bellwether" (cit. n. 21); Alexis C. Madrigal, "How Steve Jobs Ended Up Sitting in Front of a Rosetta Stone Replica with Groucho Marx Glasses On," *Atlantic*, 25 January 2013, www.theatlantic.com/technology/archive/2013/01/how-steve-jobs-ended-up-sitting-in-front-of-a-rosetta-stone-replica-with-groucho-marx-glasses-on/272530/.

[47] MacNiven, *Breakfast at Buck's* (cit. n. 25), 215–19.

[48] Letters about buying Lenin's body allegedly exchanged between MacNiven and the Kremlin hang framed on the restaurant walls; they are reproduced in Zetter, "Silicon Valley Unofficial Mayor" (cit. n. 39). MacNiven recounts this story in his *Breakfast at Buck's* (cit. n. 25), 215–16. The whole tale, or at least the reply from the Russians, is probably either a joke or a fake; a well-informed Russian friend of mine sees many clear mistakes in the alleged reply, from the address and name of the responding ministry to a typographical error in its title. He speculates that "somebody has made [the letter] up in haste, perhaps using a[n] old dispatch of the ministry of foreign affairs of the USSR as a model."

Figure 3. *Jamis MacNiven with a stack of pancakes and a selection of Buck's eccentric tchotchkes, including the supposed Yuri Gagarin space suit (center, top). (Photo courtesy of Jamis MacNiven.)*

together. It's wacky, but it's wackiness of a certain kind—a kind that appeals to the self-image of know-no-rules Valley entrepreneurs and those with the vision to fund them.

THE PATRON AS STAGE MANAGER

The owner himself is part of the show, part of the stage-set for the theater of Valley innovation, and that's certainly how MacNiven sees himself. He revels in being the

Figure 4. *View of Buck's interior. (Photo courtesy of Jamis MacNiven.)*

Figure 5. *Another view of Buck's interior. (Photo courtesy of Jamis MacNiven.)*

restaurateur for the future, and he has furnished Buck's with the props for the role. "Buck's is the Earthlinked mother ship for the New Jerusalem," the restaurant website announces.[49] MacNiven says he hasn't profited from overheard investment tips, but he's an entrepreneur himself—that status presumably is not meant to include the clutch of Bay Area restaurants he now presides over, or a desultory 2006 flutter on a travel website company.[50] Recently, he told the BBC that he had been offered a "piece" of Netscape when it started and, over time, had invested in six start-up firms, including a robotics company.[51] "I'm always coupled [in articles] with Alan Greenspan as America's pundits," he confides.[52] Bill Draper called MacNiven "the World's Most Creative Entrepreneur," recognizing the restaurateur as someone very like himself and his clients, a freebooting, can-do, out-of-the-box visionary thinker.[53] The patron-configured-as-an-entrepreneur was, so to speak, part of the restaurant's furniture, helping make VCs and their supplicants feel that this was their sort of place. In June 2013, MacNiven was one of 130 Silicon Valley visionaries invited on a twelve-hour British Airways flight from San Francisco to London, during which they were charged with the mission to come up with ideas "to change the world."[54] He's a hippie entrepreneur of a specifically California sort—the sort that loves technology rather than hates it and that is drawn to technological utopianism. A self-described "student activist" and "leftist radical" at Berkeley in the 60s—"the 60s were cool," he reminds those who need reminding—MacNiven trained as an artist and tried to make it in the New York art world. Drawn to California from the East Coast by Stewart Brand's *Whole Earth Catalog*, MacNiven (before he became a restaurateur) lived "off the grid" on forty acres in the hills near the Stanford campus.[55] That was when he worked as a builder in the Bay Area for about fifteen years. The reason Steve Jobs—exceptionally among Valley entrepreneurs—never went to Buck's was, supposedly, because of a falling-out over work that MacNiven did on Jobs's house, but he apparently got into the restaurant business through building a number of well-known area eateries.[56] Buck's is the owner's place, but it's an innovatory place, signaling welcome to the Valley's tech innovators. In Erving Goffman's vocabulary, Buck's provides the "scenery and stage props" for the theater of late modern innovation.[57]

[49] Jamis MacNiven, "Restaurant History," Buck's of Woodside (website), accessed 25 May 2018, http://buckswoodside.com/restaurant-history/.

[50] Matt Marshall, "Jamis MacNiven, of Buck's Restaurant Fame, Enters Web 2.0 Travel Fray with 'LandFrog,'" *VentureBeat*, 10 March 2006, www.venturebeat.com/2006/03/10/jamis-mcniven-of -bucks-restaurant-fame-enters-web-20-travel-fray-with-landfrog.

[51] "Inside Silicon Valley," In Business, BBC Radio 4, broadcast 17 August 2014, www.bbc.co.uk /programmes/b04d4v76.

[52] Linda Leung, "VC Feeding Frenzy Over, Buck's Keeps Cookin,'" *NetworkWorld*, IDG Communications, 31 May 2004, www.networkworld.com/article/2333511/wireless/vc-feeding-frenzy-over –buck-s-keeps-cookin-.html.

[53] Draper, *Startup Game* (cit. n. 15), 2.

[54] Peter Delevett, "UnGrounded: Silicon Valley Stars Embark on Flight 'To Change the World,'" 12 June 2013, *Mercury News* (San Jose), 12 June 2013, www.mercurynews.com/2013/06/12 /ungrounded-silicon-valley-stars-embark-on-flight-to-change-the-world/.

[55] John Markoff, "Op-Ed: The Invention of Google before Google—a Radical Mail-Order 'Catalog,'" *Los Angeles Times*, 28 March 2018, www.latimes.com/opinion/op-ed/la-oe-markoff-stewart -brand-whole-earth-catalog-20180328-story.html.

[56] MacNiven, *Breakfast at Buck's* (cit. n. 25), 68–70.

[57] Erving Goffman, *The Presentation of Self in Everyday Life* (Garden City, N.Y.: Doubleday, 1959), 22. Goffman has been the premier sociological student of face-to-face encounters, and a central role is played in *Presentation* (116–19) by an account of passages between backstage kitchen and frontstage dining room in a Shetland hotel, but his work devoted little attention to the significance of meals as face-to-face occasions.

BREAKFAST AND BANDWIDTH

What happens at Buck's? The breakfasts are *informal*. A reporter observes that "the dress code is dress-down; a suit and tie stands out like a sore thumb."[58] The suit and tie reference is both diffusely conventional and specifically substantive. For all the talk of the Valley's openness to merit, the worlds of venture capital and high-tech entrepreneurship remain notoriously masculine; 89 percent of VCs and 96 percent of senior partners in VC firms are male. East and South Asian males are significant Valley presences in both venture capital and high-tech business; Hispanics and African Americans are scarcely represented at all.[59] None of the encounters that I have seen documented of these kinds of Buck's breakfasts seem to have involved women, except perhaps as servers. So, shared cultures and systems of social recognition are bound to be core elements of the informality of the breakfast encounters; it is easier to be informal with people-of-your-own-kind than with unfamiliar others, and, as the British say, the "laddishness," as well as the geekiness, of Silicon culture lubricates that informality for males.

What actually takes place at these meetings is mostly a matter of inference and speculation; they're not, of course, minuted—and that's rather the point. The historian or sociologist learns about them through anecdote, and many of the anecdotes seem to be heavily mythologized. Most, probably the overwhelming majority, don't have any instrumental outcome at all, and the stories are all about the few that do come to something. The breakfasts are usually one on one or maybe two on one, where the two are entrepreneurs and the one is a VC. They're not supposed to be "pitches"—that is, prepared presentations that function as bids for funding—though supplicant entrepreneurs are well coached always to have at the ready the one-minute "elevator pitch," and it's hard to imagine that entrepreneurs securing a breakfast meeting come without the PowerPoint slides on their mobile digital devices or a business plan, paper or digital.[60] It's informal, but it's *formally* informal; if you're an entrepreneur, you understand that this *is* a pitch (or a pre-pitch), only with maximum deniability for the VC. A commentator who pretends to know about Buck's breakfast encounters writes that "the conversations sound disjointed," normless, conventionless: "Speakers hop from one topic to the next, interrupting each other, while the entrepreneur sketches on a napkin and the VC makes notes on a ketchup-stained placemat, to be taken back to his

---

[58] McChesney, "Checking a Tech Bellwether" (cit. n. 21).

[59] Hollie Slade, "'We Need More Women in Venture Capital' Say Female Midas Listers," *Forbes*, 26 March 2014, www.forbes.com/sites/hollieslade/2014/03/26/we-need-more-women-in-venture-capital-say-female-midas-listers/; see also Mike Peña, "Room at the Top for More Women in Venture Capital," *STVP* (Stanford Technology Ventures Program), 12 May 2014, https://ecorner.stanford.edu/articles/room-at-the-top-for-more-women-in-venture-capital/; Maria Cirino, "Lack of Women in VC Reflects Broader Bias," *Boston Globe*, 13 July 2014, http://www.bostonglobe.com/business/2014/07/12/lack-women-reflects-broader-bias/O9xjb7qGNo4toyW8dEp4TJ/story.html; Julie Carrie Wong, "Segregated Valley: The Ugly Truth about Google and Diversity in Tech," *Guardian*, 7 August 2017, https://www.theguardian.com/technology/2017/aug/07/silicon-valley-google-diversity-black-women-workers; Devin Thorpe, "Successful African-American Silicon Valley Entrepreneur Feels 'Like a Black Unicorn,'" *Forbes Magazine*, 15 January 2018, https://www.forbes.com/sites/devinthorpe/2018/01/15/successful-african-american-silicon-valley-entrepreneur-feels-like-a-black-unicorn/#618f339d29ab; and Sam Levin, "Black and Latino Representation in Silicon Valley Has Declined," *Guardian*, 3 October 2017, https://www.theguardian.com/technology/2017/oct/03/silicon-valley-diversity-black-latino-women-decline-study.

[60] For the institutions and practices of "pitching" as a face-to-face mode, see Shapin, *Scientific Life* (cit. n. 14), 276–82.

office. Slowly, a pattern emerges." The VCs are probing and probing again; what appears aimless and patternless takes shape as inquiry: "What is the science? What is its status as IP (intellectual property)? What is the competition? What is the intended product? Who is the customer? How much will it cost?"[61] Success for the entrepreneur—on this sort of occasion—would mean a promise of a further meeting, maybe for coffee and bran muffins, one on one in the VC firm's offices, maybe, and further down the road, a formal pitch to the general partners.

The breakfast meetings may be instrumental, but they are also informal and *intimate*—in the sense that supplicant and patron are just a few feet away from each other; they are constantly in face-to-face contact for an hour or so; and they are sharing a meal together—even though it's just breakfast. Venture capitalists like to say that in making an investment decision they "bet on the jockey, not on the horse"—that is, the entrepreneur and not the technology or the scientific IP. To the extent that is so, the face-to-face mode of a meal, however informal, offers the potential of rich information about the virtues, vices, capabilities, and commitments of the investable person.[62] If a visionary technoscientific future is being pitched, the high bandwidth of the occasion offers a mode of monitoring that may never be repeated over the future course of the interaction, assuming that there is a future. The VC is literally, for that hour or so, looking a future in the face. MacNiven, who has had a ringside seat for almost thirty years, was asked what advice he had for entrepreneurs hoping for access to venture funding. He said: "This is where the action is, and a lot of the action is had by face-to-face. That's what Buck's is all about."[63] What about the food? A British journalist sought the owner's views on what money eats for breakfast. Don't look at the French toast munchers, MacNiven said. "The bigger the deal, the less they eat, because you don't want your mouth full at a critical moment." The serious venture capitalists eat muffins, he says.[64] The high bandwidth of the face-to-face has for many years been supplemented by digital high bandwidth; Buck's was one of the originals of a type of Foucault's *heterotopic* spaces of late modernity,[65] claiming to be the first place in the country—and presumably the world—to offer a public Wi-Fi hotspot. This was before ubiquitous cell phones, so Buck's helped pioneer a mode of interaction that many of us now take for granted: the VC can be in an analog face-to-face relationship with the entrepreneur and in digital contact with the universe. That's a normal mode of interaction now, but Buck's was possibly one of the first to facilitate it in a public place.[66]

BREAKFAST: A QUICK BITE AND A LONG HISTORY

Breakfast at Buck's is a meal in a restaurant. We take it as obvious now that there are such places as restaurants, as we take as self-evident the conventions that obtain there

[61] Elton B. Sherwin Jr., *The Silicon Valley Way: Discover 45 Secrets for Successful Start-Ups*, 2nd ed. (1998; repr., Rocklin, Calif.: Energy House, 2010), 3; Walther, *Green Business* (cit. n. 36), 100 (for ketchup stain).

[62] For familiarity and the face-to-face mode in the VC-entrepreneur relationship, see Shapin, *Scientific Life* (cit. n. 14), 282–303.

[63] Zetter, "Silicon Valley Unofficial Mayor" (cit. n. 39).

[64] Hall, "Silicon Giants" (cit. n. 23).

[65] "The heterotopia is capable of juxtaposing in a single real place several spaces, several sites that are in themselves incompatible": Michel Foucault, "Of Other Spaces," trans. Jay Miskowiec, *Diacritics* 16 (Spring 1986): 22–7, on 25.

[66] Courtney Price, "A Visual Tour of Silicon Valley," *HuffPost*, 6 December 2017, www.huffingtonpost .com/courtney-price/a-visual-tour-of-silicon-valley_b_5638584.html.

and the mode of being-together that is routine when eating in restaurants. But recent scholarship draws attention to the historical origins of the restaurant in late eighteenth- and early nineteenth-century France and its significance as a mode of modernity.[67] There are several points to bear in mind here: the first is the *carte*, ordering off the menu (*à la carte*)—and therefore having just what you want—rather than the *table d'hôte* where all guests sat at a common table and took from common dishes what the host offered that day. The menu therefore represented a form of individualism, a manifestation of the power of the individual consumer. The second, and related, point is the form of social interaction conventional in the restaurant. It was one of the places where one could be, arguably for the first time, *alone together*. Each individual in a party is in intimate, face-to-face contact with that party's members, while disengaged from others, even those sitting at adjacent tables and perhaps even eating the same things. Eyeshot is, for the most part, irrelevant, while earshot may be problematic (as it often is for Buck's customers sitting where they can overhear other capitalists and entrepreneurs).[68] Consider the difference from, say, the cinema or theater or concert— some of which are premodern venues—where one may also be alone together, but taking in the same spectacle and reacting in a way that is sensed by, and relevant to, unfamiliar others; for example, you're all watching the same play and you hear, and may participate in, the audience's collective laughter or audible expressions of shock and horror. How- ever, unlike those other venues, the restaurant is more likely to make individuals' specific choices and body practices consequentially visible to others. Individuals in the restaurant can be part of the show in a way that they are not in the theater.

Again, Buck's is a restaurant, but there are many other restaurants, and, indeed, VCs and entrepreneurs also meet at other establishments—often for breakfast. It has al- ready been pointed out that Buck's is conveniently located, so that consideration, added (over time) to its celebrity, helps to answer the specific question, "Why Buck's?" But related, interesting, and more general questions are: "Why a restaurant?" and "Why breakfast?" One of the virtues of meeting at Buck's is, it's sometimes said, that it's "neutral territory"; it is near VCs' offices, but it *isn't* their office, and so that adds to "deniability"; you could say business isn't really happening if you decide that it isn't.[69] And as Buck's is a restaurant, neither the VC nor the entrepreneur, of course, owns it. But the neutrality comes with qualifications; the VC will be known to the owner and staff, will be familiar with the place (its menu and its décor), and will have chosen it as a meeting place. Also, the VC will be responsible for when the meeting begins and when it ends (as in "I have to go to the office now"), and will certainly—though I have seen no direct testimony on this—pick up the tab, which is what you do when you're the host. It's a neutral place that the VC is in charge of and that contains the capitalist's stage props.[70]

---

[67] Rebecca L. Spang, *The Invention of the Restaurant: Paris and Modern Gastronomic Culture* (Cambridge, Mass.: Harvard Univ. Press, 2000); see also Amy B. Trubek, "The Emergence of the Restaurant," chap. 2 in *Haute Cuisine: How the French Invented the Culinary Profession* (Philadel- phia: Univ. of Pennsylvania Press, 2000); Paul Freedman, "Restaurants," in *Food in Time and Place: The American Historical Association Companion to Food History*, ed. Freedman, Joyce E. Chaplin, and Ken Albala (Oakland: Univ. of California Press, 2014), 253–75.

[68] Hall, "Silicon Giants" (cit. n. 23), on overhearing; see also Goffman, *Presentation of Self* (cit. n. 57), e.g., 130, 138, 151, 170, 179, 213; Rachel Rich, *Bourgeois Consumption: Food, Space and Identity in London and Paris, 1850–1914* (Manchester: Manchester Univ. Press, 2011), 137–8 (on separate tables at restaurants).

[69] Piscione, *Secrets of Silicon Valley* (cit. n. 16), 154.

[70] Goffman, *Presentation of Self* (cit. n. 57), e.g., 22, 94, 96, 99, 112.

The identity of all daily European and American meals has been historically unstable—with respect to what's eaten, when the meals are taken, where and with whom, and what they signify. Dinner, for example, understood as the main meal of the day, migrated from around midday in the seventeenth century to evening by the late nineteenth century. Supper traditionally was a little something—maybe soup, maybe bread and cheese—that you took before going to bed. Even now, what's called dinner, what's supper, and, in Britain, what's "tea" varies in timing, content, mode of service, and social identity among classes and regional cultures. But the shifting identities and circumstances of dinner and supper are as nothing compared to breakfast.

For most of European history, breakfast as a routine morning meal scarcely existed—the English word was not evidently used until the middle of the fifteenth century—and a morning meal of solid food was not invariably taken, or, if it was taken, it was restricted to the very young, the very old, the invalid, and the laboring classes. They were reckoned to need proper sustenance before (in the last case) going off to hard manual work. The Romans were sometimes cited as a model of dietary discipline; they usually took nothing until supper, and even those who did eat breakfast—at eight or ten in the morning, or as late as noon—tended otherwise toward a light diet.[71] One historian describes the breakfast of the later Middle Ages as "an optional extra."[72] There were dietary writers who cautioned against the taking of breakfast for anyone not in the best health, which, in England, was said to include "the generality of People."[73] There was no standard stuff that constituted a breakfast meal, and some physicians counseled on medical grounds against eating breakfast or at least taking anything substantial.[74] And at the end of the seventeenth century, John Locke warned against accustoming young gentlemen to breakfast.[75] "Dinner" was itself a contraction of the Latin word for breaking the fast (*disjejunere*); hence, both the English "breakfast" and the French *petit déjeuner* etymologically duplicated the idea of breaking the night's fast, so encoding the anomaly of an early morning *meal*.[76]

In the Jacobean play *The Spanish Curate*, the clergyman Lopez has offended the lawyer Bartolus, but the latter offers to forgive him, marking the reconciliation by

[71] John Locke, *Some Thoughts Concerning Education* (1690; repr., London: J. and R. Tonson, 1779), 17.

[72] Bridget Ann Henisch, *Fast and Feast: Food in Medieval Society* (University Park: Pennsylvania State Univ. Press, 1976), 22–5, on 23; see also Mark Dawson, *Plenti and Grase: Food and Drink in a Sixteenth-Century Household* (Totnes, UK: Prospect, 2009), 205–6 (on breakfast not recommended save for invalids and manual laborers). But see the classic Victorian account in Thomas Wright, *A History of Domestic Manners and Sentiments in England during the Middle Ages* (London: Chapman and Hall, 1862), 425, 455.

[73] Humphrey Brooke, *Ugieine or A Conservatory of Health* (London: R. W. for G. Whittington, 1650), 123.

[74] See, for example, William Bullein, *The Government of Health* (London: Valentine Sims, 1595), 26v: "use some light things at breakfast of perfite digestion."

[75] Locke, *Some Thoughts Concerning Education* (cit. n. 71), 17.

[76] Ken Albala, "Hunting for Breakfast in Medieval and Early Modern Europe," in *The Meal: Proceedings of the Oxford Symposium on Food and Cookery 2001*, ed. Harlan W. Walker (Totnes, UK: Prospect, 2002), 20–30, on 21; Heather Arndt Anderson, "History and Social Context," chap. 1 in *Breakfast: A History* (Lanham, Md.: Rowman & Littlefield, 2013); J. C. Drummond and Anne Wilbraham, *The Englishman's Food: A History of Five Centuries of English Diet*, rev. ed. (London: J. Cape, 1958), 55; Kaori O'Connor, *The English Breakfast: The Biography of a National Meal* (London: Bloomsbury, 2006).

inviting him "to a Breakfast [which] I make but seldome, But now we will be merry."[77] This was a play that Samuel Pepys saw in 1661, and he too did not take breakfast regularly. His diary mentioned the meal just five times over ten years—and his breakfast fare showed very little stability—sometimes a bit of mackerel, sometimes a collar of brawn, certainly not eggs and bacon or porridge. Much more often—and an index of the growing importance of eating out—Pepys said that he "gave" someone their "morning draught"—wine, small beer, coffee, or chocolate. The morning draught served a social function: it was taken at a commercial establishment, someone *treated* someone else to it, news or gossip was exchanged, and deals were done or begun.[78] The far less sociable and more abstemious Isaac Newton told the physician William Stukeley that his solitary breakfast consisted only of bread and butter and boiled orange peel made into a sort of sweetened tea, which he took to dissolve phlegm.[79]

Breakfast became a more common meal during the nineteenth century, but even then, both its makeup and the social forms attending it remained unstable, varying radically between national cultures and among social classes. In nineteenth-century Boston, boardinghouse breakfasts were the chosen setting for the agreeable, varied, and informal conversations recorded in Oliver Wendell Holmes's *The Autocrat of the Breakfast-Table*: "The talks are like the breakfasts,—sometimes dipped toast, and sometimes dry. You must take them as they come."[80] American boardinghouse breakfasts attracted the scorn of European visitors and the condemnation of US health reformers; the fare was too abundant, too rich, too rapidly consumed, thereby feeding an epidemic of dyspepsia and prompting the emergence of such lighter fare as the branded breakfast cereal.[81] Mrs. Beeton's ubiquitous rule book for domestic management noted the great variety of foods making up the Victorian breakfast. She gently scolded her compatriots for neglecting due care in its preparation and consumption: "Amongst English people as a rule, breakfast, as a meal, does not hold a sufficiently important place; and with some it means, in reality, no meal at all, unless we reckon the proverbial 'cup of tea' to form one. . . . Housekeepers should make more of breakfast."[82]

While by the twentieth century breakfast became a common feature of European and American life, its identity remains far more problematic than any other daily meal. Many people don't eat breakfast at all, while others have it on the run, in the car, or pick it up on the way to the office. Breakfast was once celebrated by nutritionists and advertisers as the "Most Important Meal of the Day";[83] then, for some, it became the

---

[77] John Fletcher and Philip Massinger, *The Spanish Curate*, in *The Dramatic Works in the Beaumont and Fletcher Canon*, ed. Fredson Bowers, vol. 10 (Cambridge: Canbridge Univ. Press, 1996), 377 (play originally performed 1622).

[78] "Food," in *The Diary of Samuel Pepys*, ed. Robert C. Latham and William Matthews, 10 vols. (London: HarperCollins, 1995), 10:143–9, on 144.

[79] Quoted in David Brewster, *Memoirs of the Life, Writings, and Discoveries of Sir Isaac Newton*, 2 vols. (Edinburgh: Thomas Constable, 1855), 2:89n.

[80] Oliver Wendell Holmes Sr., *The Autocrat of the Breakfast-Table: Every Man His Own Boswell* (Boston: Phillips, Sampson, 1858), 78. The talks were first published in serial form in the early 1830s and resumed in the late 1850s.

[81] For the "reinvention of breakfast" in nineteenth-century America, see Abigail Carroll, *Three Squares: The Invention of the American Meal* (New York, N.Y.: Basic, 2013), 133–58.

[82] [Isabella Mary Beeton], *Mrs. Beeton's Cookery Book and Household Guide*, new ed. (1861; London: Ward, Lock, 1898), 240. (These remarks don't appear in earlier editions.)

[83] Charles Spence, "Breakfast: The Most Important Meal of the Day?" *Intl. J. Gastronomy and Food Sci.* 8 (2017): 1–6; Alex Mayyasi and Priceonomics, "Why Cereal Has Such Aggressive Marketing," *Atlantic*, 16 June 2016, https://www.theatlantic.com/business/archive/2016/06/how-marketers-invented-the-modern-version-of-breakfast/487130/.

"Most Productive Meal of the Day."[84] The modern American institution called the "power breakfast" feeds self-esteem as much as it fuels the body.[85] What you have for breakfast is also problematic, though maybe not quite as problematic as that newer American invention (now well migrated abroad), the weekend "brunch."[86] At Buck's, one person at a table may be having coffee and toast while another opts for the South Bay equivalent of the "Full English." For some people who do have breakfast, it's a social occasion—though, unless it's an institutional "power" or "prayer breakfast," a nonsolitary breakfast is usually one on one. Online dating apps explore the propriety of different sorts of meals for different stages of a relationship, different stations on the way to "getting to know" the other. For example, "It's Just Lunch" advertises itself by noting that "a lunch date or drink after work is the ideal first date . . . It's a no pressure, relaxed setting where you can actually talk face-to-face."[87] But the understood meaning of lunch is more stable than that of breakfast. On the one hand, breakfast is casual—in the positive sense that it signals the possibility of appropriate disengagement, and in the negative sense that proposing a breakfast date might be taken to mean a lack of proper interest: "Nobody wants to date someone who can only squeeze them in before 9 a.m." On the other hand, the informality of breakfast might signal, or be taken to signal, intimacy—the shared meal after the shared evening meal.[88] The norms of breakfast are uncertain, oscillating more between intimacy and informality than those of other meals.

Then there's the fact that it's a morning meal, and not just a morning meal, but a meal taken outside of, and prior to, a working day. On weekdays, Buck's opens at 7 a.m.—plenty of time to get to Sand Hill Road by 9—so, unlike lunch, it doesn't take time out from work, but precedes it. Anyway, this is a world in which people compete in "up with the lark" early starts to the day. By the time VCs meet entrepreneurs for breakfast, they may have been up for three hours, having already sent their texts and tweeted their tweets, gone through their emails, had a workout with their personal trainer, spent some quality time with their kids, and done a bit of tai chi or transcendental meditation. As a VC, you may take *lunch* with other VCs—at your firm or another—or with the "limited partners" who actually provide the capital you

---

[84] Nancy Trejos, "Power Breakfast, the Most Productive Meal of the Day," *USA Today*, 12 October 2015, www.usatoday.com/story/travel/hotels/2015/10/12/hotels-power-breakfasts-loews-knickerbocker/73660834/.

[85] Julia La Roche and Lisa Du, "We Experienced New York's Most Famous Power Breakfast," *Business Insider*, 23 May 2012, https://www.businessinsider.com/power-breakfast-2012-5; Alina Dizik, "Breakfast is the Real Power Hour for Success," *BBC Capital*, 28 October 2016, http://www.bbc.com/capital/story/20161027-breakfast-is-the-real-power-hour-for-success.

[86] A prominent sociologist, writing about how eating occasions give form to quotidian social life, does offer some remarks about breakfast, but omits it from the primary list of meals that structure social life: see Warde, *Practice of Eating* (cit. n. 1), 61. That's an indication of how the modern breakfast remains something of an anomaly to sociologists of food and eating.

[87] It's Just Lunch (website), accessed 16 April 2019, https://www.itsjustlunch.com/the-ijl-story.

[88] For quotation, see Jason Kessler, "Breakfast: The Crucial Date Meal," Food Republic, 3 November 2011, https://www.foodrepublic.com/2011/11/03/breakfast-the-crucial-date-meal/; "Go On Your Breakfast Dates Instead of Coffee Dates," The New Savvy, 5 June 2017, https://thenewsavvy.com/life/food-health/go-breakfast-dates-instead-coffee-dates/; "5 Reasons Why Breakfast First Dates Are Brilliant," Date My Pet, (updated) 3 February 2010, https://www.datemypet.com/5-reasons-why-breakfast-first-dates-are-brilliant; Michelle, "6 Reasons to Take Your Date to Breakfast," *#staymarried* (blog), accessed 16 April 2019, https://staymarriedblog.com/6-reasons-to-take-your-date-to-breakfast/.

manage; you may "brown-bag" it at your desk. But breakfast is *your business*, or, if you prefer, it's not business at all unless you otherwise define it. If you're a novice entrepreneur, and you're lucky enough to be offered breakfast at Buck's with a VC, you understand that you may be a long way from a pitch in their offices, a long way from lunch with the VC and the partners, and a very long way from dinner—which, if you ever get it, is likely to be a celebration of a consummated relationship. It's apparently very rare at VC-entrepreneur breakfasts for "the lawyers" to be present, or, indeed, the other "general partners" of the VC firm. Those absences signal both the early-stage nature and the informality of the occasion. Legend picks out successful interactions, yet it's likely that very few of these breakfasts lead to an investment.

There is also the local circumstance that breakfast, and, to a lesser extent lunch, has few alternatives as an occasion for the sort of interaction that's represented by the famous Buck's scenes—intimacy with informality, out of the workday, interacting one on one but in a public place, in a neutral territory; but a place in which one party can act as host, with earshot problematic but usually manageable. In the seventeenth and eighteenth centuries, the coffeehouse (celebrated by Jürgen Habermas as a site responsible for creating the modern "public sphere") was a place where you would go to hear news and to make news—even (in the celebrated case of the Royal Society) scientific news—but it was *not* the sort of place where you could be *alone together*, since the whole point of the early modern coffeehouse was its common table and common conversation.[89] At the same time, taverns, inns, or alehouses were typically places where you might engage yourself with others in common, or, if you wanted seclusion for sexual or confidential business purposes, you could secure a private room, where only the owners and their staff knew that you were there and with whom.[90]

Modern America is short of places for the sort of interactions that make breakfast at Buck's famous. It doesn't have the British pub; it does have coffeehouses, but increasingly as places where you are connected to the world but not to anyone sitting *with* you. When, years ago, Starbucks threatened to become something like a Viennese coffeehouse, where people read the news, meet with colleagues and friends, and spend the day nursing *Kaffee mit Schlag*, the national ownership took steps to move on the "loitering laptop hobos." Customers were not to confuse a business with a public space. Law enforcement was called in to make that clearly understood, and only very recent disastrous publicity about police removals of nonbuying ethnic minority customers has instigated a corporate policy change.[91] So, breakfast is that usually unremarked-on remarkable thing: an occasion that combines informality with intimacy, that isn't business

---

[89] Markman Ellis, *The Coffee House: A Cultural History* (London: Phoenix, 2004); Steve Pincus, "'Coffee Politicians Does Create': Coffeehouses and Restoration Political Culture," *J. Mod. Hist.* 67 (1995): 807–34; Brian Cowan, *The Social Life of Coffee: The Emergence of the British Coffeehouse* (New Haven, Conn.: Yale Univ. Press, 2005); Jürgen Habermas, *The Structural Transformation of the Public Sphere: An Inquiry into a Category of Bourgeois Society*, trans. Thomas Burger and Frederick Lawrence (1962; repr., Cambridge: Polity, 1989).

[90] Samuel Pepys, "Taverns, Inns and Eating Houses," in *Diary of Samuel Pepys* (cit. n. 78), 10:416–18.

[91] Hamilton Nolan, "Starbucks Now Calling the Cops on Laptop Hobos," *Gawker*, 4 October 2011, www.gawker.com/5846448/starbucks-now-calling-the-cops-on-laptop-hobos; Jacey Fortin, "A New Policy at Starbucks: People Can Sit without Buying Anything," *New York Times*, 20 May 2018, www.nytimes.com/2018/05/20/business/starbucks-customers-policy-restrooms.html. MacNiven says that he makes a point of allowing his A-list clientele to stay as long as they like.

but isn't not-business, that touches the concerns of the working day but that's usually set outside it, that's anything you want it to be or nothing at all. Its cultural and social instability and, in the case of Buck's, its relative normlessness makes it the perfect stage-setting for innovation to happen. You might well be skeptical of the stories about the breakfasts that Made the Modern World, but the breakfasts themselves are a mode of that Made Modernity.

POSTSCRIPT

As this piece entered its final editing, and as the COVID-19 pandemic of 2020 struck hard at the hospitality industry, Buck's of Woodside—like so many other American eating establishments—was closed. Rumors circulate in the Valley that it will never reopen, although, on the restaurant website, the owner says that he didn't start them. No one knows the future—for Buck's or for many other commercial eating places. Nevertheless, this treatment of the face-to-face mode of feeding and knowing now has a historical specificity that I could never have imagined when I wrote it.

Notes on Contributors

Alissa Aron holds degrees in History and Philosophy of Science from the University of Cambridge (MPhil) and Viticulture, Enology, and Vine, Wine, and Terroir Management from the Ecole Supérieure d'Agriculture d'Angers (MSc). Her research has focused on the history of the agricultural sciences. She currently works as a consumer insights consultant and researcher in the food and beverage industry.

Bradford Bouley is Assistant Professor of History at the University of California, Santa Barbara. His first book, *Pious Postmortems: Anatomy and the Making of Early Modern Saints*, was published by the University of Pennsylvania Press in 2017. He is currently working on a second book project that examines cannibalism, warfare, and the consumption of meat in early modern Rome.

Joyce E. Chaplin is the James Duncan Phillips Professor of Early American History at Harvard University. She was guest curator for "Resetting the Table: Food and Our Changing Tastes" (2019–21) at Harvard's Peabody Museum of Archaeology and Ethnology.

Carolyn Cobbold is a Research Fellow at Clare Hall, University of Cambridge. Her book *A Rainbow Palate* is being published by the University of Chicago Press. Carolyn has a PhD from Cambridge and an MSc in History of Science from Imperial College and University College London. She previously worked in risk management after earning a degree in engineering from Imperial.

Deborah Fitzgerald is the Cutten Professor of the History of Technology at the Massachusetts Institute of Technology. She is the author of *The Business of Breeding: Hybrid Corn in Illinois* (Cornell, 1990) and *Every Farm a Factory: The Industrial Ideal in American Agriculture* (Yale, 2003). She is currently writing a history of the impact of World War II on food and agriculture.

Anita Guerrini is Horning Professor in the Humanities Emerita at Oregon State University. Her most recent book, *The Courtiers' Anatomists: Animals and Humans in Louis XIV's Paris* (Chicago, 2015), won the 2018 Pfizer Prize of the History of Science Society. She is now working on a revised second edition of *Experimenting with Humans and Animals: From Galen to Animal Rights* (Johns Hopkins, 2003).

Di Lu is a Zvi Yavetz Fellow at The Zvi Yavetz School of Historical Studies, Tel Aviv University. He is currently writing a book-length transnational history of the caterpillar fungus and is also working on a project about the Sino-British trade in materia medica in the late nineteenth century.

Ted McCormick is Associate Professor of History at Concordia University, Montreal. He is the author of *William Petty and the Ambitions of Political Arithmetic* (Oxford, 2009) and has written extensively on early modern population thought. His current research focuses on technological projects and colonial labor in seventeenth-century Britain, Ireland, and the Caribbean.

Projit Bihari Mukharji is Associate Professor at the University of Pennsylvania's Department of History and Sociology of Science. His work explores the intersecting histories of knowledge and subalternity in South Asia. Mukharji is the author of *Doctoring Traditions: Ayurveda, Small Technologies, and Braided Sciences* (Chicago, 2016) and *Nationalizing the Body: The Medical Market, Print and Daktari Medicine* (Anthem, 2009).

Stefan Pohl-Valero is Associate Professor of History of Science and Medicine at the Universidad del Rosario, Bogotá, Colombia. He is the author of *Energía y cultura: Historia de la termodinámica en la España de la segunda mitad del siglo XIX* (Rosario, 2011). In 2019 he was a Fulbright Visiting Scholar at the University of North Carolina at Chapel Hill. Currently, he is writing a book on science, food, and the city (Bogotá) from 1880 to 1940, and leads the research project, "Assembling the Food Problem in Colombia during the Beginning of Development, 1940–1960."

Steven Shapin used to teach at Harvard University. His books include *The Scientific Life: A Moral History of a Late Modern Vocation* (Chicago, 2008), and he continues work on a cultural history of ideas about food and eating.

Dana Simmons is Associate Professor of History at the University of California, Riverside, and author of *Vital Minimum: Need, Science, and Politics in Modern France* (Chicago, 2015). Her current research is a history of hunger science and politics in the twentieth century.

E. C. Spary is Reader in the History of Modern European Knowledge at the University of Cambridge. Her books include *Feeding France: New Sciences of Food, 1760–1815* (Cambridge, 2014), *Eating the Enlightenment: Food and the Sciences in Paris* (Chicago, 2012), and *Utopia's Garden: French Natural History from Old Regime to Revolution*

(Chicago, 2000). She is currently at work on a study of drug taking in the reign of Louis XIV.

Ulrike Thoms is a historian who works in the fields of social and economic history, history of the body, and history of science. In her publications on food history she combines different scientific approaches and perspectives to do justice to the sociocultural totality of food. Currently, she is investigating the German education system in the context of the migration society.

Corinna Treitel is Professor of History at Washington University, St. Louis. Her publications include *Eating Nature in Modern Germany: Food, Agriculture, and Environment, 1870–2000* (Cambridge, 2017) and *A Science for the Soul: Occultism and the Genesis of the German Modern* (Johns Hopkins, 2004). Her current book project is titled *Gesundheit! Practicing German Health, 1750–2000*.

Rebecca J. H. Woods is Assistant Professor in the Department of History and at the Institute for the History and Philosophy of Science and Technology, University of Toronto. She specializes in the history of science, environment, and animals.

Benjamin Aldes Wurgaft is a writer and historian, and the author of the recent *Meat Planet: Artificial Flesh and the Future of Food* (California, 2019). He received his PhD in European Intellectual History from the University of California, Berkeley, and has written about food for *Gastronomica* and many other publications. He is currently Visiting Assistant Professor at Wesleyan University.

Anya Zilberstein is Associate Professor of History at Concordia University, Montreal. Her first book, *A Temperate Empire: Making Climate Change in Early America* (Oxford, 2016), won the Berkshire Conference Book Prize for 2016. She is working on a second book project about the history of mass feeding of people and domesticated animals since the eighteenth century and, together with Jennifer L. Anderson, editing "Empowering Appetites: The Political Economy and Culture of Food in the Early Atlantic," a special journal issue of *Early American Studies* to be published in 2021.

Index

SUGGESTIONS FOR CONTRIBUTORS TO OSIRIS

OSIRIS is devoted to thematic issues, conceived and compiled by guest editors who submit volume proposals for review by the OSIRIS Editorial Board in advance of the annual meeting of the History of Science Society in November. For information on proposal submission, please write to the Editors at pmccray@history.ucsb.edu and ss536@cornell.edu.

1. Manuscripts should be submitted electronically in Rich Text Format using Times New Roman font, 12 point, and double-spaced throughout, including quotations and notes. Notes should be in the form of footnotes, also in 12 point and double-spaced. The manuscript style should follow *The Chicago Manual of Style*, 16th ed.

2. Bibliographic information should be given in the footnotes (not parenthetically in the text), numbered using Arabic numerals. The footnote number should appear as superscript. "Pp." and "p." are not used for page references.

    a. References to books should include the author's full name; complete title of book in *italics*; place of publication; date of publication, including the original date when a reprint is being cited; and, if required, number of the particular page cited (if a direct quote is used, the word "on" should precede the page number). *Example*:

       [1] Mary Lindemann, *Medicine and Society in Early Modern Europe* (Cambridge, 1999), 119.

    b. References to articles in periodicals or edited volumes should include the author's name; title of article in quotes; title of periodical or volume in *italics*; volume number in Arabic numerals; year in parentheses; page numbers of article; and, if required, number of the particular page cited. Journal titles are abbreviated according to the journal abbreviations listed in *Isis Current Bibliography*. *Example*:

       [2] Lynn K. Nyhart, "Civic and Economic Zoology in Nineteenth-Century Germany: The 'Living Communities' of Karl Möbius," *Isis* 89 (1999): 605–30, on 611.

    c. All citations are given in full in the first reference. For succeeding citations, use an abbreviated version of the title with the author's last name. *Example*:

       [3] Nyhart, "Civic and Economic Zoology" (cit. n. 2), 612.

3. Special characters and mathematical and scientific symbols should be entered electronically.

4. A small number of illustrations, including graphs and tables, may be used in each volume. Hard copies should accompany electronic images. Images must meet the specifications of The University of Chicago Press "Artwork General Guidelines" available from the Editor.

5. Manuscripts are submitted to OSIRIS with the understanding that upon publication copyright will be transferred to the History of Science Society. That understanding precludes consideration of material that has been previously published or submitted or accepted for publication elsewhere, in whole or in part. OSIRIS is a journal of first publication.

OSIRIS is published once a year.

ISSN: 0369-7827 | E-ISSN: 1933-8287

Paperback ISBN: 978-0-226-72686-1

eISBN: 978-0-226-47974-3

Single copies are $35.00.

Address subscriptions, single issue orders, claims for missing issues, and advertising inquiries to *Osiris*, The University of Chicago Press, Journals Division, 1427 E. 60th Street, Chicago, IL 60637-2902.

Postmaster: Send address changes to *Osiris*, The University of Chicago Press Subscription Fulfillment, 1427 E. 60th Street, Chicago, IL 60637-2902.

OSIRIS is indexed in major scientific and historical indexing services, including *Biological Abstracts, Current Contexts, Historical Abstracts*, and *America: History and Life*.

Osiris

**A RESEARCH JOURNAL DEVOTED
TO THE HISTORY OF SCIENCE
AND ITS CULTURAL INFLUENCES**

A PUBLICATION OF THE
HISTORY OF SCIENCE SOCIETY

W. Patrick McCray
Co-Editor, Osiris
Department of History
University of California, Santa Barbara
Santa Barbara, CA 93106-9410 USA
pmccray@history.ucsb.edu

Suman Seth
Co-Editor, Osiris
Department of Science & Technology Studies
321 Morrill Hall
Cornell University
Ithaca, NY 14853 USA
ss536@cornell.edu